Reviews for the third edition

'A limpid and very useful account of what we know about the management of innovation. Must read for executives, scholars and students.'

Yves Doz, Timken Chaired Professor of Global Technology and Innovation,
INSEAD.

Reviews for the second edition

'This is an extraordinary synthesis of the most important things that are understood about innovation, written by some of the world's foremost scholars in this field.'

Clayton M. Christensen, Professor of Business Administration,
Harvard Business School.

'The capacity to innovate is a key source of competitive advantage; but the management of innovation is risky. The authors provide a clear, systematic and integrated framework which will guide students and practising managers alike through a complex field. Updated to address key contemporary themes in knowledge management, networks and new technology, and with an exemplary combination of research and practitioner material, this is probably the most comprehensive guide to innovation management currently available.'

Rob Goffee, Professor of Organizational Behaviour, London Business School.

'In a highly readable yet challenging text, Tidd, Bessant and Pavitt are true to their subtitle, since they do indeed achieve a rare analytical integration of technological, market and organizational change. Alive to the vital importance of context, they nonetheless reveal generic aspects to the process of innovation. Read this book and you will understand more, and with a little luck, an encounter with a rich example will resonate with experience, hopes and fears and provide a useful guide to action.'

Sandra Dawson, KPMG Professor of Management Studies and Director,
Judge Institute of Management, University of Cambridge.

'This is an excellent book. Not only is it practical and easy to read, it is also full of useful cases and examples, as well as a comprehensive reference to the current literature. I will be recommending it to my entrepreneurship students.'

Professor Sue Birley, Director, The Entrepreneurship Centre, Imperial College,
University of London, UK.

'The first edition of this book was essential reading for anyone trying to get to grips with innovation in theory and practice. This new edition, by embracing the challenges faced in the "new economy", is an ideal companion for the serious innovator. Starting from the view that anyone can develop competencies in innovation this comprehensive text provides managers with essential support as they develop their capability. The second edition contains many case illustrations illuminating both theory and practice in successful innovation and is a "must" for aspiring MBAs.'

David Birchall, Professor and Director of the Centre for Business in the Digital Age (iDE), Henley Management College, UK.

'The authors of this book have managed to capture the essence of leading-edge thinking in the management of techonological innovation and presented the multidimensional nature of the subject in an integrated manner that will be useful for the practitioner and essential reading for students and researchers in the field. This is the book we have been waiting for!'

Professor Carl W. I. Pistorius, Dean, Management of Technology Programme,
University of Pretoria, South Africa.

'Innovation has become widely recognized as a key to competitive success. Leaders of businesses of all sizes and from all industries now put sustained innovation among their top priorities and concerns – but, for many, innovation seems mysterious, unpredictable, apparently unmanageable. Yet it can be managed. This book provides a highly readable account of the best current thinking about building and sustaining innovation. It draws particular attention to important emerging issues, such as the use of networks of suppliers, customers and others outside the firm itself to stimulate innovation, and the role of knowledge and knowledge management to support and sustain it. As the authors say, there is no "one best way" to manage innovation: different situations call for different solutions. But if you want to drive innovation in your own organization, this book will help you to understand the issues that matter and the steps you can take.'

Richard J. Granger, Vice President, Technology & Innovation
Management Practice, Arthur D. Little Inc.

'Innovation has always been a challenge, but never more so nowadays in these turbulent times. This second edition of *Managing Innovation* helps address the practicalities of the challenge and places them firmly in today's new environment, where technology is changing faster and faster. Integrating the multiple aspects of innovation – and not just treating it as a technical issue – is a real benefit this book brings.'

C. John Brady, Director, McKinsey & Company Inc.

'The characteristics of doing business today – rapid change, extreme volatility and high uncertainty – mean that traditional ways of managing technology need to be radically reappraised for any company that sees technical leadership as a critical business differentiator. Through their research work and worldwide network, Joe Tidd, John Bessant and Keith Pavitt have brought together the latest thinking on innovation management, extensively illustrated with real world examples, and with pointers to how successful implementations may emerge in the future. This book is well worth reading for all who want to achieve leadership in technology management.'

David Hughes, Executive Vice President, Technology Management, Marconi plc.

'Innovation is the cornerstone of what makes businesses successful: offering something uniquely better to the consumer. Innovation, while key, is probably the most difficult (maybe even impossible) element of corporate activity to manage or plan. This book does an excellent job of setting out the specification of ways we can think about how to create innovative organizations, without prescribing a "recipe for success".'

Dr Neil MacGilp, Director, Corporate R&D, Procter & Gamble.

MANAGING INNOVATION

20140044109

MANAGING INNOVATION

Integrating Technological, Market and Organizational Change
Third Edition

Joe Tidd
Science and Technology Policy Research (SPRU), University of Sussex

John Bessant
School of Management, Cranfield University

Keith Pavitt

John Wiley & Sons, Ltd

Third edition © 2005 Joe Tidd, John Bessant and Keith Pavitt

First edition © 1997   John Wiley & Sons Ltd, The Atrium, Southern Gate, Chichester,
West Sussex PO19 8SQ, England
Telephone (+44) 1243 779777

Email (for orders and customer service enquiries): cs-books@wiley.co.uk
Visit our Home Page on www.wileyeurope.com or www.wiley.com

Reprinted August 2005, September 2005, April 2006, March 2008 and September 2008

This publication is designed to provide accurate and authoritative information in regard to the
subject matter covered. It is sold on the understanding that the Publisher is not engaged in ren-
dering professional services. If professional advice or other expert assistance is required, the services of a competent professional should be sought.

Other Wiley Editorial Offices

John Wiley & Sons Inc., 111 River Street, Hoboken, NJ 07030, USA

Jossey-Bass, 989 Market Street, San Francisco, CA 94103-1741, USA

Wiley-VCH Verlag GmbH, Boschstr. 12, D-69469 Weinheim, Germany

John Wiley & Sons Australia Ltd, 33 Park Road, Milton, Queensland 4064, Australia

John Wiley & Sons (Asia) Pte Ltd, 2 Clementi Loop #02-01, Jin Xing Distripark, Singapore
129809

John Wiley & Sons Canada Ltd, 22 Worcester Road, Etobicoke, Ontario, Canada M9W 1L1

Wiley also publishes its books in a variety of electronic formats. Some content that appears in
print may not be available in electronic books.

Library of Congress Cataloging-in-Publication Data

Tidd, Joseph, 1960–
    Managing innovation : integrating technological, market and organizational change /
Joe Tidd, John Bessant, Keith Pavitt.– 3rd ed.
        p. cm.
    Includes bibliographical references and index.
    ISBN 0-470-09326-9 (pbk. : alk. paper)
    1. Technological innovations–Management.   2. Industrial management.
3. Organizational change.   I. Bessant, J. R.   II. Pavitt, Keith.   III. Title.
        HD45.T534   2005
        658.5′14–dc22

2004026221

British Library Cataloguing in Publication Data

A catalogue record for this book is available from the British Library

ISBN: 978-0-470-09326-9 (PB)
Typeset in SNP Best-set Typesetter Ltd., Hong Kong
Printed and bound in Great Britain by TJ International Ltd, Padstow, Cornwall

This third edition of *Managing Innovation* is dedicated to our co-author, friend and colleague, Keith Pavitt, who died in December 2002. Keith was an inspiration to us, and many others. Keith's research combined empirical evidence and common sense to generate realistic and robust theoretical and practical insights. His work was based on a deep empirical understanding of innovation and firm behaviour, and contributed to the development of new data, methods and taxonomies. His contributions spanned economics, management and science and technology policy, and included insights into the structure, dynamics and management of innovation processes, the relationship between basic research and technical change, knowledge and the theory of the firm, the globalization of R&D, and science and technology policy.

Joe Tidd
John Bessant

Contents

Preface

Recent surveys confirm that whilst most managers acknowledge the importance of innovation, the majority are dissatisfied with the management of innovation in their organizations.[1] In fact the performance of innovation varies significantly between different sectors, and between firms in the same sector, suggesting that both structural and organizational factors influence the effect of innovation on performance.[2] Management research confirms that innovative firms – those that are able to use innovation to improve their processes or to differentiate their products and services – outperform their competitors, measured in terms of market share, profitability, growth or market capitalization.[3] However, the management of innovation is inherently difficult and risky: most new technologies fail to be translated into products and services, and most new products and services are not commercial successes. In short, innovation can enhance competitiveness, but it requires a different set of management knowledge and skills from those of everyday business administration.

This book aims to equip readers with the knowledge to understand, and the skills to manage, innovation at the operational and strategic levels. Specifically, the book aims to integrate the management of market, technological and organizational change to improve the competitiveness of firms and effectiveness of other organizations. The management of innovation is inherently interdisciplinary and multifunctional, but most management texts tend to emphasize a single dimension, such as the management of research and development, production and operations management, marketing management, product development or organizational development. In contrast we aim here to provide an integrative approach to the management of innovation.

Since the first edition was published in 1997 innovation has become a major driver of success in an increasing number of activities and sectors, and is no longer confined to large manufacturing firms. Services now account for around three-quarters of value and employment in the advanced economies, and innovation is increasingly central to the performance of these services in the private and public sectors.[4] The impact of the information and communication technologies (ICTs) in logistics, distribution and services has already been significant, and has resulted in a wave of new ventures, which began (rather than ended) with the Internet Bubble of the late 1990s. The completion of the initial stages of mapping the human genome promises to have a similar impact on the pharmaceutical and healthcare sectors in the near future. In both cases there is good reason to believe that the competitive and economic exploitation of these and other emerging technologies has just begun. At the same time, more innovative solutions to the challenges of sustainability are increasingly needed.[5] In parallel, research on the management of innovation has increased in depth and breadth, improving our understanding and challenging our previous assumptions. Therefore in this third edition we have taken the opportunity to update or replace about a fifth of the material, to incorporate seven years of feedback from readers and users of the first and second editions, and to improve our coverage of contemporary topics such as disruptive innovation, service innovation, innovation networks, innovation and sustainability, entrepreneurship, and intellectual property. We have also invested in the website that supports this text to increase its scale and scope, and to more fully support the needs of users (http://www.managing-innovation.com).

The latest management research and the experience of leading practitioners confirm that significant dimensions in the management of innovation are not satisfactorily addressed by management teaching or texts. For example, the management of technological innovation reaches beyond efforts to improve the efficiency of production or research and development, to include the effectiveness of technological development, that is the translation of technology into successful products and services. This suggests a competence- and knowledge-based approach to technology management, which also requires analysis of organizational structures and processes. The management of organizational innovation has shifted from an emphasis on 'change management' of structure and culture, to the design and improvement of internal processes, such as knowledge management, and external linkages and networks. In market innovation there has been a shift in emphasis from crude market segmentation and analysis of consumer behaviour, to relationship and networked marketing that demands fine targeting of product development and closer linkages with lead customers.

All this suggests that it is not sufficient to focus on a single dimension of innovation: technological, market and organizational change interact. Better management of research and development may improve the efficiency or productivity of technological innovation, but is unlikely to contribute to product effectiveness, and therefore cannot guarantee commercial or financial success. Even the most expensive and sophisticated market research will fail to identify the potential for radically new products and services. Flat organizational structures and streamlined business processes may improve efficiency of delivering today's products and services, but will not identify or deliver innovative products and services, and may become redundant due to technological or market change.[6]

In this book we aim to provide a coherent framework to integrate the management of technological, market and organizational change. We reject the 'one best way' school of management, and instead seek to identify the links between the structures and processes which support innovation, and the opportunity for, and constraints on, innovation in specific technological and market environments. We shall argue that the process of innovation management is essentially generic, although organization-, technological- and market-specific factors will constrain choices and actions. We present a number of processes that contribute to the successful management of innovation, which are based on internal knowledge and competencies, but at the same time fully exploit external sources of know-how. Contingencies such as firm size, technological complexity and environmental uncertainty will influence the precise choice of processes; for example, complexity requires increased participation in networks of suppliers and users, whereas uncertainty demands vigilance in scanning the external technological and market environment.

This book is written with the needs of postgraduate and other management students in mind, specifically those studying MBA electives or options on the management of innovation and technology, or M.Sc. courses dedicated to technology and operations management. It is also relevant to managers charged with the management of research and development, product development or organizational change. Where possible, we provide examples of good, and not so good practice, drawn from a range of sectors and countries. However, the book is designed to encourage and support organization-specific experimentation and learning, and not to substitute for it.[7] Our analysis and prescriptions are based on the systematic analysis of the latest management research and practice, and our own research, consulting and teaching experiences at SPRU – Science and Technology Policy Research, at the University of Sussex, UK, and the School of Management at Cranfield University, as well as our experience in the USA, Europe and Asia. In 2002 SPRU joined CENTRIM (Centre for Research in Innovation Management) in the £12 million purpose-built Freeman Centre at Sussex University, to create one of the greatest concentrations of researchers in the field of technology and innovation and management and policy. For details of our current teaching and research please visit us at www.sussex.ac.uk/spru and www.cranfield.ac.uk. We would appreciate your feedback.

Joe Tidd
John Bessant
Brighton, Sussex, UK, October 2004

References

1 Arthur D. Little (2004) *Innovation Excellence Study*. ADL, Boston, Mass.

2 EIRMA (2004) *Assessing R&D Effectiveness*. EIRMA working group paper no. 62. Paris.

3 Tidd, J. (2000) *From Knowledge Management to Strategic Competence: Measuring technological, market and organzational innovation*. Imperial College Press, London.

4 Tidd, J. and F.M. Hull (2003) *Service Innovation: Organizational responses to technological opportunities and market imperatives*. Imperial College Press, London.

5 Berkhout, F. and K. Green (eds) (2002) Special Issue on 'Managing Innovation for Sustainability', *International Journal of Innovation Management*, **6** (3).

6 Bessant, J. (2003) *High Involvement Innovation*. John Wiley & Sons, Ltd, Chichester.

7 Isaksen, S. and J. Tidd (2006) *Meeting the Innovation Challenge: Leadership for transformation and growth*. John Wiley & Sons, Ltd, Chichester.

About the authors

Joe Tidd

Joe Tidd is a physicist with subsequent degrees in technology policy and business administration. He is Professor of Technology and Innovation Management and Director of Studies at SPRU (Science & Technology Policy Research), University of Sussex, UK, and Visiting Professor at University College London, Copenhagen Business School and the Rotterdam School of Management. He was previously Head of the Management of Innovation Specialization and Director of the Executive MBA Programme at Imperial College, University of London.

He has worked as policy adviser to the CBI (Confederation of British Industry), responsible for industrial innovation and advanced technologies, where he developed and launched the annual CBI *Innovation Trends Survey*, and presented expert evidence to three Select Committee Enquiries held by the House of Commons and House of Lords. He was a researcher for the five-year, $5 million International Motor Vehicle Program organized by the Massachusetts Institute of Technology (MIT) in the USA, and has worked on research and consultancy projects on technology and innovation management for consultants Arthur D. Little, CAP Gemini and McKinsey, and numerous technology-based firms, including American Express Technology, Applied Materials, ASML, BOC Edwards, BT, Marconi, National Power, NKT, Nortel Networks and Petrobras. He is the winner of the Price Waterhouse Urwick Medal for contribution to management teaching and research, and the Epton Prize from the R&D Society. He has written five books and more than 70 papers on the management of technology and innovation, the most recent being *Service Innovation: Organizational responses to technological opportunities and market imperatives* (with Frank Hull), Imperial College Press, 2003, and is Managing Editor of the *International Journal of Innovation Management*. Contact: J.Tidd@sussex.ac.uk

John Bessant

John Bessant is Professor of Innovation Management at the School of Management, Cranfield University. He also holds a Fellowship of the Advanced Institute for Management Research which he was awarded in 2003. He graduated from Aston University with a degree in Chemical Engineering in 1975 and later obtained a Ph.D. for work on innovation within the chemical industry. After a spell in industry he took up full-time research and consultancy in the field of technology and innovation management working at Aston's Technology Policy Unit, the Science Policy Research Unit at Sussex University and at

Brighton University where he held the Chair in Technology Management from 1987 to 2002.

Prior to joining the faculty at Cranfield, John was Director of Brighton University's Centre for Research in Innovation Management which he set up in 1987. He oversaw its development into a research institute with a staff of 30 people working on around 50 projects for public and private sponsors in the field of effective innovation management.

He is an Honorary Professor at SPRU, Sussex University and a Visiting Fellow at several UK and international universities. He served on the Business and Management panel of the 2001 Research Assessment Exercise. In 2003 he was elected a Fellow of the British Academy of Management.

His areas of research interest include the management of discontinuous innovation, strategies for developing high involvement innovation and enabling effective inter-firm collaboration and learning in product and process innovation. He is the author of 20 books and many articles on the topic and has lectured and consulted widely around the world. He has acted as advisor to various national governments and to international bodies including the United Nations, The World Bank and the OECD. Contact: john.bessant@cranfield.ac.uk

Part I

MANAGING FOR INNOVATION

Chapter 1

Key Issues in Innovation Management

'A slow sort of country' said the Red Queen. 'Now here, you see, it takes all the running you can do to keep in the same place. If you want to get somewhere else, you must run at least twice as fast as that!'

(Lewis Carroll, Alice through the Looking Glass)

'We always eat elephants . . .' is a surprising claim made by Carlos Broens, founder and head of a successful toolmaking and precision engineering firm in Australia with an enviable growth record. Broens Industries is a small/medium-sized company of 130 employees which survives in a highly competitive world by exporting over 70% of its products and services to technologically demanding firms in aerospace, medical and other advanced markets. The quote doesn't refer to strange dietary habits but to their confidence in 'taking on the challenges normally seen as impossible for firms of our size' – a capability which is grounded in a culture of innovation in products and the processes which go to produce them.[1]

At the other end of the scale spectrum Kumba Resources is a large South African mining company which makes another dramatic claim – 'We move mountains'. In their case the mountains contain iron ore and their huge operations require large-scale excavation – and restitution of the landscape afterwards. Much of their business involves complex large-scale machinery – and their abilities to keep it running and productive depend on a workforce able to contribute their innovative ideas on a continuing basis.[2]

Innovation is driven by the ability to see connections, to spot opportunities and to take advantage of them. When the Tasman Bridge collapsed in Hobart, Tasmania, in 1975 Robert Clifford was running a small ferry company and saw an opportunity to capitalize on the increased demand for ferries – and to differentiate his by selling drinks to thirsty cross-city commuters. The same entrepreneurial flair later helped him build a company – Incat – which pioneered the wave-piercing design which helped them capture over half the world market for fast catamaran ferries. Continuing investment in innovation has helped this company from a relatively isolated island build a key niche in highly competitive international military and civilian markets (www.incat.com.au/).

But innovation is not just about opening up new markets – it can also offer new ways of serving established and mature ones. Despite a global shift in textile and clothing manufacture towards developing countries the Spanish company, Inditex (through its retail outlets under various names including Zara) have pioneered a highly flexible, fast turnaround clothing operation with over 2000 outlets in 52 countries. It was founded by Amancio Ortega Gaona who set up a small operation in the west of Spain in La Coruna – a region not previously noted for textile production – and the first store opened there in 1975. Central to the Inditex philosophy is close linkage between design, manufacture and retailing and their network of stores constantly feeds back information about trends which are used to generate new designs. They also experiment with new ideas directly on the public, trying samples of cloth or design and quickly getting back indications of what is going to catch on. Despite their global orientation, most manufacturing is still done in Spain, and they have managed to reduce the turnaround time between a trigger signal for an innovation and responding to it to around 15 days (www.inditex.com/en).

Of course, technology often plays a key role in enabling radical new options. Magink is a company set up in 2000 by a group of Israeli engineers and now part of the giant Mitsubishi concern. Its business is in exploiting the emerging field of digital ink technology – essentially enabling paper-like display technology for indoor and outdoor displays. These have a number of advantages over other displays such as liquid crystal – low-cost, high viewing angles and high visibility even in full sunlight. One of their major new lines of development is in advertising billboards – a market worth $5 bn in the USA alone – where the prospect of 'programmable hoardings' is now opened up. Magink enables high resolution images which can be changed much more frequently than conventional paper advertising, and permit billboard site owners to offer variable price time slots, much as television does at present.[3]

At the other end of the technological scale there is scope for improvement on an old product – the humble eyeglass. A chance meeting took place between an Oxford physics professor developing his own new ophthalmic lens technology (and with an interest in applying it in the developing world) and someone with a great deal of knowledge of the developing world. This has led to a new technology with the potential to transform the lives of hundreds of millions of people in the developing world – a pair of spectacles with lenses that can be adjusted by the wearer to suit their visual needs. No sight tests by opticians are required, the special lenses can be simply adjusted to accurately correct the vision of large numbers of people. Mass production of the spectacles will soon be under way, with manufacturing designed to give high quality at low cost. In the developing world, where a severe shortage of opticians is a real problem, this innovation is likely to have an impact on a larger number of people than the celebrated wind-up radio.

Innovation is of course not confined to manufactured products; examples of turnaround through innovation can be found in services and in the public and private sector.[4] For example, the Karolinska Hospital in Stockholm has managed to make radical improvements in the speed, quality and effectiveness of its care services – such as cutting waiting lists by 75% and cancellations by 80% – through innovation.[5] In banking the UK First Direct organization became the most competitive bank, attracting around 10 000 new customers each month by offering a telephone banking service backed up by sophisticated IT. A similar approach to the insurance business – Direct Line – radically changed the basis of that market and led to widespread imitation by all the major players in the sector.[6,7] Internet-based retailers such as Amazon.com have changed the ways in which products as diverse as books, music and travel are sold, whilst firms like e-Bay have brought the auction house into many living rooms.

1.1 Innovation and Competitive Advantage

What these organizations have in common is that their undoubted success derives in large measure from innovation. Whilst competitive advantage can come from size, or possession of assets, etc. the pattern is increasingly coming to favour those organizations which can mobilize knowledge and technological skills and experience to create novelty in their offerings (product/service) and the ways in which they create and deliver those offerings.[8] This is seen not only at the level of the individual enterprise but increasingly as the wellspring for national economic growth. For example, the UK Office of Science and Technology see it as 'the motor of the modern economy, turning ideas and knowledge into products and services'.[9]

Innovation contributes in several ways. For example, research evidence suggests a strong correlation between market performance and new products.[10,11] New products help capture and retain market shares, and increase profitability in those markets. In the case of more mature and established products, competitive sales growth comes not simply from being able to offer low prices but also from a variety of non-price factors – design, customization and quality.[6] And in a world of shortening product life cycles – where, for example, the life of a particular model of television set or computer is measured in months, and even complex products like motor cars now take only a couple of years to develop – being able to replace products frequently with better versions is increasingly important.[12,13] 'Competing in time' reflects a growing pressure on firms not just to introduce new products but to do so faster than competitors.[12,14]

At the same time new product development is an important capability because the environment is constantly changing. Shifts in the socio-economic field (in what people believe, expect, want and earn) create opportunities and constraints. Legislation may open up new pathways, or close down others – for example, increasing the requirements for environmentally friendly products. Competitors may introduce new products which represent a major threat to existing market positions. In all these ways firms need the capability to respond through product innovation.

Whilst new products are often seen as the cutting edge of innovation in the marketplace, process innovation plays just as important a strategic role. Being able to make something no one else can, or to do so in ways which are better than anyone else is a powerful source of advantage. For example, the Japanese dominance in the late twentieth century across several sectors – cars, motorcycles, shipbuilding, consumer electronics – owed a great deal to superior abilities in manufacturing – something which resulted from a consistent pattern of process innovation. The Toyota production system and its equivalent in Honda and Nissan led to performance advantages of around two to one over average car makers across a range of quality and productivity indicators.[15] One of the main reasons for the ability of relatively small firms like Oxford Instruments or Incat to survive in highly competitive global markets is the sheer complexity of what they make and the huge difficulties a new entrant would encounter in trying to learn and master their technologies.

Similarly, being able to offer better service – faster, cheaper, higher quality – has long been seen as a source of competitive edge. Citibank was the first bank to offer automated telling machinery (ATM) service and developed a strong market position as a technology leader on the back of this process innovation. Benetton is one of the world's most successful retailers, largely due to its, sophisticated IT-led production network, which it innovated over a 10-year period,[16] and the same model has been used to great effect by the Spanish firm Zara. Southwest Airlines achieved an enviable position as the most effective airline in the USA despite being much smaller than its rivals; its success was due to process innovation in areas like reducing airport turnaround times.[17] This model has subsequently become the template for a whole new generation of low-cost airlines whose efforts have revolutionized the once-cosy world of air travel.

Importantly we need to remember that the advantages which flow from these innovative steps gradually get competed away as others imitate. Unless an organization is able to move into further innovation, it risks being left behind as others take the lead in changing their offerings, their operational processes or the underlying models which drive their business. For example, leadership in banking has passed to others, particularly those who were able to capitalize early on the boom in information and communications technologies; in particular many of the lucrative financial services like securities and share dealing have been dominated by players with radical new models like Charles Schwab.[18] As retailers all adopt advanced IT so the lead shifts to those who are able – like Zara and Benneton – to streamline their production operations to respond rapidly to the signals flagged by the IT systems.

With the rise of the Internet the scope for service innovation has grown enormously – not for nothing is it sometimes called 'a solution looking for problems'. As Evans and

Wurster point out, the traditional picture of services being either offered as a standard to a large market (high 'reach' in their terms) or else highly specialized and customized to a particular individual able to pay a high price (high 'richness') is 'blown to bits' by the opportunities of Web-based technology. Now it becomes possible to offer both richness and reach at the same time – and thus to create totally new markets and disrupt radically those which exist in any information-related businesses.[19]

The challenge which the Internet poses is not only one for the major banks and retail companies, although those are the stories which hit the headlines. It is also an issue – and quite possibly a survival one – for thousands of small businesses. Think about the local travel agent and the cosy way in which it used to operate. Racks full of glossy brochures through which people could browse, desks at which helpful sales assistants sort out the details of selecting and booking a holiday, procuring the tickets, arranging insurance and so on. And then think about how all of this can be accomplished

---

BOX 1.1
JOSEPH SCHUMPETER – THE 'GODFATHER' OF INNOVATION STUDIES

The 'godfather' of this area of economic theory was Joseph Schumpeter who wrote extensively on the subject. He had a distinguished career as an economist and served as Minister for Finance in the Austrian Government. His argument was simple: entrepreneurs will seek to use technological innovation – a new product/service or a new process for making it – to get strategic advantage. For a while this may be the only example of the innovation so the entrepreneur can expect to make a lot of money – what Schumpeter calls 'monopoly profits'. But of course other entrepreneurs will see what he has done and try to imitate it – with the result that other innovations emerge, and the resulting 'swarm' of new ideas chips away at the monopoly profits until an equilibrium is reached. At this point the cycle repeats itself – our original entrepreneur or someone else looks for the next innovation which will rewrite the rules of the game, and off we go again. Schumpeter talks of a process of 'creative destruction' where there is a constant search to create something new which simultaneously destroys the old rules and establishes new ones – all driven by the search for new sources of profits.[20]

In his view:

[What counts is] competition from the new commodity, the new technology, the new source of supply, the new type of organization . . . competition which . . . strikes not at the margins of the profits and the outputs of the existing firms but at their foundations and their very lives.

TABLE 1.1 Strategic advantages through innovation

Mechanism	Strategic advantage	Examples
Novelty in product or service offering	Offering something no one else can	Introducing the first . . . Walkman, fountain pen, camera, dishwasher, telephone bank, on-line retailer, etc. . . . to the world
Novelty in process	Offering it in ways others cannot match – faster, lower cost, more customized, etc.	Pilkington's float glass process, Bessemer's steel process, Internet banking, on-line bookselling, etc.
Complexity	Offering something which others find it difficult to master	Rolls-Royce and aircraft engines – only a handful of competitors can master the complex machining and metallurgy involved
Legal protection of intellectual property	Offering something which others cannot do unless they pay a licence or other fee	Blockbuster drugs like Zantac, Prozac, Viagra, etc.
Add/extend range of competitive factors	Move basis of competition – e.g. from price of product to price and quality, or price, quality, choice, etc.	Japanese car manufacturing, which systematically moved the competitive agenda from price to quality, to flexibility and choice, to shorter times between launch of new models, and so on – each time not trading these off against each other but offering them all
Timing	First-mover advantage – being first can be worth significant market share in new product fields	Amazon.com, Yahoo – others can follow, but the advantage 'sticks' to the early movers
	Fast follower advantage – sometimes being first means you encounter many unexpected teething problems, and it makes better sense to watch someone else make the early mistakes and move fast into a follow-up product	Palm Pilot and other personal digital assistants (PDAs) which have captured a huge and growing share of the market. In fact the concept and design was articulated in Apple's ill-fated Newton product some five years earlier – but problems with software and especially handwriting recognition meant it flopped

Mechanism	Strategic advantage	Examples
Robust/ platform design	Offering something which provides the platform on which other variations and generations can be built	Walkman architecture – through minidisk, CD, DVD, MP3 . . . Boeing 737 – over 30 years old, the design is still being adapted and configured to suit different users – one of the most successful aircraft in the world in terms of sales. Intel and AMD with different variants of their microprocessor families
Rewriting the rules	Offering something which represents a completely new product or process concept – a different way of doing things – and makes the old ones redundant	Typewriters vs. computer word processing, ice vs. refrigerators, electric vs. gas or oil lamps
Reconfiguring the parts of the process	Rethinking the way in which bits of the system work together – e.g. building more effective networks, outsourcing and co-ordination of a virtual company, etc.	Zara, Benetton in clothing, Dell in computers, Toyota in its supply chain management
Transferring across different application contexts	Recombining established elements for different markets	Polycarbonate wheels transferred from application market like rolling luggage into children's toys – lightweight micro-scooters
Others?	Innovation is all about finding new ways to do things and to obtain strategic advantage – so there will be room for new ways of gaining and retaining advantage	Napster. This firm began by writing software which would enable music fans to swap their favourite pieces via the Internet – the Napster program essentially connected person to person (P2P) by providing a fast link. Its potential to change the architecture and mode of operation of the Internet was much greater, and although Napster suffered from legal issues followers developed a huge industry based on downloading and file sharing (see Box 1.3 for more detail on this)

at the click of a mouse from the comfort of home – and that it can potentially be done with more choice and at lower cost. Not surprisingly, one of the biggest growth areas in dot.com start-ups was the travel sector and whilst many disappeared when the bubble burst, others like lastminute.com and Expedia have established themselves as mainstream players.

Of course, not everyone wants to shop online and there will continue to be scope for the high-street travel agent in some form – specializing in personal service, acting as a gateway to the Internet-based services for those who are uncomfortable with computers, etc. And, as we have seen, the early euphoria around the dot.com bubble has given rise to a much more cautious advance in Internet-based business. The point is that whatever the dominant technological, social or market conditions, the key to creating – and sustaining – competitive advantage is likely to lie with those organizations which continually innovate.

Table 1.1 indicates some of the ways in which enterprises can obtain strategic advantage through innovation.

1.2 Types of Innovation

Before we go too much further it will be worth defining our terms. What do we mean by 'innovation'? Essentially we are talking about change, and this can take several forms; for the purposes of this book we will focus on four broad categories (the '4Ps' of innovation):[21]

- 'product innovation' – changes in the things (products/services) which an organization offers;
- 'process innovation' – changes in the ways in which they are created and delivered;
- 'position innovation' – changes in the context in which the products/services are introduced;
- 'paradigm innovation' – changes in the underlying mental models which frame what the organization does.

For example, a new design of car, a new insurance package for accident-prone babies and a new home entertainment system would all be examples of product innovation. And change in the manufacturing methods and equipment used to produce the car or the home entertainment system, or in the office procedures and sequencing in the insurance case, would be examples of process innovation.

Sometimes the dividing line is somewhat blurred – for example, a new jet-powered sea ferry is both a product and a process innovation. Services represent a particular case of this where the product and process aspects often merge – for example, is a new holiday package a product or process change?

Innovation can also take place by repositioning the perception of an established product or process in a particular user context. For example, an old-established product in the UK is Lucozade – originally developed as a glucose-based drink to help children and invalids in convalescence. These associations with sickness were abandoned by the brand owners, SmithKline Beecham, when they relaunched the product as a health drink aimed at the growing fitness market where it is now presented as a performance-enhancing aid to healthy exercise. This shift is a good example of 'position' innovation.

Sometimes opportunities for innovation emerge when we reframe the way we look at something. Henry Ford fundamentally changed the face of transportation not because he invented the motor car (he was a comparative latecomer to the new industry) nor because he developed the manufacturing process to put one together (as a craft-based specialist industry car-making had been established for around 20 years). His contribution was to change the underlying model from one which offered a handmade specialist product to a few wealthy customers to one which offered a car for Everyman at a price they could afford. The ensuing shift from craft to mass production was nothing short of a revolution in the way cars (and later countless other products and services) were created and delivered.[15] Of course making the new approach work in practice also required extensive product and process innovation – for example, in component design, in machinery building, in factory layout and particularly in the social system around which work was organized.

Recent examples of 'paradigm' innovation – changes in mental models – include the shift to low-cost airlines, the provision of online insurance and other financial services, and the repositioning of drinks like coffee and fruit juice as premium 'designer' products. Although in its later days Enron became infamous for financial malpractice it originally came to prominence as a small gas pipeline contractor which realized the potential in paradigm innovation in the utilities business. In a climate of deregulation and with global interconnection through grid distribution systems energy and other utilities like telecommunications bandwidth increasingly became commodities which could be traded much as sugar or cocoa futures.[22]

From Incremental to Radical Innovation

A second dimension to change is the degree of novelty involved. Clearly, updating the styling on our car is not the same as coming up with a completely new concept car which has an electric engine and is made of new composite materials as opposed to steel and glass. Similarly, increasing the speed and accuracy of a lathe is not the same thing as replacing it with a computer-controlled laser forming process. There are degrees

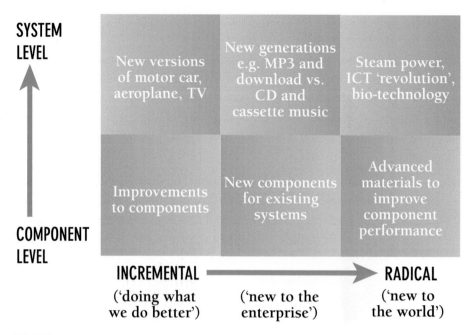

FIGURE 1.1 Dimensions of innovation

of novelty in these, running from minor, incremental improvements right through to radical changes which transform the way we think about and use them. Sometimes these changes are common to a particular sector or activity, but sometimes they are so radical and far-reaching that they change the basis of society – for example the role played by steam power in the Industrial Revolution or the ubiquitous changes resulting from today's communications and computing technologies. Figure 1.1 illustrates this continuum, highlighting the point that such change can happen at component or sub-system level or across the whole system.

Mapping Innovation Space

Each of our 4Ps of innovation can take place along an axis running from incremental through to radical change; the area indicated by the circle in Figure 1.2 is the potential innovation space within which an organization can operate. Whether it actually explores and exploits all the space is a question for innovation strategy and we will return to it later.

As far as managing the innovation process is concerned, these differences are important. The ways in which we approach incremental, day-to-day change will differ from those used occasionally to handle a radical step change in product or process. But we

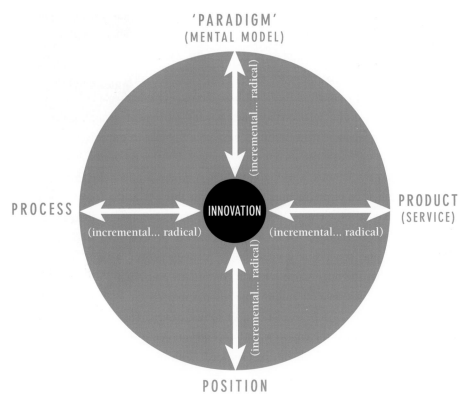

FIGURE 1.2 Innovation space

should also remember that it is the *perceived* degree of novelty which matters; novelty is very much in the eye of the beholder. For example, in a giant, technologically advanced organization like Shell or IBM advanced networked information systems are commonplace, but for a small car dealership or food processor even the use of a simple PC to connect to the Internet may still represent a major challenge.[23]

1.3 The Importance of Incremental Innovation

Although innovation sometimes involves a discontinuous shift – something completely new or a response to dramatically changed conditions – most of the time it takes place in incremental fashion. Products are rarely 'new to the world', process innovation is mainly about optimization and getting the bugs out of the system. (Ettlie suggests disruptive or new to the world innovations are only 6% to 10% of all projects labelled innovation.)[24] Studies of incremental process development (such as Hollander's famous study of Du Pont rayon plants) suggest that the cumulative gains in efficiency are often much greater over time than those which come from occasional radical changes.[25] Other examples include Tremblay's studies of paper mills, Enos on petroleum refining and Figueredo's of steel plants.[26–28]

Continuous improvement of this kind has received considerable attention in recent years, first as part of the 'total quality management' movement, reflecting the significant gains which Japanese manufacturers were able to make in improving quality and productivity through sustained incremental change.[29] But this is not new – similar principles underpin the famous 'learning curve' effect where productivity improves with increases in the scale of production; the reason for this lies in the learning and continuous incremental problem-solving innovation which accompanies the introduction of a new product or process.[30] More recent experience of deploying 'lean' thinking in manufacturing and services and increasingly between as well as within enterprises underlines further the huge scope for such continuous innovation.[15]

One way in which the continuous innovation approach can be harnessed to good effect is through the concept of platform or robust design. This is a way of creating stretch and space within the envelope and depends on being able to establish a strong basic platform or family which can be extended. Rothwell and Gardiner give several examples of such 'robust designs' which can be stretched and otherwise modified to extend the range and life of the product, including Boeing airliners and Rolls-Royce jet engines.[31] Major investments by large semiconductor manufacturers like Intel and AMD are amortized to some extent by being used to design and produce a family of devices based on common families or platforms such as the Pentium, Celeron, Athlon or Duron chipsets. Car makers are increasingly moving to produce models which although apparently different in style make use of common components and floor pans or chassis. Perhaps the most famous product platform is the 'Walkman' originally developed by Sony as a portable radio and cassette system; the platform concept has come to underpin a wide range of offerings from all major manufacturers for this market and deploying technologies like minidisk, CD, DVD and now MP3 players.

In processes much has been made of the ability to enhance and improve performance over many years from the original design concepts – in fields like steel-making and chemicals, for example. Service innovation offers other examples where a basic concept can be adapted and tailored for a wide range of similar applications without undergoing the high initial design costs – as is the case with different mortgage or insurance products.

Platforms and families are powerful ways for companies to recoup their high initial investments in R&D by deploying the technology across a number of market fields. For example, Procter & Gamble invested heavily in their cyclodextrin development for original application in detergents but then were able to use this technology or variants on it in a family of products including odour control ('Febreze'), soaps and fine fragrances ('Olay'), off-flavour food control, disinfectants, bleaches and fabric softening ('Tide', 'Bounce', etc.). They were also able to license out the technology for use in non-competing areas like industrial scale carpet care and in the pharmaceutical industry.

1.4 Innovation as a Knowledge-based Process

Innovation is about knowledge – creating new possibilities through combining different knowledge sets. These can be in the form of knowledge about what is technically possible or what particular configuration of this would meet an articulated or latent need. Such knowledge may already exist in our experience, based on something we have seen or done before. Or it could result from a process of search – research into technologies, markets, competitor actions, etc. And it could be in explicit form, codified in such a way that others can access it, discuss it, transfer it, etc. – or it can be in tacit form, known about but not actually put into words or formulae.[32]

The process of weaving these different knowledge sets together into a successful innovation is one which takes place under highly uncertain conditions. We don't know about what the final innovation configuration will look like (and we don't know how we will get there). Managing innovation is about turning these uncertainties into knowledge – but we can do so only by committing resources to reduce the uncertainty – effectively a balancing act. Figure 1.3 illustrates this process of increasing resource commitment whilst reducing uncertainty.

Viewed in this way we can see that incremental innovation, whilst by no means risk-free – is at least potentially manageable because we are starting from something we know about and developing improvements in it. But as we move to more radical options, so uncertainty is higher and at the limit we have no prior idea of what we are to develop or how to develop it! Again this helps us understand why discontinuous innovation is so hard to deal with.

A key contribution to our understanding here comes from the work of Henderson and Clark who looked closely at the kinds of knowledge involved in different kinds of innovation.[33] They argue that innovation rarely involves dealing with a single technology or market but rather a bundle of knowledge which is brought together into a configuration. Successful innovation management requires that we can get hold of and use knowledge about *components* but also about how those can be put together – what they termed the *architecture* of an innovation.

We can see this more clearly with an example. Change at the component level in building a flying machine might involve switching to newer metallurgy or composite

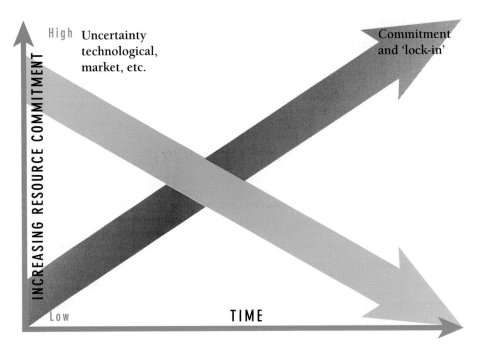

FIGURE 1.3   Innovation, uncertainty and resource commitment

materials for the wing construction or the use of fly-by-wire controls instead of control lines or hydraulics. But the underlying knowledge about how to link aerofoil shapes, control systems, propulsion systems, etc. at the *system* level is unchanged – and being successful at both requires a different and higher order set of competencies.

One of the difficulties with this is that innovation knowledge flows – and the structures which evolve to support them – tend to reflect the nature of the innovation. So if it is at component level then the relevant people with skills and knowledge around these components will talk to each other – and when change takes place they can integrate new knowledge. But when change takes place at the higher system level – 'architectural innovation' in Henderson and Clark's terms – then the existing channels and flows may not be appropriate or sufficient to support the innovation and the firm needs to develop new ones. This is another reason why existing incumbents often fare badly when major system level change takes place – because they have the twin difficulties of learning and configuring a new knowledge system and 'unlearning' an old and established one.

A variation on this theme comes in the field of 'technology fusion', where different technological streams converge, such that products which used to have a discrete identity begin to merge into new architectures. An example here is the home automation industry, where the fusion of technologies like computing, telecommunications, industrial control and elementary robotics is enabling a new generation of housing systems with integrated entertainment, environmental control (heating, air conditioning, lighting, etc.) and communication possibilities.[34,35]

Similarly, in services a new addition to the range of financial services may represent a component product innovation, but its impacts are likely to be less far-reaching (and the attendant risks of its introduction lower) than a complete shift in the nature of the service package – for example, the shift to direct-line systems instead of offering financial services through intermediaries.

Figure 1.4 highlights the issues for managing innovation. In Zone 1 the rules of the game are clear – this is about steady-state improvement to products or processes and uses knowledge accumulated around core components.

In Zone 2 there is significant change in one element but the overall architecture remains the same. Here there is a need to learn new knowledge but within an established and clear framework of sources and users – for example, moving to electronic ignition or direct injection in a car engine, the use of new materials in airframe components, the use of IT systems instead of paper processing in key financial or insurance transactions, etc. None of these involve major shifts or dislocations.

In Zone 3 we have discontinuous innovation where neither the end state nor the ways in which it can be achieved are known about – essentially the whole set of rules of the game changes and there is scope for new entrants.

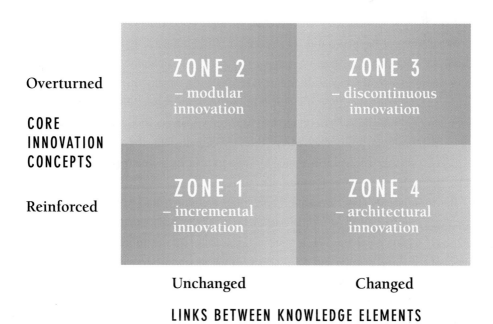

FIGURE 1.4   Component and architectural innovation

In Zone 4 we have the condition where new combinations – architectures – emerge, possibly around the needs of different groups of users (as in the disruptive innovation case). Here the challenge is in reconfiguring the knowledge sources and configurations. We may use existing knowledge and recombine it in different ways or we may use a combination of new and old. Examples might be low-cost airlines, direct line insurance, others.

1.5 The Challenge of Discontinuous Innovation

Most of the time innovation takes place within a set of rules of the game which are clearly understood, and involves players trying to innovate by doing what they have been doing (product, process, position, etc.) but better. Some manage this more effectively than others but the 'rules of the game' are accepted and do not change.[21]

But occasionally something happens which dislocates this framework and changes the rules of the game. By definition these are not everyday events but they have the capacity to redefine the space and the boundary conditions – they open up new opportunities but also challenge existing players to reframe what they are doing in the light of new conditions.[18,19,22] This is a central theme in Schumpeter's original theory of innovation which he saw as involving a process of 'creative destruction'.[20,36,37]

What seems to happen is that for a given set of technological and market conditions there is a long period of relative stability during which a continuous stream of variations around a basic innovation theme take place. Essentially this is product/process improvement along the lines of 'doing what we do, but better'. For example, the Bic ballpoint pen was originally developed in 1957 but remains a strong product with daily sales of 14 million units worldwide. Although superficially the same shape, closer inspection reveals a host of incremental changes that have taken place in materials, inks, ball technology, safety features, etc.

But these 'steady-state' innovation conditions are punctuated by occasional discontinuities – and when these occur one or more of the basic conditions (technology, markets, social, regulatory, etc.) shifts dramatically. In the process the underlying 'rules of the game' change and a new opportunity space for innovation opens up. 'Do different' conditions of this kind occur, for example, when radical change takes place along the technological frontier or when completely new markets emerge. An emerging example of this could be the replacement of the incandescent light bulb originally developed in the late nineteenth century by Edison and Swan (amongst others). This may be replaced by the solid state white light emitting diode technology patented by Nichia Chemical. This technology is 85% more energy efficient, has 16 times the life of a conventional bulb, is brighter, is more flexible in application and is likely to be subject to the scale economies associated with electronic component production. See Box 1.2 for a more detailed discussion of this.

In their pioneering work on this theme Abernathy and Utterback developed a model describing the pattern in terms of three distinct phases. Initially, under discontinuous conditions, there is what they term a 'fluid phase' during which there is high uncertainty along two dimensions:

- The target – what will the new configuration be and who will want it?
- The technical – how will we harness new technological knowledge to create and deliver this?

No one knows what the 'right' configuration of technological means and market needs will be and so there is extensive experimentation (accompanied by many failures) and fast learning by a range of players including many new entrepreneurial businesses.

Gradually these experiments begin to converge around what they call a 'dominant design' – something which begins to set up the rules of the game. This represents a convergence around the most popular (importantly not necessarily the most technologically sophisticated or elegant) solution to the emerging configuration. At this point a 'bandwagon' begins to roll and innovation options become increasingly channeled around a core set of possibilities – what Dosi calls a 'technological trajectory'.[38] It becomes increasingly difficult to explore outside this space because entrepreneurial interest and the resources which that brings increasingly focus on possibilities within the dominant design corridor.

This can apply to products or processes; in both cases the key characteristics become stabilized and experimentation moves to getting the bugs out and refining the dominant design. For example, the nineteenth-century chemical industry moved from making soda ash (an essential ingredient in making soap, glass and a host of other products) from the earliest days where it was produced by burning vegetable matter through to a sophisticated chemical reaction which was carried out on a batch process (the Leblanc process) which was one of the drivers of the Industrial Revolution. This process dominated for nearly a century but was in turn replaced by a new generation of continuous processes which used electrolytic techniques and which originated in Belgium where they were developed by the Solvay brothers. Moving to the Leblanc process or the Solvay process did not happen overnight; it took decades of work to refine and improve each process, and to fully understand the chemistry and engineering required to get consistent high quality and output.

BOX 1.2
LIVING WITH DISCONTINUOUS CHANGE

When discontinuous conditions emerge they challenge the 'rules of the game' – and both pose threats to the existing players and offer opportunities for those quick enough to take advantage of the new ones. A good example can be seen in the world of publishing. On the one hand we have an industry which was, until recently, based on very physical technologies and a complex network of specialist suppliers who contributed their particular parts of the complex puzzle of publishing. For example, copy – words or pictures – would be generated by a specialist journalist or photographer. They would then pass this on to various editors who would check, make choices about design and layout, etc. Next would come typesetting where the physical materials for printing would be made – hot metal would be cast into letters and grouped into blocks to form words and sentences within special frames. Pictures and other items would be transferred onto printing plates. The type frames or printing plates would then be fixed to presses, these would be inked and some test runs made. And finally the printed version would appear – and passed on to someone else to distribute and publish it.

Such a method might still be recognizable by Messrs Caxton and Gutenberg – the pioneers of the printing industry. But it is likely that they would not have much idea about the way in which publishing operates today – with its emphasis on IT. Now the process has changed such that a single person could undertake the whole set of operations – create text on a word processor, design and lay it out on a page-formatting program, integrate images with text and when satisfied print to either physical media or – increasingly – publish it worldwide in electronic form.

There are plenty of examples of firms which have exploited this or related opportunities. For example, Adplates is – or was until recently – a small firm in north London specializing in the production of printing plates for the advertising industry (hence the name). They used to be a small link in a long chain which began with a client and an advertising agency agreeing about an advertisement. The photo shoot and copy lines would be created and eventually the material would arrive at Adplates who would carry out the task of preparing a printing plate – which they would then pass on to a printer to use. In other words they were a small link in a long chain.

But technology has changed all that for them. They began to challenge the boundaries of the operation in which they were part – why, for example, could they not move upstream to deal directly with the client? Of course this required new skills and technology in areas like design and image and text preparation – but all of this is available on a PC. Equally, since printing has moved from hot metal to a largely digital process, they could invest in the skills and equipment to move downstream. And why should they leave it to a publisher to disseminate the material when the market and the technology in this end of the industry is changing so rapidly and opening up so many opportunities? Adplates now offers a complete service to clients from initial idea through to printing and even has its own stable of magazines and a thriving Web publishing operation.

There are winners in this game but also losers. People still think of the *Encyclopaedia Britannica* as a household name and the repository of useful reference knowledge which can be trusted. It is a well-established product – in fact the original idea came from three Scottish printers back in 1768! The brand is fine – but the business has gone through dramatic shifts and is still under threat. From a peak of sales in 1990 of around $650m. its sales have collapsed – for example, in the USA by up to 80%. The problem is not the product but the way in which it is presented – all the hard copy encyclopedias have suffered a similar fate at the hands of the CD-ROM-based versions like Encarta (which is often bundled in as part of a PC purchase).

We could go on looking at the publishing industry but the point is clear – when technology shifts dramatically it opens up major opportunities but also poses major threats to players in the industry and to those who might want to enter from outside. Under these conditions simply being an established player – even with a centuries-old brand name and an excellent product – is not enough. Indeed – as firms like Amazon.com have shown – it is at times like these that coming from outside and starting fresh may offer significant advantages.

What is going on here is clearly not conforming to a stable, big-is-beautiful model, nor is it about historically important emphasis on core competence. The foundations of a business like publishing become shaken and many of the famous names disappear whilst other unknown upstarts become major industry players – in some cases overnight! (Amazon.com was at one time worth more than double the market value of established businesses like British Airways.) Turbulence like this throws a challenge to established models of managing – not only is it a question of urgently needing to change but the very models of change management on which many traditional players rely may not be sufficient or appropriate.

The same pattern can be seen in products. For example, the original design for a camera is something which goes back to the early nineteenth century and – as a visit to any science museum will show – involved all sorts of ingenious solutions. The dominant design gradually emerged with an architecture which we would recognize – shutter and lens arrangement, focusing principles, back plate for film or plates, etc. But this design was then modified still further – for example, with different lenses, motorized drives, flash technology – and, in the case of George Eastman's work, to creating a simple and relatively 'idiot-proof' model camera (the Box Brownie) which opened up photography to a mass market. More recent development has seen a similar fluid phase around digital imaging devices.

The period in which the dominant design emerges and emphasis shifts to imitation and development around it is termed the 'transitional phase' in the Abernathy and Utterback model. Activities move from radical concept development to more focused efforts geared around product differentiation and to delivering it reliably, cheaply, with higher quality, extended functionality, etc.

As the concept matures still further so incremental innovation becomes more significant and emphasis shifts to factors like cost – which means efforts within the industries which grow up around these product areas tend to focus increasingly on rationalization, on scale economies and on process innovation to drive out cost and improve productivity. Product innovation is increasingly about differentiation through customization to meet the particular needs of specific users. Abernathy and Utterback term this the 'specific phase'.*

Finally the stage is set for change – the scope for innovation becomes smaller and smaller whilst outside – for example, in the laboratories and imaginations of research scientists – new possibilities are emerging. Eventually a new technology emerges which has the potential to challenge all the by now well-established rules – and the game is disrupted. In the camera case, for example, this is happening with the advent of digital photography which is having an impact on cameras and the overall service package around how we get, keep and share our photographs. In our chemical case this is happening with biotechnology and the emergence of the possibility of no longer needing giant chemical plants but instead moving to small-scale operations using live organisms genetically engineered to produce what we need.

Table 1.2 sets out the main elements of this model. Although originally developed for manufactured products the model also works for services – for example the early days of Internet banking were characterized by a typically fluid phase with many options and models being offered. This gradually moved to a transitional phase, build-

---

* A good example of this can be seen in the case of bicycles which went through an extended period of fluidity in design options before the dominant diamond frame emerged which has characterized the industry for the past century.[11]

TABLE 1.2 Stages in innovation life cycle

Innovation characteristic	Fluid pattern	Transitional phase	Specific phase
Competitive emphasis placed on . . .	Functional product performance	Product variation	Cost reduction
Innovation stimulated by . . .	Information on user needs, technical inputs	Opportunities created by expanding internal technical capability	Pressure to reduce cost, improve quality, etc.
Predominant type of innovation	Frequent major changes in products	Major process innovations required by rising volume	Incremental product and process innovation
Product line	Diverse, often including custom designs	Includes at least one stable or dominant design	Mostly undifferentiated standard products
Production processes	Flexible and inefficient – aim is to experiment and make frequent changes	Becoming more rigid and defined	Efficient, often capital intensive and relatively rigid

ing a dominant design consensus on the package of services offered, the levels and nature of security and privacy support, the interactivity of website, etc. The field has now become mature with much of the competition shifting to marginal issues like relative interest rates.

The pattern can be seen in many studies and its implications for innovation management are important. In particular it helps us understand why established organizations often find it hard to deal with discontinuous change. Organizations build capabilities around a particular trajectory and those who may be strong in the later (specific) phase of an established trajectory often find it hard to move into the new one. (The example of the firms which successfully exploited the transistor in the early 1950s is a good case in point – many were new ventures, sometimes started by enthusiasts in their garage, yet they rose to challenge major players in the electronics industry like Raytheon.[39]) This is partly a consequence of sunk costs and commitments to existing technologies and markets and partly because of psychological and institutional barriers.[40] They may respond but in slow fashion – and they may make the mistake of giving responsibility for the new development to those whose current activities would be threatened by a shift.[41]

Importantly, the 'fluid' or 'ferment' phase is characterized by *co-existence* of old and new technologies and by rapid improvements of both.[41,42] (It is here that the so-called 'sailing ship' effect can often be observed, in which a mature technology accelerates in its rate of improvement as a response to a competing new alternative – as was the case with the development of sailing ships in competition with newly emerging steamship technology.[43,44]

Whilst some research suggests existing incumbents do badly, we need to be careful here. Not all existing players do badly – many of them are able to build on the new trajectory and deploy/leverage their accumulated knowledge, networks, skills and financial assets to enhance their competence through building on the new opportunity.[42†] Equally whilst it is true that new entrants – often small entrepreneurial firms – play a strong role in this early phase we should not forget that we see only the successful players. We need to remember that there is a strong ecological pressure on new entrants which means only the fittest or luckiest survive.

It is more helpful to suggest that there is something about the ways in which innovation is *managed* under these conditions which poses problems. Good practice of the 'steady-state' kind described above is helpful in the mature phase but can actively militate against the entry and success in the fluid phase of a new technology.[46] How do enterprises pick up signals about changes if they take place in areas where they don't normally do research? How do they understand the needs of a market which doesn't exist yet but which will shape the eventual package which becomes the dominant design? If they talk to their existing customers the likelihood is that those customers will tend to ask for more of the same, so which new users should they talk to – and how do they find them?

The challenge seems to be to develop ways of managing innovation not only under 'steady-state' but also under the highly uncertain, rapidly evolving and changing conditions which result from a dislocation or discontinuity. The kinds of organizational behaviour needed here will include things like agility, flexibility, the ability to learn fast, the lack of preconceptions about the ways in which things might evolve, etc. – and these are often associated with new small firms. There are ways in which large and established players can also exhibit this kind of behaviour but it does often conflict with their normal ways of thinking and working.

Extensive studies have shown the power of shifting technological boundaries in creating and transforming industry structures – for example, in the case of the typewriter, the computer and the automobile. Such transformations happen relatively often – no industry is immune (see Box 1.3 for an example).

Worryingly the source of the technology which destabilizes an industry often comes from outside that industry. So even those large incumbent firms which take time and resources to carry out research to try and stay abreast of developments in their field

---

[†] For example, Microsoft was able to manage the shift towards Web-based services and towards PDA/mobile phones by extending its operating system and leveraging its marketing strength.[45]

BOX 1.3
THE DIMMING OF THE LIGHT BULB

In the Beginning . . .

God said let there be light. And for a long time this came from a rather primitive but surprisingly effective method – the oil lamp. From the early days of putting simple wicks into congealed animal fats, through candles to more sophisticated oil lamps people have been using this form of illumination. Archaeologists tell us this goes back at least 40 000 years so there has been plenty of scope for innovation to improve the basic idea! Certainly by the time of the Romans domestic illumination – albeit with candles – was a well-developed feature of civilized society.

Not a lot changed until the late eighteenth century when the expansion of the mining industry led to experiments with uses for coal gas – one of which was as an alternative source of illumination. One of the pioneers of research in the coal industry – Humphrey Davy – invented the carbon arc lamp and ushered in a new era of safety within the mines – but also opened the door to alternative forms of domestic illumination and the era of gas lighting began.

But it was not until the middle of the following century that researchers began to explore the possibilities of using a new power source and some new physical effects. Experiments by Joseph Swann in England and Farmer in the USA (amongst others) led to the development of a device in which a tiny metal filament enclosed within a glass envelope was heated to incandescence by an electric current. This was the first electric light bulb – and it still bears more than a passing resemblance to the product found hanging from millions of ceilings all around the world.

By 1879 it became clear that there was significant commercial potential in such lighting – not just for domestic use. Two events occurred during that year which were to have far-reaching effects on the emergence of a new industry. The first was that the city of Cleveland – although using a different lamp technology (carbon arc) – introduced the first public street lighting. And the second was that patents were registered for the incandescent filament light bulb by Joseph Swann in England and one Thomas Edison in the USA.

Needless to say the firms involved in gas supply and distribution and the gas lighting industry were not taking the threat from electric light lying down and they responded with a series of improvement innovations which helped retain gas lighting's popularity for much of the late nineteenth century. Much of what happened over the next 30 years is a good example of what is sometimes called the 'sailing

continues overleaf

BOX 1.3   (*continued*)

ship effect'. That is, just as in the shipping world the invention of steam power did not instantly lead to the disappearance of sailing ships but instead triggered a whole series of improvement in that industry, so the gas lighting industry consolidated its position through incremental product and process innovations.

But electric lighting was also improving and the period 1886–1920 saw many important breakthroughs and a host of smaller incremental performance improvements. In a famous and detailed study (carried out by an appropriately named researcher called Bright) there is evidence to show that little improvements in the design of the bulb and in the process for manufacturing it led to a fall in price of over 80% between 1880 and 1896.[46] Examples of such innovations include the use of gas instead of vacuum in the bulb (1913 Langmuir) and the use of tungsten filament.

Innovation theory teaches us that after an invention there is a period in which all sorts of designs and ideas are thrown around before finally a 'dominant design' settles out and the industry begins to mature. So it was with the light bulb; by the 1920s the basic configuration of the product – a tungsten filament inside a glass gas-filled bulb – was established and the industry began to consolidate. It is at this point that the major players with whom we associate the industry – Philips, General Electric (GE), Westinghouse – become established.

Technological Alternatives

Although the industry then entered a period of stability in the marketplace there was still considerable activity in the technology arena. Back in the nineteenth century Henri Becquerel invented the fluorescent lamp and in 1911 Claude invented the neon lamp – both inventions which would have far-reaching effects in terms of the industry and its segmentation into different markets.

The neon lamp started a train of work based on forming different glass tubes into shapes for signs and in filling them with a variety of gases with similar properties to neon but which gave different colours.

The fluorescent tube was first made commercially by Sylvania in the USA in 1938 following extensive development work by both GE and Westinghouse. The technology had a number of important features including low power consumption and long life – factors which led to their widespread use on office and business environments although less so in the home. By the 1990s this product had matured alongside the traditional filament bulb and a range of compact and shaped fittings were available from the major lighting firms.

Meanwhile, in Another Part of the World . . .

Whilst neon and fluorescent tubes were variations on the same basic theme of lights, a different development began in a totally new sector in the 1960s. In 1962 work on the emerging solid state electronics area led to the discovery of a light emitting diode – LED – a device which would, when a current passed through it, glow in red or green colour. These lights were bright and used little power; they were also part of the emerging trend towards miniaturization. They quickly became standard features in electronic devices and today the average household will have hundreds of LEDs in orange, green or red to indicate whether devices such as TV sets, mobile phones or electric toothbrushes are on and functioning.

Development and refinement of LEDs took place in a different industry for a different market and in particular one line of work was followed in a small Japanese chemical company supplying LEDs to the major manufacturers like Sony. Nichia Chemical began a programme of work on a type of LED which would emit blue light – something much more difficult to achieve and requiring complex chemistry and careful process control. Eventually they were successful and in 1993 produced a blue LED based on gallium arsenide technology. The firm then committed a major investment to development of both product and process technology, amassing around 300 patents along the way. Their research culminated in the development in 1995 of a white light LED – using the principle that white light is made up of red, green and blue light mixed together.

So what? The significance of Shuji Nakamura's invention may not be instantly apparent – and at present the only products which can be bought utilizing it are small high power torches. But think about the significance of this discovery. White LEDs offer the following advantages:

- 85% less power consumption
- 16 times brighter than normal electric lights
- Tiny size
- Long life – tests suggest the life of an LED could be 100 000 hours – about 11 years
- Can be packaged into different shapes, sizes and arrangements
- Will follow the same economies of scale in manufacturing that led to the continuing fall in the price of electronic components so will become very cheap very quickly.

continues overleaf

BOX 1.3   (continued)

If people are offered a low-cost, high-power, flexible source of white light they are likely to adopt it – and for this reason the lighting industry is feeling some sense of threat. The likelihood is that the industry as we know it will be changed dramatically by the emergence of this new light source – and whilst the names may remain the same they will have to pay a high price for licensing the technology. They may try and get around the patents – but with 300 already in place and the experience of the complex chemistry and processing which go into making LEDs Nichia have a long head start. When Dr Nakamura left Nichia Chemical for a chair at University of California, Santa Barbara, sales of blue LEDs and lasers were bringing the firm more than $200 m. a year and the technology is estimated to have earned Nichia nearly $2 bn.

Things are already starting to happen. Many major cities are now using traffic lights which use the basic technology to make much brighter green and red lights since they have a much longer life than conventional bulbs. One US company, Traffic Technology Inc., has even offered to give away the lights in return for a share of the energy savings the local authority makes! Consumer products like torches are finding their way into shops and online catalogues whilst the automobile industry is looking at the use of LED white light for interior lighting in cars. Major manufacturers such as GE are entering the market and targeting mass markets such as street lighting and domestic applications, a market estimated to be worth $12 bn in the USA alone.

---

may find that they are wrong-footed by the entry of something which has been developed in a different field. The massive changes in insurance and financial services which have characterized the shift to online and telephone provision were largely developed by IT professionals often working outside the original industry.[6] In extreme cases we find what is often termed the 'not invented here' – NIH – effect, where a firm finds out about a technology but decides against following it up because it does not fit with their perception of the industry or the likely rate and direction of its technological development. Famous examples of this include Kodak's rejection of the Polaroid process or Western Union's dismissal of Bell's telephone invention. In a famous memo dated 1876 the board commented, 'this "telephone" has too many shortcomings to be seriously considered as a means of communication. The device is inherently of no value to us.'

1.6 Christensen's Disruptive Innovation Theory

Although major advances or breakthroughs along the *technological* frontier can disrupt the rules of the game they are not the only mechanism. For example, Box 1.4 gives some examples where the technological leaders in industrial sectors found themselves in deep trouble as a result of changes in the ways *existing* technological knowledge was deployed.

The influential work of Clayton Christensen drew attention to cases where the *market* was the effective trigger point. He studied a number of industries in depth and particularly focused on the hard disk drive sector because it represented an industry where a number of generations of dominant design could be found within a relatively short history.[47]

BOX 1.4

TECHNOLOGICAL EXCELLENCE MAY NOT BE ENOUGH . . .

In the 1970s Xerox was the dominant player in photocopiers, having built the industry from its early days when it was founded on the radical technology pioneered by Chester Carlsen and the Battelle Institute. But despite their prowess in the core technologies and continuing investment in maintaining an edge it found itself seriously threatened by a new generation of small copiers developed by new entrants including several Japanese players. Despite the fact that Xerox had enormous experience in the industry and a deep understanding of the core technology it took them almost eight years of mishaps and false starts to introduce a competitive product. In that time Xerox lost around half its market share and suffered severe financial problems. As Henderson and Clark put it, in describing this case, 'apparently modest changes to the existing technology . . . have quite dramatic consequences'.[33]

In similar fashion in the 1950s the electronics giant RCA developed a prototype portable transistor-based radio using technologies which it had come to understand well. However, it saw little reason to promote such an apparently inferior technology and continued to develop and build its high range devices. By contrast Sony used it to gain access to the consumer market and to build a whole generation of portable consumer devices – and in the process acquired considerable technological experience which enabled them to enter and compete successfully in higher value, more complex markets.[40]

His distinctive observation was that with each generation almost all of the previously successful players in what was a multimillion dollar market failed to make the transition effectively and were often squeezed out of the market or into bankruptcy (see Table 1.3). In 1976 there were 17 major firms in the industry; by 1995 of these only IBM remained a player. During that period 129 firms had entered the industry – but 109 exited. Yet these were not non-innovative firms – quite the reverse. They were textbook examples of good practice, ploughing a high percentage of sales back into R&D, working closely with lead users to understand their needs and develop product innovations alongside them, delivering a steady stream of continuous product and process innovations and systematically exploring the full extent of the innovation space defined by their market. So what explains why such apparently smart firms fail?

The answer was not their failure to cope with a breakthrough in the technological frontier – indeed, all of the technologies which were involved in the new dominant designs for each generation were well-established and many of them had originated in the laboratories of the existing (and later disrupted) incumbents. What was changing was the emergence of new *markets* with very different needs and expectations. Generally these involved players who were looking for something simpler and cheaper to meet a very different set of needs – essentially outside or at the fringes of the mainstream.

For example the pioneers of the personal computer (Apple, Atari, Commodore, etc.) in the mid-1970s were trying to make a machine for the home and hobby market – but for a fraction of the price and with much less functionality than the existing mainstream mini-computer market where high capacity, fast access disk drives were required. Messrs Jobs, Wozniak and colleagues would be quite satisfied with something much less impressive technically but available to fit the tight budget of the kind of hobbyists to whom their product was initially addressed. The trouble was that they were not taken seriously as an alternative market prospect by the established suppliers of disk drives.

TABLE 1.3 Changing shape of US disk drive industry (derived from Christensen[47])

Time frame	1970	1975	1980	1984	1990
Dominant size inches	14	8	5.25	3.5	2.5
Main market applications	Mainframes	Mini computers	Desktops	Laptops	Advanced laptops
Main manufacturers	IBM Plug compatible manufacturers CDC	Shugart Priam Quantum Micropolis Ampex	Seagate Computer Memories International Memories	Rodime Conner Peripherals	Seagate Quantum Western Digital

In essence the existing players were too good at working with their mainstream users and failed to see the longer-term potential in the newly emerging market. Their systems for picking up signals about user needs and feeding these into the product development process were all geared around a market for machines for running sophisticated engineering and financial applications software. And their success in meeting these needs helped their businesses to grow through keeping up with that industry. We shouldn't be surprised at this – new markets do not emerge in their full scale or with clearly identifiable needs but start out as messy, uncertain and risky places with small size and dubious growth prospects. The early days of the PC industry were characterized by enthusiasm amongst a group of nerds and geeks running small and highly speculative ventures. These hardly represented a serious alternative market to the multibillion dollar business of supplying the makers of mainstream mini-computers. As Steve Jobs described their attempts to engage interest, 'So we went to Atari and said, "Hey, we've got this amazing thing, even built with some of your parts, and what do you think about funding us? Or we'll give it to you. We just want to do it. Pay our salary, we'll come work for you." And they said, "No." So then we went to Hewlett-Packard, and they said, "Hey, we don't need you. You haven't got through college yet." '[48]

But while these markets appeared irrelevant to mainstream players their requirements gave the outline specification for what would become a new dominant design based on a significantly different price/performance configuration. As the new market grew so the technology around delivering the dominant design matured and became more reliable and capable – as we would predict using the Abernathy and Utterback model. Eventually it became able to meet not only the needs of the new market but also those of the original business – but from a position of much more attractive price/performance. At this point the makers of mini-computers began to see significant benefits in using drives which were based on a different dominant design but which would still give them the functionality they needed – only much more cheaply.

It is here that market *disruption* emerges – what began as a fringe business has moved into the mainstream and eventually changes the rules under which the mainstream operates. By the time the established suppliers of disk drives to the mainstream industry woke up to what was happening the best they could do was to imitate but from a position of being far behind the learning curve. Not surprisingly in many cases they failed to make the grade and withdrew or went bankrupt.

Importantly the new players who rewrote rule book for one generation found their markets disrupted in turn by a later generation of players doing the same thing to them. This underlines the point that it is not stupid firms who suffer this kind of disruption – rather it is the fact that the recipe for success in following a new dominant design becomes one which shapes the signals firms perceive about future opportunities and the ways in which they allocate resources to them. Riding along on one particular bandwagon makes the enterprise vulnerable in its ability to jump on to the next one when it starts to roll.

The pattern of disruptive innovation can be seen in a variety of industries – for example mini-mills disrupting the market for integrated large-scale steel producers or manufacturers of mechanical excavators finding their world challenged by a new breed of smaller, simpler hydraulic equipment. In later work Christensen and Raynor have extended this powerful market-linked analysis to deal with two dimensions of discontinuity – where disruption occurs because of a new bundle of performance parameters competing against existing markets and where it competes against non-consumption. Effectively the latter case is about creating completely new markets.[40]

The key challenge which organizations find difficult to deal with in these cases is not technological advance but rather a change in the technology/needs configuration for new and mainstream markets. The 'innovator's dilemma' in the title of Christensen's first book refers to the difficulties established players have in simultaneously managing the steady-state (sustaining) and the discontinuous (disruptive) aspects.

At its heart this powerful theory is a challenge to the ways in which we approach managing innovation. Sustaining conditions require innovation but along very different tracks – and involving very different networks – to disruptive conditions. The track record of existing players to ride both horses is poor but they face the need to deal with this innovator's dilemma. Either they surrender the ground to newcomers or they spin off new ventures and become newcomers themselves. A third option involving balancing the two – ambidextrous capability – is a tough challenge but one we pose throughout the book.

1.7 Other Sources of Discontinuity

This problem – of managing both the discontinuous and the steady state – emerges frequently and can be triggered not only by radical technology or significant market change. For example, it can come from dramatic breakthroughs in technology or by clever use of existing technology in a new configuration for a newly emerging market. It can come from reframing a business model – such as has happened with the 'reinvention' of the airline industry around low-cost models. Or it can come from an external shock forcing change on an industry or sector – as is often the case in wartime.

Table 1.4 gives some examples of such triggers for discontinuity. Common to these from an innovation management point of view is the need to recognize that under discontinuous conditions (which thankfully don't emerge every day) we need different

TABLE 1.4 Sources of discontinuity

Triggers/ sources of discontinuity	Explanation	Problems posed	Examples (of good and bad experiences)
New market emerges	Most markets evolve through a process of growth, segmentation, etc. But at certain times completely new markets emerge which can not be analysed or predicted in advance or explored through using conventional market research/analytical techniques	Established players don't see it because they are focused on their existing markets May discount it as being too small or not representing their preferred target market – fringe/cranks dismissal Originators of new product may not see potential in new markets and may ignore them – e.g. text messaging	Disk drives, excavators, mini-mills[47] Mobile phone/SMS where market which actually emerged was not the one expected or predicted by originators
New technology emerges	Step change takes place in product or process technology – may result from convergence and maturing of several streams (e.g. industrial automation, mobile phones) or as a result of a single breakthrough (e.g. LED as white light source)	Don't see it because beyond the periphery of technology search environment Not an extension of current areas but completely new field or approach Tipping point may not be a single breakthrough but convergence and maturing of established technological streams, whose combined effect is underestimated Not invented here effect – new technology represents a different basis for delivering value – e.g. telephone vs. telegraphy	Ice harvesting to cold storage[49] Valves to solid state electronics[39] Photos to digital images

continues overleaf

TABLE 1.4 (continued)

Triggers/ sources of discontinuity	Explanation	Problems posed	Examples (of good and bad experiences)
New political rules emerge	Political conditions which shape the economic and social rules may shift dramatically – for example, the collapse of communism meant an alternative model – capitalist, competition – as opposed to central planning – and many ex-state firms couldn't adapt their ways of thinking	Old mindset about how business is done, rules of the game, etc. are challenged and established firms fail to understand or learn new rules	Centrally planned to market economy e.g. former Soviet Union Apartheid to post-apartheid South Africa – inward and insular to externally linked[50] Free trade/globalization results in dismantling protective tariff and other barriers and new competition basis emerges[50,51]
Running out of road	Firms in mature industries may need to escape the constraints of diminishing space for product and process innovation and the increasing competition of industry structures by either exit or by radical reorientation of their business	Current system is built around a particular trajectory and embedded in a steady-state set of innovation routines which militate against widespread search or risk taking experiments	Medproducts[52] Kodak Encyclopaedia Britannica[19] Preussag[53] Mannesmann
Sea change in market sentiment or behaviour	Public opinion or behaviour shifts slowly and then tips over into a new model – for example, the music industry is in the midst of a (technology-enabled) revolution in delivery systems from buying records, tapes and CDs to direct download of tracks in MP3 and related formats	Don't pick up on it or persist in alternative explanations – cognitive dissonance – until it may be too late	Apple, Napster, Dell, Microsoft vs. traditional music industry[54]

Deregulation/ shifts in regulatory regime	Political and market pressures lead to shifts in the regulatory framework and enable the emergence of a new set of rules – e.g. liberalization, privatization or deregulation	New rules of the game but old mindsets persist and existing player unable to move fast enough or see new opportunities opened up	Old monopoly positions in fields like telecommunications and energy were dismantled and new players/ combinations of enterprises emerged. In particular, energy and bandwidth become increasingly viewed as commodities. Innovations include skills in trading and distribution – a factor behind the considerable success of Enron in the late 1990s as it emerged from a small gas pipeline business to becoming a major energy trader[22] – unquantifiable chances may need to be taken
Fractures along 'fault lines'	Long-standing issues of concern to a minority accumulate momentum (sometimes through the action of pressure groups) and suddenly the system switches/tips over – for example, social attitudes to smoking or health concerns about obesity levels and fast-foods	Rules of the game suddenly shift and then new pattern gathers rapid momentum wrong-footing existing players working with old assumptions. Other players who have been working in the background developing parallel alternatives may suddenly come into the limelight as new conditions favour them	McDonalds and obesity Tobacco companies and smoking bans Oil/energy and others and global warming Opportunity for new energy sources like wind-power – c.f. Danish dominance[55]
Unthinkable events	Unimagined and therefore not prepared for events which – sometimes literally – change the world and set up new rules of the game	New rules may disempower existing players or render competencies unnecessary	9/11
Business model innovation	Established business models are challenged by a reframing, usually by a new entrant who redefines/reframes the problem and the consequent 'rules of the game'	New entrants see opportunity to deliver product/service via new business model and rewrite rules – existing players have at best to be fast followers	Amazon.com Charles Schwab Southwest and other low cost airlines[22,54,56]

continues overleaf

TABLE 1.4 *(continued)*

Triggers/ sources of discontinuity	*Explanation*	*Problems posed*	*Examples (of good and bad experiences)*
Shifts in 'techno-economic paradigm' – systemic changes which impact whole sectors or even whole societies	Change takes place at system level, involving technology and market shifts. This involves the convergence of a number of trends which result in a 'paradigm shift' where the old order is replaced	Hard to see where new paradigm begins until rules become established Existing players tend to reinforce their commitment to old model, reinforced by 'sailing ship' effects	Industrial Revolution[38,57,58] Mass production
Architectural innovation	Changes at the level of the system architecture rewrite the rules of the game for those involved at component level	Established players develop particular ways of seeing and frame their interactions – for example who they talk to in acquiring and using knowledge to drive innovation – according to this set of views Architectural shifts may involve reframing but at the component level it is difficult to pick up the need for doing so – and thus new entrants better able to work with new architecture can emerge	Photo-lithography in chip manufacture[33,59]

approaches to organizing and managing innovation. If we try and use established models which work under steady-state conditions we find – as is the reported experience of many – we are increasingly out of our depth and risk being upstaged by new and more agile players.

1.8 Innovation Is Not Easy . . .

Although innovation is increasingly seen as a powerful way of securing competitive advantage and a more secure approach to defending strategic positions, success is by no means guaranteed. The history of product and process innovations is littered with examples of apparently good ideas which failed – in some cases with spectacular consequences. For example:

- In 1952 Ford engineers began working on a new car to counter the mid-size models offered by GM and Chrysler – the 'E' car. After an exhaustive search for a name involving some 20 000 suggestions the car was finally named after Edsel Ford, Henry Ford's only son. It was not a success; when the first Edsels came off the production line Ford had to spend an average of $10 000 per car (twice the vehicle's cost) to get them roadworthy. A publicity plan was to have 75 Edsels drive out on the same day to local dealers; in the event the firm only managed to get 68 to go, whilst in another live TV slot the car failed to start. Nor were these teething troubles; by 1958 consumer indifference to the design and concern about its reputation led the company to abandon the car – at a cost of $450 m. and 110 847 Edsels.[60]

- During the latter part of the Second World War it became increasingly clear that there would be a big market for long-distance airliners, especially on the transatlantic route. One UK contender was the Bristol Brabazon, based on a design for a giant long-range bomber which was approved by the Ministry of Aviation for development in 1943. Consultation with BOAC, the major customer for the new airliner, was 'to associate itself closely with the layout of the aircraft and its equipment' but not to comment on issues like size, range and payload! The budget rapidly escalated, with the construction of new facilities to accommodate such a large plane and, at one stage, the demolition of an entire village in order to extend the runway at Filton, near Bristol. Project control was weak and many unnecessary features were included – for example, the mock-up contained 'a most magnificent ladies' powder room with wooden aluminium-painted mirrors and even receptacles for the various lotions and powders used by the modern young lady'. The prototype took six and a half years to build and involved major technical crises with wings and engine design; although it flew well in tests the character of the post-war aircraft market was very different from that envisaged by the technologists. Consequently in 1952, after flying less than 1000 miles, the project was abandoned at considerable cost to the taxpayer. The parallels with the Concorde project, developed by the same company on the same site a decade later, are hard to escape.[61]

- During the late 1990s revolutionary changes were going on in mobile communications involving many successful innovations – but even experienced players can get their fingers burned. Motorola launched an ambitious venture which aimed to offer mobile communications from literally anywhere on the planet – including the middle of the Sahara Desert or the top of Mount Everest! Achieving this involved a $7 bn project to put 88 satellites into orbit, but despite the costs Iridium – as the venture was known – received investment funds from major backers and the network was established. The trouble was that, once the novelty had worn off, most people realized that they did not need to make many calls from remote islands or at the North Pole and that their needs were generally well met with less exotic mobile networks based around large cities and populated regions. Worse, the handsets for Iridium were large and clumsy because of the complex electronics and wireless equipment they had to contain – and the cost of these hi-tech bricks was a staggering $3000! Call charges were similarly highly priced. Despite the incredible technological achievement which this represented the take-up of the system never happened, and in 1999 the company filed for Chapter 11 bankruptcy. Its problems were not over – the cost of maintaining the satellites safely in orbit was around $2m. per month. Motorola who had to assume the responsibility had hoped that other telecoms firms might take advantage of these satellites, but after no interest was shown they had to look at a further price tag of $50m. to bring them out of orbit and destroy them safely! Even then the plans to allow them to drift out of orbit and burn up in the atmosphere were criticized by NASA for the risk they might pose in starting a nuclear war, since any pieces which fell to earth would be large enough to trigger Russian anti-missile defences since they might appear not as satellite chunks but Moscow-bound missiles!

- A survey of 14 000 organizations purchasing computer software carried out for the UK Department of Trade and Industry suggested that between 80 and 90% of projects failed to meet their performance goals, around 80% were delivered late and over budget, around 40% failed or were abandoned and only 10–20% fully met their success criteria.

- Whilst the Internet was seen as a seedbed for an enormous number of new ventures, the experience of the 'dot.coms' has not all been rosy. Some firms like Amazon and Yahoo! saw their share prices surge upwards on initial flotation – but for them and many others the bubble burst. New players were ill-equipped to survive and only a handful of the original start-ups remain – but even large and established players were hit hard. For example, the giant telecommunications player BT lost 60% of its market value, whilst Marconi eventually went under.

Of course, not all failures are as dramatic or as complete as these; for most organizations the pattern is one of partial success but with problems. For example, studies of product innovation consistently point to a high level of 'failure' between initial idea and having a successful product in the marketplace. Actual figures range from 30% to as high as 95%; an accepted average is 38%.[62] But this shouldn't surprise us – after all, innovation is by its nature a risky business and like omelettes eventual success will involve broken eggs. And we need to remember that there is a great deal of uncertainty in innovation, made up of technical, market, social, political and other factors, with the result that the odds are not too good for success unless the process is managed carefully. Even the best-managed firms still make mistakes – for example, the success story of 3 M's 'Post-it' notes is actually a somewhat chequered history where the innovation might have failed at several points.[63,64] And, as Perez points out, the pattern of riding on technology-driven bubbles which eventually burst, with dramatic consequences, is not a new one.[58]

The key point is to ensure that experiments are well designed and controlled so as to minimize the incidence of failure and to ensure that where it does occur lessons are learned to avoid falling into the same trap in the future.

1.9 . . . But It Is Imperative

Faced with what is clearly a risky and uncertain process many organizations could be forgiven for deciding not to innovate, even though the possible rewards are attractive. However, that approach – of doing nothing – is rarely an option, especially in turbulent and rapidly changing sectors of the economy. In essence, unless organizations are prepared to renew their products and processes on a continuing basis, their survival chances are seriously threatened.

In the mid-1980s a study by Shell suggested that the average corporate survival rate for large companies was only about half as long as that of a human being. Since then the pressures on firms have increased enormously from all directions – with the inevitable result that life expectancy is reduced still further. Many studies look at the changing composition of key indices and draw attention to the demise of what were often major firms and in their time key innovators. For example, Foster and Kaplan point out that of the 500 companies originally making up the Standard and Poor 500 list in 1857, only 74 remained on the list through to 1997.[18] Of the top 12 companies which made up the Dow Jones index in 1900 only one – General Electric – survives today. Even apparently robust giants like IBM, GM or Kodak can suddenly display worrying signs of mortality, whilst for small firms the picture is often considerably worse since they lack the protection of a large resource base.

Some firms have had to change dramatically to stay in business. For example, a company founded in the early nineteenth century, which had Wellington boots and toilet paper amongst its product range, is now one of the largest and most successful in the world in the telecommunications business. Nokia began life as a lumber company, making the equipment and supplies needed to cut down forests in Finland. It moved through into paper and from there into the 'paperless office' world of IT – and from there into mobile telephones.

Another mobile phone player – Vodafone Airtouch – grew to its huge size by merging with a firm called Mannesman which, since its birth in the 1870s, has been more commonly associated with the invention and production of steel tubes! Tui is the company which now owns Thomson the travel group in the UK, and is the largest European travel and tourism services company. Its origins, however, lie in the mines of old Prussia where it was established as a public sector state lead mining and smelting company![53]

Nor is this only a problem for individual firms; as Utterback's study indicates, whole industries can be undermined and disappear as a result of radical innovation which rewrites the technical and economic rules of the game. Two worrying conclusions emerge from his work; first, that many innovations which destroy the existing order originate from newcomers and outsiders to a particular industry, and second, that a significant number of the original players survive such transformations.[49]

So the question is not one of whether or not to innovate but rather of *how* to do so successfully. What lessons can we learn from research and experience about success and failure, and is there any pattern to these which might be used to guide future action?

In a process as uncertain and complex as innovation, luck plays a part. There are cases where success comes by accident – and sometimes the benefits arising from one lucky break are enough to cover several subsequent failures. But real success lies in being able to repeat the trick – to manage the process consistently so that success, whilst never guaranteed, is more likely. And this depends on understanding and managing the process such that little gets left to chance. Research suggests that success is based on the ability to learn and repeat these behaviours; it's similar to the golfer Gary Player's comment that 'the more I practise, the luckier I get . . .'

So what do we have to manage? We suggest that innovation is a core process concerned with renewing what the organization offers (its products and/or services) and the ways in which it generates and delivers these. Whether the organization is concerned with bricks, bread, banking or baby care, the underlying challenge is still the same. How to obtain a competitive edge through innovation – and through this survive and grow? (This is as much a challenge for non-profit organizations – in police work, in health care, in education the competition is still there, and the role of innovation still one of getting a better edge to dealing with problems of crime, illness or illiteracy.)

At this generic level we suggest that organizations have to manage four phases making up the innovation process. They have to:

- Scan and search their environments (internal and external) to pick up and process signals about potential innovation. These could be needs of various kinds, or opportunities arising from research activities somewhere, or pressures to conform to legislation, or the behaviour of competitors – but they represent the bundle of stimuli to which the organization must respond.

- Strategically select from this set of potential triggers for innovation – those things which the organization will commit resources to doing. Even the best resourced organization cannot do everything, so the challenge lies in selecting those things which offer the best chance of developing a competitive edge.

- Resource the option – providing (either by creating through R&D or acquiring through technology transfer) the knowledge resources to exploit it. This might be a simple matter of buying off the shelf, or exploiting the results of research already carried out – or it might require extensive search to find the right resources. It is also not just about embodied knowledge, but about the surrounding bundle of knowledge – often in tacit form – which is needed to make the technology work.

- Implement the innovation, growing it from an idea through various stages of development to final launch – as a new product or service in the external marketplace or a new process or method within the organization.

- A fifth – optional – phase is to reflect upon the previous phases and review experience of success and failure – in order to learn about how to manage the process better, and to capture relevant knowledge from the experience.

Of course there are countless variations on this basic theme in terms of how organizations actually carry this out. And much depends on where they start from – their particular contingencies. For example, large firms may structure the process much more extensively than smaller firms who work on an informal basis. And firms in knowledge-intensive sectors like pharmaceuticals will concentrate more on formal R&D – often committing sizeable amounts of their income back to this activity – whereas others like clothing will emphasize closer links with their customers as a source of innovation. Non-profit organizations may be more concerned with reducing costs and improving quality, whereas private-sector firms may worry about market share.

Networks of firms may have to operate complex co-ordination arrangements to ensure successful completion of joint projects – and to devise careful legal frameworks to ensure that intellectual property rights are respected.

But at heart the process is the same basic sequence of activity. Innovation management is about learning to find the most appropriate solution to the problem of consistently managing this process, and doing so in the ways best suited to the particular circumstances in which the organization finds itself. Therefore particular solutions to the general problem of managing this core process will be firm-specific. (We will look at this process view of innovation in more detail in the following chapter.)

We suggest that there are three key questions in innovation management which form the basis of this book:

1. How do we structure the innovation process appropriately?
2. How do we develop effective behavioural patterns (routines) which define how it operates on a day-to-day basis?
3. How do we adapt or develop parallel ones to deal with the different challenges of 'steady-state' and discontinuous innovation?

A great deal of research on the management of innovation has attempted to identify some form of 'best practice', but most of these studies have been based on the experience of particular contexts. For example, the dominant models of technology management are derived from the experience of US high-technology firms, whereas many of the 'rules' for product development are based on research on the practice of Japanese manufacturers of consumer durables. However, there is unlikely to be 'one best way' to manage innovation, as industries differ in terms of technological and market opportunity, and firm-specific features constrain management options.

For this reason in this book we reject the 'one best way' school of management, and instead seek to explore the links between the structures, processes and culture of an organization, the opportunity for and characteristics of technological innovation, and the competitive and market environment in which the organization operates.

1.10 New Challenges, Same Old Responses?

Constant revolutionizing of production, uninterrupted disturbance of all social conditions, everlasting uncertainty . . . all old-established national industries have been destroyed or are daily being destroyed. They are dislodged by new industries . . . whose products are consumed not only at home, but in every quarter of the globe. In place of old wants satisfied by the production of the country, we find new wants . . . the intellectual creativity of individual nations become common property.

(K. Marx and F. Engels, The Manifesto of the Communist Party, 1848)

The quote demonstrates that uncertainty, globalization and innovation are not new, and that the only certainty about tomorrow's environment is that it will be just as uncertain as today's. This flash of the blindingly obvious reminds us of a major difficulty in managing innovation – the fact that we are doing so against a constantly shifting backdrop. And it is clear that some trends in the current environment are converging to create conditions which many see as rewriting the rules of the competitive game.

Certainly there are big changes taking place in the environment in which we have to try and manage innovation, and in this final section we will look briefly at some of the major forces underpinning such change. Our view is, however, that whilst there is no room for complacency, we should also not be in a hurry to throw away the basic principles on which this book is based – they will certainly need adapting and configuring to dramatically new circumstances but underneath the innovation management puzzle is what it always was – a challenge to accumulate and deploy knowledge resources in strategically effective fashion.

For example, we have already looked at the challenge of discontinuous innovation. History suggests that although the technological and market shifts are dramatic the basic innovation management issues remain. In particular, organizations need to search actively and widely; developing sensitive antennae and a strong future orientation are important activities.

Similarly it is becoming clear that under current competitive conditions in many sectors protectable competitive advantage comes increasingly from knowledge – because what firms know and have is hard to copy and requires others to go through a similar learning process.[65–67] But in such a turbulent environment it is inevitable that some knowledge assets become redundant and others need to be acquired quickly. This places emphasis on strategic management of the knowledge base, and of developing effective mechanisms for resourcing technological knowledge. In the latter case it is likely that the generation of relevant knowledge will increasingly take place outside the firm and will need capabilities to ensure that technology transfer can be absorbed and deployed quickly and effectively.

The difficulties of a firm like Kodak illustrate the problem. Founded around 100 years ago the basis of the business was the production and processing of film and the sales and service associated with mass-market photography. Whilst the latter set of competencies are still highly relevant (even though camera technology has shifted) the move away from wet physical chemistry in the dark (coating emulsions onto film and paper) to digital imaging represents a profound change for the firm. It needs – across a global operation and a workforce of thousands – to let go of old competencies which are unlikely to be needed in the future whilst at the same time to rapidly acquire and absorb cutting edge new technologies in electronics and communication. Although they are making strenuous efforts to shift from being a manufacturer of film to becoming a key player in the digital imaging industry the response from the stock markets suggests some scepticism as to their ability to do so.

The concept of component and architectural innovation is relevant here – organizations need to develop the ability to see which parts of their activity are affected by technological change and to react accordingly.[33] In some cases change at the component level opens up new opportunities – for example, new materials or propulsion systems like fuel cells may open up new options for vehicle assemblers but will not necessarily challenge their core operations. But the shift to peer-to-peer networking and downloading (Box 1.5) as an alternative way of creating and distributing music via the Internet poses challenges to the whole system of music production and publishing and may require a much more significant response.

In other words, even under conditions of high uncertainty and apparent rewriting of the rules, the basic themes around which this book is structured remain constant. Successful innovation depends on being able to look widely and ahead and develop strategic approaches based on an understanding of the knowledge aspects.

In the following pages we look briefly at some of the powerful forces shaping the competitive environment and arguably rewriting the rules of the game. These are:
- globalization of markets and technology supply;
- the emergence of technologies enabling a 'virtual' mode of working;
- the growing concern about sustainability;
- the rise of networking as a business model.

Innovation in a Global Environment

A key challenge in the current environment is that the stage on which the innovation game is played out has expanded enormously. Whereas technological development was confined to a few nations in the early twentieth century it has expanded to the point where it is generated and used globally – and where the challenges are those of being a global player. This has always been the theme of innovation strategy in multinational corporations, but it now becomes the issue for small enterprises. Even a local firm is no longer insulated; increasingly, large firms are looking to source components, handle administrative processes and manage distribution on a global basis. For example, in the automobile industry components for a car made in Germany may be sourced from as far afield as Brazil or South Africa, whilst in the airline business most of the data processing required to handle reservations and billing is done in the West Indies. Software production for Citigroup is handled by a 'software factory' in Bangalore, India. This

BOX 1.5
THE CHANGING NATURE OF THE MUSIC INDUSTRY

One of the less visible but highly challenging aspects of the Internet is the impact it has had – and is having – on the entertainment business. This is particularly the case with music. At one level its impacts could be assumed to be confined to providing new 'e-tailing' channels through which you can obtain the latest CD of your preference – for example from Amazon.com or CD-Now or 100 other websites. These innovations increase the choice and tailoring of the music purchasing service and demonstrate some of the 'richness/reach' economic shifts of the new Internet game.

But beneath this updating of essentially the same transaction lies a more fundamental shift – in the ways in which music is created and distributed and in the business model on which the whole music industry is currently predicated. In essence the old model involved a complex network in which songwriters and artists depended on A&R (artists and repertoire) to select a few acts, production staff who would record in complex and expensive studios, other production staff who would oversee the manufacture of physical discs, tapes and CDs and marketing and distribution staff who would ensure the product was publicized and disseminated to an increasingly global market.

Several key changes have undermined this structure and brought with it significant disruption to the industry. Old competencies may no longer be relevant whilst acquiring new ones becomes a matter of urgency. Even well-established names like Sony find it difficult to stay ahead whilst new entrants are able to exploit the economics of the Internet. At the heart of the change is the potential for creating, storing and distributing music in digital format – a problem which many researchers have worked on for some time. One solution, developed by one of the Fraunhofer Institutes in Germany, is a standard based on the Motion Picture Experts Group (MPEG) level 3 protocol – MP3. MP3 offers a powerful algorithm for managing one of the big problems in transmitting music files – that of compression. Normal audio files cover a wide range of frequencies and are thus very large and not suitable for fast transfer across the Internet – especially with a population who may only be using relatively slow modems. With MP3 effective compression is achieved by cutting out those frequencies which the human ear cannot detect – with the result that the files to be transferred are much smaller.

As a result MP3 files can be moved across the Internet quickly and shared widely. Various programs exist for transferring normal audio files and inputs – such as CDs – into MP3 and back again.

continues overleaf

BOX 1.5   (*continued*)

What does this mean for the music business? In the first instance aspiring musicians no longer need to depend on being picked up by A&R staff from major companies who can bear the costs of recording and production of a physical CD. Instead they can use home recording software and either produce a CD themselves or else go straight to MP3 – and then distribute the product globally via newsgroups, chatrooms, etc. In the process they effectively create a parallel and much more direct music industry which leaves existing players and artists on the sidelines.

Such changes are not necessarily threatening. For many people the lowering of entry barriers has opened up the possibility of participating in the music business – for example, by making and sharing music without the complexities and costs of a formal recording contract and the resources of a major record company. There is also scope for innovation around the periphery – for example in the music publishing sector where sheet music and lyrics are also susceptible to lowering of barriers through the application of digital technology. Journalism and related activities become increasingly open – now music reviews and other forms of commentary become possible via specialist user groups and channels on the Web, whereas before they were the province of a few magazine titles. Compiling popularity charts – and the related advertising – is also opened up as the medium switches from physical CDs and tapes distributed and sold via established channels to new media such as MP3 distributed via the Internet.

As if this were not enough, the industry is also challenged from another source – the sharing of music between different people connected via the Internet. Although technically illegal this practice of sharing between people's record collections has always taken place – but not on the scale which the Internet threatens to facilitate. Much of the established music industry is concerned with legal issues – how to protect copyright and how to ensure that royalties are paid in the right proportions to those who participate in production and distribution. But when people can share music in MP3 format and distribute it globally, the potential for policing the system and collecting royalties becomes extremely difficult to sustain.

It has been made much more so by another technological development – that of person-to-person or P2P networking. Sean Fanning, an 18-year-old student with the nickname 'the Napster', was intrigued by the challenge of being able to enable his friends to 'see' and share between their own personal record collections. He argued that if they held these in MP3 format then it should be possible to set up some kind of central exchange program which facilitated their sharing.

The result – the Napster.com site – offered sophisticated software which enabled P2P transactions. The Napster server did not actually hold any music on its files – but every day millions of swaps were made by people around the world exchanging their music collections. Needless to say this posed a huge threat to the established music business since it involved no payment of royalties. A number of high-profile lawsuits followed but whilst Napster's activities have been curbed the problem did not go away. There are now many other sites emulating and extending what Napster started – sites such as Gnutella take the P2P idea further and enable exchange of many different file formats – text, video, etc. In Napster's own case the phenomenally successful site concluded a deal with entertainment giant Bertelsman which paved the way for subscription-based services which provide some revenue stream to deal with the royalty issue.

Expectations that legal protection would limit the impact of this revolution have been dampened by a US Court of Appeal ruling which rejected claims that P2P violated copyright law. Their judgment said, 'History has shown that time and market forces often provide equilibrium in balancing interests, whether the new technology be a player piano, a copier, a tape recorder, a video recorder, a PC, a karaoke machine or an MP3 player.'[68]

Significantly the new opportunities opened up by this were seized not by music industry firms but by computer companies, especially Apple. In parallel with the launch of their successful i-Pod personal MP3 player they opened a site called itunes which offered users a choice of thousands of tracks for download at 99c each. In its first weeks of operation it recorded 1m. hits and has gone on to be the market leader in an increasingly populated field, having notched up over 50m. downloads since opening in mid-2003.

forces a reappraisal of positioning in global economic terms, whether at the level of individual enterprises within global value chains[69] or at the national economy level. For example, a recent report by Michael Porter and colleagues for the UK government concluded that

the UK currently faces a transition to a new phase of economic development. We find that the competitiveness agenda facing UK leaders in government and business reflects the challenges of moving from a location competing on relatively low costs of doing business to a location competing on unique value and innovation. This transition requires investments in different elements of the business environment, upgrading of company strategies, and the creation and strengthening of new types of institutions.[70]

One of the key enablers of this distribution is information and communications technology (ICT) which – as we saw above – radically changes the balance of richness and reach involved in all kinds of information-based businesses. In the case of design, for example, a firm like IBM can now work on a 24-hour day by mobilizing design teams in the UK, the USA and Japan with each team handing over after its 'shift' to the next time zone where the work will be continued. This has two effects – first it radically compresses the time in which the design of new components or equipment takes place, and second it brings to bear different and complementary knowledge sets. But in order to make such systems work a new form of network/global management is required, one which addresses some of the underlying national cultural characteristics as well as the departmental or functional ones.[40,41]

The production of knowledge has become far more global – although R&D is still a heavy investment item in major industrialized countries, there is an acceleration across the newly-industrializing world. Similarly the number of scientists and engineers is increasing faster in Asia than elsewhere and this is likely to fuel further innovation-led growth in that region. For example, the number of engineering degrees awarded in 1998 in Europe was 159 000, in the USA 62 000 and in Asia around 280 000.[71]

Consequently the major challenge to innovation management is one of managing the same basic principles but on a much bigger stage. With trade liberalization and the opening of markets has come a massive upsurge in overall activity and the number of players in the game. (It is estimated, for example, that the entire volume of world trade which took place during 1950 is now transacted in a single day!) Competition has intensified and much of it is being driven by innovation in products, services and processes. The response of successful firms is increasingly likely to involve some measure of networking and collaboration.

Innovation in a Virtual World

One of the defining symbols of the early twenty-first-century environment for innovation is the Internet. Born out of informal exchanges and a desire amongst scientists to share and collaborate more effectively, this has grown into a framework for change which bears comparison with the advent of the railways in the nineteenth century.[72] It has fuelled – and been fuelled by – the rise in the power and versatility of ICT, and it has generated an enormous user base – estimates vary, but from a figure of around 35 million users in the late 1990s there are now probably over 1 billion people with access to the Internet around the world.

Mobile telephones provide a similar example of huge growth and penetration. There are currently around 600 m. units/year sold, and markets in developed countries close to saturation. Even in developing countries there is a high access rate – for example, 'telephone ladies' in Bangladesh rent out by the minute so even the poorest citizens have access.[73]

Such developments – and their parallel and complementary versions inside organizations, across private networks and using different media – wireless, cable, satellite, etc. – create a communications and participation revolution which, one might expect, has all the characteristics of a discontinuous shift in the innovation environment. Yet if we analyse this we can see the same forces for innovation at work as were operating centuries ago. On the 'technology push' side the range of opportunity created by ICT developments is enormous – it has become 'a solution looking for problems'. But similar characteristics were present when steam power first became widely available and reliable. And – as the glut of failed Internet start-ups demonstrates – simply having the technological means is no guarantee of business success – innovation, as always, is about effective coupling of needs and means within a strategic framework.

Similarly on the demand side, there are forces at work which are acting to pull innovations through and to shape and direct the pace and nature of change. Not surprisingly, much of the impact has come in areas which are essentially information rich in terms of their content and delivery – for example, services like banking and insurance have been heavily hit by new developments. Two useful concepts in this connection are those of 'richness' and 'reach' – terms coined by the Boston Consulting Group to help think about where the impacts of the e-revolution are likely to be felt. Richness refers to the content of an information service – how customized and deep it is – whereas reach refers to the extent to which it can be offered to a population. Normally there is a trade-off – you can have rich services but they tend to be high price and reach only a few people with the means to access them – for example, a personalized bank or a tailored travel package via a personal consultant. Equally, low cost services with high reach tend to be characterized by a 'one size fits all' mentality and to compete on the basis of low cost. What the ICT revolution does is shift the balance between these two so that rich services are available but with global reach – and a new economics emerges.[19,74]

This is a seductive argument and there are certainly good examples where industries or sectors have been transformed by the new balance of richness and reach – in addition to banking and insurance we can think of travel (last-minute.com), publishing (Amazon), retailing (QXL, e-Bay) and many others. But there are clear limits to the extent to which even revolutionary changes in the availability of service delivery options will lead to discontinuity. Not all sectors are information rich and consumers still consume goods as well as services. For much of the retail end of the e-revolution there is still the problem of the 'last mile' – getting the physical goods delivered to particular households. These goods have to be manufactured and although the co-ordination and control may become increasingly subject to ICT innovation, it will still be necessary to store and move physical goods around. And in hospitals automated medicine still can't help with the growing demands of care especially amongst an increasingly aged population.

In other words, the innovation management picture remains surprisingly constant. There will certainly be differences – for example, we will need to consider:

- Very high velocity interactions;
- Very rich potential connectivity involving many different players;
- Global orientation where distance becomes irrelevant.

But the underlying problem remains one of picking up – and making sense of – signals about triggers for innovation, and then managing the process of change effectively.

Innovation and Sustainability

Of increasing relevance in the innovation agenda is the concern being expressed about sustainability. Issues here include:

- Global warming and the threats posed by climate change.
- Environmental pollution and the pressure towards 'greener' products and services.
- Population growth and distribution, with accompanying problems of urban concentration.
- Declining availability of energy and pressure to find renewable and alternative resources.
- Health and related issues of access to basic standards of care, clean water, simple public hygiene, etc.

Such concerns are not new – there was, for example, an extensive debate during the 1970s around 'the limits to growth' in which a variety of 'doomsday' scenarios were predicted.[75,76] Although enormously relevant, the resolution of such concerns owed much to an underlying innovation process which helped deal with some of the more urgent problems and opened up new possible directions to ameliorate others. In similar fashion, the sustainability agenda today poses challenges but also opens up significant innovation opportunities. We can see these distributed across our range of innovation types, for example involving:

- New or more sustainable products and services such as fuel cells, solar power systems, biodegradable waste, organic foods, low-impact transportation systems, etc.
- New or more sustainable processes such as low-energy processing, minimal impact mining operations, electronic rather than physical transaction processing, etc.
- New or extended markets built on exploiting a growing concern with sustainability issues – for example 'clean and green' foodstuffs, furniture made with Forestry Stewardship Council certification, eco-tourism, etc.
- New business models reframing existing arrangements to emphasize sustainability – for example, ethical investment services, environmentally responsible retailing (B&Q, IKEA, Body Shop), socially responsible business promotions (such as the Co-op and its support for 'fair-trade' products), etc.

Beyond these new opportunities lies a second powerful driver for innovation around sustainability – its potential for creating discontinuous conditions. As we saw earlier in the chapter there are periods when the 'rules of the game' change and this often threatens existing incumbents and opens up opportunities for new entrants to particular sectors. Trends such as those outlined above can build for some time and suddenly flip as social attitudes harden or new information emerges. The shift in perception of smoking from recreation to health hazard and the recent concerns about fast foods as a major contributor to high obesity levels are examples and have had marked impacts on the rate and pattern of innovation in their industries.

Sustainability issues are often linked to regulation and such legislation can add additional force to changing the rules of the game – for example, the continuing effects of clean air and related environmental pollution legislation have had enormous and cumulative effects on industries involved in chemicals, materials processing, mining and transportation, both in terms of products and processes. Current directives such as those of the European Union around waste and recycling mean that manufacturers are increasingly having to take into account the long-term use and disposal of their products as well as their manufacture and sales – and this is forcing innovation in both products, processes and administrative models (such as whole life costing).[77]

Discontinuities open up new opportunities as well as challenging existing arrangements, and the other side of this sustainability coin is the potential for new growth markets in, for example, alternative energy sources, green products and services and new transportation or construction systems.

Innovation linked to issues of sustainability often has major systems-level implications and emphasizes the need to manage in integrated fashion. Such innovations arise from concerns in, and need to be compatible with, complex social, political and cultural contexts and there is a high risk of failure if these demand-side elements are neglected. For example, the wind power industry is an old one, originally going back to the windmill technologies of the medieval times. It expanded significantly during the opening up of the United States and Australia, and significant acceleration of innovation in various aspects of product design took place. But although there is now another wave of technological innovation and market growth associated with exploiting wind power on a large scale, the leaders in this have not been the USA (despite extensive R&D investment) but rather Denmark where the development followed a simpler, smaller-scale approach matched to meeting energy needs of small and local communities. As Douthwaite points out this has enabled the Danish industry to develop significant competence through interacting with a growing user base and building technological sophistication from the bottom up.[55]

In similar fashion the development of 'appropriate technologies' essentially involves matching local demand-side conditions by configuring specific solutions, often involving established technologies. Examples include the clockwork radio, intermediate technology pumps, tractors and other machinery and micro-credit investment banking.

No Firm is an Island – The Challenge of Networking

Innovation could once have been seen as the province of a few heroic individuals who pioneered ideas into action – and certainly many of the great nineteenth-century names conform to this stereotype. Of course, even then it was actually a linked system with sources of finance, of marketing, etc. being part of the puzzle. But the twentieth century – as Freeman observed – was essentially the era of organized R&D and the rise of the firm as the unit of innovation.[78] We can think of particular names and innovations in this context – Bell Labs, 3M, Pilkington, Ford, Hewlett-Packard. Here the role of champions is still important, but the stage on which they act is essentially defined by the firm. But in the twenty-first century the game has moved on again and it's now very clearly a multiplayer one. Innovation involves trying to deal with an extended and rapidly advancing scientific frontier, fragmenting markets flung right across the globe, political uncertainties, regulatory instabilities – and a set of competitors who are increasingly coming from unexpected directions. The response has to be one of spreading the net wide and trying to pick up and make use of a wide set of knowledge signals – in other words, learning to manage innovation at the *network* level.

This is something which Roy Rothwell foresaw in his pioneering work on models of innovation with a gradual move away from thinking about (and organizing) a linear science/technology push or demand pull process to one which saw increasing *interactivity* – at first across the firm with cross-functional teams and other boundary-spanning activities and increasingly outside the firm in its links with others. His vision of the 'fifth generation' innovation is essentially the one in which we now need to operate – with rich and diverse network linkages accelerated and enabled by an intensive set of information and communication technologies.[79]

A key driver of this is the division of labour effect whereby firms increasingly question their core competencies and purpose and configure networks accordingly. For example, one of the most successful firms of the twentieth century – General Electric – reconfigured its business in aircraft engines by thinking about 'selling power by the hour' – and as a result moved away from manufacturing activities like grinding turbine blades and into outsourcing these areas of competence. It increasingly became a co-ordinator and began to explore how it could provide financing and other necessary support services – with the result that it is now largely a service business offering a turnkey package to airlines who equally see their needs as buying power for lifting their aircraft rather than shopping for jet engines.

Similar examples include the running shoe firm Nike which sees its competencies in design and marketing rather than in manufacturing, and Dell which has built a business out of configuring computers to individual needs but which makes extensive use of outsourcing and the management of complementary networks.

Even the biggest and most established innovators are recognizing this shift. Procter & Gamble spend around $2 bn each year on what used to be termed R&D – but these days they use the phrase 'Connect and Develop' instead and have set themselves the ambitious goal of sourcing much of their idea input from outside the company. As Nabil Sakkab, Senior Vice President of Research and Development commented recently, 'The future of R&D is C&D – collaborative networks that are in touch with the 99% of research that we don't do ourselves. P&G plans to keep leading innovation and this strategy is crucial for our future growth.' Similar stories can be told for firms like IBM, Cisco, Intel – they are all examples of what Henry Chesborough calls the move towards 'open innovation' where links and connections become as important as actual production and ownership of knowledge.[80]

Third, there is a recognition that networks may not simply be one end of the traditional spectrum between doing everything in-house (vertical integration) and of outsourcing everything to suppliers (with the consequent transaction costs of managing them).[81] It is possible to argue for a 'third way' which builds on the theory of systems and that networks have emergent properties – the whole is greater than the sum of the parts. This does not mean that the benefits flow without effort – on the contrary, unless participants in a network can solve the problems of co-ordination and management they risk being suboptimal. But there is growing evidence of the benefits of networking as a mode of operation in innovation.[82–84] We pick up this theme in more detail in Chapter 8.

For example, participating in innovation networks can help firms bump into new ideas and creative combinations – even in mature businesses. It's well known in studies of creativity that the process involves making associations – and sometimes the unexpected conjunction of different perspectives can lead to surprising results. And the same seems to be true at the organizational level; studies of networks indicate that getting together in such fashion can help open up new and productive territory. For instance, recent developments in the use of titanium components in Formula 1 engines have been significantly advanced by lessons learned about the moulding process from a company producing golf clubs.[85]

Another way in which networking can help innovation is in providing support for shared learning. Much process innovation is about configuring and adapting what has been developed elsewhere and applying it – for example, in the many efforts which firms have been making to adopt world class manufacturing (and increasingly service) practice. Whilst it is possible to go it alone in this process, an increasing number of firms are seeing the value in using networks to give them some extra traction on the learning process.

These principles also underpin an increasing number of policy initiatives aimed at getting firms to work together on innovation-related learning. For example, the UK's Society of Motor Manufacturers and Traders has run the successful Industry Forum for many years helping a wide range of firms adopt and implement process innovations around world class manufacturing. This model has been rolled out (with government support) to sectors as diverse as ceramics, aerospace, textiles and tourism. Many Regional Development Agencies now try and use networks and clusters as a key aid to helping stimulate economic growth through innovation. And the same principles can be to help diffuse innovative practices along supply chains; companies like IBM and BAe Systems have made extensive efforts to make 'supply chain learning' the next key thrust in their supplier development programmes.[86]

The importance of such networking is not simply firm to firm – it is also about building rich linkages within the national system of innovation. Government policy to support innovation is increasingly concerned with enabling better connections between elements – for example, between the many small firms with technological needs and the major research and technology institutes, universities, etc. which might be able to meet these needs.[87]

Innovation is about taking risks and deploying what are often scarce resources in projects which may not succeed. So another way in which networking can help is by helping spread the risk and in the process extending the range of things which might be tried. This is particularly useful in the context of smaller firms where resources are scarce – and it is one of the key features behind the success of many industrial clusters. The case of the Italian furniture industry is one in which a consistently strong export performance has been achieved by firms with an average size of fewer than 20 employees. Keeping their position at the frontier in terms of performance has come through sustained innovation in design and quality — enabled by a network-based approach. This isn't an isolated case — one of the most respected research institutes in the world for textiles is CITER, based in Emilia Romagna. Unlike so many world class institutions this was not created in top-down fashion but evolved from the shared innovation concerns of a small group of textile producers who built on the network model to share risks and resources. Their initial problems with dyeing and with computer-aided design helped them gain a foothold in terms of innovation in their processes and in the years since its founding in 1980 it has helped its 500 (mostly small firm) members develop a strong innovation capability.

Long-lasting innovation networks can create the capability to ride out major waves of change in the technological and economic environment. Michael Best's fascinating account of the ways in which the Massachusetts economy managed to reinvent itself several times is one which places innovation networking at its heart.[88]

The implications for innovation management are again that the underlying questions remain the same — how to identify triggers and develop coherent strategic responses — the difference is that the unit to be managed is now a co-operative federation of players. The levers will need to be different and the routines may need to evolve — a major challenge but one with potentially high pay-offs.

1.11 Outline of the Book

The layout of the book is as follows. Chapter 2 looks at the core process of innovation and at variations in the way in which different organizations handle it in response to different contingencies. It also looks at the question of how organizations manage the operation of that process — and the behavioural patterns (routines) which they learn and develop to do so effectively. Drawing on research on success and failure in innovation the chapter provides a framework for categorizing these behaviour patterns into five clusters of enabling routines:

- Providing a supportive strategic framework.
- Developing pro-active linkages.
- Creating effective enabling mechanisms for the innovation process to operate.
- Building an innovative organizational context.
- Learning and capability development for innovation management.

Part II explores the first of these clusters — the creation of a strategic context for innovation. Chapter 3 considers the significance of an innovation strategy in conditions of complexity, continuous change and consequent uncertainty, and contrasts the rational and incrementalist approaches. It develops the three elements of innovation strategy proposed by David Teece and Gary Pisano: market and national positions, technological paths and organizational processes. In Chapter 4 we address the question of how the firm's national and market environment shapes its innovation strategy, in particular the effects of the home country, competencies, economic inducements and institutions. In Chapter 5 we show that marked differences amongst sectors are also central to corporate choices about technological trajectories, firm-specific competencies and innovation strategies, and identify five broad technological trajectories that firms can follow, each of which has distinct implications for the tasks of innovation strategy. We also identify three key technologies (biotechnology, materials and IT) where rapid advances lead to major shifts in technological trajectories, and where it is increasingly important to distinguish the microelectronics revolution (making and using electronic chips) from the more important information revolution (making and using software).

Part III is concerned with the enabling routines for building effective linkages outside the organization. Innovation does not take place in a vacuum, and research has consistently shown that successful organizations understand and work with different actors in their environment. Chapter 7 focuses particularly on the market-related linkages, looking at how markets are defined, explored and understood – and how this knowledge is communicated and updated throughout the organization. It also looks at how understanding of buyer behaviour can be used to support the launch of innovations – whether to an external market (for example, in launching a new consumer product) or an internal market (for example, managing the change process associated with introducing new machinery or systems). Chapter 8 looks at linkages of a different kind, associated with developing collaborations, networks and strategic alliances.

Part IV is concerned with the routines and mechanisms for enabling and implementing innovation. These include the particular structures for decision-making throughout the life of an innovation project, the arrangements for project monitoring and management and the mechanisms whereby change is planned and introduced to the organization. Every organization needs to do these things as part of managing innovation; research indicates that some do it better than others. Chapter 9 explores the different ways in which organizations operationalize these aspects of innovation. Chapter 10 looks at the special case of starting up an innovative new venture, building on the growing research base of work in this field. It highlights the ways in which organizations can move beyond their current range of technologies, products and processes, and the learning processes involved in doing so effectively. In particular the chapter focuses on internal corporate ventures and on the establishment of new technology-based firms.

In Part V the emphasis shifts to exploring the organizational context in which innovation takes place. Much is written about the need for loose, organic and flexible organizations – typified by Tom Peters' concept of 'thriving on chaos' – which offer considerable individual freedom to innovate. But there is a need to balance these models with some element of formality and control, and to ensure a clear sense of strategic direction. Chapter 11 looks at the different elements which influence the way in which innovation takes place, and the choices available to manage under different conditions. Issues explored include organizational structures, team working, participation, training and development, motivation and the development of a creative climate within the organization. The chapter also looks briefly at how these different elements can contribute to some form of corporate learning process which helps develop and accumulate competence – the 'learning organization'. A key theme in this chapter is that there is no 'best' model for organizing innovation; the key task is to find the most appropriate fit for a particular set of contingencies. Chapter 12 examines the special case of building a new organization for innovation, looking at the example of innovative small firms.

The book concludes by bringing together key themes. In particular we argue that success in innovation management is not a matter of doing one or two things exceptionally well but one of good all-round performance in the areas highlighted above. But there are also no simple and standard solutions to the problem of how to do this; organizations are like people and come in widely varying shapes, sizes and personalities. As we argue throughout the book, there is a need for each organization to find its own particular answers to the general puzzle of innovation management. There are general recipes available which can be adapted, and the discussion in the body of the book provides an indication of how this can be and has been done to advantage.

Innovation is particularly about learning, both in the sense of acquiring and deploying knowledge in strategic fashion and also in acquiring and reinforcing patterns of behaviour which help this competence-building learning to happen. Managing innovation is particularly about identifying and enabling the development of behaviour patterns – routines – which make such learning possible.

One important aspect of learning is structured reflection on the organization's current position as an input into its next strategic development. Taking stock – auditing – can be a powerful aid to organizational development, and the final chapter looks at the ways in which what we know about innovation management can be integrated into an audit framework.

Throughout the book we will also try and reflect the influence of two key challenges on the way in which we think about managing innovation – dealing with it under discontinuous conditions ('beyond the steady state') and as an inter-organizational, networked phenomenon ('beyond boundaries').

1.12 Summary and Further Reading

Few other texts cover the technological, market and organizational aspects of innovation in integrated fashion. Peter Drucker's *Innovation and Entrepreneurship*[89] provides a more accessible introduction to the subject, but perhaps relies more on intuition and experience than on empirical research. Since we published the first edition in 1997 a number of interesting texts have been published. Paul Trott's *Innovation Management and New Product Development* (now in its second edition) particularly focuses on the management of product development,[90] books by Bettina von Stamm[91] and Margaret Bruce[92] have a strong design emphasis and Tim Jones' book targets practitioners in particular.[7] Brockhoff *et al.*[93] and Sundbo and Fugelsang[94] provide some largely European views, while John Ettlie's *Managing Technological Innovation*,[24] is based on the experience of American firms, mainly from manufacturing, as are Mascitelli[95] and Schilling.[96] A few books explore the implications for a wider developing country context, notably Forbes and Wield,[97] and a number look at public policy implications.[98,99] Mark Dodgson's *The Management of Technological Innovation*,[100] has a strong historical and international perspective.

There are several compilations and handbooks covering the field, the best known being *Strategic Management of Technology and Innovation*, now in its fourth edition and containing a wide range of key papers and case studies, though with a very strong US emphasis.[45] A more international flavour is present in Dodgson and Rothwell,[16] and Shavinina.[101] The work arising from the Minnesota Innovation Project also provides a good overview of the field and the key research themes contained within it.[102]

Case studies of innovation provide a rich resource for understanding the workings of the process in particular contexts. Good compilations include those of Baden-Fuller[6] and Pitt, Nayak and Ketteringham[103] and Von Stamm[104] whilst other books link theory to case examples – e.g. Tidd and Hull[105] with its focus on service innovation. Several books cover the experiences of particular companies including 3M, Corning, DuPont and others.[64,106–108] Internet-related innovation is well covered in a number of books mostly oriented towards practitioners – for example, Evans and Wurster,[19] Loudon,[109] Oram,[110] Alderman[111] and Pottruck and Pearce.[74]

Most other texts tend to focus on a single dimension of innovation management. In the 'The nature of the innovative process', Giovanni Dosi adopts an evolutionary economics perspective and identifies the main issues in the management of technological innovation.[112] On the subject of organizational innovation, Jay Galbraith and E. Lawler[113] summarize recent thinking on organizational structures and processes, although a more critical account is provided by Wolfe (1994) in 'Organizational innovation: review, critique and suggested research', *Journal of Management Studies*, **31** (3), 405–432. For a review of the key issues and leading work in the field of organizational change and learning see M. D. Cohen and L. S. Sproull (eds), *Organizational Learning* (Sage, London, 1996).

Most marketing texts fail to cover the specific issues related to innovative products and services, although a few specialist texts exist which examine the more narrow problem of marketing so-called 'high-technology' products – for example, Jolly and Moore.[114–115] Helpful coverage of the core issues are to be found in the chapter, 'Securing the future' in Gary Hamel and C. K. Prahalad's *Competing for the Future* (Harvard Business School Press, 1994) and the chapter 'Learning from the market', in Dorothy Leonard's *Wellsprings of Knowledge* (Harvard Business School Press, 1995). There are also extensive insights into adoption behaviour drawn from a wealth of studies drawn together by Everett Rogers and colleagues.[116]

Particular themes in innovation are covered by a number of books and journal special issues; for example, services,[117] networks and clusters,[88,118] sustainability,[119] and discontinuous innovation.[18,40,120]

References

1 www.broens.com.au
2 De Jager, B. *et al.* (2004) 'Enabling continuous improvement – an implementation case study', *International Journal of Manufacturing Technology Management*, **15**, 315–324.
3 Port, O. (2004) 'The signs they are a changing', *Business Week*, September 10, 24.
4 Wagstyl, S. (1996) 'Innovation zealots a cut above the rest', *Financial Times*.
5 Kaplinsky, R., F. den Hertog and B. Coriat (1995) *Europe's Next Step*. Frank Cass, London.
6 Baden-Fuller, C. and M. Pitt (1996) *Strategic Innovation*. Routledge, London.
7 Jones, T. (2002) *Innovating at the Edge*. Butterworth Heinemann, London.
8 Kay, J. (1993) *Foundations of Corporate Success: How business strategies add value*. Oxford University Press, Oxford.
9 Office of Science and Technology (2000) *Excellence and Opportunity: A science policy for the 21st century*. London.
10 Souder, W. and J. Sherman (1994) *Managing New Technology Development*. McGraw-Hill, New York.
11 Tidd, J. (ed.) (2000) *From Knowledge Management to Strategic Competence: Measuring technological, market and organizational innovation*. London, Imperial College Press.
12 Stalk, G. and T. Hout (1990) *Competing against Time: How time-based competition is reshaping global markets*. Free Press, New York.
13 Walsh, V. *et al.* (1992) *Winning by Design: Technology, product design and international competitiveness*. Basil Blackwell, Oxford.
14 Rosenau, M. *et al.* (eds) (1996) *The PDMA Handbook of New Product Development*. John Wiley & Sons, Inc., New York.
15 Womack, J. and D. Jones (1996) *Lean Thinking*. Simon & Schuster, New York.
16 Dodgson, M. and R. Rothwell (eds) (1995) *The Handbook of Industrial Innovation*. Edward Elgar, London.

17 Pfeffer, J. (1994) *Competitive Advantage through People*. Harvard Business School Press, Boston, Mass.

18 Foster, R. and S. Kaplan (2002) *Creative Destruction*. Harvard University Press, Cambridge.

19 Evans, P. and T. Wurster (2000) *Blown to Bits: How the new economics of information transforms strategy*. Harvard Business School Press, Cambridge, Mass.

20 Schumpeter, J. (1950) *Capitalism, Socialism and Democracy*, 3rd edn. Harper & Row, New York.

21 Francis, D. and J. Bessant (2005) 'Targeting innovation and implications for capability development', *Technovation*, **25**(3), 171–183.

22 Hamel, G. (2000) *Leading the Revolution*. Harvard Business School Press, Boston, Mass.

23 Garcia, R. and R. Calantone (2002) 'A critical look at technological innovation typology and innovativeness terminology: A literature review', *Journal of Product Innovation Management*, **19**, 110–132.

24 Ettlie, J. (1999) *Managing Innovation*. John Wiley & Sons, Inc., New York.

25 Hollander, S. (1965) *The Sources of Increased Efficiency: A study of Dupont rayon plants*. MIT Press, Cambridge, Mass.

26 Tremblay, P. (1994) 'Comparative Analysis of Technological Capability and Productivity Growth in the Pulp and Paper Industry in Industrialised and Industrialising Countries', PhD thesis, University of Sussex.

27 Enos, J. (1962) *Petroleum Progress and Profits: A history of process innovation*. MIT Press, Cambridge, Mass.

28 Figueiredo, P. (2002) 'Does technological learning pay off? Inter-firm differences in technological capability-accumulation paths and operational performance improvement', *Research Policy*, **31**, 73–94.

29 Bessant, J. (2003) *High Involvement Innovation*. John Wiley and Sons, Ltd, Chichester.

30 Bell, R.M. and D. Scott-Kemmis (1990) *The Mythology of Learning-by-doing in World War 2 Airframe and Ship Production*. Science Policy Research Unit, University of Sussex.

31 Rothwell, R. and P. Gardiner (1985) 'Invention, Innovation and Re-innovation and the role of the user', *Technovation*, **3**, 176–186.

32 Nonaka, I., S. Keigo and M. Ahmed (2003) *Continuous Innovation: The power of tacit knowledge*, in Shavinina, L. (ed.), *International Handbook of Innovation*. Elsevier, New York.

33 Henderson, R. and K. Clark (1990) 'Architectural innovation: the reconfiguration of existing product technologies and the failure of established firms', *Administrative Science Quarterly*, **35**, 9–30.

34 Tidd, J. (1994) *Home Automation: Market and technology networks*. Whurr Publishers, London.

35 Kodama, F. (1992) 'Technology fusion and the new R&D', *Harvard Business Review*, July–August.

36 Abernathy, W. and K. Clark (1985) 'Mapping the winds of creative destruction', *Research Policy*, **14**, 3–22.

37 Boisot, M. (1995) 'Is your firm a creative destroyer? Competitive learning and knowledge flows in the technological strategies of firms', *Research Policy*, **24**, 489–506.

38 Dosi, G. (1982) 'Technological paradigms and technological trajectories', *Research Policy*, **11**, 147–162.

39 Braun, E. and S. Macdonald (1980) *Revolution in Miniature*. Cambridge University Press, Cambridge.

40 Christensen, C. and M. Raynor (2003) *The Innovator's Solution: Creating and sustaining successful growth*. Harvard Business School Press, Boston, Mass.

41 Foster, R. (1986) *Innovation – the attacker's advantage*. Pan Books, London.

42 Tushman, M. and P. Anderson (1987) 'Technological discontinuities and organizational environments', *Administrative Science Quarterly*, **31** (3), 439–465.

43 Gilfillan, S. (1935) *Inventing the Ship*. Follett, Chicago.

44 Cooper, A. and D. Schendel (1988) 'Strategic responses to technological threats', in Tushman, M. and Moore, W. (eds), *Readings in the Management of Innovation*. HarperCollins, London.

45 Burgelman, R., C. Christensen and S. Wheelwright (eds) (2004) *Strategic Management of Technology and Innovation*, 4th edn. McGraw Hill Irwin, Boston, Mass.

46 Bright, A. (1949) *The Electric Lamp Industry: Technological change and economic development from 1800 to 1947*. Macmillan, New York.

47 Christensen, C. (1997) *The Innovator's Dilemma*. Harvard Business School Press, Cambridge, Mass.

48 Cited at http://web.mit.edu/randy/www/words.html

49 Utterback, J. (1994) *Mastering the Dynamics of Innovation*. Harvard Business School Press, Boston, Mass., 256.

50 Barnes, J. *et al.* (2001) 'Developing manufacturing competitiveness in South Africa'. *Technovation*, **21** (5).

51 Kaplinsky, R., M. Morris and J. Readman (2003) 'The globalisation of product markets and immiserising growth: lessons from the South African furniture industry', *World Development*, **30** (7), 1159–1178.

52 Bessant, J. and D. Francis (2004) 'Developing parallel routines for product innovation', in *11th PDMA Product Development Conference*, Dublin, Ireland. EIASM, Brussels.

53 Francis, D., J. Bessant and M. Hobday (2003) 'Managing radical organisational transformation', *Management Decision*, **41** (1), 18–31.

54 Prahalad, C. (2004) 'The blinders of dominant logic', *Long Range Planning*, **37** (2), 171–179.

55 Douthwaite, B. (2002) *Enabling Innovation*. Zed Books, London.

56 Day, G. and P. Schoemaker (2004) 'Driving through the fog: managing at the edge'. *Long Range Planning*, **37** (2), 127–142.

57 Freeman, C. and C. Perez (1989) 'Structural crises of adjustment: business cycles and investment behaviour', in Dosi, G. (ed.), *Technical Change and Economic Theory*. Frances Pinter, London.

58 Perez, C. (2002) *Technological Revolutions and Financial Capital*. Edward Elgar, Cheltenham.

59 Henderson, R. (1994) 'The evaluation of integrative capability: innovation in cardiovascular drug discovery', *Industrial and Corporate Change*, **3** (3), 607–630.

60 Bryson, B. (1994) *Made in America*. Minerva, London.

61 Gilbert, J. (1975) *The World's Worst Aircraft*. Coronet, London.

62 Crawford, M. and C. Di Benedetto (1999) *New Products Management*. McGraw-Hill/Irwin, New York.

63 Gundling, E. (2000) *The 3M Way to Innovation: Balancing people and profit*. Kodansha International, New York.

64 Graham, M. and A. Shuldiner (2001) *Corning and the Craft of Innovation*. Oxford University Press, Oxford.

65 Teece, D. (1998) 'Capturing value from knowledge assets: the new economy, markets for know-how, and intangible assets', *California Management Review*, **40** (3), 55–79.

66 Quinn, J. (1992) *Intelligent Enterprise: A knowledge and service-based paradigm for industry*. Free Press, New York.

67 Quintas, P., P. Lefrere and G. Jones (1997) 'Knowledge management: a strategic agenda', *Long Range Planning*, **13** (3), 387.

68 *Personal Computer World* (2004), 32.

69 Morris, M. (2001) 'Creating value chain co-operation', IDS Special Bulletin, *The Value of Value Chains*, **32** (3).

70 DTI (2003) *Competing in the Global Economy: The innovation challenge*. Department of Trade and Industry, London.

71 Gann, D. (2004) 'Think, play, do: the business of innovation', *Inaugural Lecture*. Imperial College, London.

72 Berners-Lee, T. (2000) *Weaving the Web: The original design and ultimate destiny of the World Wide Web by its inventor*. Harper Business, New York.

73 *The Economist* (2004) 'March of the mobiles', 15, 25 September.

74 Pottruck, D. and T. Pearce (2000) *Clicks and Mortar*. Jossey-Bass, San Francisco.

75 Meadows, D. and J. Forrester (1972) *The Limits to Growth*. Universe Books, New York.

76 Freeman, C. et al. (eds) (1973) *Thinking about the Future: A critique of the limits to growth*. Universe Books, New York.

77 Dunphy, D., A. Griffiths and S. Benn (2003) *Organizational Change for Corporate Sustainability*. Routledge, London.

78 Freeman, C. and L. Soete (1997) *The Economics of Industrial Innovation*, 3rd edn. MIT Press, Cambridge.

79 Rothwell, R. (1992) 'Successful industrial innovation: critical success factors for the 1990s'. *R&D Management*, **22** (3), 221–239.

80 Chesborough, H. (2003) *Open Innovation: The new imperative for creating and profiting from technology*. Harvard Business School Press, Boston, Mass.

81 Williamson, O. (1975) *Markets and Hierarchies*. Free Press, New York.

82 Grandori, A. and G. Soda (1995) 'Inter-firm networks: antecedents, mechanisms and forms', *Organization Studies*, **16** (2), 183–214.

83 Danish Technological Institute (1991) *Network Co-operation – achieving SME competitiveness in a global economy*. Danish Technological Institute, Aarhus.

84 AIM (2004) *i-works: How high value innovation networks can boost UK productivity*. ESRC/EPSRC Advanced Institute of Management Research, London.

85 Mariotti, F. and R. Delbridge (2005) 'A portfolio of ties: managing knowledge transfer and learning within network firms', *Academy of Management Review* (forthcoming).

86 Bessant, J., R. Kaplinsky and R. Lamming (2003) 'Putting supply chain learning into practice', *International Journal of Operations and Production Management*, **23** (2), 167–184.

87 Bessant, J. (1999) 'The rise and fall of "Supernet": a case study of technology transfer policy for smaller firms', *Research Policy*, **28** (6), 601–614.

88 Best, M. (2001) *The New Competitive Advantage*. Oxford University Press, Oxford.

89 Drucker, P. (1985) *Innovation and Entrepreneurship*. Harper & Row, New York.

90 Trott, P. (2004) *Innovation Management and New Product Development*, 2nd edn. Prentice-Hall, London.

91 Von Stamm, B. (2003) *Managing Innovation, Design and Creativity*. John Wiley & Sons, Ltd, Chichester.

92 Bruce, M. and J. Bessant (eds) (2001) *Design in Business*. Pearson Education, London.

93 Brockhoff, K., A. Chakrabarti and J. Hauschildt (eds) (1999) *The Dynamics of Innovation*. Springer, Heidelberg.

94 Sundbo, J. and L. Fugelsang (eds) (2002) *Innovation as Strategic Reflexivity*. London, Routledge.

95 Mascitelli, R. (1999) *The Growth Warriors: Creating sustainable global advantage for America's technology industries*. Technology Perspectives, Northridge, California.

96 Schilling, M. (2005) *Strategic Management of Technological Innovation*. McGraw-Hill, New York.

97 Forbes, N. and D. Wield (2002) *From Followers to Leaders*. Routledge, London.

98 Branscomb, L. and J. Keller (eds) (1999) *Investing in Innovation: Creating a research and innovation policy that works*. MIT Press, Cambridge, Mass.

99 Dodgson, M. and J. Bessant (1996) *Effective Innovation Policy*. International Thomson Business Press, London.

100 Dodgson, M. (2000) *The Management of Technological Innovation*. Oxford University Press, Oxford.

101 Shavinina, L. (2003) *International Handbook on Innovation*. Elsevier, New York.

102 Van de Ven, A. (1999) *The Innovation Journey*. Oxford University Press, Oxford.

103 Nayak, P. and J. Ketteringham (1986) *Breakthroughs: How leadership and drive create commercial innovations that sweep the world*. Mercury, London.

104 Von Stamm, B. (2003) *The Innovation Wave*. John Wiley & Sons, Ltd, Chichester.

105 Tidd, J. and F. Hull (eds) (2003) *Service Innovation: Organizational responses to technological opportunities and market imperatives*. Imperial College Press, London.

106 Leifer, R. *et al.* (2000) *Radical Innovation*. Harvard Business School Press, Boston, Mass.

107 Kanter, R. (ed) (1997) *Innovation: Breakthrough thinking at 3M, DuPont, GE, Pfizer and Rubbermaid*. Harper Business, New York.

108 Kelley, T., J. Littman and T. Peters (2001) *The Art of Innovation: Lessons in Creativity from Ideo, America's leading design firm*. Currency, New York.

109 Loudon, A. (2001) *Webs of Innovation*. FT.Com, London.

110 Oram, A. (ed) (2001) *Peer-to-peer: Harnessing the power of disruptive technologies*. O'Reilly, Sebastopol, California.

111 Alderman, J. (2001) *Sonic Boom*. Fourth Estate, London.

112 Dosi, G. *et al.* (1988) *Technical Change and Economic Theory*. Pinter Publishers, London.

113 Galbraith, J. and E. Lawler (1988) *Organizing for the Future*. Jossey-Bass, San Francisco.

114 Jolly, V. (1997) *Commercialising New Technologies: Getting from mind to market*. Harvard Business School Press, Boston, Mass.

115 Moore, G. (1999) *Crossing the Chasm: Marketing and selling high-tech products to mainstream customers*. Harper Business, New York.

116 Rogers, E. (1995) *Diffusion of Innovations*. Free Press, New York.

117 Boden, M. and I. Miles (ed) (1998) *Services and the Knowledge-led Economy*. Continuum, London.

118 Cooke, P. and K. Morgan (1991) *The Intelligent Region: Industrial and institutional innovation in Emilia-Romagna*. University of Cardiff, Cardiff.

119 Dodgson, M. and A. Griffiths (2004) 'Sustainability and innovation', Special Issue, *Innovation Management, Policy and Practice*; Berkhout, F. and K. Green (eds) (2002) Special Issue on 'Managing innovation for sustainability', *International Journal of Innovation Management*, **6** (3).

120 Day, G. and P. Schoemaker (2000) *Wharton on Managing Emerging Technologies*. John Wiley & Sons, Inc., New York.

Chapter 2

Innovation as a Management Process

One of America's most successful innovators was Thomas Alva Edison who during his life registered over 1000 patents. Products for which his organization was responsible include the light bulb, 35 mm cinema film and even the electric chair. Edison appreciated better than most that the real challenge in innovation was not invention – coming up with good ideas – but in making them work technically and commercially. His skill in doing this created a business empire worth, in 1920, around $21.6 bn. He put to good use an understanding of the interactive nature of innovation, realizing that both technology push (which he systematized in one of the world's first organized R&D laboratories) and demand pull need to be mobilized.

His work on electricity provides a good example of this; Edison recognized that although the electric light bulb was a good idea it had little practical relevance in a world where there was no power point to plug it into. Consequently, his team set about building up an entire electricity generation and distribution infrastructure, including designing lamp stands, switches and wiring. In 1882 he switched on the power from the first electric power generation plant in Manhattan and was able to light up 800 bulbs in the area. In the years that followed he built over 300 plants all over the world.[1]

As Edison realized, innovation is more than simply coming up with good ideas; it is the *process* of growing them into practical use. Definitions of innovation may vary in their wording, but they all stress the need to complete the development and exploitation aspects of new knowledge, not just its invention. Box 2.1 gives some examples.

If we only understand part of the innovation process, then the behaviours we use in managing it are also likely to be only partially helpful – even if well intentioned and executed. For example, innovation is often confused with invention – but the latter is only the first step in a long process of bringing a good idea to widespread and effective use. Being a good inventor is – to contradict Emerson – no guarantee of commercial success and no matter how good the better mousetrap idea, the world will only beat a path to the door if attention is also paid to project management, market development, financial management, organizational behaviour, etc.* Box 2.2 gives some examples which highlight the difference between invention and innovation.

* Ralph Waldo Emerson, 'If a man has good corn, or wood, or boards, or pigs to sell, or can make better chairs or knives, crucibles or church organs than anybody else, you will find a broad-beaten road to his home, though it be in the woods.'

BOX 2.1
WHAT *IS* INNOVATION?

One of the problems in managing innovation is variation in what people understand by the term, often confusing it with invention. In its broadest sense the term comes from the Latin – *innovare* – meaning 'to make something new'. Our view, shared by the following writers, assumes that innovation is a process of turning opportunity into new ideas and of putting these into widely used practice.

'Innovation is the successful exploitation of new ideas' – Innovation Unit, UK Department of Trade and Industry (2004).

'Industrial innovation includes the technical, design, manufacturing, management and commercial activities involved in the marketing of a new (or improved) product or the first commercial use of a new (or improved) process or equipment' – Chris Freeman (1982) *The Economics of Industrial Innovation*, 2nd edn. Frances Pinter, London.

'. . . Innovation does not necessarily imply the commercialization of only a major advance in the technological state of the art (a radical innovation) but it includes also the utilization of even small-scale changes in technological know-how (an improvement or incremental innovation)' – Roy Rothwell and Paul Gardiner (1985) 'Invention, innovation, re-innovation and the role of the user', *Technovation*, **3**, 168.

'Innovation is the specific tool of entrepreneurs, the means by which they exploit change as an opportunity for a different business or service. It is capable of being presented as a discipline, capable of being learned, capable of being practised' – Peter Drucker (1985), *Innovation and Entrepreneurship*. Harper & Row, New York.

'Companies achieve competitive advantage through acts of innovation. They approach innovation in its broadest sense, including both new technologies and new ways of doing things' – Michael Porter (1990) *The Competitive Advantage of Nations*. Macmillan, London.

'An innovative business is one which lives and breathes "outside the box". It is not just good ideas, it is a combination of good ideas, motivated staff and an instinctive understanding of what your customer wants' – Richard Branson (1998) DTI Innovation Lecture.

BOX 2.2
INVENTION AND INNOVATION

In fact, some of the most famous inventions of the nineteenth century were invented by men whose names are forgotten; the names which we associate with them are of the entrepreneurs who brought them into commercial use. For example, the vacuum cleaner was invented by one J. Murray Spengler and originally called an 'electric suction sweeper'. He approached a leather goods maker in the town who knew nothing about vacuum cleaners but had a good idea of how to market and sell them – a certain W. H. Hoover. Similarly, a Boston man called Elias Howe produce the world's first sewing machine in 1846. Unable to sell his ideas despite travelling to England and trying there, he returned to the USA to find one Isaac Singer had stolen the patent and built a successful business from it. Although Singer was eventually forced to pay Howe a royalty on all machines made, the name which most people now associate with sewing machines is Singer not Howe. And Samuel Morse, widely credited as the father of modern telegraphy, actually invented only the code which bears his name; all the other inventions came from others. What Morse brought was enormous energy and a vision of what could be accomplished; to realize this he combined marketing and political skills to secure state funding for development work, and to spread the concept of something which for the first time would link up people separated by vast distances on the continent of America. Within five years of demonstrating the principle there were over 5000 miles of telegraph wire in the USA. and Morse was regarded as 'the greatest man of his generation'.[1]

2.1 Innovation as a Core Business Process

Chapter 1 set out a view of innovation as the core process within an organization associated with renewal – with refreshing what it offers the world and how it creates and delivers that offering. Viewed in this way innovation is a generic activity associated with survival and growth. And at this level of abstraction we can see the underlying process as common to all firms. At its heart the process involves:

- *Searching* – scanning the environment (internal and external) for, and processing relevant signals about, threats and opportunities for change.
- *Selecting* – deciding (on the basis of a strategic view of how the enterprise can best develop) which of these signals to respond to.

- *Implementing* – translating the potential in the trigger idea into something new and launching it in an internal or external market. Making this happen is not a single event but requires attention to:
  - *Acquiring* the knowledge resources to enable the innovation (for example, by creating something new through R&D, market research, etc., acquiring knowledge from elsewhere via technology transfer, strategic alliance, etc.).
  - *Executing* the project under conditions of uncertainty which require extensive problem-solving.
  - *Launching* the innovation and managing the process of initial adoption.
  - *Sustaining* adoption and use in the long term – or revisiting the original idea and modifying it – reinnovation.
- *Learning* – enterprises have (but may not always take) the opportunity to learn from progressing through this cycle so that they can build their knowledge base and can improve the ways in which the process is managed.

This process is shown in Figure 2.1.

The challenge facing any organization is to try and find ways of managing this process to provide a good solution to the problem of renewal. Different circumstances lead to many different solutions – for example, large science-based firms like the phar-

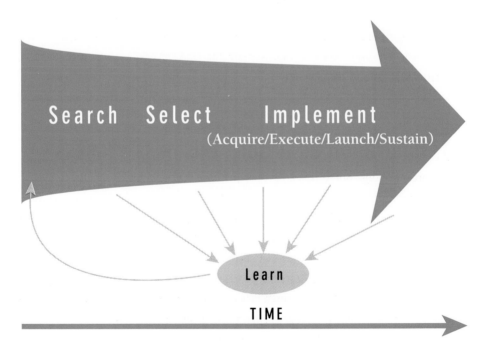

FIGURE 2.1   Simple representation of the innovation process

maceutical companies will tend to create solutions which have heavy activities around formal R&D, patent searching, etc. whilst small engineering subcontractors will emphasize rapid implementation capability. Retailers may have relatively small R&D commitments in the formal sense but stress scanning the environment to pick up new consumer trends, and they are likely to place heavy emphasis on marketing. Consumer goods producers may be more concerned with rapid product development and launch, often with variants and repositioning of basic product concepts. Heavy engineering firms involved in products like power plant are likely to be design-intensive, and critically dependent on project management and systems integration aspects of the implementation phase. Public sector organizations have to configure it to cope with strong external political and regulatory influences.

Despite these variations the underlying pattern of phases in innovation remains constant. In this chapter we want to explore the *process* nature of innovation in more detail, and to look at the kinds of variations on this basic theme. But we also want to suggest that there is some commonality around the things which are managed and the influences which can be brought to bear on them in successful innovation. These 'enablers' represent the levers which can be used to manage innovation in any organization. Once again, how these enablers are actually put together varies between firms, but they represent particular solutions to the general problem of managing innovation. Exploring these enablers in more detail is the basis of the following chapters in the book.

Central to our view is that innovation management is a *learned* capability. Although there are common issues to be confronted and a convergent set of recipes for dealing with them, each organization must find its own particular solution and develop this in its own context. Simply copying ideas from elsewhere is not enough; these must be adapted and shaped to suit particular circumstances.

Variations on a Theme

Innovations vary widely, in scale, nature, degree of novelty and so on – and so do innovating organizations. But at this level of abstraction it is possible to see the same basic process operating in each case. For example, developing a new consumer product will involve picking up signals about potential needs and new technological possibilities, developing a strategic concept, coming up with options and then working those up into new products which can be launched into the marketplace.

In similar fashion deciding to install a new piece of process technology also follows this pattern. Signals about needs – in this case internal ones, such as problems with the current equipment – and new technological means are processed and provide an input to developing a strategic concept. This then requires identifying an existing option, or inventing a new one which must then be developed to such a point that it can be implemented, i.e. launched, by users within the enterprise – effectively a group of internal customers. The same principles of needing to understand their needs and to prepare the marketplace for effective launch will apply as in the case of product innovation.

Services may appear different because they are often less tangible – but the same underlying model applies. The process whereby an insurance or financial services company launches a new product will follow a path of signal processing, strategic concept, product and market development and launch. What is developed may be less tangible than a new television set, but the underlying structure to the process is the same.[2] We should also recognize that increasingly what we call manufacturing includes a sizeable service component with core products being offered together with supporting services – a website, a customer information or help-line, updates, etc. Indeed for many complex product systems – such as aircraft engines – the overall package is likely to have a life in excess of 30 or 40 years and the service and support component may represent a significant part of the purchase. At the limit such manufacturers are recognizing that their users actually want to buy some service attribute which is embodied in the product – so aero engine manufacturers are offering 'power by the hour' rather than simply selling engines.[3]

Similarly the huge growth in 'outsourcing' of key business processes – IT, call centre management, human resources administration, etc. – although indicative of a structural shift in the economy has at its heart the same innovation drivers. Even if companies are being 'hollowed out' the challenges facing the outsourcer and its client remain those of process innovation. The underlying business model of outsourcing is based on being able to do something more efficiently than the client and thereby creating a business margin – but achieving this depends critically on the ability to re-engineer and then continuously improve on core business processes. And over time the attractiveness of one outsourcer over another increasingly moves from simply being able to execute outsourced standard operations more efficiently and towards being able to offer – or to co-evolve with a client – new products and services.

The distinction between commercial and not-for profit organizations may also blur when considering innovation. Whilst private sector firms may compete for the attentions of their markets through offering new things or new ways of delivering them, public sector and non-profit organizations use innovation to help them compete against the challenges of delivering healthcare, education, law and order, etc. They are similarly preoccupied with process innovation (the challenge of using often scarce resources more effectively or becoming faster and more flexible in their response to a diverse environment) and with product innovation – using combinations of new and existing knowledge to deliver new or improved 'product concepts' – such as decentralized healthcare, community policing or micro-credit banking.[4]

Size Matters

Another important influence on the particular ways in which innovation is managed is the size of the organization. Typically smaller organizations possess a range of advantages – such as agility, rapid decision-making – but equally limitations such as resource constraints. These mean that developing effective innovation management will depend on creating structures and behaviours which play to these – for example, keeping high levels of informality to build on shared vision and rapid decision-making but possibly to build network linkages to compensate for resource limitations.

But we need to be clear that small organizations differ widely. In most economies small firms account for 95% or more of the total business world and within this huge number there is enormous variation, from micro-businesses like hairdressing and accounting services through to high-technology start-ups. (We explore this theme in more detail in Chapter 5.) Once again we have to recognize that the generic challenge of innovation can be taken up by businesses as diverse as running a fish and chip shop through to launching a nanotechnology spin-out with millions of pounds in venture capital – but the particular ways in which the process is managed are likely to differ widely.

National, Regional, Local Context

Regional and national systems of innovation vary widely. By innovation system we mean the range of actors – government, financial, educational, labour market, science and technology infrastructure, etc. – which represent the context within which organizations operate their innovation process. In some cases there is clear synergy between these elements which create the supportive conditions within which innovation can flourish – for example, the regional innovation led clusters of Baden-Wurttemberg in Germany, Cambridge in the UK, Silicon Valley and Route 128 in the USA, or the island of Singapore.[5,6] Increasingly effective innovation management is being seen as a challenge of connecting to and working with such innovation systems – and this again has implications for how we might organize and manage the generic process (see Box 2.3).

Networks and Systems

As we saw in Chapter 1, one of the emerging features of the twenty-first-century innovation landscape is that it is much less of a single enterprise activity. For a variety of reasons it is increasingly a multiplayer game in which organizations of different shapes and sizes work together in networks. These may be regional clusters, or supply chains or product development consortia or strategic alliances which bring competitors and

> ## BOX 2.3
> ## THE POWER OF REGIONAL INNOVATION SYSTEMS
>
> Michael Best's fascinating account of the ways in which the Massachusetts economy managed to reinvent itself several times is one which underlines the importance of innovation systems. In the 1950s the state suffered heavily from the loss of its traditional industries of textiles and shoes but by the early 1980s the 'Massachusetts miracle' led to the establishment of a new high-tech industrial district. It was a resurgence enabled in no small measure by an underpinning network of specialist skills, high-tech research and training centres (the Boston area has the highest concentration of colleges, universities, research labs and hospitals in the world) and by the rapid establishment of entrepreneurial firms keen to exploit the emerging 'knowledge economy'. But in turn this miracle turned to dust in the years between 1986 and 1992 when around one third of the manufacturing jobs in the region disappeared as the minicomputer and defence-related industries collapsed. Despite gloomy predictions about its future, the region built again on its rich network of skills, technology sources and a diverse local supply base which allowed rapid new product development to emerge again as a powerhouse in high technology such as special purpose machinery, optoelectronics, medical laser technology, digital printing equipment and biotech.[6]

customers into a temporary collaboration to work at the frontier of new technology application. Although the dynamics of such networks are significantly different from those operating in a single organization and the controls and sanctions much less visible, the underlying innovation process challenge remains the same – how to build shared views around trigger ideas and then realize them. Throughout the book we will look at the particular issues raised in trying to manage innovation beyond the boundaries of the organization.

Project-based Organizations

For many enterprises the challenge is one of moving towards project-based organization – whether for realizing a specific project (such as construction of a major facility like an airport or a hospital) or for managing the design and build around complex product systems like aero engines, flight simulators or communications networks. Project organization of this kind represents an interesting case, involving a *system* which brings together many different elements into an integrated whole, often involving different firms, long timescales and high levels of technological risk.[7]

Increasingly they are associated with innovations in project organization and management – for example, in the area of project financing and risk sharing. Although such projects may appear very different from the core innovation process associated with, for example, producing a new soap powder for the mass market, the underlying process is still one of careful understanding of user needs and meeting those. The involvement of users throughout the development process, and the close integration of different perspectives will be of particular importance, but the overall map of the process is the same.

Do Better/Do Different

It's not just the sector which moderates the way the innovation process operates. An increasing number of authors draw attention to the need to take the degree of novelty in an innovation into account. At a basic level the structures and behaviours needed to help enable incremental improvements will tend to be incorporated into the day-to-day standard operating procedures of the organization. More radical projects may require more specialized attention – for example, arrangements to enable working across functional boundaries. At the limit the organization may need to review the whole bundle of routines which it uses for managing innovation when it confronts discontinuous conditions and the 'rules of the game' change.

As we saw in Chapter 1, we can think of innovation in terms of two complementary modes. The first can be termed 'doing what we do but better' – a 'steady state' in which innovation happens but within a defined envelope around which our 'good practice' routines can operate. This contrasts with 'do different' innovation where the rules of the game have shifted (due to major technological, market or political shifts, for example) and where managing innovation is much more a process of exploration and co-evolution under conditions of high uncertainty. A number of writers have explored this issue and conclude that under turbulent conditions firms need to develop capabilities for managing both aspects of innovation.[8–11]

Once again the generic model of the innovation process remains the same. Under 'do different' conditions, organizations still need to search for trigger signals – the difference is that they need to explore in much less familiar places and deploy peripheral vision to pick up weak signals early enough to move. They still need to make strategic choices about what they will do – but they will often have vague and incomplete information and the decision-making involved will thus be much more risky – arguing for a higher tolerance of failure and fast learning. Implementation will require much higher levels of flexibility around projects – and monitoring and review may need to take place against more flexible criteria than might be applied to 'do better' innovation types.

For established organizations the challenge is that they need to develop the capability to manage both kinds of innovation. Much of the time they will need robust systems for dealing with 'do better' but from time to time they risk being challenged by new entrants better able to capitalize on the new conditions opened up by discontinuity – unless they can develop a 'do different' capability to run in parallel. New entrants don't have this problem when riding the waves of a discontinuous shift – for example, exploiting opportunities opened up by a completely new technology. But they in turn will become established incumbents and face the challenge later if they do not develop the capacity to exploit their initial advantage through 'do better' innovation process and also build capability for dealing with the next wave of change by creating a 'do different' capability.

The challenge is thus – as shown in Figure 2.2 – to develop an ambidextrous capability for managing both kinds of innovation within the same organization. We will return to this theme repeatedly in the book, exploring the additional or different challenges posed when innovation has to be managed beyond the steady state.

Table 2.1 lists some of the wide range of influences around which organizations need to configure their particular versions of the generic innovation process.

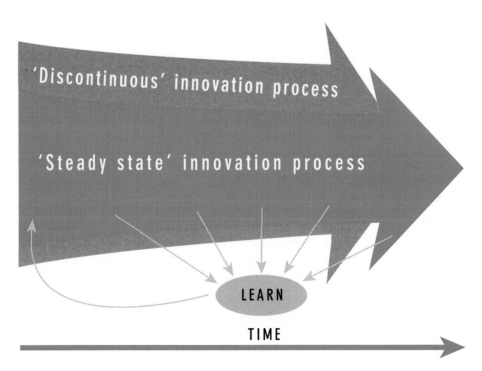

FIGURE 2.2   'Do better' and 'do different' innovation processes

TABLE 2.1 How context affects innovation management

Context variable	Modifiers to the basic process	Example references discussing these
Sector	Different sectors have different priorities and characteristics – for example, scale-intensive, science-intensive	2, 12
Size	Small firms differ in terms of access to resources, etc. and so need to develop more linkages	13–17
National systems of innovation	Different countries have more or less supportive contexts in terms of institutions, policies, etc.	5, 18, 19
Life cycle (of technology, industry, etc.)	Different stages in life-cycle emphasize different aspects of innovation – for example, new technology industries versus mature established firms	20–23
Degree of novelty- continuous vs. discontinuous innovation	'More of the same' improvement innovation requires different approaches to organization and management to more radical forms. At the limit firms may deploy 'dual structures' or even split or spin off in order to exploit opportunities	8, 24–26
Role played by external agencies such as regulators	Some sectors – e.g. utilities, telecommunications and some public services – are heavily influenced by external regimes which shape the rate and direction of innovative activity. Others – like food or healthcare – may be highly regulated in certain directions	26, 27

2.2 Evolving Models of the Process

The importance of understanding innovation as a process is that this understanding shapes the way in which we try and manage it. This has changed a great deal over time. Early models (both explicit and, more important, the implicit mental models whereby people managed the process) saw it as a linear sequence of functional activities. Either new opportunities arising out of research gave rise to applications and refinements which eventually found their way to the marketplace ('technology push') or else the market signalled needs for something new which then drew through new solutions to the problem ('need pull', where necessity becomes the mother of invention).

The limitations of such an approach are clear; in practice innovation is a coupling and matching process where interaction is the critical element.[28–30] Sometimes the 'push' will dominate, sometimes the 'pull', but successful innovation requires interaction between the two. The analogy to a pair of scissors is useful here; without both blades it is difficult to cut.

One of the key problems in managing innovation is that we need to make sense of a complex, uncertain and highly risky set of phenomena. Inevitably we try and simplify these through the use of mental models – often reverting to the simplest linear models to help us explore the management issues which emerge over time. Prescriptions for structuring the process along these lines abound; for example, one of the most cited models for product innovation is due to Booz, Allen and Hamilton.[31] Many variations exist on this theme – for example, Robert Cooper's work suggests a slightly extended view with 'gates' between stages which permit management of the risks in the process.[32] There is also a British Standard (BS 7000) which sets out a design-centred model of the process.[33]

Much recent work recognizes the limits of linear models and tries to build more complexity and interaction into the frameworks. For example, the Product Development Management Association (PDMA) offers a detailed guide to the process and an accompanying toolkit.[34] Increasingly there is recognition of some of the difficulties around what is often termed the 'fuzzy front end' where uncertainty is highest, but there is still convergence around a basic process structure as a way of focusing our attention.[35] The balance needs to be struck between simplifications and representations which help thinking – but just as the map is not the same as the territory it represents so they need to be seen as frameworks for thinking, not as descriptions of the way the process actually operates.

Most innovation is messy, involving false starts, recycling between stages, dead ends, jumps out of sequence, etc. Various authors have tried different metaphors – for example, seeing the process as a railway journey with the option of stopping at different stations, going into sidings or even, at times, going backwards – but most agree that there is still some sequence to the basic process.[36,37] In an important programme of case-study-based research looking at widely different innovation types, Van de Ven and colleagues explored the limitations of simple models of the process. They drew attention to the complex ways in which innovations actually evolve over time, and derived some important modifiers to the basic model:

- Shocks trigger innovations – change happens when people or organizations reach a threshold of opportunity or dissatisfaction.
- Ideas proliferate – after starting out in a single direction, the process proliferates into multiple, divergent progressions.
- Setbacks frequently arise, plans are over-optimistic, commitments escalate, mistakes accumulate and vicious cycles can develop.

- Restructuring of the innovating unit often occurs through external intervention, personnel changes or other unexpected events.
- Top management plays a key role in sponsoring – but also in criticizing and shaping – innovation.
- Success criteria shift over time, differ between groups and make innovation a political process.
- Innovation involves learning, but many of its outcomes are due to other events which occur as the innovation develops – making learning often 'superstitious' in nature.

They suggest that the underlying structure can be represented by the metaphor of an 'innovation journey', which has key phases of initiation, development and implementation/termination. But the progress of any particular innovation along this will depend on a variety of contingent circumstances; depending on which of these apply, different specific models of the process will emerge.

Roy Rothwell was for many years a key researcher in the field of innovation management, working at SPRU at the University of Sussex. In one of his later papers he provided a useful historical perspective on this, suggesting that our appreciation of the nature of the innovation process has been evolving from such simple linear models (characteristic of the 1960s) through to increasingly complex interactive models (Table 2.2). His 'fifth-generation innovation' concept sees innovation as a multi-actor process which requires high levels of integration at both intra- and inter-firm levels and which is increasingly facilitated by IT-based networking.[38] Whilst his work did not explicitly mention the Internet, it is clear that the kinds of innovation management challenge posed by the emergence of this new form fit well with the model. Although such fifth-generation models and the technologies which enable them appear complex, they still involve the same basic process framework.[39]

TABLE 2.2 Rothwell's five generations of innovation models

Generation	Key features
First and second	Simple linear models – need pull, technology push
Third	Coupling model, recognizing interaction between different elements and feedback loops between them
Fourth	Parallel model, integration within the firm, upstream with key suppliers and downstream with demanding and active customers, emphasis on linkages and alliances
Fifth	Systems integration and extensive networking, flexible and customized response, continuous innovation

2.3 Consequences of Partial Understanding of the Innovation Process

Mental models are important because they help us frame the issues which need managing – but therein also lies the risk. If our mental models are limited then our approach to managing is also likely to be limited. For example, if we believe that innovation is simply a matter of coming up with a good invention – then we risk managing that part of the process well but failing to consider or deal with other key issues around actually taking that invention through technological and market development to successful adoption.

Examples of such 'partial thinking' here include:

- Seeing innovation as a linear 'technology push' process (in which case all the attention goes into funding R&D with little input from users) or one in which the market can be relied upon to pull through innovation.
- Seeing innovation simply in terms of major 'breakthroughs' – and ignoring the significant potential of incremental innovation. In the case of electric light bulbs, the original Edison design remained almost unchanged in concept, but incremental product and process improvement over the 16 years from 1880 to 1896 led to a fall in price of around 80%.[40]
- Seeing innovation as a single isolated change rather than as part of a wider system (effectively restricting innovation to component level rather than seeing the bigger potential of architectural changes).[41]
- Seeing innovation as product or process only, without recognizing the interrelationship between the two.

Table 2.3 provides an overview of the difficulties which arise if we take a partial view of innovation.

2.4 Can We Manage Innovation?

It would be hard to find anyone prepared to argue against the view that innovation is important and likely to be more so in the coming years. But that still leaves us with the big question of whether or not we can actually *manage* what is clearly an enormously complex and uncertain process.

There is certainly no easy recipe for success. Indeed, at first glance it might appear that it is impossible to manage something so complex and uncertain. There are prob-

TABLE 2.3 Problems of partial views of innovation

If innovation is only seen as . . .	. . . the result can be
Strong R&D capability	Technology which fails to meet user needs and may not be accepted
The province of specialists	Lack of involvement of others, and a lack of in the R&D laboratory key knowledge and experience input from other perspectives
Understanding and meeting customer needs	Lack of technical progression, leading to inability to gain competitive edge
Advances along the technology frontier	Producing products or services which the market does not want or designing processes which do not meet the needs of the user and whose implementation is resisted
The province only of large firms	Weak small firms with too high a dependence on large customers Disruptive innovation as apparently insignificant small players seize new technical or market opportunities
Only about 'breakthrough' changes	Neglect of the potential of incremental innovation. Also an inability to secure and reinforce the gains from radical change because the incremental performance ratchet is not working well
Only about strategically targeted projects	May miss out on lucky 'accidents' which open up new possibilities
Only associated with key individuals	Failure to utilize the creativity of the remainder of employees, and to secure their inputs and perspectives to improve innovation
Only internally generated	The 'not invented here' effect, where good ideas from outside are resisted or rejected
Only externally generated	Innovation becomes simply a matter of filling a shopping list of needs from outside and there is little internal learning or development of technological competence
Only concerning single firms	Excludes the possibility of various forms of inter-organizational networking to create new products, streamline shared processes, etc.

lems in developing and refining new basic knowledge, problems in adapting and applying it to new products and processes, problems in convincing others to support and adopt the innovation, problems in gaining acceptance and long-term use, and so on. Since so many people with different disciplinary backgrounds, varying responsibilities and basic goals are involved, the scope for differences of opinion and conflicts over ends and means is wide. In many ways the innovation process represents the place where Murphy and his associated band of lawmakers hold sway, where if anything can go wrong, there's a very good chance that it will!

But despite the uncertain and apparently random nature of the innovation process, it *is* possible to find an underlying pattern of success. Not every innovation fails, and some firms (and individuals) appear to have learned ways of responding and managing it such that, while there is never a cast-iron guarantee, at least the odds in favour of successful innovation can be improved. We are using the term *'manage'* here not in the sense of designing and running a complex but predictable mechanism (like an elaborate clock) but rather that we are creating conditions within an organization under which a successful resolution of multiple challenges under high levels of uncertainty is made more likely.

One indicator of the possibility of doing this comes from the experiences of organizations which have survived for an extended period of time. Whilst most organizations have comparatively modest lifespans there are some which have survived at least one and sometimes multiple centuries. Looking at the experience of these '100 club' members – firms like 3M, Corning, Procter & Gamble, Reuters, Siemens, Philips and Rolls-Royce – we can see that much of their longevity is down to having developed a capacity to innovate on a continuing basis. They have learned – often the hard way – how to manage the process (both in its 'do better' and 'do different' variants) so that they can sustain innovation.[11,42–44]

It is important to note the distinction here between 'management' and managers. We are not arguing here about who is involved in taking decisions or directing activity, but rather about what has to be done. Innovation is a *management* question, in the sense that there are choices to be made about resources and their disposition and co-ordination. Close analysis of many technological innovations over the years reveals that although there are technical difficulties – bugs to fix, teething troubles to be resolved and the occasional major technical barrier to surmount – the majority of failures are due to some weakness in the way the process is managed. Success in innovation appears to depend upon two key ingredients – technical resources (people, equipment, knowledge, money, etc.) *and* the capabilities in the organization to manage them.

This brings us back to the concept of *routines*, mentioned in Chapter 1. Organizations develop particular ways of behaving which become 'the way we do things around here' as a result of repetition and reinforcement. These patterns reflect an underlying set of shared beliefs about the world and how to deal with it, and form part of the organization's culture – 'the way we do things in this organization'. They emerge as a result of repeated experiments and experience around what appears to work well – in other words, they are learned. Over time the pattern becomes more of an automatic response to particular situations, and the behaviour becomes what can be termed a 'routine'.

This does not mean that it is necessarily repetitive, only that its execution does not require detailed conscious thought. The analogy can be made with driving a car; it is possible to drive along a stretch of motorway whilst simultaneously talking to someone else, eating or drinking, listening to, and concentrating on, something on the radio or planning what to say at the forthcoming meeting. But driving is not a passive behaviour; it requires continuous assessment and adaptation of responses in the light of other traffic behaviour, road conditions, weather and a host of different and unplanned factors. We can say that driving represents a behavioural routine in that it has been learned to the point of being largely automatic.

In the same way an organizational routine might exist around how projects are managed, or new products researched. For example, project management involves a complex set of activities such as planning, team selection, monitoring and execution of tasks, replanning, coping with unexpected crises, and so on. All of these have to be integrated – and offer plenty of opportunities for making mistakes. Project management is widely recognized as an organizational skill, which experienced firms have developed to a high degree but which beginners can make a mess of. Firms with good project management routines are able to codify and pass them on to others via procedures and systems. Most important, the principles are also transmitted into 'the way we run projects around here' by existing members passing on the underlying beliefs about project management behaviour to new recruits.

Over time organizational behaviour routines create and are reinforced by various kinds of artefacts – formal and informal structures, procedures and processes which describe 'the way we do things around here', and symbols which represent and characterize the underlying routines. It could be in the form of a policy – for example, 3M is widely known for its routines for regular and fast product innovation. They have enshrined a set of behaviours around encouraging experimentation into what they term 'the 15% policy' in which employees are enabled to work on their own curiosity-driven agenda for up to 15% of their time.[43,45] These routines are firm-specific – for example, they result from an environment in which the costs of product development experimentation are often quite low.

Levitt and March describe routines as involving established sequences of actions for undertaking tasks enshrined in a mixture of technologies, formal procedures or strategies, and informal conventions or habits.[46] Importantly, routines are seen as evolving in the light of experience that works – they become the mechanisms that 'transmit the lessons of history'. In this sense, routines have an existence independent of particular personnel – new members of the organization learn them on arrival, and most routines survive the departure of individual routines. Equally, they are constantly being adapted and interpreted such that formal policy may not always reflect the current nature of the routine – as Augsdorfer points out in the case of 3M.[47]

For our purposes the important thing to note is that routines are what makes one organization different from another in how they carry out the same basic activity. We could almost say they represent the particular 'personality' of the firm. Each enterprise learns its own particular 'way we do things around here' in answer to the same generic questions – how it manages quality, how it manages people, etc. 'How we manage innovation around here' is one set of routines which describes and differentiates the responses which organizations make to the question of structuring and managing the generic model described above.

It follows that some routines are better than others in coping with the uncertainties of the outside world, in both the short and the long term. And it is possible to learn from others' experience in this way; the important point is to remember that routines are firm-specific and must be learned. Simply copying what someone else does is unlikely to help, any more than watching someone drive and then attempting to copy them will make a novice into an experienced driver. There may be helpful clues which can be used to improve the novice's routines, but there is no substitute for the long and experience-based process of learning. Box 2.4 gives some examples where change has been introduced without this learning perspective.

BOX 2.4
FASHION STATEMENTS VS. BEHAVIOURAL CHANGE IN ORGANIZATIONS

The problem with routines is that they have to be learned – and learning is difficult. It takes time and money to try new things, it disrupts and disturbs the day-to-day working of the firm, it can upset organizational arrangements and require efforts in acquiring and using new skills. Not surprisingly most firms are reluctant learners – and one strategy which they adopt is to try and short-cut the process by borrowing ideas from other organizations.

Whilst there is enormous potential in learning from others, simply copying what seems to work for another organization will not necessarily bring any benefits and may end up costing a great deal and distracting the organization from finding its own ways of dealing with a particular problem. The temptation to copy gives rise to the phenomenon of particular approaches becoming fashionable – something which every organization thinks it needs in order to deal with its particular problems.

Over the past 20 years we have seen many apparent panaceas for the problems of becoming competitive. Organizations are constantly seeking for new answers to old problems, and the scale of investment in the new fashions of management thinking have often been considerable. The *original* evidence for the value of these tools and techniques was strong, with case studies and other reports testifying to their proven value within the context of origin. But there is also extensive evidence to suggest that these changes do not always work, and in many cases lead to considerable dissatisfaction and disillusionment.

Examples include:

- advanced manufacturing technology (AMT – robots, flexible machines, integrated computer control, etc.);[48,49]
- total quality management (TQM);[50,51]
- business process re-engineering (BPR);[52,53]
- benchmarking best practice;[54]
- quality circles;[55,56]
- networking/clustering;[57,58]
- knowledge management.[59]

What is going on here demonstrates well the principles behind behavioural change in organisations. It is not that the original ideas were flawed or that the initial evidence was wrong. Rather it was that other organisations assumed they could simply be copied, without the need to adapt them, to customize them, to modify and change them to suit their circumstances. In other words, there was no learning, and no progress towards making them become routines, part of the underlying culture within the firm. Chapter 3 picks up this theme in the context of thinking about strategy.

Successful innovation management routines are not easy to acquire. Because they represent what a particular firm has learned over time, through a process of trial and error, they tend to be very firm-specific. Whilst it may be possible to identify the kinds of thing which 3M, Toyota, Hewlett-Packard or others have learned to do, simply copying them will not work. Instead each firm has to find its own way of doing these things – in other words, developing its own particular routines.

In the context of innovation management we can see the same hierarchical relationship in developing capability as there is in learning to drive. Basic skills are behaviours associated with things like planning and managing projects or understanding customer needs. These simple routines need to be integrated into broader abilities which taken together make up an organization's capability in managing innovation. Table 2.4 gives some examples.

One last point concerns the negative side of routines. They represent, as we have seen, embedded behaviours which have become reinforced to the point of being almost second nature – 'the way we do things around here'. Therein lies their strength, but

TABLE 2.4 Core abilities in managing innovation

Basic ability	Contributing routines
Recognizing	Searching the environment for technical and economic clues to trigger the process of change
Aligning	Ensuring a good fit between the overall business strategy and the proposed change – not innovating because it is fashionable or as a knee-jerk response to a competitor
Acquiring	Recognizing the limitations of the company's own technology base and being able to connect to external sources of knowledge, information, equipment, etc. Transferring technology from various outside sources and connecting it to the relevant internal points in the organization
Generating	Having the ability to create some aspects of technology in-house – through R&D, internal engineering groups, etc.
Choosing	Exploring and selecting the most suitable response to the environmental triggers which fit the strategy and the internal resource base/external technology network
Executing	Managing development projects for new products or processes from initial idea through to final launch Monitoring and controlling such projects
Implementing	Managing the introduction of change – technical and otherwise – in the organization to ensure acceptance and effective use of innovation
Learning	Having the ability to evaluate and reflect upon the innovation process and identify lessons for improvement in the management routines
Developing the organization	Embedding effective routines in place – in structures, processes, underlying behaviours, etc.

also their weakness. Because they represent ingrained patterns of thinking about the world, they are resilient – but they can also become barriers to thinking in different ways. Thus core capabilities can become core rigidities – when the 'way we do things round here' becomes inappropriate, but when the organization is too committed to the old ways to change.[60] So it becomes important, from the standpoint of innovation management, not only to build routines but also to recognize when and how to destroy them and allow new ones to emerge. This is a particularly important issue in the context of managing discontinuous innovation; we return to it in Chapter 5, in the context of strategy.

Our argument in this book is that successful innovation management is primarily about building and improving effective routines. Learning to do this comes from recognizing and understanding effective routines (whether developed in-house or observed in another enterprise) and facilitating their emergence across the organization.

2.5 Successful Innovation and Successful Innovators

Before we move to look at examples of successful routines for innovation management, we should pause for a moment and define what we mean by 'success'.

We have already seen that one aspect of this question is the need to measure the overall process rather than its constituent parts. Many successful inventions fail to become successful innovations, even when well planned.[61-63] Equally, innovation alone may not always lead to business success. Although there is strong evidence to connect innovation with performance, success depends on other factors as well. If the fundamentals of the business are weak, then all the innovation in the world may not be sufficient to save it. This argues for strategically focused innovation as part of a 'balanced scorecard' of results measurement.[64,65]

We also need to consider the time perspective. The real test of innovation success is not a one-off success in the short term but sustained growth through continuous invention and adaptation. It is relatively simple to succeed once with a lucky combination of new ideas and receptive market at the right time – but it is quite another thing to repeat the performance consistently. Some organizations clearly feel able to do the latter to the point of presenting themselves as innovators – for example, 3M, Sony, IBM, Samsung and Philips, all of whom currently use the term in their advertising campaigns and stake their reputations on their ability to innovate consistently.

In our terms, success relates to the overall innovation process and its ability to contribute consistently to growth. This question of measurement – particularly its use to help shape and improve management of the process – is one to which we will return in Chapter 13.

2.6 What Do We Know About Successful Innovation Management?

The good news is that there is a knowledge base on which to draw in attempting to answer this question. Quite apart from the wealth of experience (of success and failure) reported by organizations involved with innovation, there is a growing pool of knowledge derived from research. Over the past 80 years or so there have been many studies of the innovation process, looking at many different angles. Different innovations, different sectors, firms of different shapes and sizes, operating in different countries, etc. have all come under the microscope and been analysed in a variety of ways. Table 2.5

TABLE 2.5 Examples of innovation studies

Study name	Key focus	Further reference
Project SAPPHO	Success and failure factors in matched pairs of firms, mainly in chemicals and scientific instruments	70
Wealth from knowledge	Case studies of successful firms – all were winners of the Queen's Award for Innovation	71
Post-innovation performance	Looked at these cases 10 years later to see how they fared	72
Project Hindsight	Historical reviews of US government funded work within the defence industry looking back over 20 years (from 1966) at key projects and success/failure factors	73
TRACES	As Project Hindsight but with 50-year review and exploring civilian projects as well. Main aims were to identify sources of successful innovation and management factors influencing success	74
Industry and technical progress	Survey of UK firms to identify why some were apparently more innovative than others in the same sector, size range, etc. Derived a list of managerial factors which comprised 'technical progressiveness'	75
Minnesota studies	Detailed case studies over an extended period of 14 innovations. Derived a 'road map' of the innovation process and the factors influencing it at various stages	37
Project NEWPROD and replications	Long-running survey of success and failure in product development	76
Stanford Innovation Project	Case studies of (mainly) product innovations, emphasis on learning	77
Lilien and Yoon	Literature review of major studies of success and failure	62
Rothwell	25-year retrospective review of success and failure studies and models of innovation process	38
Mastering the dynamics of innovation	Five retrospective in-depth industry-level cases	21
Sources of innovation	Case studies involving different levels and types of user involvement	29
Product Development Management Association	Handbook distilling key elements of good practice from a range of success and failure studies in product development	34
Ernst	Extensive literature review of success factors in product innovation	63
Interprod	International study (17 countries) collecting data on the factors influencing new product success and failure	78, 79

Study name	Key focus	Further reference
Christensen	Industry level studies of disruptive innovation – includes disk drives, mechanical excavators, steel mini-mills	10, 24
Eisenhardt and Brown	Detailed case studies of five semiconductor equipment firms	80
Revolutionizing product development	Case studies of product development	81
Winning by design	Case studies of product design and innovation	82
Innovation audits	Various frameworks synthesizing literature and reported key factors	83–85
Radical innovation	Review of radical innovation practices in case study firms	25
Rejuvenating the mature business	Review of mature businesses in Europe and their use of innovation to secure competitive advantage	86
Innovation Wave	Case studies of manufacturing and service innovations based on experiences at the London Business School Innovation Exchange	87

gives some examples of the research which underpins what we know about successful innovation management.

From this knowledge base it is clear that there are no easy answers and that innovation varies enormously – by scale, type, sector, etc. Nonetheless, there does appear to be some convergence around our two key points:

- Innovation is a process, not a single event, and needs to be managed as such.
- The influences on the process can be manipulated to affect the outcome – that is, it *can* be managed.

Most important, the research base highlights the concept of success routines which are learned over time and through experience. For example, successful innovation correlates strongly with how a firm selects and manages projects, how it co-ordinates the inputs of different functions, how it links up with its customers, etc. Developing an integrated set of routines is strongly associated with successful innovation management, and can give rise to distinctive competitive ability – for example, being able to introduce new products faster than anyone[66] or being able to use new process technology better.[67,68]

The other critical point to emerge from research is that innovation needs managing in an *integrated* way; it is not enough just to manage or develop abilities in some of these areas. One metaphor which helps draw attention to this is to see managing the process in sporting terms; success is more akin to winning a multi-event group of activities (like the pentathlon) than to winning a single high performance event like the 100 metres.[69]

There are many examples of firms which have highly developed abilities for managing part of the innovation process but which fail because of a lack of ability in others. For example, there are many with an acknowledged strength in R&D and the generation of technological innovation – but which lack the abilities to relate these to the marketplace or to end-users. Others may lack the ability to link innovation to their business strategy; for example, many firms invested in advanced manufacturing technologies – robots, computer aided design, computer controlled machines, etc. – during the late twentieth century, but most surveys suggest that only half of these investments really paid off. For the other half the problem was an inability to match the 'gee whiz' nature of a glamorous technology to their particular needs, and the result was what might be called 'technological jewellery' – visually impressive but with little more than a decorative function.

The concept of capability in innovation management also raises the question of how it is developed over time. This must involve a learning process. It is not sufficient simply to have experiences (good or bad); the key lies in evaluating and reflecting upon them and then developing the organization in such a way that the next time a similar challenge emerges the response is ready. Such a cycle of learning is easy to prescribe but very often missing in organizations – with the result that there often seems to be a great deal of repetition in the pattern of mistakes, and a failure to learn from the misfortunes of others. For example, there is often no identifiable point in the innovation process where a post-mortem is carried out, taking time to try and distil useful learning for next time. In part this is because the people involved are too busy, but it is also because of a fear of blame and criticism. Yet without this pause for thought the odds are that the same mistakes will be repeated.[88,89] (We will return to this theme in Chapter 9.)

2.7 Roadmaps for Success

Successful innovators acquire and accumulate technical resources and managerial capabilities over time; there are plenty of opportunities for learning – through doing, using, working with other firms, asking the customers, etc. – but they all depend upon the readiness of the firm to see innovation less as a lottery than as a process which can be continuously improved.

From the various studies of success and failure in innovation it is possible to construct checklists and even crude blueprints for effective innovation management. A number of models for auditing innovation have been developed in recent years, which provide a framework against which to assess performance in innovation management. Some of these involve simple checklists, others deal with structures, others with the operation of particular sub-processes.[83,85,90] (We will return to the theme of innovation audits and their role in helping develop capability in Chapter 13.)

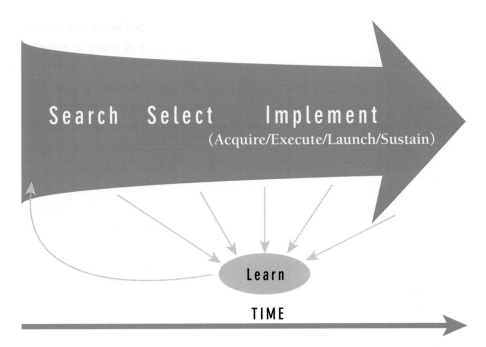

FIGURE 2.3   Innovation process model

For our purposes in exploring innovation management throughout the rest of the book it will be helpful to build our own simple model and use it to focus attention on key aspects of the innovation management challenge. At its heart we have the generic process described earlier which sees innovation as a core set of activities distributed over time. (Of course, as we noted earlier, innovation in real life does not conform neatly to this simple representation – and it is rarely a single event but rather a cycle of activities repeated over time.) The key point is that a number of different actions need to take place as we move through the phases of this model and associated with each are some consistent lessons about effective innovation management routines (see Figure 2.3, which is a reminder of Figure 2.1).

Search Phase

The first phase in innovation involves detecting signals in the environment about potential for change. These could take the form of new technological opportunities, or changing requirements on the part of markets; they could be the result of legislative pressure or competitor action. Most innovations result from the interplay of several forces, some coming from the need for change pulling through innovation and others from the push which comes from new opportunities.

Given the wide range of signals it is important for successful innovation management to have well-developed mechanisms for identifying, processing and selecting information from this turbulent environment. Chapter 9 explores enabling routines associated with successful scanning and processing of relevant signals.

Organizations don't, of course, search in infinite space but rather in places where they expect to find something helpful. Over time their search patterns become highly focused and this can – as we have seen – sometimes represent a barrier to more radical forms of innovation. A key challenge in innovation management relates to the clear understanding of what factors shape the 'selection environment' and the development of strategies to ensure their boundaries of this are stretched.

Selection Phase

Innovation is inherently risky, and even well-endowed firms cannot take unlimited risks. It is thus essential that some selection is made of the various market and technological opportunities, and that the choices made fit with the overall business strategy of the firm, and build upon established areas of technical and marketing competence. The purpose of this phase is to resolve the inputs into an innovation concept which can be progressed further through the development organization.

Three inputs feed this phase. The first is the flow of signals about possible technological and market opportunities available to the enterprise. The second input concerns the current technological base of the firm – its distinctive technological competence.[91] By this we mean what it knows about terms of its product or service and how that is produced or delivered effectively. This knowledge may be embodied in particular products or equipment, but is also present in the people and systems needed to make the processes work. The important thing here is to ensure that there is a good fit between what the firm currently knows about and the proposed changes it wants to make.

This is not to say that firms should not move into new areas of competence; indeed there has to be an element of change if there is to be any learning. But rather there needs to be a balance and a development *strategy*. This raises the third input to this phase – the fit with the overall business. At the concept stage it should be possible to relate the proposed innovation to improvements in overall business performance. Thus if a firm is considering investing in flexible manufacturing equipment because the business is moving into markets where increased customer choice is likely be critical, it will make sense. But if it is doing so in a commodity business where everyone wants exactly the same product at the lowest price, then the proposed innovation will not underpin the strategy – and will effectively be a waste of money. Getting close alignment between the overall strategy for the business and the innovation strategy is critical at this stage.

In similar fashion many studies have shown that product innovation failure is often caused by firms trying to launch products which do not match their competence base.[92]

This knowledge base need not be contained within the firm; it is also possible to build upon competencies held elsewhere. The requirement here is to develop the relationships needed to access the necessary complementary knowledge, equipment, resources, etc. Strategic advantage comes when a firm can mobilize a set of internal and external competencies – what Teece calls 'the appropriability regime' – which make it difficult for others to copy or enter the market.[93] (This theme is picked up in more depth in Chapter 4, and Chapter 9 explores in more detail some of the key routines associated with managing the strategic selection of innovation projects and building a coherent and robust portfolio.)

Implementing

Having picked up relevant trigger signals and made a strategic decision to pursue some of them, the next key phase is actually turning those potential ideas into some kind of reality – a new product or service, a change in process, a shift in business model, etc. In some ways this implementation phase can be seen as one which gradually pulls together different pieces of knowledge and weaves them into an innovation. At the early stages there is high uncertainty – details of technological feasibility, of market demand, of competitor behaviour, of regulatory and other influences, etc. – all of these are scarce and strategic selection has to be based on a series of 'best guesses'. But gradually over the implementation phase this uncertainty is replaced by knowledge acquired through various routes and at an increasing cost. Technological and market research helps clarify whether or not the innovation is technically possible or if there is a demand for it and if so, what are its characteristics. As the innovation develops so a continuing thread of problem-finding and -solving – getting the bugs out of the original concept – takes place, gradually building up relevant knowledge around the innovation. Eventually it is in a form which can be launched into its intended context – internal or external market – and then further knowledge about its adoption (or otherwise) can be used to refine the innovation.

We can explore the implementation phase in a little more detail by considering three core elements – acquiring knowledge resources, executing the project and launching and sustaining the innovation.

Acquiring Knowledge Resources

This phase involves combining new and existing knowledge (available within and outside the organization) to offer a solution to the problem. It involves both generation of technological knowledge (via R&D carried out within and outside the organization) and technology transfer (between internal sources or from external sources).

As such it represents a first draft of a solution, and is likely to change considerably in its development. The output of this stage in the process is both forward to the next stage of detailed development, and back to the concept stage where it may be abandoned, revised or approved.

'Invention' is used here to denote the first combination of ideas around a concept; the concept may be one articulated by market research, triggered by competitor action or emerging from R&D work in-house or externally. The key point is that it is here that the innovation moves from a collection of ideas, conscious or unconscious, to some physical reality. Much depends at this stage on the nature of the new concept. If it involves an incremental modification to an existing design, there will be little activity within the invention stage. By contrast, if the concept involves a totally new concept, there is considerable scope for creativity.

Whilst individuals may differ in terms of their preferred creative style, there is strong evidence to support the view that everyone has the latent capability for creative problem-solving.[94] Unfortunately, a variety of individual inhibitions and external social and environmental pressures combine and accumulate over time to place restrictions on the exercise of this creative potential. The issue in managing this stage is thus to create the conditions under which this can flourish and contribute to effective innovation.

Another problem with this phase is the need to balance the open-ended environmental conditions which support creative behaviour with the somewhat harsher realities involved elsewhere in the innovation process. As with concept testing and development, it is worth spending time exploring ideas and potential solutions rather than jumping on the first apparently workable option.

The challenge in effective R&D is not simply one of putting resources into the system; it is how those resources are used. Effective management of R&D requires a number of organizational routines, including clear strategic direction, effective communication and 'buy-in' to that direction, and integration of effort across different groups.

But not all firms can afford to invest in R&D; for many smaller firms the challenge is to find ways of using technology generated by others or to complement internally generated core technologies with a wider set drawn from outside (see Chapter 8). This places emphasis on the strategy system discussed above – the need to know which to carry out where and the need for a framework to guide policy in this area. Firms can survive even with no in-house capability to generate technology – but to do so they need to have a well-developed network of external sources which can supply it, and the ability to put that externally acquired technology to effective use.

It also requires abilities in finding, selecting and transferring technology in from outside the firm. This is rarely a simple shopping transaction although it is often treated as such; it involves abilities in selecting, negotiating and appropriating the benefits from such technology transfer.[95] Chapter 9 explores some key routines which are associated with acquiring the knowledge resources to enable innovation.

Executing the Project

This phase forms the heart of the innovation process. Its inputs are a clear strategic concept and some initial ideas for realizing the concept. Its outputs are both a developed innovation and a prepared market (internal or external), ready for final launch. This is fundamentally a challenge in project management under uncertain conditions. As we will see in Chapter 9, the issue is not simply one of ensuring certain activities are completed in a particular sequence and delivered against a time and cost budget. The lack of knowledge at the outset and the changing picture as new knowledge is brought in during development means that a high degree of flexibility is required in terms of overall aims and subsidiary activities and sequencing. Much of the process is about weaving together different knowledge sets coming from groups and individuals with widely different functional and disciplinary backgrounds. And the project may involve groups who are widely distributed in organizational and geographical terms – often belonging to completely separate organizations. Consequently the building and managing of a project team, of communicating a clear vision and project plan, of maintaining momentum and motivation, etc. are not trivial tasks.

One way of representing the development stage is as a funnel, moving gradually from broad exploration to narrow focused problem-solving and hence to final (and successful) innovation. Unfortunately the apparent rational progress implied in this model is often not borne out in practice; instead various problems emerge, such as the lack of input (or sometimes too much input) from key functions, lack of communication between functions, conflicting goals, etc.

It is during this stage that most of the time, costs and commitment are incurred, and it is characterized by a series of problem-solving loops dealing with expected and unexpected difficulties in the technical and market areas. Although we can represent it as a parallel process, in practice effective management of this stage requires close interaction between marketing-related and technical activities. For example, product development involves a number of functions, ranging from marketing, through design and development to manufacturing, quality assurance and finally back to marketing. Differences in the tasks which each of these functions performs, in the training and experience of those working there and in the timescales and operating pressures under which they work all mean that each of these areas becomes characterized by a different working culture. Functional divisions of this kind are often exaggerated by location, where R&D and design activities are grouped away from the mainstream production and sales operations – in some cases on a completely different site.

Separation of this kind can lead to a number of problems in the overall development process. Distancing the design function from the marketplace can lead to inappropriate designs which do not meet the real customer needs, or which are 'over-engineered', embodying a technically sophisticated and elegant solution which exceeds the actual requirement (and may be too expensive as a consequence). This kind of phenomenon is often found in industries which have a tradition of defence contracting, where work has been carried out on a cost-plus basis involving projects which have emphasized technical design features rather than commercial or manufacturability criteria.

Similarly, the absence of a close link with manufacturing means that much of the information about the basic 'make-ability' of a new design either does not get back to the design area at all or else does so at a stage too late to make a difference or to allow the design to be changed. There are many cases in which manufacturing has wrestled with the problem of making or assembling a product which requires complex manipulation, but where minor design change – for example, relocation of a screw hole – would considerably simplify the process. In many cases such an approach has led to major reductions in the number of operations necessary – simplifying the process and often, as an extension, making it more susceptible to automation and further improvements in control, quality and throughput.

In similar fashion, many process innovations fail because of a lack of involvement on the part of users and others likely to be affected by the innovation. For example, many IT systems, whilst technically capable, fail to contribute to improved performance because of inadequate consideration of current working patterns which they will disrupt, lack of skills development amongst those who will be using them, inadequately specified user needs, and so on.

Although services are often less tangible, the underlying difficulties in implementation are similar. Different knowledge sets need to be brought together at key points in the process of creating and deploying new offerings. For example, developing a new insurance or financial service product requires technical input on the part of actuaries, accountants, IT specialists, etc. – but this needs to be combined with information about customers and key elements of the marketing mix – the presentation, the pricing, the positioning, etc. of the new service. Knowledge of this kind will lie particularly with marketing and related staff – but their perspective must be brought to bear early enough in the process to avoid creating a new service which no one actually wants to buy.

The 'traditional' approach to this stage was a linear sequence of problem-solving, but much recent work in improving development performance (especially in compressing the time required) involves attempts to do much of this concurrently or in overlapping stages. Useful metaphors for these two approaches are the relay race and the rugby team.[96] These should be seen as representing two poles of a continuum; as we shall see (in Chapter 5 and again in Chapter 9) the important issue is to choose an appropriate level of parallel development.

Launching the Innovation

In parallel with the technical problem-solving associated with developing an innovation, there is also a set of activities associated with preparing the market into which it will be launched. Whether this market is a group of retail consumers or a set of internal users of a new process, the same requirement exists for developing and preparing this market for launch, since it is only when the target market makes the decision to adopt the innovation that the whole innovation process is completed. The process is again one of sequentially collecting information, solving problems and focusing efforts towards a final launch. In particular it involves collecting information on actual or antici-pated customer needs and feeding this into the product development process, whilst simultaneously preparing the marketplace and marketing for the new product. It is essential throughout this process that a dialogue is maintained with other functions involved in the development process, and that the process of development is staged via a series of 'gates' which control progress and resource commitment.

A key aspect of the marketing effort involves anticipating likely responses to new product concepts and using this information to design the product and the way in which it is launched and marketed. This process of analysis builds upon knowledge about various sources of what Thomas calls 'market friction'.[97]

Buyer behaviour is a complex subject, but there are several key guidelines which emerge to help shape market development for a new product. The first is the underlying process of adoption of something new; typically this involves a sequence of awareness, interest, trial, evaluation and adoption. Thus simply making people aware, via advertising, etc. of the existence of a new product, will not be sufficient; they need to be drawn into the process through the other stages. Converting awareness to interest, for example, means forging a link between the new product concept and a personal need (whether real or induced via advertising). Chapter 7 deals with this issue in greater depth.

Successful implementation of internal (process) innovations also requires skilled change management. This is effectively a variation on the marketing principles outlined above, and stresses communication, involvement and intervention (via training, etc.) to minimize resistance to change – again essentially analogous to Thomas's concept of 'market friction'. Chapter 7 discusses this theme in greater detail, whilst Chapter 9 presents some key enabling routines for the implementation phase.

Understanding user needs has always been a critical determinant of innovation success and one way of achieving this is by bringing users into the loop at a much earlier stage. The work of Eric von Hippel and others has shown repeatedly that early involvement and allowing them to play an active role in the innovation process leads to better adoption and higher quality innovation. It is, effectively, the analogue of the early involvement/parallel working model mentioned above – and with an increasingly powerful set of tools for simulation and exploration of alternative options there is growing scope for such an approach.[29,98,99]

Where there is a high degree of uncertainty – as is the case with discontinuous innovation conditions – there is a particular need for adaptive strategies which stress the co-evolution of innovation with users, based on a series of 'probe and learn' experimental approaches. The role here for early and active user involvement is critical.[100]

Learning and Reinnovation

An inevitable outcome of the launch of an innovation is the creation of new stimuli for restarting the cycle. If the product/service offering or process change fails, this offers valuable information about what to change for next time. A more common scenario is what Rothwell and Gardiner call 're-innovation'; essentially building upon early success but improving the next generation with revised and refined features. In some cases, where the underlying design is sufficiently 'robust' it becomes possible to stretch and reinnovate over many years and models.[101]

But although the opportunities emerge for learning and development of innovations and the capability to manage the process which created them, they are not always taken up by organizations. Amongst the main requirements in this stage is the willingness to learn from completed projects. Projects are often reviewed and audited, but these reviews may often take the form of an exercise in 'blame accounting' and in trying to cover up mistakes and problems. The real need is to capture all the hard-won lessons, from both success and failure, and feed these through to the next generation. Nonaka and Kenney provide a powerful argument for this perspective in their comparison of product innovation at Apple and at Canon.[102] Much of the current discussion around the theme of knowledge management represents growing concern about the lack of such 'carry-over' learning – with the result that organizations are often 'reinventing the wheel' or repeating previous mistakes.

Learning can be in terms of technological lessons learned – for example, the acquisition of new processing or product features – which add to the organization's technological competence. But learning can also be around the capabilities and routines needed for effective product innovation management. In this connection some kind of structured audit framework or checklist is useful.

Chapter 9 explores some key routines for enabling learning.

2.8 Key Contextual Influences

So far we have been considering the core generic innovation process as a series of stages distributed over time and have identified key challenges which emerge in their effective management. But the process doesn't take place in a vacuum – it is subject to a range of internal and external influences which shape what is possible and what actually emerges. Roy Rothwell distinguishes between what he terms 'project related factors' – essentially those which we have been considering so far – and 'corporate conditions' which set the context in which the process is managed.[38] For the purposes of the book we will consider three sets of such contextual factors:

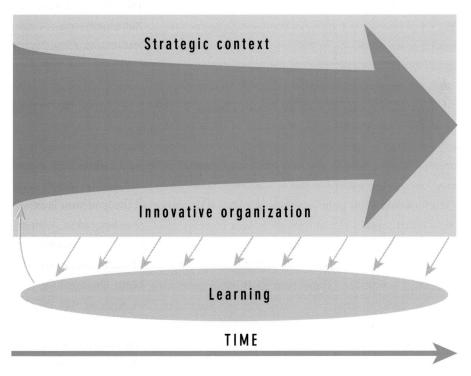

FIGURE 2.4 Influences on the innovation process

- The strategic context for innovation.
- The innovativeness of the organization.
- The connection between the organization and key elements in its external environment.

2.9 Beyond the Steady State

The model we have been developing in this chapter is very much about the world of repeated, continuous innovation where there is the underlying assumption that we are 'doing what we do but better'. This is not necessarily only about incremental innovation – it is possible to have significant step changes in product/service offering, process, etc. – but these still take place within an established envelope. The 'rules of the game' in terms of technological possibilities, market demands, competitor behaviour, political context, etc. are fairly clear and although there is scope for pushing at the edges the space within which innovation happens is well defined.

Central to this model is the idea of learning through trial and error to build effective routines which can help improve the chances of successful innovation. Because we get a lot of practice at such innovation it becomes possible to talk about a 'good' (if not 'best') practice model for innovation management which can be used to audit and guide organizational development.

But we need to also take into account that innovation is sometime *discontinuous* in nature. Things happen – as we saw in Chapter 1 – which lie outside the 'normal' frame and result in changes to the 'rules of the game'. Under these conditions doing more of the same 'good practice' routines may not be enough and may even be inappropriate to dealing with the new challenges. Instead we need a different set of routines – not to use instead of but as well as those we have developed for 'steady-state' conditions. It is likely to be harder to identify and learn these, in part because we don't get so much practice – it is hard to make a routine out of something which happens only occasionally. But we can observe some of the basic elements of the complementary routines which are associated with successful innovation management under discontinuous conditions. These tend to be associated with highly flexible behaviour involving agility, tolerance for ambiguity and uncertainty, emphasis on fast learning through quick failure, etc. – very much characteristics that are often found in small entrepreneurial firms. Table 2.6 lists some outline elements of the two complementary models.

As we will see throughout the book, a key challenge in managing innovation is the ability to create ways of dealing with both sets of challenges – and if possible to do so

TABLE 2.6 Different innovation management archetypes

	Type 1 – Steady-state archetype	Type 2 – Discontinuous-innovation archetype
Interpretive schema – how the organization sees and makes sense of the world	There is an established set of 'rules of the game' by which other competitors also play. Particular pathways in terms of search and selection environments and technological trajectories exist and define the 'innovation space' available to all players in the game	No clear 'rules of the game' – these emerge over time but cannot be predicted in advance. Need high tolerance for ambiguity – seeing multiple parallel possible trajectories. 'Innovation space' defined by open and fuzzy selection environment. Probe and learn experiments needed to build information about emerging patterns and allow dominant design to emerge.
	Strategic direction is highly – path dependent	Highly path-independent
Strategic decision-making	Makes use of decision-making processes which allocate resources on the basis of risk management linked to the above 'rules of the game'. (Does the proposal fit the business strategic directions? Does it build on existing competence base?) Controlled risks are taken within the bounds of the 'innovation space'. Political coalitions are significant influences maintaining the current trajectory	High levels of risk taking since no clear trajectories – emphasis on fast and lightweight decisions rather than heavy commitment in initial stages. Multiple parallel bets, fast failure and learning as dominant themes. High tolerance of failure but risk is managed by limited commitment. Influence flows to those prepared to 'stick their neck out' – entrepreneurial behaviour
Operating routines	Operates with a set of routines and structures/procedures which embed them which are linked to these 'risk rules' – for example, stage gate monitoring and review for project management. Search behaviour is along defined trajectories and uses tools and techniques for R&D, market research, etc. which assume a known space to be explored – search and selection environment. Network building to support innovation – e.g. user involvement, supplier partnership, etc. – is on basis of developing close and strong ties	Operating routines are open-ended, based around managing emergence. Project implementation is about 'fuzzy front end', light touch strategic review and parallel experimentation. Probe and learn, fast failure and learn rather than managed risk. Search behaviour is about peripheral vision, picking up early warning through weak signals of emerging trends. Linkages are with heterogeneous population and emphasis less on established relationships than on weak ties

in 'ambidextrous' fashion, maintaining close links between the two rather than spinning off completely separate ventures.

2.10 Beyond Boundaries

As we noted in Chapter 1, part of the management challenge around innovation in the twenty-first century is learning to deal with the process at an *inter*-organizational level. Innovation involves an increasingly large and diverse set of players arranged in various kinds of network, and managing across these boundaries represents a new set of issues and requires new and complementary routines to help deal with them. We argue that the underlying model of innovation presented in this chapter still offers a relevant framework around which to help think about and develop suitable routines for managing the process. Throughout the remainder of the book we will look at these core innovation themes but also at the ways in which this 'beyond the boundaries' challenge is being met and dealt with.

2.11 Summary and Further Reading

A number of writers have looked at innovation from a process perspective and some good examples can be found in references 103–109. Case studies provide a good lens through which this process can be seen and there are several useful collections including references 25, 45, 87, 110–113. Some books cover company histories in detail and give an insight into the particular ways in which firms develop their own bundles of routines – for example, references 42, 43, 114. Autobiographies of key innovation leaders provide a similar – if sometimes personally biased – insight into this.[115–118] In addition several websites – such as the Product Development Management Association (www.pdma.org) – carry case studies on a regular basis.

Many books and articles focus on particular aspects of the process – for example, on technology strategy,[119–122] on product or service development,[2,34,76,81] on process innovation,[53,123,124] on technology transfer,[125] on implementation[126–130] and on learning.[60,131–133]

For a good review and critique from the academic standpoint which raises a number of issues concerned with managing innovation see references 37, 105, 134–136.

References

1 Bryson, B. (1994) *Made in America*. Minerva, London.

2 Tidd, J. and F. Hull (eds) (2003) *Service Innovation: Organizational responses to technological opportunities and market imperatives*. Imperial College Press, London.

3 Davies, A. (1997) *Complex Product Systems: Europe's competitive advantage?* COPS Research Centre, Brighton.

4 Albury, D. (2004) *Innovation in the Public Sector*. Strategy Unit, Cabinet Office, London.

5 Nelson, R. (1993) *National Innovation Systems: A comparative analysis*. Oxford University Press, New York.

6 Best, M. (2001) *The New Competitive Advantage*. Oxford University Press, Oxford.

7 Gann, D. and A. Salter (2000) 'Innovation in project-based, service-enhanced firms: the construction of complex products and systems', *Research Policy*, **29**, 955–972.

8 Hamel, G. (2000) *Leading the Revolution*. Harvard Business School Press, Boston, Mass.

9 Tushman, M. and C. O'Reilly (1996) *Winning through Innovation*. Harvard Business School Press, Boston, Mass.

10 Christensen, C. and M. Raynor (2003) *The Innovator's Solution: Creating and sustaining successful growth*. Harvard Business School Press, Boston, Mass.

11 Foster, R. and S. Kaplan (2002) *Creative Destruction*. Harvard University Press, Cambridge, Mass.

12 Pavitt, K. (1984) 'Sectoral patterns of technical change: Towards a taxonomy and a theory', *Research Policy*, **13**, 343–373.

13 Hoffman, K. *et al.* (1997) 'Small firms, R&D, technology and innovation in the UK', *Technovation*, **18**, 39–55.

14 Oakey, R. (1991) 'High technology small firms: Their potential for rapid industrial growth', *International Small Business Journal*, **9**, 30–42.

15 OECD (1993) 'Small and medium-sized enterprise: Technology and competitiveness', Organisation for Economic Co-operation and Development, Paris.

16 Rothwell, R. and M. Dodgson (1993) 'SMEs: Their role in industrial and economic change', *International Journal of Technology Management*, Special issue on small firms, 8–22.

17 Voss, C. (1999) *Made in Europe 3: The small company study*. London Business School/IBM Consulting, London.

18 Lundvall, B. (1990) *National Systems of Innovation: Towards a theory of innovation and interactive learning*. Frances Pinter, London.

19 Pavitt, K. (2000) *Technology, Management and Systems of Innovation*. Edward Elgar, London.

20 Abernathy, W. and J. Utterback (1978) 'Patterns of industrial innovation', *Technology Review*, **80**, 40–47.

21 Utterback, J. (1994) *Mastering the Dynamics of Innovation*. Harvard Business School Press, Boston, Mass., 256.

22 Tushman, M. and P. Anderson (1987) 'Technological discontinuities and organizational environments', *Administrative Science Quarterly*, **31** (3), 439–465.

23 Perez, C. (2002) *Technological Revolutions and Financial Capital*. Edward Elgar, Cheltenham.

24 Christenson, C. (1997) *The Innovator's Dilemma*. Harvard Business School Press, Cambridge, Mass.

25 Leifer, R. *et al.* (2000) *Radical Innovation*. Harvard Business School Press, Boston, Mass.

26 IPTS (1998) *The Impact of EU Regulation on Innovation of European Industry*. IPTS/European Union, Seville.

27 Whitley, R. (2000) 'The institutional structuring of innovation strategies: business systems, firm types and patterns of technical change in different market economies', *Organization Studies*, **21** (5), 855–886.

28 Freeman, C. and L. Soete (1997) *The Economics of Industrial Innovation*. 3rd edn. MIT Press, Cambridge, Mass.

29 Von Hippel, E. (1988) *The Sources of Innovation*. MIT Press, Cambridge, Mass.

30 Coombs, R., P. Saviotti and V. Walsh (1985) *Economics and Technological Change*. Macmillan, London.

31 Booz, Allen and Hamilton Consultants (1982) *New Product Management for the 1980s*. Booz, Allen and Hamilton Consultants.

32 Cooper, R. (2003) 'Profitable product innovation', in Shavinina, L. (ed), *International Handbook of Innovation*. Elsevier, New York.

33 Hollins, G. and W. Hollins (1990) *Total Design*. Pitman, London.

34 Rosenau, M. *et al.* (eds) (1996) *The PDMA Handbook of New Product Development*. John Wiley & Sons, Inc., New York.

35 Koen, P.A. *et al.* (2001) 'New concept development model: providing clarity and a common language to the "fuzzy front end" of innovation', *Research Technology Management*, **44** (2), 46–55.

36 Souder, W. and J. Sherman (1994) *Managing New Technology Development*. McGraw-Hill, New York.

37 Van de Ven, A., H. Angle and M. Poole (1989) *Research on the Management of Innovation*. Harper & Row, New York.

38 Rothwell, R. (1992) 'Successful industrial innovation: critical success factors for the 1990s', *R&D Management*, **22** (3), 221–239.

39 Dodgson, M., D. Gann and A. Salter (2002) 'The intensification of innovation', *International Journal of Innovation Management*, **6** (1), 53–83.

40 Bright, A. (1949) *The Electric Lamp Industry: technological change and economic development from 1800 to 1947*. Macmillan, New York.

41 Henderson, R. and K. Clark (1990) 'Architectural innovation: the reconfiguration of existing product technologies and the failure of established firms', *Administrative Science Quarterly*, **35**, 9–30.

42 Graham, M. and A. Shuldiner (2001) *Corning and the Craft of Innovation*. Oxford University Press, Oxford.

43 Gundling, E. (2000) *The 3M Way to Innovation: Balancing people and profit*. Kodansha International, New York.

44 De Geus, A. (1996) *The Living Company*. Harvard Business School Press, Boston, Mass.

45 Kanter, R. (ed) (1997) *Innovation: Breakthrough thinking at 3M, DuPont, GE, Pfizer and Rubbermaid*. Harper Business, New York.

46 Levitt, B. and J. March (1988) 'Organisational learning', *Annual Review of Sociology*, **14**, 319–340.

47 Augsdorfer, P. (1996) *Forbidden Fruit*. Avebury, Aldershot.

48 Ettlie, J. (1988) *Taking Charge of Manufacturing*. Jossey-Bass, San Francisco.

49 Bessant, J. (1991) *Managing Advanced Manufacturing Technology: The challenge of the fifth wave*. NCC-Blackwell, Oxford/Manchester.

50 Povey, B. (1996) *Business Process Improvement*. University of Brighton, Brighton.

51 Knights, D. and D. McCabe (1999) '"Are there no limits to authority?": TQM and organizational power', *Organizational Studies*, Spring.

52 Grover, V. and S. Jeong (1995) 'The implementation of business process re-engineering', *Journal of Management Information Systems*, **12** (1), 109–144.

53 Davenport, T. (1995) 'Will participative makeovers of business processes succeed where re-engineering failed?', *Planning Review*, 24.

54 Camp, R. (1989) *Benchmarking – The Search for Industry Best Practices that Lead to Superior Performance*. Quality Press, Milwaukee, WI.

55 Dale, B. and S. Hayward (1984) *A Study of Quality Circle Failures*. UMIST.

56 Lillrank, P. and N. Kano (1990) *Continuous Improvement; Quality control circles in Japanese industry*. University of Michigan Press, Ann Arbor.

57 Swann, P., M. Prevezer and D. Stout (eds) (1998) *The Dynamics of Industrial Clustering*. Oxford, Oxford University Press.

58 Humphrey, J. and H. Schmitz (1996) 'The Triple C approach to local industrial policy', *World Development*, **24** (12), 1859–1877.

59 Malhotra, Y. (2004) 'Why knowledge management systems fail', in Koenig, M. and Srikantaiah, T. (eds), *Knowledge Management Lessons Learned*. American Society for Information Science and Technology, New York, 87–112.

60 Leonard-Barton, D. (1995) *Wellsprings of Knowledge: Building and sustaining the sources of innovation*. Harvard Business School Press, Boston, Mass., 335.

61 Robertson, A. (1974) *The Lessons of Failure*. Macdonald, London.

62 Lilien, G. and E. Yoon (1989) 'Success and failure in innovation – a review of the literature', *IEEE Transactions on Engineering Management*, **36** (1), 3–10.

63 Ernst, H. (2002) 'Success factors of new product development: a review of the empirical literature', *International Journal of Management Reviews*, **4** (1), 1–40.

64 Kaplan, R. and D. Norton (1996) 'Using the balanced scorecard as a strategic management system', *Harvard Business Review*, January–February.

65 Mills, J. *et al.* (2002) *Creating a Winning Business Formula*. Cambridge University Press, Cambridge.

66 Smith, P. and D. Reinertsen (1991) *Developing Products in Half the Time*. Van Nostrand Reinhold, New York.

67 Tidd, J. (1989) *Flexible Automation*. Frances Pinter, London.

68 Brown, S. *et al.* (2000) *Strategic Operations Management*. Butterworth Heinemann, Oxford.

69 Goffin, K. and R. Pfeiffer (2001) 'Competing in the innovation pentathlon', *Innovation: Management, Policy and Practice*, **4** (1/3), 143–150.

70 Rothwell, R. (1977) 'The characteristics of successful innovators and technically progressive firms', *R&D Management*, **7** (3), 191–206.

71 Langrish, J. *et al.* (1972) *Wealth from Knowledge*. Macmillan, London.

72 Georghiou, L. *et al.* (1986) *Post-innovation Performance*. Macmillan, Basingstoke.

73 Sherwin, C. and S. Isenson (1967) 'Project hindsight', *Science*, **156**, 571–577.

74 Isenson, R. (1968) *Technology in Retrospect and Critical Events in Science (Project TRACES)*. Illinois Institute of Technology/National Science Foundation.

75 Carter, C. and B. Williams (1957) *Industry and Technical Progress*. Oxford University Press, Oxford.

76 Cooper, R. (2001) *Winning at New Products*. 3rd edn. Kogan Page, London.

77 Maidique, M. and B. Zirger (1985) 'The new product learning cycle', *Research Policy*, **14** (6), 299–309.

78 Ledwith, A. (2004) 'Management of new product development in small Irish electronics firms', PhD thesis, *CENTRIM*, University of Brighton, Brighton.

79 Souder, W. and S. Jenssen (1999) 'Management practices influencing new product success and failure in the US and Scandinavia', *Journal of Product Innovation Management*, **16**, 183–204.

80 Eisenhardt, K. and S. Brown (1997) 'The art of continuous change: linking complexity theory and time-paced evolution in relentlessly shifting organizations', *Administrative Science Quarterly*, **42** (1), 1–34.

81 Wheelwright, S. and K. Clark (1992) *Revolutionising Product Development*. Free Press, New York.

82 Walsh, V. *et al.* (1992) *Winning by Design: Technology, product design and international competitiveness*. Basil Blackwell, Oxford.

83 Chiesa, V., P. Coughlan and C. Voss (1996) 'Development of a technical innovation audit', *Journal of Product Innovation Management*, **13** (2), 105–136.

84 Design Council (2002) *Living Innovation*. Design Council/Department of Trade and Industry, London. website: www.livinginnovation.org.uk

85 Francis, D. (2001) *Developing Innovative Capability*. University of Brighton, Brighton.

86 Baden-Fuller, C. and J. Stopford (1995) *Rejuvenating the Mature Business*. Routledge, London.

87 Von Stamm, B. (2003) *The Innovation Wave*. John Wiley & Sons, Ltd, Chichester.

88 Leonard-Barton, D. (1992) 'The organisation as learning laboratory', *Sloan Management Review*, **34** (1), 23–38.

89 Rush, H., T. Brady and M. Hobday (1997) *Learning Between Projects in Complex Systems*. Centre for the Study of Complex Systems, University of Brighton.

90 Johne, A. and P. Snelson (1988) 'Auditing product innovation activities in manufacturing firms', *R&D Management*, **18** (3), 227–233.

91 Hamel, G. and C. Prahalad (1994) *Competing for the Future*. Harvard Business School Press, Cambridge, Mass.

92 Cooper, R. and E. Kleinschmidt (1990) *New Products: The key factors in success*. American Marketing Association, Chicago.

93 Teece, D. (1998) 'Capturing value from knowledge assets: the new economy, markets for know-how, and intangible assets', *California Management Review*, **40** (3), 55–79.

94 Bessant, J. (2003) *High Involvement Innovation*. John Wiley & Sons, Ltd, Chichester.

95 Dodgson, M. and J. Bessant (1996) *Effective Innovation Policy*. International Thomson Business Press, London.

96 Clark, K. and T. Fujimoto (1992) *Product Development Performance*. Harvard Business School Press, Boston, Mass.

97 Thomas, R. (1993) *New Product Development: Managing and forecasting for strategic success*. John Wiley & Sons, Inc., New York.

98 Dodgson, M., D. Gann and A. Salter (2005) *Think, Play, Do: Technology and organization in the emerging innovation process*. Oxford University Press, Oxford.

99 Schrage, M. (2000) *Serious Play: How the world's best companies simulate to innovate*. Harvard Business School Press, Boston, Mass.

100 Moore, G. (1999) *Crossing the Chasm; Marketing and selling high-tech products to mainstream customers*. Harper Business, New York.

101 Rothwell, R. and P. Gardiner (1985) 'Invention, innovation, re-innovation and the role of the user', *Technovation*, **3**, 167–186.

102 Nonaka, I. and M. Kenney (1991) 'Towards a new theory of innovation management', *Journal of Engineering and Technology Management*, **8**, 67–83.

103 Brockhoff, K., A. Chakrabarti and J. Hauschildt (eds) (1999) 'The Dynamics of Innovation'. Springer, Heidelberg.

104 Jelinek, M. and J. Litterer (1994) 'Organising for technology and innovation', in Souder, W. and Sherman, J. (eds), *Managing New Technology Development*. McGraw-Hill, New York.

105 Loveridge, R. and M. Pitt (1990) *The Strategic Management of Technological Innovation*. John Wiley & Sons, Ltd, Chichester.

106 Trott, P. (1998) *Innovation Management and New Product Development*. FT-Pitman, London.

107 Ettlie, J. (1999) *Managing Innovation*. John Wiley & Sons, Inc., New York.

108 Jones, T. (2002) *Innovating at the Edge*. Butterworth Heinemann, London.

109 Hargadon, A. (2003) *How Breakthroughs Happen*. Harvard Business School Press, Boston, Mass.

110 Baden-Fuller, C. and M. Pitt (1996) *Strategic Innovation*. Routledge, London.

111 Burgelman, R. and R. Rosenbloom (1993) *Research on Technological Innovation: Management and policy*. JAI Press, Greenwich, Conn.

112 Gallagher, M. and S. Austin (1997) *Continuous Improvement Casebook*. Kogan Page, London.

113 Weisberg, R. (2003) 'Case studies of innovation: ordinary thinking, extraordinary outcomes', in Shavinina, L. (ed.), *International Handbook of Innovation*. Elsevier, New York.

114 Kelley, T., J. Littman and T. Peters (2001) *The Art of Innovation: Lessons in creativity from Ideo, America's leading design firm*. Currency, New York.

115 Groves, A. (1999) *Only the Paranoid Survive*. Bantam Books, New York.

116 Dyson, J. (1997) *Against the Odds*. Orion, London.

117 Welch, J. (2001) *Jack! What I've learned from leading a great company and great people*. Headline, New York.

118 Gates, B. (1996) *The Road Ahead*. Viking, New York.

119 Burgelman, R., C. Christensen and S. Wheelwright (eds) (2004) *Strategic Management of Technology and Innovation*. Fourth edn. McGraw Hill Irwin, Boston.

120 Dodgson, M. (ed) (1990) *Technology and the Firm: strategies, management and public policy*. Longman, Harlow.

121 Ford, D. and M. Saren (1996) 'Technology strategy for business', in Bessant, J. and Preece, D. (eds), *Management of Technology and Innovation*. International Thomson Business Press, London.

122 Adler, P. (1989) 'Technology strategy: A guide to the literature', *Research in Technological Innovation, Management and Policy*, **4**, 25–151.

123 Pisano, G. (1996) *The Development Factory: Unlocking the potential of process innovation*. Harvard Business School Press, Boston, Mass.

124 Zairi, M. (1999) *Process Innovation Management*. Butterworth Heinemann, London.

125 Saad, M. (2000) *Development through Technology Transfer*. Intellect Publishers, Bristol.

126 Bennett, D. and M. Kerr (1996) 'A systems approach to the implementation of total quality management', *Total Quality Management*, **7**, 631–665.

127 Leonard-Barton, D. (1988) 'Implementation as mutual adaptation of technology and organization', *Research Policy*, **17**, 251–267.

128 Voss, C. (1986) 'Implementation of advanced manufacturing technology', in Voss, C. (ed.), *Managing Advanced Manufacturing Technology*. IFS Publications, Kempston.

129 Fleck, J. (1994) 'Learning by trying', *Research Policy*, **23**, 637–652.

130 Voss, C. (1988) 'Success and failure in AMT', *International Journal of Technology Management*, **3** (3), 285–297.

131 Cohen, W. and D. Levinthal (1990) 'Absorptive capacity: A new perspective on learning and innovation', *Administrative Science Quarterly*, **35** (1), 128–152.

132 Boisot, M. (1995) 'Is your firm a creative destroyer? competitive learning and knowledge flows in the technological strategies of firms', *Research Policy*, **24**, 489–506.

133 Ayas, K. (1997) *Design for Learning for Innovation*. Eburon, Delft.

134 King, N. (1992) 'Modelling the innovation process: an empirical comparison of approaches', *Journal of Occupational and Social Psychology*, **65**, 89–100.

135 Clark, P. (2002) *Organizational Innovations*. Sage, London.

136 Van de Ven, A. (1999) *The Innovation Journey*. Oxford University Press, Oxford.

Part II

TAKING A STRATEGIC APPROACH

> A great deal of business success depends on generating new knowledge and on having the capabilities to react quickly and intelligently to this new knowledge . . . I believe that strategic thinking is a necessary but overrated element of business success. If you know how to design great motorcycle engines, I can teach you all you need to know about strategy in a few days. If you have a PhD in strategy, years of labor are unlikely to give you the ability to design great new motorcycle engines.
>
> (Richard Rumelt, 1996, *California Management Review*, **38**, 110, on the continuing debate about the causes of Honda's success in the US motorcycle market)

The above quotation from a distinguished professor of strategy appears on the surface not to be a strong endorsement of his particular trade. In fact, it offers indirect support for the central propositions of this section of our book:

1. Firm-specific knowledge – including the capacity to exploit it – is an essential feature of competitive success.
2. An essential feature of corporate strategy should therefore be an innovation strategy, the purpose of which is deliberately to accumulate such firm-specific knowledge.
3. An innovation strategy must cope with an external environment that is complex and ever-changing, with considerable uncertainties about present and future developments in technology, competitive threats and market (and non-market) demands.
4. Internal structures and processes must continuously balance potentially conflicting requirements:
   (a) to identify and develop specialized knowledge within technological fields, business functions and product divisions;
   (b) to exploit this knowledge through integration across technological fields, business functions and product divisions.

In Chapter 3, we address the question, 'What is the appropriate framework for understanding innovation strategy?' Given complexity, continuous change and consequent uncertainty, we conclude that the so-called rational approach to innovation strategy is less likely to be effective than the incrementalist approach that stresses continuous adjustment in the light of new knowledge and learning. We also conclude that the approach pioneered by Michael Porter correctly identifies the nature of the competitive threats and opportunities that emerge from advances in technology, and rightly stresses the importance of developing and protecting firm-specific technology in order to enable firms to position themselves against the competition. But it underestimates the power of technology to change the rules of the competitive game by modifying industry boundaries, developing new products and shifting barriers to entry. It also overestimates the capacity of corporate management to identify and predict the important changes outside the firm, and to implement radical changes in competencies and organizational practices within the firm. We therefore adopt the three elements of inno-vation strategy proposed by David Teece and Gary Pisano: market and national *pos-itions*, technological *paths* and organizational *processes*.

In Chapter 4, we address the question, 'How does the firm's national and market environment shape its innovation strategy?' We first show that the *home country* pos-itions of even global firms have a strong influence on their innovation strategies. The national influences can be grouped into three categories: *competencies* (workforce edu-cation, research), *economic inducement mechanisms* (local demand and input prices, com-petitive rivalry) and *institutions* (methods of funding, controlling and managing business firms). However, managements still have ample influence over their firms' innovation strategies, and firms can benefit from foreign systems of innovation through a variety of market mechanisms.

We also show that firms can obtain information to position themselves compared to their competitors through an increasing range of sources (including so-called bench-marking). *Information* about what competitors are doing must be clearly distinguished from the *competence* to keep up with competitors, which requires a much greater cor-porate investment in R&D and reverse engineering activities. Firms maintain their innovative leads over their competitors through a variety of often complementary mechanisms, the relative importance of which varies from industry to industry: for example, patent protection is recognized as a more effective means of protection against imitators in the drug industry than in the paper and pulp industry.

In Chapter 5, we address the question, 'How does the nature of technology shape the firm's innovation strategy?' We show that marked differences amongst sectors are also central to corporate choices about *technological trajectories, firm-specific competencies* and *innovation strategies*. We identify five broad technological trajectories that firms can follow, each of which has distinct implications for the tasks of innovation strategy. We also identify three key technologies (biotechnology, materials and IT) where rapid advances lead to major shifts in technological trajectories, and where it is increasingly important to distinguish the microelectronics revolution (making and using electronic chips) from the more important information revolution (making and using software).

C. K. Prahalad and Gary Hamel have had a major influence on management think-ing by showing that the capacity to open up new product markets requires distinctive core competencies, coupled with methods of corporate organization and evaluation that explicitly recognize the importance of these competencies, and top management visions that identify future opportunities. Experience shows that, along some techno-logical trajectories, the opportunities for product diversification are abundant but uncer-tain, whilst along others they hardly exist at all. It also shows that companies also need background competencies to co-ordinate and integrate changes coming from outside the firm, and that corporate visions can be wrong.

In Chapter 6, we address the question 'What are the linkages across functional areas that are essential for a successful innovation strategy?' We identify three areas where continuous flows of information and knowledge across functional boundaries are crucial. First, the problems of deciding the organizational and geographical *location of R&D activities* reflect conflicting needs: (1) the advantages of concentration – geographical for costly programmes to develop and launch major innovations, and organizational to orchestrate networks with public (including university) research activities; (2) the advantages of decentralization, in order to be responsive to the demands of specific product or geographic markets.

Second, links between the corporate technical function and the corporate *resource allocation function* should be developed to ensure that resource allocation decisions reflect the dual nature of corporate investments in R&D: as a *business investment*, and as an *investment in learning* or *strategic positioning*. For the latter, conventional financial techniques are not appropriate, and effective decisions depend on success in mobilizing a range of professional and functional competencies from across the corporation.

Third, different *corporate strategic* styles – defined along two dimensions: the relative importance *of financial control versus entrepreneurship*, and of *centralization versus decentralization* – are appropriate for different technologies, depending on the range of opportunities that they open, and the costs of experimentation. In other words, corporate technological competencies not only serve corporate strategy, they often determine its main directions.

This section of the book has more to say about large firms, where deliberate innovation strategies are of great importance. However, positions, paths and processes also matter in small firms, and they are therefore discussed towards the end of Chapters 4–6. The purpose of innovation strategy in small firms is the same as in large ones: to create a firm-specific innovative advantage. But the balance of advantages and disadvantages is very different. Small firms in general have the organizational advantages of ease of communication, rapid decision-making and flexibility. They also have the disadvantage of either narrow or shallow technological competencies, and a greater reliance for innovation on suppliers and customers, and for strategies and competencies on the experience and qualifications of their senior managers. As with large firms, small firms vary greatly in their technological trajectories and innovation strategies. Unfortunately, most research attention has been devoted to the minority of small firms who emerge from high-technology opportunities and sometimes become very big. We know very little about innovation strategies in the rest, except that they are nearly all now trying to cope with IT.

Chapter 3

Developing the Framework for an Innovation Strategy

Innovation plays an important and dual role, as both a major source of uncertainty and change in the environment, and a major competitive resource within the firm. In this chapter, we develop what we think is the most useful framework for defining and implementing corporate innovation strategy.

We begin by summarizing the well-known debate in corporate strategy between 'rationalist' and 'incrementalist' approaches to the characteristics of technological innovation; we conclude that the latter approach is more useful, given the inevitable complexities and uncertainties in the innovation process. We then describe and evaluate Michael Porter's pioneering framework that links innovation strategy to overall corporate strategy; we conclude that its major strength is in linking the firm's technology strategy to its market and competitive position. But it both underestimates the power of technological change to upset established market and competitive conditions, and overestimates the influence that managers actually have over corporate choice in technology strategy. For this reason, we propose that the most useful framework so far is the one developed by David Teece and Gary Pisano. It gives central importance to *the dynamic capabilities of firms*, and distinguishes three elements of corporate innovation strategy: (1) competitive and national *positions*; (2) technological *paths*; (3) organizational and managerial *processes*. These will then be discussed in detail in the subsequent three chapters.

3.1 'Rationalist' or 'Incrementalist' Strategies for Innovation?

The long-standing debate between 'rational' and 'incremental' strategies is of central importance to the mobilization of technology and to the purposes of corporate strategy. We begin by reviewing the main terms of the debate, and conclude that the supposedly clear distinction between strategies based on 'choice' or on 'implementation' breaks down when firms are making decisions in complex and fast-changing competitive environments. Under such circumstances, formal strategies must be seen as part of a wider process of continuous learning from experience and from others to cope with complexity and change.

Notions of corporate strategy first emerged in the 1960s. A lively debate has continued since then amongst the various 'schools' or theories. Here we discuss the two most influential: the 'rationalist' and the 'incrementalist'. The main protagonists are Ansoff[1] of the rationalist school and Mintzberg[2] amongst the incrementalists. An excellent summary of the terms of the debate can be found in Whittington,[3] and a face-to-face debate between the two in the *Strategic Management Journal* in 1991.

Rationalist Strategy

'Rationalist' strategy has been heavily influenced by military experience, where strategy (in principle) consists of the following steps: (1) describe, understand and analyse the environment; (2) determine a course of action in the light of the analysis; (3) carry out the decided course of action. This is a 'linear model' of rational action: appraise, determine and act. The corporate equivalent is SWOT: the analysis of corporate strengths and weaknesses in the light of external opportunities and threats. This approach is intended to help the firm to:

- Be conscious of trends in the competitive environment.
- Prepare for a changing future.
- Ensure that sufficient attention is focused on the longer term, given the pressures to concentrate on the day-to-day.
- Ensure coherence in objectives and actions in large, functionally specialized and geographically dispersed organizations.

However, as John Kay has pointed out, the military metaphor can be misleading.[4] Corporate objectives are different from military ones: namely, to establish a distinctive competence enabling them to satisfy customers better than the competition – and not to mobilize sufficient resources to destroy the enemy. Excessive concentration on the 'enemy' (i.e. corporate competitors) can result in strategies emphasizing large commitments of resources for the establishment of monopoly power, at the expense of profitable niche markets and of a commitment to satisfying customer needs.

More important, as Box 3.1 shows, professional experts, including managers, have difficulties in appraising accurately their real situation, essentially for two reasons. First, their external environment is both *complex*, involving competitors, customers, regulators and so on, and *fast-changing*, including technical, economic, social and political change. It is therefore difficult enough to understand the essential features of the

---

BOX 3.1

'STRATEGIZING IN THE REAL WORLD' BY WILLIAM STARBUCK[5]

'The war in Vietnam is going well and will succeed.' (R. MacNamara, 1963)

'I think there is a world market for about five computers.' (T. Watson, 1948)

'Gaiety is the most outstanding feature of the Soviet Union.' (J. Stalin, 1935)

'Prediction is very difficult, especially about the future.' (N. Bohr)

'I cannot conceive of any vital disaster happening to this vessel.'

(Captain of *Titanic*, 1912)

The above quotes are from the paper by Starbuck,[5] in which he criticizes formal strategic planning:

> First, formalization undercuts planning's contributions. Second, nearly all managers hold very inaccurate beliefs about their firms and market environments. Third, no-one can forecast accurately over the long term . . . However, planners can make strategic planning more realistic and can use it to build healthier, more alert and responsive firms. They can make sensible forecasts and use them to foster alertness; exploit distinctive competencies, entry barriers and proprietary information; broaden managers' horizons and help them develop more realistic beliefs; and plan in ways that make it easier to change strategy later. (p. 77)

---

present, let alone to predict the future. Second, managers in large firms disagree on their firms' strengths and weaknesses in part because their knowledge of what goes on *inside* the firm is imperfect.

As a consequence, internal corporate strengths and weaknesses are often difficult to identify before the benefit of practical experience, especially in new and fast-changing technological fields. For example:

- In the 1960s, the oil company Gulf defined its distinctive competencies as producing energy, and so decided to purchase a nuclear energy firm. The venture was unsuccessful, in part because the strengths of an oil company in finding, extracting, refining and distributing oil-based products, i.e. geology and chemical processing technologies, logistics, consumer marketing, were largely irrelevant to the design, construction and sale of nuclear reactors, where the key skills are in electro-mechanical technologies and in selling to relatively few, but often politicized electrical utilities.[6]

- In the 1960s and 1970s, many firms in the electrical industry bet heavily on the future of nuclear technology as a revolutionary breakthrough that would provide virtually costless energy. Nuclear energy failed to fulfil its promise, and firms only recognized later that the main revolutionary opportunities and threats for them came from the virtually costless storage and manipulation of information provided by improvements in semiconductor and related technologies.[7]

- In the 1980s, analysts and practitioners predicted that the 'convergence' of computer and communications technologies through digitalization would lower the barriers to entry of mainframe computer firms into telecommunications equipment, and vice versa. Many firms tried to diversify into the other market, often through acquisitions or alliances, e.g. IBM bought Rohm, AT&T bought NCR. Most proved unsuccessful, in part because the software requirements in the telecommunications and office markets were so different.[8]

- The 1990s similarly saw commitments in the fast-moving fields of ICT (information and communication technology) where initial expectations about opportunities and complementarities have been disappointed (see Box 3.2). For example, the investments of major media companies in the Internet in the late 1990s took more than a decade to prove profitable: problems remain in delivering products to consumers and in getting paid for them, and advertising remains ineffective.[9] There have been similar disappointments so far in development of 'e-entertainment'.[10]

- The Internet Bubble, which began in the late 1990s but had burst by 2000, placed wildly optimistic and unrealistic valuations on new ventures utilizing e-commerce. In particular, most of the new e-commerce businesses selling to consumers which floated on the US and UK stock exchanges between 1998 and 2000 subsequently lost around 90% of their value, or were made bankrupt. Notorious failures of that period include Boo.com in the UK, which attempted to sell sports clothing via the Internet, and Pets.com in the USA, which attempted to sell pet food and accessories.

Incrementalist Strategy

Given these conditions, 'incrementalists' argue that the complete understanding of complexity and change is impossible: our ability both to comprehend the present and to predict the future is therefore inevitably limited. As a consequence, successful practitioners – engineers, doctors and politicians, as well as business managers – do not, in general, follow strategies advocated by the rationalists, but incremental strategies which explicitly recognize that the firm has only very imperfect knowledge of its environment, of its own strengths and weaknesses, and of the likely rates and directions of change in the future. It must therefore be ready to adapt its strategy in the light of new information and understanding, which it must consciously seek to obtain. In such circumstances the most efficient procedure is to:

BOX 3.2
THE LIMITS OF RATIONAL STRATEGIZING

Jonathan Sapsed's thought-provoking analysis of corporate strategies of entry into new digital media[11] concludes that the rationalist approach to strategy in emerging industries is prone to failure. Because of the intrinsic uncertainty in such an area, it is impossible to forecast accurately and predict the circumstances on which rationalist strategy, e.g. as recommended by Porter will be based. Sapsed's book includes case studies of companies that have followed the classical rational approach and subsequently found their strategies frustrated.

An example is Pearson, the large media conglomerate, which conducted a SWOT (identify strengths, weaknesses, opportunities, threats) analysis in response to developments in digital media. The strategizing showed the group's strong assets in print publishing and broadcasting, but perceived weaknesses in new media. Having established its 'gaps' in capability Pearson then searched for an attractive multimedia firm to fill the gap. It expensively acquired Mindscape, a small Californian firm. The strategy failed with Mindscape being sold for a loss of £212 m. four years later, and Pearson announcing exit from the emerging market of consumer multimedia.

The strategy failed for various reasons. First, unfamiliarity with the technology and market; second, a misjudged assessment of Mindscape's position; and third, a lack of awareness of the multimedia activities already within the group. The formal strategy exercises that preceded action were prone to misinterpretation and misinformation. The detachment from operations recommended by rationalist strategy exacerbated the information problems. The emphasis of rational strategy is not on assessing information arising from operations, but places great credence in detached, logical thought.

Sapsed argues that whilst formal strategizing is limited in what it can achieve, it may be viewed as a form of therapy for managers operating under uncertainty. It can enable disciplined thought on linking technologies to markets, and direct attention to new information and learning. It focuses minds on products, financial flows and anticipating options in the event of crisis or growth. Rather than determining future action, it can prepare the firm for unforeseen change.

1. Make deliberate steps (or changes) towards the stated objective.
2. Measure and evaluate the effects of the steps (changes).
3. Adjust (if necessary) the objective and decide on the next step (change).

This sequence of behaviour goes by many names, such as incrementalism, trial and error, 'suck it and see', muddling through and learning. When undertaken deliberately, and based on strong background knowledge, it has a more respectable veneer, such as:

- Symptom → diagnosis → treatment → diagnosis → adjust treatment → cure (for medical doctors dealing with patients)
- Design → development → test → adjust design → retest → operate (for engineers making product and process innovations)

Corporate strategies that do not recognize the complexities of the present, and the uncertainties associated with change and the future, will certainly be rigid, will probably be wrong, and will potentially be disastrous if they are fully implemented. But this is not a reason for rejecting analysis and rationality in innovation management. On the contrary, under conditions of complexity and continuous change, it can be argued that 'incrementalist' strategies are more rational (that is, more efficient) than 'rationalist' strategies. Nor is it a reason for rejecting all notions of strategic planning. The original objectives of the 'rationalists' for strategic planning – set out above – remain entirely valid. Corporations, and especially big ones, without any strategies will be ill-equipped to deal with emerging opportunities and threats: as Pasteur observed '. . . chance favours only the prepared mind'.[12]

Implications for Management

This debate has two sets of implications for managers. The first concerns the practice of corporate strategy, which should be seen as a form of corporate *learning, from analysis and experience, how to cope more effectively with complexity and change.* The implications for the processes of strategy formation are the following:

- Given uncertainty, explore the implications of a *range* of possible future trends.
- Ensure broad participation and informal channels of communication.
- Encourage the use of multiple sources of information, debate and skepticism.
- Expect to change strategies in the light of new (and often unexpected) evidence.

The second implication is that *successful management practice is never fully reproducible.* In a complex world, neither the most scrupulous practising manager nor the most rigorous management scholar can be sure of identifying – let alone evaluating – all the necessary ingredients in real examples of successful management practice. In addition, the conditions of any (inevitably imperfect) reproduction of successful management practice will differ from the original, whether in terms of firm, country, sector, physical conditions, state of technical knowledge, or organizational skills and cultural norms.

Thus, in conditions of complexity and change – in other words, the conditions for managing innovation – there are no easily applicable recipes for successful management practice. This is one of the reasons why there are continuous swings in management fashion (see Box 3.3). Useful learning from the experience and analysis of others necessarily requires the following:

1. *A critical reading of the evidence underlying any claims to have identified the factors associated with management success.* Compare, for example, the explanations for the success of Honda in penetrating the US motorcycle market in the 1960s, given (1) by the Boston Consulting Group: exploitation of cost reductions through manufacturing investment and production learning in deliberately targeted and specific market segments;[13] and (2) by Richard Pascale: flexibility in product-market

---

BOX 3.3
SWINGS IN MANAGEMENT FASHION

'**Upsizing.** After a decade of telling companies to shrink, management theorists have started to sing the praises of corporate growth.' (Feature title from *The Economist*, 10 February 1996, p. 81)

'**Fire and forget?** Having spent the 1990s in the throes of restructuring, re-engineering and downsizing, American companies are worrying about corporate amnesia.' (Feature title from *The Economist*, 20 April 1996, pp. 69–70)

Above are two not untypical examples of swings in management fashion and practice that reflect the inability of any recipe for good management to reflect the complexities of the real thing, and to put successful experiences in the past in the context of the function, firm, country, technology, etc. More recently, a survey of 475 global firms by Bain and Co. showed that the proportion of companies using management tools associated with *business process re-engineering*, *core competencies* and *total quality management* has been declining since the mid-1990s. But they still remain higher than the more recently developed tools associated with *knowledge management*, which have been less successful, especially outside North America. ('Management fashion: fading fads', *The Economist*, 22 April 2000, pp. 72–73)

strategy in response to unplanned market signals, high-quality product design, manufacturing investment in response to market success.[14] The debate has recently been revived, although not resolved, in the California Management Review.[15]

2. A careful comparison of the context of successful management practice, with the context of the firm, industry, technology and country in which the practice might be reused. For example, one robust conclusion from management research and experience is that major ingredients in the successful implementation of innovation are effective linkages amongst functions within the firm, and with outside sources of relevant scientific and marketing knowledge. Although very useful to management, this knowledge has its limits. As we shall show later in Chapter 5, conclusions from a drug firm that the key linkages are between university research and product development are profoundly misleading for an automobile firm, where the key linkages are amongst product development, manufacturing and the supply chain. And even within each of these industries, important linkages may change over time. In the drug industry, the key academic disciplines are shifting from chemistry to include more biology. And in automobiles, computing and associated skills have become important for the development of 'virtual prototypes', and for linkages between product development, manufacturing and the supply chain.[16]

The management tools and techniques described in this book represent only very imperfectly the complexities and changes of the real world. As such, they can be no more than aids to systematic thinking, and to collective learning based on analysis and experience. Especially in conditions of complexity and change, *tacit knowledge* of individuals and groups (i.e. know-how that is based on experience, and that cannot easily be codified and reproduced) is of central importance, whether in the design of automobiles and drugs, or in the strategic management of innovation.

3.2 Technology and Competitive Analysis

In the early 1980s, Michael Porter made a major contribution to the analysis of innovation in corporate strategy, by explicitly linking technology to the 'five forces' driving industry competition, and to the choice amongst a number of 'generic strategies' that must be made by the firm. His approach situates the firm's technological activities in a wider context of industry competition, and he develops a systematic SWOT analysis, based on competitive forces and firms' internal choices. His approach has been very influential, and it illustrates both the strengths and the weaknesses of the 'rationalist' approach to technology strategy, which is why we shall discuss it now. Despite criticism by many academics, it remains the most dominant approach to strategy, in both the business schools and in practice.

The 'Five Forces' Driving Industry Competition

For Porter, the unit of analysis is the *industry* producing similar products. Profitable opportunities are fewer in mature industries. There are five forces driving industry competition, each of which generates opportunities and threats:

1. Relations with suppliers.
2. Relations with buyers.
3. New entrants.
4. Substitute products.
5. Rivalry amongst established firms.

According to Porter, '(t)he goal of competitive strategy . . . is to find a position in an industry where a company can best defend itself against these competitive forces or can influence them in its favour'.[17] Technological change can influence all the five forces, as is shown in Box 3.4.

Generic Market Strategies for Firms

According to Porter, there are also four generic market strategies from which firms must choose:

1. Overall cost leadership.
2. Product differentiation.
3. Cost focus.
4. Differentiation focus.

As is shown in Table 3.1, the choice of product strategy has direct and obvious implications for the choice of technology strategy, in particular for priorities in product and process development. Thus, in consumer durable goods markets like automobiles, consumer electronics and 'white (kitchen) goods', we can observe a range of products with different trade-offs between performance and price, with each aiming at specific market segments, and each requiring different choices in the balance between product and process innovation. Porter insists on the importance of these choices: he argues that firms that get 'stuck in the middle' between cost and product quality will have low profits.

BOX 3.4

THREATS AND OPPORTUNITIES FROM CHANGES IN TECHNOLOGY IN PORTER'S COMPETITIVE ANALYSIS

Potential Entrants and Substitute Products

- Threats of new entrants can be *increased* through reduced economies of scale (e.g. telecommunications, publishing), and through substitute products (e.g. mini-computers, aluminium for steel cans).
- They can be *decreased* through 'lock-in' to technological standards (e.g. Microsoft), and through patent and other legal protection (e.g. most major ethical drugs).

Power of Suppliers over Buyers

- Can be *increased* by innovations that are more essential to the firm's inputs (e.g. microprocessors into computers).
- Can be *decreased* by innovations that reduce technological dependence on suppliers (engineering materials).

Rivalry amongst Established Firms

- Rival firms can establish a monopoly position through innovation (e.g. Polaroid in instant photography), or destroy a monopoly position through imitation (US General Electric in brain scanners).

TABLE 3.1   Porter's generic technology strategies

	Cost leadership	Differentiation	Cost focus	Differentiation focus
Product development	Lower material inputs	Enhance quality Enhance features	Minimum features	Niche market
	Ease of manufacture			
		Deliverability		
	Improve logistics			
Process development	Learning curve Economies of scale	Precision Quality control	Minimize costs	Precision Quality control
		Response time		Response time

Innovation 'Leadership' versus 'Followership'

Finally, according to Porter, firms must also decide between two market strategies:

1. Innovation 'leadership' – where firms aim at being first to market, based on technological leadership. This requires a strong corporate commitment to creativity and risk-taking, with close linkages both to major sources of relevant new knowledge, and to the needs and responses of customers.
2. Innovation 'followership' – where firms aim at being late to market, based on imitating (learning) from the experience of technological leaders. This requires a strong commitment to competitor analysis and intelligence, to reverse engineering (i.e. testing, evaluating and taking to pieces competitors' products, in order to understand how they work, how they are made and why they appeal to customers), and to cost cutting and learning in manufacturing.

However, in practice the distinction between 'innovator' and 'follower' is much less clear. For example, a study of the product strategies of 2273 firms found that market pioneers continue to have high expenditures on R&D, but that this subsequent R&D is most likely to be aimed at minor, incremental innovations. A pattern emerges where pioneer firms do not maintain their historical strategy of innovation leadership, but instead focus on leveraging their competencies in minor incremental innovations. Conversely, late entrant firms appear to pursue one of two very different strategies. The first is based on competencies other than R&D and new product development – for example, superior distribution or greater promotion or support. The second, more interesting strategy, is to focus on major new product development projects in an effort to compete with the pioneer firm.[18]

3.3 Assessment of Porter's Framework

The strengths of Porter's framework are precisely those we should expect from someone whose initial training was in industrial economics: a deep understanding of the competitive environment in which the business firm operates, and in which it must consciously try to *position* itself in its innovation strategy, as well as in other dimensions of corporate policy. Competitive rivalry provides the essential incentive for innovation. Technology can help give a firm a distinctive competence, enabling it to provide goods and services better than competitors. However, technology can never be completely monopolized. Knowledge always leaks out, and competencies can always be imitated, unless they are continually renewed. A distinctive competence must be sustainable over the long term. This can be decided by the firm only after a careful analysis of competitors and the market conditions in which it is embedded. In later analysis Porter also identifies the major influence of home country conditions on the innovation strategies of firms in global markets.[19] The main elements of the strategic task of *positioning* corporate technology strategy in the context of competition and markets will be discussed in Chapter 4.

Porter also argues that innovation strategy should aim to repel competitive threats, both from incumbents in the industry and potential new entrants, including new and substitute products based on new technological opportunities. Box 3.5 describes such a case: how radical change in semiconductor technology completely transformed the structure and competitive conditions in the US computer industry, over a period of 20 years, from a few integrated oligopolistic firms supplying essentially the office machinery market, to a disintegrated structure with many new entrants and a few remaining incumbents.

BOX 3.5

TECHNOLOGICAL CHANGE IN THE COMPUTER INDUSTRY SINCE 1970[20]

The experience over the past 25 years of the US computer industry is a spectacular example of the power of technological change to transform completely the structure and competitive conditions in an industry. In the early 1970s, the industry was dominated by a few mainframe producers, some of whom were fully vertically integrated from basic circuitry through to distribution. Barriers to entry were high, suppliers relatively weak and customers had a limited range of choice. By the early 1990s, the industry had literally disintegrated, with independent firms (most of whom are new entrants since the 1970s) competing at each stage from basic circuitry to distribution. The main destabilizing factor has been the rapid rate of technical improvement in the microprocessor – the computer on a chip. This has drastically reduced the costs of computing, thereby lowering barriers to entry to the users of microprocessors and opening a whole range of potential applications outside mainframes (of which the personal computer is one of the most spectacular), and thereby creating a whole range of new opportunities for firms in systems and applications software. Recent developments in wireless technologies have further extended the reach of the computer and application of microprocessors, with ever-smarter hand-held devices and the blurring of the distinctions between mobile computing, cell phones and personal digital assistants (PDAs).

However, this example also reveals the essential weaknesses of Porter's framework for analysis and action. As Martin Fransman has pointed out, technical personnel in firms like IBM in the 1970s were well aware of trends in semiconductor technology, and their possible effects on the competitive position of mainframe producers.[21] IBM in fact made at least one major contribution to developments in the revolutionary new technology: RISC microprocessors. Yet, in spite of this knowledge, none of the established firms proved capable over the next 20 years of achieving the primary objective of strategy, as defined by Porter: '. . . to find a position . . . where a company can best defend itself against these competitive forces or can influence them in its favour'.

Like many mainstream industrial economics, Porter's framework underestimates the power of technological change to transform industrial structures, and overestimates the power of managers to decide and implement innovation strategies. Or, to put it another way, it underestimates the importance of *technological trajectories*, and of the firm-specific *technological and organizational competencies* to exploit them. Large firms in mainframe computers could not control the semiconductor trajectory. Although they had the necessary technological competencies, their organizational competencies were geared to selling expensive products in a focused market, rather than a proliferating range of cheap products in an increasing range of (as yet) unfocused markets.

These shortcomings of Porter's framework in its treatment of corporate technology and organization led it to underestimate the constraints on individual firms in choosing their innovation strategies. In particular:

- Firm *size* influences the choice between 'broad front' and 'focused' technological strategies. Large firms typically have 'broad front' strategies whilst small firms are 'focused'.
- The firm's *established product base* and related technological competencies will influence the range of technological fields and industrial sectors in which it can hope to compete in future. Chemical-based firms do not diversify into making electronic products, and vice versa. It is very difficult (but not impossible, see, for example, the case of Nokia in Chapter 11) for a firm manufacturing traditional textiles to have an innovation strategy to develop and make computers.[22]
- The *nature of its products and customers* will strongly influence its degree of choice between quality and cost. Compare food products, where there are typically a wide range of qualities and prices, with ethical drugs and passenger aeroplanes where product quality (i.e. safety) is rigidly regulated by Government legislation (e.g. FAR, JAR) or agencies (e.g. FAA, CAA). Food firms therefore have a relatively wide range of potential innovation strategies to be chosen from amongst those described by Porter. The innovation strategies of drug and aircraft firms, on the other hand, inevitably require large-scale expenditures on product development and rigorous testing.

In addition, technological opportunities are always emerging from advances in knowledge, so that:

- Firms and technologies do not fit tidily into preordained and static industrial structures. In particular, firms in the chemical and electrical–electronic industries are typically active in a number of product markets, and also create new ones, like personal computers. Really new innovations (as distinct from radical or incremental), which involve some discontinuity in the technological or marketing base of a firm are actually very common, and often evolve into new businesses and product lines, such as the Sony Walkman or Canon Laserjet printers.[23]

- Technological advances can increase opportunities for profitable innovation in so-called mature sectors. See, for example, the opportunities generated over the past 15 years by applications of IT in marketing, distribution and coordination in such firms as Benetton and Hotpoint.[24] See also the increasing opportunities for technology-based innovation in traditional service activities like banking, following massive investments in IT equipment and related software competencies.[25]

- Firms do not become stuck in the middle as Porter predicted. John Kay has shown that firms with medium costs and medium quality compared to the competition achieve higher returns on investment than those with either low–low or high–high strategies.[26] Furthermore, some firms achieve a combination of high quality and low cost compared to competitors and this reaps high financial returns. These and related issues of product strategy will be discussed in Chapter 7.

There is also little place in Porter's framework for the problems of *implementing* a strategy:

- Organizations which are large and specialized must be capable of learning and changing in response to new and often unforeseen opportunities and threats. This does not happen automatically, but must be consciously managed. In particular, the continuous transfer of knowledge and information across functional and divisional boundaries is essential for successful innovation. Studies confirm that the explicit management of competencies across different business divisions can help to create radical innovations, but that such interactions demand attention to leadership roles, team composition and informal networks.[27]

- Elements of Porter's framework (Box 3.3) have been contradicted as a result of organizational and related technological changes. The benefits of non-adversarial relations with both suppliers and customers have become apparent. Instead of bargaining in what appears to be a zero-sum game, co-operative links with customers and suppliers can increase competitiveness, by improving both the value of innovations to customers, and the efficiency with which they are supplied.[28] According to a survey of innovation strategies in Europe's largest firms, just over 35% replied that the technical knowledge they obtain from their suppliers and customers is very important for their own innovative activities.[29] The role of collaboration will be discussed more fully in Chapter 8.

Finally, Porter returned to the subject of strategy in the mid-1990s.[30] He finally recognizes the importance of the path-dependent nature of corporate activity by stressing the importance of 'fit' (i.e. coherence and balance) between the various elements of what firms have done in the past, and what they plan to do in the future (e.g. in their customers, products, forms of organization). He also makes a more questionable distinction between 'operational effectiveness' (i.e. doing things better) and 'strategy' (i.e. doing things that others cannot do), arguing that the latter is always essential. He has nothing to say about the contemporary problems of adopting business methods based on radical improvements in ICT technology (e.g. IT-based supply chains or accounting methods) that are necessary for corporate survival, but which will by themselves never be a source of sustainable competitive advantage. More generally, Porter seems uncomfortable in dealing with the inevitably high level of uncertainty found in emerging industries.

Clayton Christensen provides a recent and balanced summary of the relative merits of the rational versus incremental approaches to strategy:

> core competence, as used by many mangers, is a dangerously inward-looking notion. Competitiveness is far more about doing what customers value, than doing what you think you're good at . . . the problem with the core competence/not your core competence categorization is that what might seem to be a non core activity today might become an absolutely critical competence to have mastered in a proprietary way in the future, and vice versa . . . emergent processes should dominate in circumstances in which the future is hard to read and it is not clear what the right strategy should be . . . the deliberate strategy process should dominate once a winning strategy has become clear, because in those circumstances effective execution often spells the difference between success and failure.[31]

3.4 The Dynamic Capabilities of Firms

David Teece and Gary Pisano[32] integrate the various dimensions of innovation strategy identified above into what they call the 'dynamic capabilities' approach to corporate strategy, which underlines the importance of dynamic change and corporate learning:

> This source of competitive advantage, dynamic capabilities, emphasizes two aspects. First, it refers to the shifting character of the environment; second, it emphasizes the key role of strategic management in appropriately adapting, integrating and re-configuring internal and external organisational skills, resources and functional competencies towards a changing environment. (p. 537)

To be strategic, a capability must be honed to a user need (so that there are customers), unique (so that the products/services can be priced without too much regard for the competition), and difficult to replicate (so that profits will not be competed away). (p. 539)

We advance the argument that the strategic dimensions of the firm are its managerial and organisational *processes*, its present *position*, and the *paths* available to it. By managerial *processes* we refer to the way things are done in the firm, or what might be referred to as its 'routines', or patterns of current practice and learning. By *position*, we refer to its current endowment of technology and intellectual property, as well as its customer base and upstream relations with suppliers. By *paths* we refer to the strategic alternatives available to the firm, and the attractiveness of the opportunities which lie ahead. (pp. 537–541, our italics)

The role of position, paths and processes in the strategic management of innovation is examined in the next three chapters.

3.5 Innovation Strategy in Small Firms

Much of the above analysis has been directed to the problems of managing innovation in large, complex organizations where deliberate management action is necessary to co-ordinate or integrate specialized resources and skills. Like their large counterparts, small firms also need to concern themselves with their market position, their technological trajectories and competence-building, and their organizational processes. However, the challenges to management present themselves in somewhat different ways. In this section we summarize the key differences between innovation in small and large firms. At the end of the three subsequent chapters, we shall explore their implications for the strategic management of small firms in defining their positions, paths and processes. The discussion of small firms will necessarily be short, given the lack of systematic research on the majority of firms that are not particularly innovative, but which must necessarily cope with changing technology that impacts their business, as IT does today.

As Table 3.2 shows, debates about the role of small firms (<500 employees) in technological innovation excite strong opinions based on weak empirical evidence. The evidence also shows that, compared to large innovating firms, small innovating firms have the following characteristics:

- Similar *objectives* – *to develop* and combine technological and other competencies to provide goods and services that satisfy customers better than alternatives, and that are difficult to imitate.
- Organizational strengths – ease of communication, speed of decision-making, degree of employee commitment and receptiveness to novelty. This is why small firms often do not need the formal strategies that are used in large firms to ensure communication and co-ordination.

TABLE 3.2   Misleading assertions about innovation in small firms

Misleading assertions	What the evidence shows
'Small firms make most of the major innovations'	It depends on the product and the technology
'Small firms make few innovations since they do so little R&D'	They do lots of 'informal', 'part-time' and non-measured R&D, and produce a share of total innovations roughly proportionate to their output and employment
'Small firms are much more innovative than large firms, since they account for a higher share of innovations than of R&D'	Not if you include unmeasured, 'part-time' R&D
'New small firms create a lot of employment'	They also lose a lot, since they have high birth *and* death rates

- Technological weaknesses – specialized range of technological competencies, inability to develop and manage complex systems, inability to fund long-term and risky programmes.
- *Different* sectors – small firms make a greater contribution to innovation in certain sectors, such as machinery, instruments and software, than in chemicals, electronics and transport.

We shall discuss the implications of these and other characteristics of small firms[33] for innovation strategy in subsequent chapters.

3.6 Summary and Further Reading

Innovation involves complexity and change, whether in the firm's technology, its organization or its economic environment. As a consequence, technological opportunities and threats are often difficult to identify, innovation strategies difficult to define, and outcomes difficult to predict. There are therefore no management *recipes* and tools that guarantee success. In all cases a capacity to learn from experience and analysis is essential. None the less, both research and experience point to three essential ingredients in corporate innovation strategies:

1. The *position* of the firm, compared to its competitors, in terms of its product, processes and technologies; and in terms of the national system of innovation in which it is embedded.

2. The technological *paths* open to the firm, given its accumulated competencies, and the emerging opportunities that these enable it to exploit.

3. The organizational *processes* followed by the firm, in order to integrate strategic learning across functional and divisional boundaries.

Two streams of literature are relevant to innovation and strategy. The first, from mainstream strategic management, is in general weak with respect to innovation, but a few writers consider some of the issues. For example, Porter's paper 'The technological dimension of competitive strategy' in *Research on Technological Innovation, Management and Policy*, Vol. 1 (UAI Press, London, 1983) epitomizes the rational planning approach to strategy, whereas the case for incrementalism is made by Henry Mintzberg's 'Rethinking strategic planning' in *Long Range Planning*, **27** (3), 12–30 (1994). Richard Whittington's *What is Strategy and Does it Matter?* (Routledge, London, 1994) presents both cases. Chapter 21 of John Kay's book *Foundations of Corporate Success: How businesses add value* (OUP, Oxford, 1993) is a masterly survey of the strengths and weaknesses of the strategic management literature. More recent contributions include C. Baden-Fuller and M. Pitt, *Strategic Innovation* (Routledge, 1996); A. Chandler, P. Hagstrom and O. Solvell, *The Dynamic Firm* (OUP, 1998); and R. Stacey, *Strategic Management and Organizational Dynamics: The challenge of complexity* (Prentice Hall, 3rd edn, 2000).

Second, there is a growing literature on technology strategy. Three books of edited papers contain excellent contributions: Mark Dodgson (ed.) *Technology Strategy and the Firm: Management and public policy* (Longman, Harlow, 1989); Mel Horwitch (ed.) *Technology and the Modern Corporation: A strategic perspective* (Pergamon, 1986); and Ray Loveridge and Martyn Pitt (eds) *The Strategic Management of Technological Innovation* (John Wiley & Sons, Ltd, Chichester, 1990). Paul Adler provides a comprehensive, but now dated review in 'Technology strategy: A guide to the literatures', in Rosenbloom, R. and Burgelman, R. A. (eds) *Research on Technological Innovation, Management and Policy* (Vol. 4, JAI Press, 1989, pp. 25–80); Goodman and Lawless, and Dussuage, Hart and Ramanantsoa identify a number of useful frameworks in *Technology and Strategy* (OUP, 1994) and *Strategic Technology Management* (John Wiley & Sons, Ltd, Chichester, 1994) respectively. UMIST maintains a website of tools and techniques for technology strategy at http://info.mcc.ac.uk/UMIST-CROMTEC/itm/itm.htm. More recent contributions include R. Burgelman, M. Maidique and S. Wheelwright, *Strategic Management of Technology and Innovation*, (Irwin, 2nd edn, 1996); M. Tushman and P. Anderson (eds) *Managing Strategic Innovation and Change: A collection of readings* (OUP, 1997); M. Dodgson, *The Management of Technological Innovation* (OUP, 2000); and K. Pavitt and E. Steinmueller, 'Technology in corporate strategy: change, continuity and the information revolution', chapter in Pettigrew, A. Thomas, H. and Whittington, R. (eds) *Handbook of Strategy and Management* (Sage Publications, 2001).

References

1 Ansoff, I. (1965) 'The firm of the future', *Harvard Business Review*, Sept.–Oct., 162–178.

2 Mintzberg, H. (1987) 'Crafting strategy', *Harvard Business Review*, July–August, 66–75. See also the interview with Mintzberg in *The Academy of Management Executive* (2000), **14** (3), 31–42.

3 Whittington, R. (1994) *What is Strategy and Does it Matter?* Routledge, London.

4 Kay, J. (1993) *Foundations of Corporate Success: How business strategies add value.* Oxford University Press, Oxford.

5 Starbuck, W.H. (1992) 'Strategizing in the real world', *International Journal of Technology Management*, special publication on 'Technological foundations of strategic management', **8** (1/2), 77–85.

6 Howard, N. (1983) 'A novel approach to nuclear fusion', *Dun's Business Month (dmi)*, **123**, 72, 76, November.

7 Berton, L. (1974) 'Nuclear energy stocks, set to explode', *Financial World (two)*, **141** (16 Jan.), 8–11; Freeman, C. (1984) 'Prometheus unbound', *Futures*, **16**, 495–507.

8 Duysters, G. (1995) *The Evolution of Complex Industrial Systems: The dynamics of major IT sectors.* MERIT, University of Maastricht, Maastricht; *The Economist* (1996) 'Fatal attraction: Why AT&T was led astray by the lure of computers', management brief, 23 March; Von Tunzelmann, N. (1997) 'Technological accumulation and corporate change in the electronics industry', in Gambardella, A. and Malerba, F. (eds) *The Organization of Scientific and Technological Research in Europe*, Cambridge University Press (1999).

9 See 'The failure of new media', *The Economist*, 19 August 2000, 59–60.

10 See 'A survey of e-entertainment' *The Economist*, 7 October 2000.

11 Sapsed, J. (2001) *Restricted Vision: Strategizing under uncertainty.* Imperial College Press, London.

12 Pasteur, L. (1854) Address given on the inauguration of the Faculty of Science, University of Lille, 7 December. Reproduced in *Oxford Dictionary of Quotations*. Oxford University Press.

13 Boston Consulting Group (1975) *Strategy Alternatives for the British Motorcycle Industry.* HMSO, London.

14 Pascale, R. (1984) 'Perspectives on strategy: the real story behind Honda's success', *California Management Review*, **26**, 47–72.

15 Mintzberg, H. *et al.* (1996) 'The "Honda effect" revisited', *California Management Review*, **38**, 78–117.

16 Lee, G. (1995) 'Virtual prototyping on personal computers', *Mechanical Engineering*, **117** (July), 70–73.

17 Porter, M. (1980) *Competitive Strategy.* Free Press, New York.

18 Robinson, W. and Chiang, J. (2002) 'Product development strategies for established market pioneers, early followers, and late entrants', *Strategic Management Journal*, **23**, 855–866.

19 Porter, M. (1990) *The Competitive Advantage of Nations.* Macmillan, London.

20 Manasian, D. (1993) 'The computer industry: reboot system and start again', Computer Industry Survey. *The Economist*, 27 February; *The Economist* (1993) 'What went wrong at IBM', 16 January, 23–25; Snell, I. (1994) 'R&D configurational strategies in the M-form: the case of Hewlett Packard', M.Sc. dissertation, Science Policy Research Unit, University of Sussex, UK; Steinmueller, E. (1995) 'The US software industry: an interpretative history', in Mowery, D. (ed.), *The International Computer Software Industry.* Oxford University Press, Oxford, 15–52.

21 Fransman, M. (1994) 'Information, knowledge, vision and theories of the firm', *Industrial and Corporate Change*, **3**, 713–757.

22 Patel, P. and K. Pavitt (1998) 'The wide (and increasing) spread of technological competencies in the world's largest firms: a challenge to conventional wisdom', in Chandler, A., Hagstrom, P. and Solvell, O. (eds), *The Dynamic Firm*, Oxford University Press, Oxford.

23 Garcia, R. and R. Calantone (2002) 'A critical look at technological innovation typology and innovativeness terminology: a literature review', *Journal of Product Innovation Management*, **19**, 110–132.

24 Baden-Fuller, C. and J. Stopford (1994) *Rejuvenating the Mature Business: The competitive challenge*, Harvard Business School Press, Boston, Mass.; Belussi, F. (1989) 'Benetton – a case study of corporate strategy for innovation in traditional sectors', in Dodgson, M. (ed.), *Technology Strategy and the Firm: Management and public policy.* Longman, London, 116–133.

25 Barras, R. (1990) 'Interactive innovation in financial and business services: the vanguard of the service revolution', *Research Policy*, **19**, 215–238.

26 Kay, J. (1996) 'Oh Professor Porter, whatever did you do?', *Financial Times*, 10 May, 17.

27 Tidd, J. (1995) 'The development of novel products through intra- and inter-organizational networks: the case of home automation', *Journal of Product Innovation Management*, **12** (4), 307–322; McDermott, C. and G. O'Connor (2002) 'Managing radical innovation: An overview of emergent strategy issues', *Journal of Product Innovation Management*, **19**, 424–438.

28 Lamming, R. (1993) *Beyond Partnership.* Prentice-Hall, Hemel Hempstead.

29 Arundel, A., G. van de Paal and L. Soete (1995) *Innovation Strategies of Europe's Largest Industrial Firms.* PACE Report, MERIT, University of Limbourg, Maastricht.

30 Porter, M. (1996) 'What is strategy?', *Harvard Business Review*, Nov.–Dec., 61–78.

31 Christensen, C. and M. Raynor (2003) *The Innovator's Solution: Creating and sustaining successful growth.* Harvard Business School Press, Boston, Mass.

32 Teece, D. and G. Pisano (1994) 'The dynamic capabilities of firms: An introduction', *Industrial and Corporate Change*, **3**, 537–556.

33 Rothwell, R. and M. Dodgson (1990) 'Innovation and size of firm', in *The Handbook of Industrial Innovation*, Edward Elgar, Aldershot. MERIT, University of Maastricht, Maastricht; *The Economist* (1996) 'Fatal attraction: why AT&T was led astray by the lure of computers', management brief, 23 March; Von Tunzelmann, N. (1997) 'Technological accumulation and corporate change in the electronics industry', in Gambardella, A. and Malerba, F. (eds) *The Organization of Scientific and Technological Research in Europe*, Cambridge University Press (1999).

Positions: The National and Competitive Environment

Two features of the firm's environment have a major influence on its innovation strategy: first, the *national system of innovation* in which the firm is embedded, and which in part defines its range of choices in dealing with opportunities and threats; and second, its *market position* compared to *competing firms*, which in part defines the innovation-based opportunities and threats that it faces.

4.1 National Systems of Innovation

On 14 May 1990, the cover story in *Business Week* was entitled 'The stateless corporation'. The main thrust of the analysis was that global firms now rely hardly at all on their home – or any other one – country for their operations, since they compete and increasingly produce in global markets. On the other hand, analysts like Porter[1] and his colleagues[2] have shown that business firms – and even the largest ones competing in global markets – are strongly influenced in their choice of technological strategies by the conditions existing in their home countries. This is because even global firms draw on mainly one – or perhaps two – countries for their strategic skills and expertise in formulating and executing their innovation strategies. As we shall see in Chapter 6, only about 12% of the innovative activities of the world's largest 500, technologically active firms were located outside their home countries in the 1990s, compared to about 25% of their production and much larger shares of sales.[3] As a consequence, we find that the technological strengths and weaknesses of countries are reflected in their major firms.

In Table 4.1 we see that, in the 1990s, the largest numbers of European firms amongst the technical leaders were to be found in the technological fields of industrial and fine chemicals, and defence-related technologies (i.e. aerospace), which are fields of national technological strength, whilst the reverse is the case in electronic capital and consumer goods. Japanese firms predominate in consumer electronics and motor vehicle technologies, and US firms in fine chemicals and in raw materials-based (i.e.

TABLE 4.1 Nationalities of top 20 firms patenting in the USA in 1992–96

Broad technological field	Europe	USA	Japan	South Korea	Canada
Industrial chemicals	9	9	2		
Defence-related technologies	9	11			
Fine chemicals	7	12	1		
Telecommunications	6	6	7		1
Mechanical engineering	5	9	6		
Composite materials	4	9	7		
Motor vehicles	4	4	12		
Electrical machinery	4	7	8	1	
Raw material-related technologies	4	16			
Electronic capital goods and components	2	8	9	1	
Electronic consumer goods	2	3	14	1	

Source: P. Patel, *Large Firm Database*, SPRU, University of Sussex. Based on data provided by the US Patent Office.

TABLE 4.2 Trends in business-funded R&D as percentage of GDP

	1967	1971	1975	1979	1983	1987	1991	1995	1996	1997	1998
Switzerland	1.78	1.67	1.67	1.74	1.67	1.92	1.79		1.80		
USA	0.99	0.97	0.98	1.05	1.31	1.37	1.60	1.54	1.65	1.71	1.80
Japan	0.83	1.09	1.12	1.19	1.59	1.82	2.13	1.90	1.99	2.06	
Germany	0.94	1.13	1.11	1.32	1.48	1.80	1.58	1.36	1.35	1.38	1.42
Sweden	0.71	0.80	0.96	1.11	1.45	1.73	1.69	2.32		2.56	
Denmark	0.34	0.39	0.41	0.42	0.53	0.66	0.86	0.88	0.99	1.00	
Finland	0.30	0.44	0.44	0.53	0.73	0.99	1.10	1.33		1.67	
Canada	0.40	0.38	0.33	0.39	0.46	0.57	0.58	0.68		0.73	0.75
UK	1.00	0.81	0.80	0.82	0.86	1.02	0.99	0.92	0.88	0.87	
Netherlands	1.12	1.02	0.97	0.86	0.89	1.11	0.91	0.86	0.94		

Source: OECD.

oil, gas and food) and defence-related technologies, again reflecting the technological strengths of their home countries.

The strategic importance to corporations of home countries' technological competencies would matter little if they were all more or less the same. But Table 4.1 and other data[4] show that they are not. Patterns of sectoral specialization differ greatly: for example, the Japanese pattern of strengths and weaknesses is almost the opposite of that in the USA. In addition, Table 4.2 illustrates that countries differ in both the level and the rate of increase in the resources devoted by business firms to innovative activities. It shows, for selected OECD countries, trends over 31 years in business-funded R&D as a share of GDP (gross domestic product). As we have already seen, R&D captures corporate innovative activities only imperfectly. But it is one of the best available indicators of aggregate innovative investments, and international differences significantly influence national economic growth and trade performance.[5]

Three major conclusions emerge from Table 4.2. First, and contrary to what many observers continue to assume, Europe and Japan did not progressively and smoothly catch up with the USA, which was the technological leader in the period after the Second World War. Switzerland has always been amongst the leaders and remains so. As early as 1971, Germany and Japan overtook the USA and progressively increased their lead until the late 1980s. This was reflected in the relative performance in R&D and sales growth of the large firms based in these countries.[6] Since then, the trend has changed, with the shares of business-funded R&D in GDP stabilizing in Japan, declining in Germany, and increasing in the USA. At the same time, the three Scandinavian countries have continued to increase their shares, with the growth of major firms in pharmaceuticals and telecommunications.

Second, the other early leaders of the late 1960s – the UK and the Netherlands – have not reacted to the growing competition like the USA in the 1990s. The share of business-funded R&D in GDP in both countries declined considerably in the 1970s, and has not recovered to earlier levels.

Third, the rate of increase in corporate commitment to R&D in a country is not closely related to its industrial structure. Compare Finland and Canada, both of whose economies rely heavily on natural resources; Finland's R&D expenditures have increased even more rapidly than Japan's as a share of GDP, whilst Canada's increased only slightly.

A recent study of the innovation capabilities of European countries based on two Community Innovation Surveys (which are conducted every four years by all nation states within the EU) and other data estimated the effects of different macro and micro factors on innovation. Table 4.3 provides a summary of the results. Using patents as

TABLE 4.3 European national systems of innovation and innovation capability

NIS variable	Regression coefficient on:	
	Patents granted	Sales of new products
Public R&D expenditure	+0.839	
Firm expenditure on R&D		+0.421
Gross domestic product (GDP)	+0.691	+0.310
Openness of national economy	+0.319	−0.454
Availability of venture capital	+0.200	
Presence of SMEs	−0.146	+0.621
External sources of innovation		+0.688
Presence of innovative firms		+0.591

Source: Derived from Faber, J. and A. B. Hesen (2004) 'Innovation capabilities of European nations', *Research Policy*, **33**, 193–207.

an indicator of innovation, innovation at the national level is positively influenced by the size of the economy, foreign competition in the domestic market, public expenditure on R&D and the availability of venture capital; it is negatively influenced by the presence of a relatively large number of small and medium-sized firms, high company tax and a high level of economic prosperity. Using relative sales of innovative products as an indicator of innovation, firm level effects become more evident: national innovation is positively influenced by the size of the economy, R&D expenditure of firms, use of external sources of innovation and the presence of small and medium-sized firms, but negatively influenced by economic prosperity and foreign competition in the home market. Put another way, macro-economic conditions in a country and the structure of the national economy have significant effects on innovation, measured by patenting and sales of innovative products. At the national level, the innovative activities of firms appear to have a stronger influence on sales of innovative products, than patenting.

Thus, the national systems of innovation in which a firm is embedded matter greatly, since they strongly influence both the direction and the vigour of its own innovative activities. Several approaches have been taken on the nature and impact of such national systems.[7] Our own is to identify the main national factors that influence the rate and direction of technological innovation in a country: more specifically, the national market *incentives and pressures* to which firms have to respond, their *competencies in production and research*, and the *institutions of corporate governance*.[8] We shall now discuss each of these in turn.

Incentives and Pressures: National Demand and Competitive Rivalry

Patterns of national demands   Those concerned to explain international patterns of innovative activities have long recognized the important influence of local demand and price conditions on patterns of innovation in local firms.[9] Strong local 'demand pull' for certain types of product generates innovation opportunities for local firms, especially when the demand depends on face-to-face interactions with customers.

In Table 4.4 we identify the main factors that influence local demands for innovation, and give some examples. In addition to the obvious examples of local buyers' tastes, we identify:

- Local (private and public) investment activities, which create innovative opportunities for local suppliers of machinery and production inputs, where competence is accumulated mainly through experience in designing, building and operating machinery.
- Local production input prices, where international differences can help generate very different pressures for innovation (e.g. the effects of different petrol prices on the design and related competencies in automobiles in the USA and Europe). High

TABLE 4.4 Factors influencing national demands for innovation

Factors in	Examples
Local buyers' tastes	• Quality food and clothing in France and Italy • Reliable machinery in Germany
Private investment activities	• Automobile and other downstream investments stimulating innovation in computer-aided design and robots in Japan, Italy, Sweden and Germany
Public investment activities	• Railways in France • Medical instruments in Sweden • Coal-mining machinery in the UK (<1979)
Input prices	• Labour-saving innovations in the USA • Europe–USA differences in automobile technology • Environmental technology in Scandinavia • Synthetic fertilizers in Germany
Local natural resources	• Innovations in oil and gas, mineral ores, and food and agriculture in North America, Scandinavia and Australia

prices can also generate pressure for substitute products, like synthetic fertilizers in Germany at the beginning of the twentieth century.

• Local natural resources, which create opportunities for innovation in both upstream extraction and downstream processing.

A more subtle, but increasing significant influence is the role of social concerns and pressure about the environment, safety and governance. For example, nuclear power as a technological innovation has evolved in very different ways in countries like the USA, UK, France and Japan. Similarly, innovation in genetically modified crops and foods has taken radically different paths in the USA and Europe, mainly due to public concerns and pressure. Box 4.1 discusses some of the issues related to managing sustainable innovation.

Competitive rivalry Innovation is always difficult and often upsetting to established interests and habits, so that local demands by themselves do not create the necessary conditions for innovation. Both case studies and statistical analysis show that competitive rivalry stimulates firms to invest in innovation and change, since their very existence will be threatened if they do not.[10] A comparison by Lacey Thomas of public policies towards the pharmaceutical industries in Britain and France has shown that the former was more successful in creating a demanding local competitive environment conducive to the emergence of British firms amongst the world leaders.[11] German strength in chemicals is based on three large and technologically dynamic firms, BASF,

BOX 4.1
MANAGING INNOVATION FOR SUSTAINABILITY

In their review of the field, Frans Berkhout and Ken Green argue that:

> "technological and organizational innovation stands at the heart of the most popular and policy discourses about sustainability. Innovation is regarded as both a cause and solution . . . yet, very little attempt has been made in the business and environment, environmental management and environmental policy literatures to systematically draw on the concepts, theories and empirical evidence developed over the past three decades of innovation studies."

They identify a number of limitations in the innovation literature, and suggest potential ways to link innovation and sustainability research, policy and management:

1. A focus on managers, the firm, or the supply chain is too narrow. Innovation is a distributed process across many actors, firms and other organizations, and is influenced by regulation, policy and social pressure. We advocate a similar argument in Chapter 8, where we examine the nature and role of innovation networks.
2. A focus on a specific technology or product is inappropriate. Instead the unit of analysis must be on technological systems or regimes, and their evolution rather than management. This is consistent with our discussion here on the influence of national systems of innovation, and the role of technological trajectories in Chapter 5.
3. The assumption that innovation is the consequence of coupling technological opportunity and market demand is too limited. It needs to include the less obvious social concerns, expectations and pressures. These may appear to contradict stronger but misleading market signals. This is a characteristic of disruptive innovation, which we discussed in Chapters 1 and 2.

They present empirical studies of industrial production, air transportation and energy to illustrate their arguments, and conclude that 'greater awareness and interaction between research and management of innovation, environmental management, corporate social responsibility and innovation and the environment will prove fruitful'.

Source: From *International Journal of Innovation Management*, **6** (3), Special Issue on Managing Innovation for Sustainability, edited by F. Berkhout and K. Green (2002).

Bayer and Hoechst, rather than on one super-large firm. Similarly, Table 4.1 shows that Japanese strength in consumer electronics is based on numerous technologically active firms rather than a few giants. Relatively smaller size also reduces the severity of the task of management to maintain corporate entrepreneurship. This is because managers can spend more time familiarizing themselves with the innovative potentialities of the various businesses, and can thereby avoid the dangers of managing divisions purely through financial indicators (see Chapter 6).

Thus although corporate policy-makers in large firms might often be tempted in the short term to avoid strong competition – and to reap extra monopoly profits – by merging with their competitors, the long-term costs could be considerable. Public policy-makers should be persuaded by the evidence that creating gigantic firms does not increase innovation, quite the contrary, and therefore take countervailing measures. Lack of competitive rivalry makes firms less fit to compete on global markets through innovation.

Competencies in Production and Research

Local demand opportunities and competitive pressures will not result in innovation unless firms have the competencies that enable them to respond. Corporate and national competencies in *production* and in *research* are essential.

There are significant differences amongst apparently similar European countries in the level of *production* competencies. In the 1980s and 1990s, Professor Prais and his colleagues made detailed comparisons of the level of education of the workforce in five European countries (France, the Federal Republic of Germany, the Netherlands, Switzerland and the UK). Although similar at the top of the educational pyramid, the differences amongst countries in the proportion without any qualifications were particularly striking, varying from less than about a quarter in Germany and Switzerland to nearly two-thirds in Britain. More detailed case studies comparing British and German firms show that these differences in skills make an important competitive difference, with productivity in the German firms being as much as twice as high, with equal or superior product quality, because German workers are more skilled in repairs and in learning new techniques more quickly.[12]

Behind the differences in vocational skills are important international differences in basic education, which will become critical with the growth of the knowledge economy. These have recently been revealed in an OECD survey, some of the results of which are summarized in Table 4.5. The first three columns compare three major dimensions of literacy and the fourth gives the share of the population with only rudimentary levels of skills, judged inadequate for coping in modern society. A number of conclusions emerge from this table and related sources:

TABLE 4.5 Rankings in international literacy survey* (population aged 16–64, 1994–98)

Country	Prose skills* (USA = 100)	Document skills* (USA = 100)	Quantitative skills* (USA = 100)	% of population with quantitative skills at level 1[†]
Sweden	110	113	111	7.5
Denmark	100	110	108	9.6
Czech Republic	98	106	108	15.7
Norway	105	111	108	8.5
Germany	101	106	107	14.4
Netherlands	103	107	105	10.5
Finland	105	108	104	10.4
Canada	102	104	102	16.6
USA	100	100	100	20.7
Hungary	88	93	98	33.8
UK	97	100	97	21.8
Ireland	97	97	96	22.6
Portugal	81	82	84	48.0
Chile	81	82	76	50.1

* Literacy: Understanding and employing printed information in daily activities at work, at home, and in the community, and developing knowledge and potential.
* Prose: Understanding text.
* Document: Locating and using written information.
* Quantitative: Applying arithmetic operations to numbers in written material.
Country rankings: Sorted according to Quantitative Index. USA = 100.
[†] Level 1 skills: Lowest on five-point scale. Considered insufficient for coping with modern society.
Source: *Literacy in the Information Age: Final report of the International Adult Literacy Survey*, Paris, OECD, 2000, Tables 2.1 and 2.2.

- The survey confirms the higher level of skills in Germany than in the UK, but not to such a striking extent, probably because the survey includes inhabitants of the old German Democratic Republic, who are not as well educated as their West German counterparts, who were the subject of the earlier studies by Prais.
- In spite of a well-developed system of higher education, the US ranking is not the highest, in part because of a relatively large share of the population with Level 1 (i.e. lowest level) skills. The same is true of the UK.
- The Czech Republic and (to a lesser degree) Hungary have skills comparable to those in Western European countries which, combined with their currently low salary levels, may help explain their attractiveness to international investors.
- The Scandinavian countries have the highest levels of skills, and the smallest proportions of the population with lowest level skills, which may partly explain their relatively high levels of adoption of Internet-related technologies.[13]

National competencies in *research* are also an important input into firms' technological capabilities. Especially in large firms, R&D laboratories actively seek support, knowledge and skills from national basic research activities, especially in universities. The knowledge they seek is mainly tacit and person-embodied, which explains why language and distance are real barriers to co-operation and why the firms generally prefer to deal with domestic universities.[14] It also explains why the differing national levels of production of basic research, measured in Table 4.6 as the number of papers per head of population, are similar to the national levels of investment in technology, measured in Table 4.2. This is because technology in dynamic business firms demands high-quality investment in national basic research. The apparent exception is Japan, but this simply reflects the lag in the basic research system, which is now expanding rapidly – in quantity and quality – in response to the world frontier technological investments by business firms.[15]

TABLE 4.6 Comparative performance of national systems of basic research

Country	Papers per 1000 population in 1993
Switzerland	1.471
Sweden	1.297
Denmark	1.074
Finland	0.964
Netherlands	0.962
UK	0.912
USA	0.886
Norway	0.817
Ireland	0.631
France	0.621
Germany	0.569
Singapore	0.452
Japan	0.416
Italy	0.362
Spain	0.329
Taiwan	0.225
Portugal	0.121
South Korea	0.068
India	0.016

Source: Lattimore, R. and J. Ravesz (1996) *Australian Science: Performance from published papers.* Bureau of Industry Economics, Report 96/3, Australian Government Printing Office, Canberra. Table 4.10. Based on data provided by the Institute of Scientific Information.

These differences in national endowments of research and production competencies influence managers in their search to identify technological fields and related product markets where specific national systems of innovation are likely to be most supportive to corporate innovative activities. For example, firms in the UK and USA are particularly strong in software and pharmaceuticals, both of which require strong basic research and graduate skills, but few production skills; they are therefore particularly well matched to local skill structures. Similarly, Japanese strength in consumer electronics and automobiles is particularly well matched to its local strength in production skills, as are the German strengths in mechanical engineering.

In many countries, national advantages in natural resources and traditional industries have been fused with related competencies in broad technological fields that then become the basis for technological advantage in new product fields. Figure 4.1 illustrates how this happened in Denmark, Sweden and Switzerland. It shows, with some inevitable simplification, how strong technological competencies emerged over time in specific fields in the three countries. In all cases, linkages with established fields of strength were the basis of local technological accumulation. This accumulation reinforced corporate and national competencies and created the potential for entry and

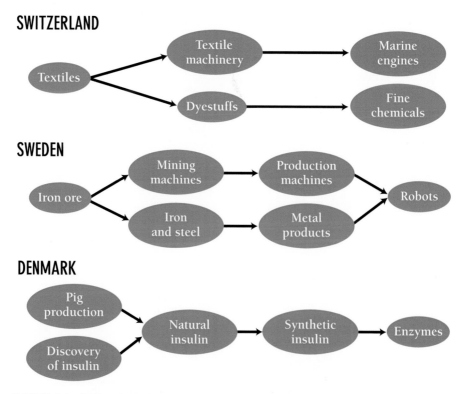

FIGURE 4.1   *Technological accumulation in three countries*

competitiveness in new product fields. Firm-specific investments in technology and related basic research and training in universities led to the mastery of broad technological fields with multiple potential applications: metallurgy and materials in Sweden, machinery in Switzerland and Sweden, and chemistry and (more recently) biology in Switzerland and Denmark.[16] Another example is the development of chemical engineering in the USA, in response to the challenges and opportunities of refining petrol.[17]

Institutions: Finance, Management and Corporate Governance

Firms' innovative behaviours are strongly influenced by the competencies of their managers and the ways in which their performance is judged and rewarded (and punished). Methods of judgment and reward vary considerably amongst countries, according to their national systems of *corporate governance*: in other words, the systems for exercising and changing corporate ownership and control. In broad terms, we can distinguish two systems: one practised in the USA and UK, and the other in Japan, Germany and its neighbours, such as Sweden and Switzerland. In his book *Capitalism against Capitalism*, Michel Albert calls the first the 'Anglo-Saxon' and the second the 'Nippon–Rhineland' variety.[18] A lively debate continues about the essential characteristics and performance of the two systems, in terms of innovation and other performance variables. Table 4.7 is based on a variety of sources, and tries to identify the main differences that affect innovative performance.

TABLE 4.7 The effects of corporate governance on innovative activities

Characteristics	Anglo-Saxon	Nippon–Rhineland
Ownership	Individuals, pension funds, insurers	Companies, individuals, banks
Control	Dispersed, arm's length	Concentrated, close and direct
Management	Business schools (USA), accountants (UK)	Engineers with business training
Evaluation of R&D investments	Published information	Insider knowledge
Strengths	• Responsive to radically new technological opportunities • Efficient use of capital	• Higher priority to R&D than to dividends for shareholders • Remedial investment in failing firms
Weakness	• Short-termism • Inability to evaluate firm-specific intangible assets	• Slow to deal with poor investment choices • Slow to exploit radically new technologies

In the UK and the USA, corporate ownership (shareholders) is separated from corporate control (managers), and the two are mediated through an active stock market. Investors can be persuaded to hold shares only if there is an expectation of increasing profits and share values. They can shift their investments relatively easily. On the other hand, in countries with governance structures like those of Germany or Japan, banks, suppliers and customers are more heavily locked into the firms in which they invest. Until the 1990s, countries strongly influenced by German and Japanese traditions persisted in investing heavily in R&D in established firms and technologies, whilst the US system has since been more effective in generating resources to exploit radically new opportunities in IT and biotechnology.

During the 1980s, the Nippon–Rhineland model seemed to be performing better. As we saw in Table 4.2, aggregate R&D expenditures were on a healthy upward trend, and so were indicators of aggregate economic performance. Since then, there have been growing doubts. The aggregate technological and economic indicators have been performing less well. Japanese firms have proved unable to repeat in telecommunications, software, microprocessors and computing their technological and competitive successes in consumer electronics.[19] German firms have been slow to exploit radically new possibilities in IT and biotechnology,[20] and there have been criticisms of expensive and unrewarding choices in corporate strategy, like the entry of Daimler Benz into aerospace.[21] At the same time, US firms appear to have learned important lessons, especially from the Japanese in manufacturing technology, and to have reasserted their eminence in IT and biotechnology. The 1990s have also seen sustained increases in productivity in US industry. According to *The Economist* in 1995, in a report entitled 'Back on top?', one professor at the Harvard Business School believed that people will look back at this period as 'a golden age of entrepreneurial management in the USA'.[22]

However, some observers have concluded that the strong US performance in innovation cannot be satisfactorily explained simply by the combination of entrepreneurial management, a flexible labour force, and a well-developed stock exchange. The US experience has not been repeated in the other Anglo-Saxon country with apparently similar characteristics – the UK (see Tables 4.2, 4.5 and 4.6). They argue that the groundwork for US corporate success in exploiting IT and biotechnology was laid initially by the US Federal Government, with the large-scale investments by the Defense Department in California in electronics, and by the National Institutes of Health in the scientific fields underlying biotechnology.[23] In addition, we should not write off Germany and Japan too soon. The former is now dealing with the dirt and inefficiency of the former East Germany[24] (the inclusion of which in official statistics is one reason for the German decline in the 1990s in business R&D as a share of GDP in Table 4.2). Japanese firms like Sony are world leaders in exploiting in home electronics the opportunities opened up by advances in digital technology. And Scandinavian countries are now well ahead of the rest of the world (including the USA) in mobile telephony,[25] as well as in more general indicators of skills and knowledge (see Tables 4.2, 4.5 and 4.6). The jury is still out.

Learning from Foreign Systems of Innovation

Firms have at least three reasons for monitoring and learning from the development of technological, production and organizational competencies of national systems of innovation other than those in which they are embedded themselves, and especially from those that are growing and strong:

1. They will be the sources of firms with a strong capacity to compete through innovation. For example, beyond Japan, other East Asian countries are developing strong innovation systems. In particular, business firms in South Korea and Taiwan now spend more than 2% of GDP on R&D, which puts them up with the advanced OECD countries. By the early 1990s, Taiwan was granted more patents in the USA than Sweden, and together with South Korea, is catching up fast with Italy, the Netherlands and Switzerland. Other Asian countries like Malaysia are also developing strong technological competencies. Following the collapse of the Russian Empire, we can also anticipate the re-emergence of strong systems of innovation in the Czech Republic and Hungary (see Table 4.5).

2. They are also potential sources of improvement in the corporate management of innovation, and in national systems of innovation. However, as we shall see below, understanding, interpreting and learning general lessons from foreign systems of innovation is a difficult task. Effectiveness in innovation has become bound up with wider national and ideological interests, which makes it more difficult to separate fact from belief. Both the business press and business education are dominated by the English language and Anglo-Saxon examples: very little is available in English on the management of innovation in Germany; and much of the information about the management of innovation in Japan has been via interpretations of researchers from North America.

3. Finally, firms can benefit more specifically from the technology generated in foreign systems of innovation. Table 4.8 shows that a high proportion of large European firms attach great importance to foreign sources of technical knowledge, whether obtained through affiliated firms (i.e. direct foreign investment) and joint ventures, links with suppliers and customers, or reverse engineering. In general, they find it is more difficult to learn from Japan than from North America and elsewhere in Europe, probably because of greater distances – physical, linguistic and cultural. Perhaps more surprising, European firms find it most difficult to learn from foreign

publicly funded research. This is because effective learning involves more subtle linkages than straightforward market transactions: for example, the membership of informal professional networks. This public knowledge is often seen as a source of potential world innovative advantage, and we shall see in Chapter 6 that firms are increasingly active in trying to access foreign sources. In contrast, knowledge obtained through market transactions and reverse engineering enables firms to catch up, and keep up, with competitors. East Asian firms have been very effective over the past 25 years in making these channels an essential feature of their rapid technological learning (see Box 4.2).

TABLE 4.8 Outside sources of technical knowledge for large European firms: percentage judging the source as very important

	Home country	Other Europe	North America	Japan
Affiliated firms	48.9	42.9	48.2	33.6
Joint ventures	36.6	35.0	39.7	29.4
Independent suppliers	45.7	40.3	30.8	24.1
Independent customers	51.2	42.2	34.8	27.5
Public research	51.1	26.3	28.3	12.9
Reverse engineering	45.3	45.9	40.0	40.0

Source: Arundel, A., G. van der Paal and L. Soete (1995) *Innovation Strategies of Europe's Largest Industrial Firms*. PACE Report, MERIT, University of Limbourg, Maastricht. Reproduced by permission of Anthony Arundel.

BOX 4.2
TECHNOLOGY STRATEGIES OF LATECOMER FIRMS IN EAST ASIA

The spectacular modernization in the past 25 years of the East Asian 'dragon' countries – Hong Kong, South Korea, Singapore and Taiwan – has led to lively debate about its causes. Michael Hobday has provided important new insights into how business firms in these countries succeeded in rapid learning and technological catch-up, in spite of underdeveloped domestic systems of science and technology, and of lack of technologically sophisticated domestic customers.

Government policies provided the favourable general economic climate: export orientation; basic and vocational education, with strong emphasis on industrial needs; and a stable economy, with low inflation and high savings. However, of major importance were the strategies and policies of specific business firms for the effective assimilation of foreign technology.

The main mechanism for catching up was the same in electronics, footwear, bicycles, sewing machines and automobiles, namely the 'OEM' (original equipment manufacture) system. OEM is a specific form of subcontracting, where firms in catching-up countries produce goods to the exact specification of a foreign transnational company (TNC) normally based in a richer and technologically more advanced country. For the TNC, the purpose is to cut costs, and to this end offers assistance to the latecomer firms in quality control, choice of equipment, and engineering and management training.

OEM began in the 1960s, and became more sophisticated in the 1970s. The next stage in the mid-1980s was ODM (own design and manufacture), where the latecomer firms learned to design products for the buyer. The last stage was OBM (own brand manufacture) when latecomer firms market their own products under their own brand name (e.g. Samsung, Acer) and compete head-on with the leaders.

For each stage of catching up, the company's technology position must be matched with a corresponding market position, as is shown below:

Stage	Technology position	Market position
1.	Assembly skills	Passive importer pull
	Basic production	Cheap labour
	Mature products	Distribution by buyers
2.	Incremental process change	Active sales to foreign buyer
	Reverse engineering	Quality and cost-based
3.	Full production skills	Advanced production sales
	Process innovation	International marketing department
	Product design	Markets own design
4.	R&D	Product marketing push
	Product innovation	Own brand product range and sales
5.	Frontier R&D	Own brand push
	R&D linked to market needs	In-house market research
	Advanced innovation	Independent distribution

Source: Hobday, M. (1995) *Innovation in East Asia: The challenge to Japan.* Edward Elgar, Guildford.

The slow but significant internationalization of R&D is also a means of firms learning from foreign systems of innovation. There are many reasons why multinational companies choose to locate R&D outside their home country, including regulatory regime and incentives, lower cost or more specialist human resources, proximity to lead

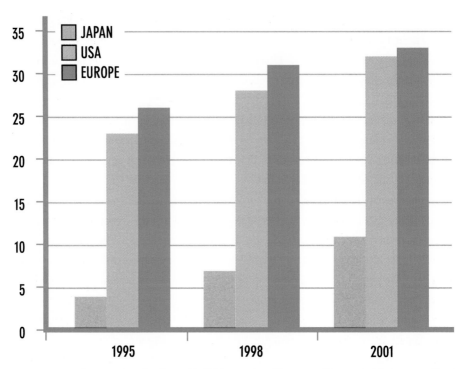

FIGURE 4.2   Internationalization of R&D by region (% expenditure outside home nation)

Source: Derived from Edler, J., F. Meyer-Krahmer and G. Reger (2002) 'Changes in the strategic management of technology', *R&D Management*, **32** (2), 149–164.

suppliers or customers, but in many cases a significant motive is to gain access to national or regional innovation networks. Overall, the proportion of R&D expenditure made outside the home nation has grown from less than 15% in 1995, to 22% by 2001. However, some countries are more advanced in internationalizing their R&D than others (Figure 4.2). In this respect European firms are the most internationalized, and the Japanese the least.

4.2 Coping with Competitors

In defining their innovation strategies, firms must be in a position to answer four questions about their competitors' innovative resources compared to their own:
* How do they compare in terms of size and composition?
* How efficiently are they used?
* How effectively do we learn from their knowledge and experience?
* How do we maintain our own innovative advantages?

In this context, it is important to distinguish knowledge of what technological developments are being undertaken by competitors, from knowledge of how the innovations can be made to work in practice. As we shall now see, the former can be obtained relatively quickly through *information-gathering* activities, whilst the latter requires extensive investments of resources and time in *benchmarking* and *learning* activities. Finally, in the light of the evidence available to it, corporate management must evaluate the market *position* of its innovation strategy in the light of intentions and competencies of competitors, and of the characteristics of its innovations: in particular, should it aim at innovation leadership or followership?

Large firms are surprisingly well informed about the technological activities of their competitors. According to a study by Mansfield, large US firms typically know about rivals' product development characteristics and plans within 6 to 12 months.[26] There are numerous methods of obtaining information about competitors' innovative strategies, all of which are imperfect, and some of which are of doubtful morality or legality. The sources based on publicly available literature are summarized in Table 4.9. The range and ease of access to these and other sources has been increasing rapidly with development of the Internet, with sites such as Yet2.com which are used by leading companies such as 3M, Boeing, Hitachi, NEC, Philips, Siemens and Toshiba to showcase their technologies and intellectual property, and more specialist patent databases such as Delphian.com.

Comparisons of Effectiveness through Benchmarking

'Benchmarking' goes beyond the routine collection of publicly available information. It consists of comparisons amongst competitor companies on specific dimensions of corporate performance – beyond financial performance – with the purpose of identifying and catching up with best practice. The approach started in the 1980s, when US firms found that their loss of market share to Japanese firms reflected underlying deficiencies in both manufacturing and product development. A pioneer was Xerox, who made systematic comparisons with Japanese competitors and found huge deficiencies in performance, as measured by the frequency of assembly line rejects, defects per machine, and the costs and time required for product development.[27] The improvement resulting from conscious policies to overcome these deficiencies has led to both higher customer satisfaction and higher financial returns on assets.

In some cases, groups of firms have funded academics and consultants to make the inter-firm comparisons. Probably the best known and most influential has been the International Motor Vehicle Programme at MIT. Its favourable conclusions in the late 1980s about production efficiency in Japanese automobile companies, compared to US

TABLE 4.9 Public information sources on corporate innovative activities

Nature of information	Sources	Strengths	Limits
Corporate R&D expenditures	• Company Annual Reports • *R&D Scoreboard* (June each year. UK firms and international comparisons) www.innovation.gov.uk/finance/ • European Commission (1997) *Second European Report on S&T Indicators*, EUR 17639	• Easy access	• No detail of projects • Misses innovative activities outside R&D
Corporate patents and scientific publications	• US Patent Office (USPO) www.almaden.ibm.com/cs/ibm_pat_srvr/homepage.html; www.uspto.gov/web/offices/ • European Patent Office www.Austria.EU.net/epo/ • Private consultants (CHI, Derwent) www.chiresearch.com/ • Patel and Vega (1998), *Technology Strategies of Large European Firms*, P.R.Patel@sussex.ac.uk • Web of Science (Science Citation Index) wos.mimas.ac.uk	• Comparisons possible in great detail • Identifies possible entrants as well as incumbents • Citations and joint authorship identify networks and alliances	• Choosing relevant patent classes • Dealing with firms with several names • Non-patented innovations
Public announcements and press analysis	• Conferences • Media • Trade press • Corporate web sites • Product catalogues	• Direct and detailed signal of corporate intentions	• Distortions for financial or marketing reasons

and European counterparts, had major effects, both in mobilizing laggard firms to do better, and in developing new concepts – such as *lean production* compared to mass production – which are now common currency amongst both managers and academics. A similar approach to benchmarking has been applied to the semiconductor industry in a programme of studies at Berkeley.[28]

In Europe, periodic surveys of performance and problems in manufacturing have been undertaken since the 1980s.[29] And after 1993, Chris Voss and his colleagues at London Business School made detailed and comprehensive benchmarking surveys of manufacturing and design in companies in Britain and other European countries. Some of their findings are summarized in Box 4.3.

BOX 4.3
BRITAIN'S MANUFACTURING AND DESIGN, ACCORDING TO CHRIS VOSS AND COLLEAGUES[30]

- Only 2% of the manufacturing sites can be said to be truly world class – but another 42% have most of the practices in place to become world class.
- The food industry is very strong, particularly in its logistics; aerospace/automotive industry sites, on the other hand, consistently lag in terms of both practice and performance.
- Purchasing power can be a force for good: companies which supply certain types of large organization (such as food retailers and the computer industry) have significantly better practice and/or performance.
- Leaders (the top 10% of companies on the scale) are eight times more likely to use external benchmarks to improve their business than laggards (the lower 10%).
- There are no quick fixes – the leading companies are better than the laggards because they have adopted best practices and improved their performance at every level.
- Compared to firms in Britain, those in Germany made higher investments and had higher productivity growth, but not enough to achieve lower costs than British plants.
- In manufacturing practice and performance both British and German leaders rank with the best in the world, but Britain has a long tail of low achievers.
- However, design in the British firms consistently lags the German firms.

According to another British survey of benchmarking by Coopers & Lybrand, managers claim that benchmarking is practised in 78% of the top 1000 companies, and achieves positive results.[31] It is more common in customer-facing areas (customer service, marketing and sales, logistics) than in manufacturing, and is least common in product development and R&D. This balance probably reflects the greater degree of difficulty in obtaining accurate information on competitors' performance in the manufacturing and R&D functions. Customer satisfaction can be assessed by surveys, and some of them are in the public domain (e.g. the surveys of US automobile users by J.D. Power Associates; consumer journals like *Which?*). In quality management various awards have long been established which allow firms to benchmark their quality systems, such as the European Quality Award, the Baldridge Award in the USA and the Deming Prize in Japan. More recently, specialized organizations have now emerged to collect and compare performance indicators in manufacturing and product development.

Whatever the corporate function concerned, all companies stress the importance of being willing to commit substantial corporate resources to benchmarking activities. This is because, as we shall now see, imitation and learning are neither cheap nor easy.

Learning and Imitating

Whilst information on competitors' innovations is relatively cheap and easy to obtain, corporate experience shows that knowledge of how to replicate competitors' product and process innovations is much more costly and time-consuming to acquire. Such imitation typically costs between 60 and 70% of the original, and typically takes three years to achieve.[32]

These conclusions are illustrated by the examples of Japanese and Korean firms, where very effective imitation has been sustained by heavy and firm specific investments in education, training and R&D.[33] They are confirmed by a large-scale survey of R&D managers in US firms in the 1980s. As Table 4.10 shows, these managers reported that the most important methods of learning about competitors' innovations were independent R&D, reverse engineering and licensing, all of which are expensive compared to reading publications and the patent literature. Useful and usable knowledge does not come cheap. A similar and more recent survey of innovation strategy in more than 500 large European firms also found that nearly half reported the great importance, for their own innovative activities, of the technical knowledge they accumulated through the reverse engineering of competitors' products.[34]

More formal approaches to technology intelligence gathering are less widespread, and the use of different approaches varies by company and sector (Figure 4.3). For example, in the pharmaceutical sector, where much of the knowledge is highly

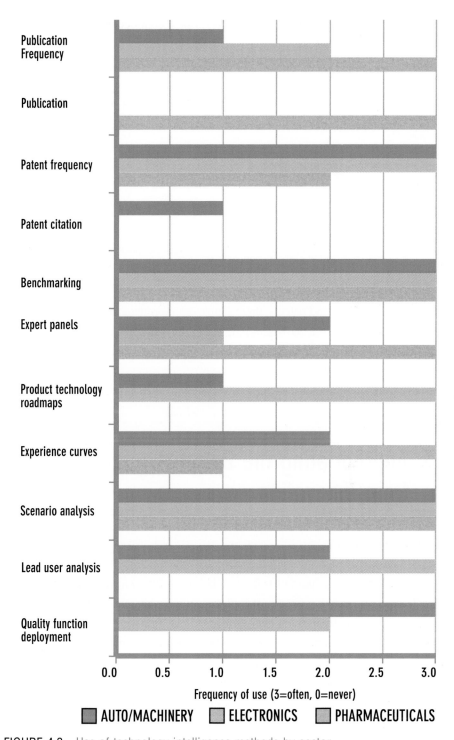

FIGURE 4.3 Use of technology intelligence methods by sector

Source: Derived from: Lichtenthaler, E. (2004) 'Technology intelligence processes in leading European and North American multinationals', *R&D Management*, **34** (2), 121–134.

TABLE 4.10 The effectiveness of methods of learning about competitors' innovations in large US firms

Method of learning	Overall sample means*	
	Processes	Products
Independent R&D	4.76	5.00
Reverse engineering	4.07	4.83
Licensing	4.58	4.62
Hiring employees from innovating firm	4.02	4.08
Publications or open technical meetings	4.07	4.07
Patent disclosures	3.88	4.01
Consultations with employees of the innovating firm	3.64	3.64

* Range: 1 = not at all effective; 7 = very effective.
Source: Levin, R. *et al.* (1987) 'Appropriating the returns from industrial research and development', *Brookings Papers on Economic Activity*, **3**, 783–820. Reproduced by permission of The Brookings Institution.

codified in publications and patents, these sources of information are scanned routinely, and the proximity to the science base is reflected in the widespread use of expert panels. In electronics, product technology roadmaps are commonly used, along with the lead users (see Chapter 7 for a discussion of lead users). Surprisingly (according to this study of 26 large firms), long-established and proven methods such as Delphi-studies, S-curve analysis and patent citations are not in widespread use.

4.3 Appropriating the Benefits from Innovation

Technological leadership in firms does not necessarily translate itself into economic benefits.[35] Table 4.11 reproduces some examples of technology leaders and followers, some of which in each category turned out to be competitive winners and others losers. David Teece argues that the capacity of the firm to appropriate the benefits of its investment in technology depends on two factors: (i) the firm's capacity to translate its technological advantage into commercially viable products or processes; (ii) the firm's capacity to defend its advantage against imitators. Thus, effective patent protection enabled Pilkington to defend its technological breakthrough in glass-making, and stopped Kodak imitating Polaroid's instant photography. Lack of commitment of complementary assets in production and marketing resulted in the failure of EMI and Xerox to reap commercial benefits from their breakthroughs in medical scanning and personal computing technologies (see also Chapter 7). De Havilland's pioneering Comet jet passenger aircraft paid the price of revealing the effects of metal fatigue on high-altitude

TABLE 4.11 Technological leaders and followers: competitive outcomes

	Technology leaders	Technology followers
Competitive winners	Pilkington (float glass)	Matsushita (VHS)
		IBM (personal computer?)
Competitive losers	EMI (scanner)	Kodak (instant photos)
	Xerox (PC)	
	De Havilland (Comet)	

Source: Based on Teece (1986).[35] Reprinted with kind permission of Elsevier Science-NL, Sara Burgerhartstraat 25, 1055 KV Amsterdam, The Netherlands.

flight, the details of which were immediately available to all competitors. In video recorders, Matsushita succeeded against the more innovative Sony in imposing its standard, in part because of a more liberal licensing policy towards competitors.

Some of the factors that enable a firm to benefit commercially from its own technological lead can be strongly shaped by its management: for example, the provision of complementary assets to exploit the lead. Other factors can be influenced only slightly by the firm's management, and depend much more on the general nature of the technology, the product market and the regime of intellectual property rights: for example, the strength of patent protection. We identify below nine factors which influence the firm's capacity to benefit commercially from its technology:

1. Secrecy.
2. Accumulated tacit knowledge.
3. Lead times and after-sales service.
4. The learning curve.
5. Complementary assets.
6. Product complexity.
7. Standards.
8. Pioneering radical new products.
9. Strength of patent protection.

We begin with those over which management has some degree of discretion for action, and move on to those where its range of choices is more limited.

1. *Secrecy* is considered an effective form of protection by industrial managers, especially for process innovations. However, it is unlikely to provide absolute protection, because some process characteristics can be identified from an analysis of the final product, and because process engineers are a professional community, who talk to each other and move from one firm to another, so that information and knowledge

inevitably leak out.[36] Moreover, there is evidence that, in some sectors, firms that share their knowledge with their national system of innovation outperform those that do not, and that those that interact most with global innovation systems have the highest innovative performance.[37] Specifically, firms that regularly have their research (publications and patents) cited by foreign competitors are rated more innovative than others, after controlling for the level of R&D. In some cases this is because sharing knowledge with the global system of innovation may influence standards and dominant designs (see below), and can help attract and maintain research staff, alliance partners and other critical resources.

2. *Accumulated tacit knowledge* can be long and difficult to imitate, especially when it is closely integrated in specific firms and regions. Examples include product design skills, ranging from those of Benetton and similar Italian firms in clothing design, to those of Rolls-Royce in aircraft engines.

3. *Lead times and after-sales service* are considered by practitioners as major sources of protection against imitation, especially for product innovations. Taken together with a strong commitment to product development, they can establish brand loyalty and credibility, accelerate the feedback from customer use to product improvement, generate learning curve cost advantages (see below) and therefore increase the costs of entry for imitators. Based on the survey of large European firms, Table 4.12 shows that there are considerable differences amongst sectors in product development lead times, reflecting differences both in the strength of patent protection and in product complexity.

4. *The learning curve* in production generates both lower costs, and a particular and powerful form of accumulated and largely tacit knowledge that is well recognized by practitioners. In certain industries and technologies (e.g. semiconductors, continuous processes), the first-comer advantages are potentially large, given the major possibilities for reducing unit costs with increasing cumulative production. However, such 'experience curves' are not automatic, and require continuous investment in training, and learning.

5. *Complementary assets.* The effective commercialization of an innovation very often depends on assets (or competencies) in production, marketing and after-sales to complement those in technology. As we have seen above, EMI did not invest in them to exploit its advances in electronic scanning. On the other hand, Teece argues that strong complementary assets enabled IBM to catch up in the personal computer market.[38]

6. *Product complexity.* However, Teece was writing in the mid-1980s, and IBM's performance in personal computers has been less than impressive since then. Previously, IBM could rely on the size and complexity of their mainframe computers as an effective barrier against imitation, given the long lead times required to design and build copy products. With the advent of the microprocessor and standard soft-

TABLE 4.12 Inter-industry differences in product development lead times

Industry	% of firms noting >5 years for development and marketing of alternative to a significant product innovation
All	11.0
Pharmaceuticals	57.5
Aerospace	26.3
Chemicals	17.2
Petroleum products	13.6
Instruments	10.0
Automobiles	7.3
Machinery	5.7
Electrical equipment	5.3
Basic metals	4.2
Utilities	3.7
Glass, cement and ceramics	0
Plastics and rubber	0
Food	0
Telecommunications equipment	0
Computers	0
Fabricated metals	0

Source: Arundel et al. (1995).[34] Reproduced by permission of Anthony Arundel.

ware, these technological barriers to imitation disappeared and IBM was faced in the late 1980s with strong competition from IBM 'clones', made in the USA and in East Asia. Boeing and Airbus have faced no such threat to their positions in large civilian aircraft, since the costs and lead times for imitation remain very high (see Table 4.12). Product complexity is recognized by managers as an effective barrier to imitation.

7. *Standards.* The widespread acceptance of a company's product standard widens its own market and raises barriers against competitors. Carl Shapiro and Hal Varian have written the standard (so far) text on the competitive dynamics of the Internet economy,[39] where standards compatibility is an essential feature of market growth, and 'standards wars' an essential feature of the competitive process (see Box 4.4). However, they point out that such wars have been fought in the adoption of each new generation of radically new technology: for example, over the width of railway gauges in the nineteenth century, the provision of electricity through direct or alternating current early in the twentieth, and rival technical systems for colour television more recently. Amongst other things they conclude that the market leader

BOX 4.4
STANDARDS AND 'WINNER TAKES ALL' INDUSTRIES

Charles Hill has gone so far as to argue that standards competition creates 'winner takes all industries'.[40] This results from so-called 'increasing returns to adoption', where the incentive for customers to adopt a standard increases with the number of users who have already adopted it, because of the greater availability of complementary and compatible goods and services (e.g. content programmes for video recorders, and computer application programs for operating systems). While the experiences of Microsoft and Intel in personal computers give credence to this conclusion, it does not always hold. The complete victory of the VHS standard has not stopped the loser (Sony) from a successful business in the video market, based on its rival's standard.[41] Similarly, IBM has not benefited massively (some would say at all), compared to its competitors, from the success of its own personal computer standard.[42] In both cases, rival producers have been able to copy the standard and to prevent 'winner takes all', because the costs to producers of changing to other standards have been relatively small. This can happen when the technology of a standard is licensed to rivals, in order to encourage adoption. It can also happen when technical differences between rival standards are relatively small. When this is the case (e.g. in TV and mobile phones) the same firms will often be active in many standards.

normally has the advantage in a standards war, but this can be overturned through radical technological change, or a superior response to customers' needs. Competing firms can adopt either 'evolutionary' strategies minimizing switching costs for customers (e.g. backward compatibility with earlier generations of the product), or 'revolutionary' strategies based on greatly superior performance–price characteristics, such that customers are willing to accept higher switching costs. Standards wars are made less bitter and dramatic when the costs to the losers of adapting to the winning standard are relatively small (see Box 4.4).

A recent review by Fernando Suarez of the literature on standards criticized much of the research as being 'ex-post', and therefore offering few insights into the 'ex-ante' dynamics of standards formation most relevant to managers.[43] It identifies that both firm level and environmental factors influence standards-setting:

- *Firm-level factors*: technological superiority, complementary assets, installed base, credibility, strategic manoeuvering, including entry timing, licensing, alliances, managing market expectations.

- *Environmental factors*: regulation, network effects, switching costs, appropriability regime, number of actors and level of competition versus cooperation. The appropriability regime refers to the legal and technological features of the environment which allow the owner of a technology to benefit from the technology. A strong or tight regime makes it more difficult for a rival firm to imitate or acquire the technology.

Different factors will have an influence at different phases of the standards process. In the early phases, aimed at demonstrating technical feasibility, factors such as the technological superiority, complementary assets and credibility of the firm are most important, combined with the number and nature of other firms and appropriability regime. In the next phase, creating a market, strategic manoeuvering and regulation are most important. In the decisive phase, the most significant factors are the installed base, complementary assets, credibility and influence of switching costs and network effects. However, in practice it is not always easy to trace such ex-ante factors to ex-post success in successfully establishing a standard (Table 4.13). This is one reason that increasingly collaboration is occurring earlier in the standards process, rather than the more historical 'winner takes all' standards battles in the later stages.[44] Research in the telecommunications and other complex technological environments where system-wide compatibility is necessary, confirms that early advocates of standards via alliances are more likely to create standards and achieve dominant positions in the industry network (see also Box 4.5 on Ericsson and the GSM standard).[45] Contrast the failure of Philips and Sony to establish their respec-

TABLE 4.13 Cases of standardization and pioneering technology

Standard	Outcome	Key actors and technology
Betamax	failure	Sony, pioneering technology
VHS	success	Matsushita and JVC alliance, follower technology
CD	success	Sony and Philips alliance for hardware, Columbia and Polygram for content
DCC	failure	Philips digital evolution of analogue cassette
Minidisc	failure	Sony competitor to DCC, re-launched after DCC withdrawn, limited subsequent success
MS-DOS	success	Microsoft and IBM
Navigator	mixed	Netscape was a pioneer and early standard for Internet browsers, but Microsoft's Explorer is fast over-taking this position

Source: Derived from Chiesa, V. and G. Toletti (2003) 'Standards-setting in the multimedia sector', *International Journal of Innovation Management*, 7 (3), 281–308.

BOX 4.5

STANDARDS, INTELLECTUAL PROPERTY AND FIRST-MOVER ADVANTAGES:
THE CASE OF GSM

The development of the global system for mobile communications (GSM) standard began around 1982. Around 140 patents formed the essential intellectual property behind the GSM standard. In terms of numbers of patents, Motorola dominated with 27, followed by Nokia (19) and Alcatel (14). Philips also had an initial strong position with 13 essential patents, but later made a strategic decision to exit the mobile telephony business. Ericsson was unusual in that it held only four essential patents for GSM, but later became the market leader. One reason for this was that Ericsson wrote the original proposal for GSM. Another reason is that it was second only to Philips in its position in the network of alliances between relevant firms. Motorola continued to patent after the basic technical decisions had been agreed, whereas the other firms did not. This allowed Motorola greater control over which markets GSM would be made available, and also enabled it to influence licensing conditions and to gain access to others' technology. Subsequently, virtually all the GSM equipment was supplied by companies which participated in the cross-licensing of this essential intellectual property: Ericsson, Nokia, Siemens, Alcatel and Motorola, together accounting for around 85% of the market for switching systems and stations, a market worth US $100bn.

As the GSM standard moved beyond Europe, North American suppliers such as Nortel and Lucent began to license the technology to offer such systems, but never achieved the success of the five pioneers. Most recently Japanese firms have licensed the technology to provide GSM based systems. Royalties for such technology can be high, representing up to 29% of the cost of a GSM handset.

Source: Bekkers R., G. Duysters and B. Verspagen (2002) 'Intellectual property rights, strategic technology agreements and market structure', *Research Policy*, **31**, 1141–1161.

---

tive analogue video standards, and subsequent recordable digital media standards, compared to the success of VHS, CD and DVD standards which were the result of early alliances. Where strong appropriability regimes exist, compatibility standards may be less important than customer interface standards, which help to 'lock-in' customers.[46] Apple's graphic user interface is a good example of this trade-off.

8. *Pioneering radical new products.* It is not necessarily a great advantage to be a technological leader in the early stages of the development of radically new products, when the product performance characteristics, and features valued by users, are not

always clear, either to the producers or to the users themselves. Especially for consumer products, valued features emerge only gradually through a process of dynamic competition, that involves a considerable amount of trial, error and learning by both producers and users. New features valued by users in one product can easily be recognized by competitors and incorporated in subsequent products. This is why market leadership in the early stages of the development of personal computers was so volatile, and why pioneers are often displaced by new entrants.[47] In such circumstances, product development must be closely coupled with the ability to monitor competitors' products and to learn from customers. According to research by Tellis and Golder, pioneers in radical consumer innovations rarely succeed in establishing long-term market positions. Success goes to so-called 'early entrants' with the vision, patience and flexibility to establish a mass consumer market.[48] As a result, studies suggest that the success of product pioneers ranges between 25% (for consumer products) and 53% (for higher technology products), depending on the technological and market conditions. For example, studies of the PIMS (Profit Impact of Market Strategy) database indicate that (surviving) product pioneers tend to have higher quality and a broader product line than followers, whereas followers tend to compete on price, despite having a cost disadvantage. A pioneer strategy appears more successful in markets where the purchasing frequency is high, or distribution important (e.g. fast-moving consumer goods), but confer no advantage where there are frequent product changes or high advertising expenditure (e.g. consumer durables).[49]

9. *Strength of patent protection* can, as we have already seen in the examples described above, be a strong determinant of the relative commercial benefits to innovators and imitators. Table 4.14 summarizes the results of the surveys of the judgments of managers in large European and US firms about the strength of patent protection. The firms' sectors are ordered according to the first column of figures, showing the strength of patent protection for product innovations for European firms. On the whole, European firms value patent protection more than their US counterparts. However, with one exception (cosmetics), the variations across industry in the strength of patent protection are very similar in Europe and the USA. Patents are judged to be more effective in protecting product innovations than process innovations in all sectors except petroleum refining, probably reflecting the importance of improvements in chemical catalysts for increasing process efficiency. It also shows that patent protection is rated more highly in chemical-related sectors (especially drugs) than in other sectors. This is because it is more difficult in general to 'invent round' a clearly specified chemical formula than round other forms of invention.

Radically new technologies are now posing new problems for the protection of intellectual property, including the patenting system. The number of patents granted

TABLE 4.14   Inter-industry differences in the effectiveness of patent protection, according to large European and US firms*

Industry	Products		Processes	
	Europe	USA	Europe	USA
Drugs	4.8	4.6	4.3	3.5
Plastic materials	4.8	4.6	3.4	3.3
Cosmetics	4.6	2.9	3.9	2.1
Plastic products	3.9	3.5	2.9	2.3
Motor vehicle parts	3.9	3.2	3.0	2.6
Medical instruments	3.8	3.4	2.1	2.3
Semiconductors	3.8	3.2	3.7	2.3
Aircraft and parts	3.8	2.7	2.8	2.2
Communications equipment	3.6	2.6	2.4	2.2
Steel mill products	3.5	3.6	3.5	2.5
Measuring devices	3.3	2.8	2.2	2.6
Petroleum refining	3.1	3.1	3.6	3.5
Pulp and paper	2.6	2.4	3.1	1.9

* Range: 1 = not at all effective; 5 = very effective.
Note: Some industries omitted because of lack of Europe–USA comparability.
Sources: Arundel *et al.* (1995)[34] and Levin *et al.* (1987).[32] Reproduced by permission of Anthony Arundel.

to protect software technology is growing in the USA, and so are the numbers of financial institutions getting involved in patenting for the first time.[50] Debate and controversy surround important issues, such as the possible effects of digital technology on copyright protection,[51] the validity of patents to protect living organisms, and the appropriate breadth of patent protection in biotechnology.[52]

Finally, we should note that firms can use more than one of the above nine factors to defend their innovative lead. For example, in the pharmaceutical industry secrecy is paramount during the early phases of research, but in the later stages of research patents become critical. Complementary assets such as global sales and distribution become more important at the later stages. Despite all the merger and acquisitions in this sector, these factors, combined with the need for a significant critical mass of R&D, have resulted in relatively stable international positions of countries in pharmaceutical innovation over a period of some 70 years (Figure 4.4). By any measure, firms in the USA have dominated the industry since the 1940s, followed by a second division consisting of Switzerland, Germany, France and the UK. Some of the methods are mutually exclusive: for example, secrecy precludes patenting, which requires disclosure of

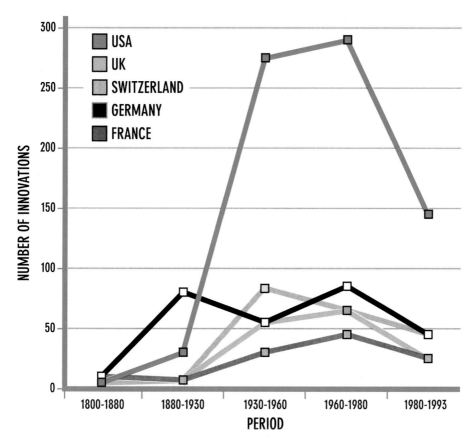

FIGURE 4.4 Innovative positions of countries in the pharmaceutical industry, 1800–1993

Source: Derived from Achilladelis, B. and N. Antonakis (2001) 'The dynamics of technological innovation: the case of the pharmaceutical industry', *Research Policy*, **30**, 535–588.

information, although it can precede patenting. However, firms typically deploy all the useful means available to them to defend their innovations against imitation.[53]

In some cases the advantages of pioneering technology, intellectual property and standards combine to create a sustainable market position (Box 4.5).

4.4 Positioning of Small Firms

Even more than in large firms, the opportunities for innovation in small firms are strongly influenced by the 'system of innovation' in which they are embedded. Table 4.15, which is based on a comprehensive survey in France, shows that the frequency of innovation in firms with fewer than 100 employees is much lower than in larger

TABLE 4.15   Frequency and sources of innovation, by firm size (France, 1993/94)

Firm size	Innovating firms (%)	Sources of innovation (%)					
		Own R&D	Part-time R&D	Outside R&D	Licences	Machine suppliers	Material suppliers
20–49	55	16	25	10	5	26	18
50–99	66	19	25	10	5	23	16
100–199	70	21	25	11	5	22	16
200–499	80	24	24	12	6	20	15
500–1999	86	26	23	13	6	19	14
2000+	96	25	21	14	6	18	14

Sources: 'L'innovation technologique', Min. de l'Industrie (1994); Kaminski, P. (1994) 'Le cas particulier de la micro-entreprise', INSEE, France.

firms. They also rely more heavily in-house on informal rather than formal R&D. Smaller firms also make less frequent use of outside sources of knowledge (R&D and licences) than larger firms, reflecting their limited capacity to absorb outside knowledge. Above all, small firms depend more than others for their innovations on their suppliers of machinery and materials, in which the innovations are embodied.

These conclusions are reinforced and extended by a survey in Canada,[54] covering 1500 small industrial firms distinguished by their strong growth performance (their average annual sales were $6.6m., and their average employment 44 people). According to their managers:

- Only 9.3% performed their own R&D, 10.4% introduced their product innovations and 5.4% their own process innovations.
- The three major sources of innovation were customers, suppliers and internal management – formal R&D was considered much less important.
- About 55% of firms introduced innovations from one of these sources.
- The main factors contributing to growth are the skills in management, labour and marketing.
- Their distinctive competence is in product quality, flexibility and customer services.

Thus, the opportunities for innovation in small firms are strongly influenced by the innovativeness of their suppliers. We shall also see in Chapter 5 that small firms in some important sectors (like machinery) are also strongly influenced by the innovativeness of their customers. In both cases personal contacts with, and close geographical proximity to, suppliers and customers reinforce and augment the effectiveness of innovation in small firms. So do the quality and skills of the local labour force. As a consequence, a small firm's innovativeness is strongly conditioned by the national and regional context in which it finds itself embedded. Examples of the regional concentration of innovative small firms include not only Silicon Valley in northern California, but also the small machinery firms linked to large firms like Robert Bosch and Daimler Benz in Baden-Württemberg, and the 'industrial districts' producing textiles in Italy.[55] In addition, small-firm clusters are now emerging in the provision of software-based services.

4.5 Summary and Further Reading

In formulating and executing their innovation strategies, business firms cannot ignore the national systems of innovation in which they are embedded. Through their strong influences on demand and competitive conditions, the provision of human resources, and forms of corporate governance, national systems of innovation both open opportunities and impose constraints on what firms can do.

However, although firms' technological strategies are *influenced* by their own national systems of innovation, they are not *determined* by them. In an analysis of more than 400 of the world's largest firms, only part of the variance between firms in their share of corporate resources devoted to innovative activities could be explained by their products and by their country of origin: a considerable share of the differences reflects the discretionary decisions by managers on the proportion of corporate resources to be devoted to innovative activities.[56]

In deciding how to cope with competitors' technology, there are a variety of useful and accessible public sources of information on firms' innovative activities. Benchmarking is more rewarding, since it makes detailed comparisons between specific firms and identifies specific factors for improvement. It is also more difficult and costly, given problems of access, comparability and interpretation. Learning (i.e. assimilating knowledge) from competitors is both essential for the firm's own innovative activities, and costly since it requires extensive investment in R&D, reverse engineering and related training. There are no readily applicable recipes for translating a technological lead into commercial success. This depends in part on what management itself does, by way of investing in complementary assets in production, marketing and after-sales. It also depends on a variety of factors that make the pioneering innovation more or less difficult to imitate, and over which management can sometimes have very little influence.

There are a number of texts which describe and compare different systems of national innovation policy, including *National Innovation Systems* (Oxford University Press, 1993), edited by Richard Nelson; *National Systems of Innovation* (Pinter, London, 1992) edited by B.-A. Lundvall; and *Systems of Innovation: Technologies, institutions and organisations* (Pinter, 1997) edited by Charles Edquist. The former is stronger on US policy, the other two on European, but all have an emphasis on public policy rather than corporate strategy. Bo Carlsson (ed.) and his colleagues have developed the notion of sectoral systems of innovation in *Technological Systems and Economic Performance: The case of factory automation* (Kluwer, 1995); Michael Porter's *The Competitive Advantage of Nations* (Macmillan, London, 1990) provides a useful framework in which to examine the direct impact on corporate behaviour of innovation systems. At the other extreme, David Landes' *Wealth and Poverty of Nations* (Little Brown, 1998), takes a broad (and stimulating) historical view of the subject.

The literature on analysing and benchmarking competitors' innovative capability is less well developed. One of the pioneering texts is Camp's (1989) *Benchmarking: The search for industry best practices that lead to superior performance* (Quality Press, Milwaukee, 1989), and more recent are the excellent series of publications by Chris Voss and his colleagues, M. Zairi *Benchmarking for Best Practice* (Butterworth-Heinemann, 1996) and J. Tidd (ed.) *From Knowledge Management to Strategic Competence: Measuring technological, market and organizational innovation* (Imperial College Press, 2000). Published comparisons of corporate innovative activities are becoming increasingly frequent, for example the annual comparisons of *The R&D Scoreboard* (Table 4.9).

The best analysis of the strategic advantages of being a technological leader or follower remains Teece's 1986 paper. The surveys by Arundel, Levin and their colleagues are rich sources of information on corporate experience in assimilating technological knowledge and on how to maintain innovative leads. Comprehensive and balanced reviews of the arguments and evidence for product leadership versus follower positions is provided by G. J. Tellis and P. N. Golder *Will and Vision: How latecomers grow to dominate markets* (McGraw-Hill, 2002) and D. A. Shepherd and M. S. Shanley *New Venture Strategies: Timing of entry, demand uncertainty and technological uncertainty* (Sage, 1998). Fernando Suarez provides an excellent up-to-date overview of the factors influencing technology standards in 'Battles for technological dominance: an integrative framework' (*Research Policy*, 2004, **33**, 271–286).

Finally there has been a recent spate of papers and books on the implications of the Internet (and of ICT more generally) for corporate market positioning strategies. On product development and market positioning, these include M. Iansiti, 'Mastering the rapids: managing product development in turbulent environments', *California Management Review*, **38** (1), Fall 1995, 37–58; D. B. Yoffie and M. A. Cusumano: 'Judo strategy. The competitive dynamics of Internet time', *Harvard Business Review*, Jan.–Feb. 1999, 71–81; M. Iansiti and A. MacCormack, 'Developing products on internet time', *Harvard Business Review*, Sept.–Oct. 1997, 108–117; *The Economist*, 25 May 1996, 'Survey of the software industry'; M. A. Cusumano, Y. Mylonadis and R. S. Rosenbloom: 'Strategic manoeuvring and mass-market dynamics: the triumph of VHS over Beta', *Business History Review*, **66**, Spring 1992, 51–94.

On open source software: J. B. De Long and A. H. Froomkin: 'Beating Microsoft at its own game', *Harvard Business Review*, Jan.–Feb. 2000, 159–164; *The Economist*, 12 June 1999, 'Business: venture communism: open-source software: the next high-tech listings', 92–94; *The Economist*, 11 July 1998, 'Business: revenge of the hackers: the software is free', 81–84; R. Garud and A. Kumaraswary, 'Changing competitive dynamics in networks industries. An exploration of Sun Microsystems' open source strategy', *Strategic Management Journal*, **14** (5), July 1993, 351.

References

1 Porter, M. (1990) *The Competitive Advantage of Nations*. Macmillan, London.
2 Nelson, R. (ed.) (1993) *National Innovation Systems*. Oxford University Press, Oxford.
3 Cantwell, J. (1992) 'The internationalisation of technological activity and its implications for competitiveness', in Granstrand, O., Hakanson, L. and Sjolander, S. (eds) *Technology Management and International Business*, John Wiley & Sons, Ltd, Chichester, 75–95; Patel, P. (1995) 'The localised production of global technology', *Cambridge Journal of Economics*, **19**, 141–153.
4 Patel, P. and K. Pavitt (1994) 'Uneven (and divergent) technological accumulation amongst advanced countries: evidence and a framework of explanation', *Industrial and Corporate Change*, **3**, 759–787.
5 Fagerberg, J. (1987) 'A technology gap approach to why growth rates differ', in Freeman, C. (ed.), *Output Measurement in Science and Technology: Essays in honour of Y. Fabian*. North-Holland, Amsterdam, 33–45; Fagerberg, J. (1988) 'International competitiveness', *Economic Journal*, **98**, 355–374.
6 Franko, L. (1989) 'Global corporate competition: who's winning, who's losing, and the R&D factor as one reason why', *Strategic Management Journal*, **10**, 449–474.
7 Freeman, C. (1988) 'Japan: a new national system of innovation?', in Dosi, G. *et al.* (eds), *Technical Change and Economic Theory*, Pinter, London, 330–348; Nelson, R. (ed.) (1993) *National Innovation Systems*. Oxford University Press, Oxford; Lundvall, B.-A. (ed.) (1992) *National Systems of Innovation*. Pinter, London.
8 Patel, P. and K. Pavitt (1994) 'National innovation systems: why they are important, and how they might be measured and compared', *Economics of Innovation and New Technology*, **3**, 77–95.
9 Vernon, R. (1966) 'International investment and international trade in the product cycle', *Quarterly Journal of Economics*, **80**, 190–207; Dunning, J. (1980) 'Towards an eclectic theory of international production: some empirical tests', *Journal of International Business Studies*, **1**, 9–31.
10 Patel, P. and K. Pavitt (1994) 'National innovation systems: why they are important, and how they might be measured and compared', *Economics of Innovation and New Technology*, **3**, 77–95; Porter, M. (1990) *The Competitive Advantage of Nations*. Macmillan, London.
11 Thomas, L. (1994) 'Implicit industrial policy: the triumph of Britain and the failure of France in global pharmaceuticals' *Industrial and Corporate Change*, **3**, 451–489.
12 Prais, S. (1993) 'Economic performance and education: the nature of Britain's deficiencies', *Proceedings of the British Academy*, **84**, 151–207.
13 See OECD (2000) *Internet Technology Outlook, 2000: ICTs, E-commerce and the information economy*. Paris, www.oecd.org/dsti/sti/stat-ana/.

14 Faulkner, W. and J. Senker (1995) *Knowledge Frontiers*. Clarendon Press, Oxford; Hicks, D. (1995) 'Published papers, tacit competencies and corporate management of the public/private character of knowledge', *Industrial and Corporate Change*, **4**, 401–424; Jaffe, A. (1989) 'Real effects of academic research', *American Economic Review*, **79**, 957–970; Narin, F., K. Hamilton and D. Olivastro (1998) 'The increasing linkage between U.S. technology and public science', *Research Policy*, **26**, 317.

15 Martin, B. (1992) 'The position of British science', *Nature*, **355**, 760; and 'Struggling to keep up appearances', *New Scientist*, 7 November; Nathan, R. (1996) 'Japan hints at a shift to the life sciences', *Nature*, **383**, 7.

16 Carlsson, B. and S. Jacobsson (1994) 'Technological systems and economic policy: the diffusion of factory automation in Sweden', *Research Policy*, **23**, 235–248; Laursen, K. (1997) 'Horizontal diversification in the Danish national system of innovation: the case of pharmaceuticals', *Research Policy*, **25**, 1121.

17 Landau, R. and N. Rosenberg (1992) 'Successful commercialisation in the chemical process industries', in Rosenberg, N., Landau, R. and Mowery, D. (eds), *Technology and the Wealth of Nations*. Stanford University Press, Stanford, 73–119.

18 Albert, M. (1992) *Capitalism against Capitalism*. Whurr, London.

19 Fransman, M. (1995) *Japan's Computer and Communications Industry*. Oxford University Press, Oxford.

20 Albach, H. (1996) 'Global competitive strategies for scienceware products', in Koopmann, G. and Scharrer, H. (eds), *The Economics of High Technology Competition and Co-operation in Global Markets*. Nomos, Baden-Baden, 203–217.

21 *The Economist* (1995) 'Dismantling Daimler-Benz', 18 November, 99–100.

22 *The Economist* (1995) 'Back on top. A survey of American business', 16 September.

23 Gordon, P. (1996) 'Industrial districts and the globalisation of innovation: regions and networks in the new economic space', in Vence-Deza, X. and Metcalfe, J. (eds), *Wealth from Diversity*. Kluwer, Dordrecht, 103–134; Computer Science and Telecommunications Board (1999) *Funding a Revolution: Government support for computing research*, National Research Council, Washington, DC.

24 Harding, R. and W. Paterson (eds) (2000) *The Future of the German Economy: an end to the miracle?*, Manchester University Press, Manchester.

25 See *The Economist* (1999) 'The world in your pocket: a survey of telecommunications', 9 October.

26 Mansfield, E. (1985) 'How rapidly does new industrial technology leak out?', *Journal of Industrial Economics*, **34**, 217–223.

27 Camp, R. (1989) *Benchmarking – the Search for Industry Best Practices That Lead to Superior Performance*. Quality Press, Milwaukee, Wis.

28 Altshuler, A. *et al.* (1984) *The Future of the Automobile: The report of MIT's international automobile program*. MIT Press, Cambridge, Mass.; Womack, J., D. Jones and D. Roos (1990) *The Machine that Changed the World*. Macmillan, New York; Leachman, R. and D. Hodges (1996) 'Benchmarking semiconductor manufacturing', *IEEE Transactions on Semiconductor Manufacturing*, **9**, 158–169.

29 De Meyer, A. and B. Pycke (1997) 'Separating the fads from the facts: trends in manufacturing action programmes and competitive priorities from 1986 till 1996', in Miller, J., de Meyer, A. and Nakane, J. (eds), *Benchmarking Global Manufacturing: Understanding international suppliers, customers, and competitors*. Irwin, Ill.

30 Voss, C. and K. Blackmon (1996) 'The impact of national and parent company origin on world-class manufacturing: findings from Britain and Germany', *International Journal of Oper-*

ations and Production Management, **16**, 96–112; Hanson, P. and C. Voss (1993) *Made in Britain*. IBM Consulting, London; P. Hanson *et al.* (1996) *Made in Europe 2*. IBM Consulting, London.

31 Coopers & Lybrand (1994) *Survey of Benchmarking in the UK: Executive summary*. London.

32 Levin, R. *et al.* (1987) 'Appropriating the returns from industrial research and development', *Brookings Papers on Economic Activity*, **3**, 783–820; Mansfield, E., M. Schwartz and S. Wagner (1981) 'Imitation costs and patents: an empirical study', *Economic Journal*, **91**, 907–918.

33 Kim, L. (1993) 'National system of industrial innovation: dynamics of capability building in Korea', and Odagiri, H. and A. Goto (1993) 'The Japanese system of innovation: past, present and future', in Nelson, R. (ed.), *National Innovation Systems*, Oxford University Press, Oxford, 357–383, 76–114.

34 Arundel, A., G. van de Paal and L. Soete (1995) *Innovation Strategies of Europe's Largest Industrial Firms* (PACE Report). MERIT, University of Limbourg, Maastricht.

35 Teece, D. (1986) 'Profiting from technological innovation: implications for integration, collaboration, licensing and public policy', *Research Policy*, **15**, 285–305.

36 Von Hippel, E. (1987) 'Co-operation between rivals: informal know-how training', *Research Policy*, **16**, 291–302.

37 Spencer, J. (2003) 'Firms' knowledge-sharing strategies in the global innovation system: empirical evidence from the flat panel display industry', *Strategic Management Journal*, **24**, 217–233.

38 Teece, D. (1986) 'Profiting from technological innovation: implications for integration, collaboration, licensing and public policy', *Research Policy*, **15**, 285–305.

39 Shapiro, C. and H. Varian (1998) *Information Rules: A strategic guide to the network economy*. Harvard Business School Press, Boston, Mass.

40 Hill, C. (1997) 'Establishing a standard: competitive strategy and technological standards in winner-take-all industries', *Academy of Management Executive*, **11**, 7–25.

41 Rosenbloom, R. and M. Cusumano (1987) 'Technological pioneering and competitive advantage: the birth of the VCR industry', *California Management Review*, **24**, 51–76.

42 Chesbrough, H. and D. Teece (1996) 'When is virtual virtuous? Organizing for innovation', *Harvard Business Review*, Jan.–Feb., 65–73.

43 Suarez, F. (2004) 'Battles for technological dominance: an integrative framework', *Research Policy*, **33**, 271–286.

44 Chiesa, V., R. Manzini, and G. Toletti (2002) 'Standards-setting processes: evidence from two case studies', *R&D Management*, **32** (5), 431–450.

45 Soh, P. and E. Roberts (2003) 'Networks of innovators: a longitudinal perspective', *Research Policy*, **32**, 1569–1588.

46 Sahay, A. and D. Riley (2003) 'The role of resource access, market conditions, and the nature of innovation in the pursuit of standards in the new product development process', *Journal of Product Innovation Management*, **20**, 338–355.

47 Steffens, J. (1994) *Newgames: Strategic competition in the PC revolution*. Pergamon Press, Oxford.

48 Tellis, G. and P. Golder (1996) 'First to market, first to fail? Real causes of enduring market leadership', *Sloan Management Review*, Winter, 65–75; Tellis, G. and P. Golder (2002) *Will and Vision: How latecomers grow to dominate markets*. McGraw-Hill.

49 Lambkin, M. (1992) 'Pioneering new markets. A comparison of market share winners and losers', *International Journal of Research on Marketing*, 5–22; Robinson, W. (2002) 'Product development strategies for established market pioneers, early followers and late entrants', *Strategic Management Journal*, **23**, 855–866.

50 *The Economist* (2000) 'The knowledge monopolies: patent wars', 8 April, 95–99; (1996) 'A dose of patent medicine', 10 February, 93–94.

51 *The Economist* (1999) 'Digital rights and wrongs', 17 July, 99–100.

52 Mazzolini, R. and R. Nelson (1998) 'The benefits and costs of strong patent protection: a contribution to the current debate', *Research Policy*, **26**, 405.

53 Bertin, G. and S. Wyatt (1988) *Multinationals and Industrial Property: The control of the world's technology.* Harvester-Wheatsheaf, Hemel Hempstead.

54 Baldwin, K. (1994) *Innovation: The key to success in small firms.* Statistics Canada, Ottawa.

55 Cooke, P. and K. Morgan (1997) *The Associational Economy: Firms, regions and innovation.* Oxford University Press, Oxford.

56 Patel, P. and K. Pavitt (1998) 'The wide (and increasing) spread of technological competencies in the world's largest firms: a challenge to conventional wisdom', in Chandler, A., Hagstrom, P. and Solvell, O. (eds), *The Dynamic Firm.* Oxford University Press, Oxford.

Chapter 5

Paths: Exploiting Technological Trajectories

Firms' strategies are strongly constrained by their current position and by the specific opportunities open to them in future: in other words, they are path-dependent.[1] At any point in time, two sets of constraints make path-dependency in corporate innovation strategy inevitable: those of the present and likely future state of technological knowledge, and those of the limits of corporate competence.

Critics of technological determinism notwithstanding, pure technological development does have its own internal logic, which helps define where firms will find innovative opportunities. Thus, we can marvel at the rapid rate of improvement in the performance–price ratio of the electronic chip and at the economic and social changes it has made possible. But we can also be frustrated that our personal computers can rarely be made to run independently for more than three hours, or that battery-driven cars are so heavy, limited in range and slow to be refuelled: in spite of extensive private investments, existing knowledge of battery technology has not enabled us to do much better. The energy density of gasoline fuel (i.e. energy generated per unit weight) remains 100 times higher than electric batteries.[2] Similarly, we can speculate that a set of technologies that could convert deep-mined coal into oil and gas at the same price, and with lesser adverse environmental consequences than existing supplies, would have economic, social and political effects at least equal to those of the microchip. But it will remain speculation, since the present state of knowledge does not enable it to be done.[3]

In addition to the constraints of knowledge, there are those of competence: in other words, of what specific firms are capable of learning and exploiting. As we have seen in Chapter 3, innovation requires improvements and changes in the operation of complex technical and organizational systems. This involves trial, error and learning. Learning tends to be incremental, since major step changes in too many parameters both increase uncertainty and reduce the capacity to learn. As a consequence, firms' learning processes are path-dependent, with the directions of search strongly conditioned by the competencies accumulated for the development and exploitation of their existing product base.[4] Moving from one path of learning to another can be costly, even impossible, given cognitive limits – think of the problems of learning a foreign language from scratch.

Furthermore, firms cannot easily jump from one major path to another through hiring individuals with the required competencies. Corporate competencies are rarely those of an individual, and most often those of specialized, interdependent and co-ordinated groups, where tacit technical and organizational knowledge accumulated through experience are of central importance. This is why firms perform most of their innovative activities in-house.[5] And even when competencies come from outside the firm as part of a corporate acquisition, different practices and cognitive structures may make their assimilation costly or impossible. For example, it is no accident that electrical firms find it much easier to master and exploit semiconductor technology than chemical firms: the fields of technological competencies required are much closer.[6]

From the fact of path dependency has emerged the notion of technological trajectory, first proposed by Nelson and Winter[7] and later extended by Dosi.[8] As we have seen above, it can be applied equally to a technology, constrained by knowledge limits, and to a firm, constrained by limits in competence. In Chapter 4, we have shown that it can also be applied to a country, which will often have more than one trajectory. These overlapping categories are inevitable, since technologies develop in firms, which themselves are situated in sectors and countries.

5.1 Major Technological Trajectories

We focus here on firms and broad technological trajectories. This is because firms and industrial sectors differ greatly in their underlying technologies. For example, designing and making an automobile is not the same as designing and making a therapeutic drug, or a personal computer. We are dealing not with one *technology*, but with several *technologies*, each with its historical pattern of development, skill requirements and strategic implications. It is therefore a major challenge to develop a framework, for integrating changing technology into strategic analysis, that deals effectively with corporate and sectoral diversity. We describe below the framework that one of us has developed over the past 10 or more years to encompass diversity.[9] It has been strongly influenced by the analyses of the emergence of the major new technologies over the past 150 years by Chris Freeman and his colleagues,[10] and by David Mowery and Nathan Rosenberg.[11]

A number of studies have shown marked, similar and persistent differences amongst industrial sectors in the sources and directions of technological change. They can be summarized as follows:

- *Size of innovating firms*: typically *big* in chemicals, road vehicles, materials processing, aircraft and electronic products; and *small* in machinery, instruments and software.

- *Type of product made*: typically *price sensitive* in bulk materials and consumer products; and *performance sensitive* in ethical drugs and machinery.
- *Objectives of innovation*: typically *product* innovation in ethical drugs and machinery; *process* innovation in steel; and *both* in automobiles.
- *Sources of innovation*: *suppliers* of equipment and other production inputs in agriculture and traditional manufacture (like textiles); *customers* in instrument, machinery and software; *in-house* technological activities in chemicals, electronics, transport, machinery, instruments and software; and *basic research* in ethical drugs.
- *Locus of own innovation*: *R&D laboratories* in chemicals and electronics; *production engineering departments* in automobiles and bulk materials; *design offices* in machine-building; and *Systems Departments* in service industries (e.g. banks and supermarket chains).

Five Major Technological Trajectories

In the face of such diversity there are two opposite dangers. One is to generalize about the nature, source, directions and strategic implications of innovation on the basis of experience in one firm or in one sector. In this case, there is a strong probability that many of the conclusions will be misleading or plain wrong. The other danger is to say that all firms and sectors are different, and that no generalizations can be made. In this case, there can be no cumulative development of useful knowledge. In order to avoid these twin dangers, one of us distinguished five major technological trajectories, each with its distinctive nature and sources of innovation, and with its distinctive implications for technology strategy and innovation management. This was done on the basis of systematic information on more than 2000 significant innovations in the UK, and of a reading of historical and case material. In Table 5.1 we identify for each trajectory its typical core sectors, its major sources of technological accumulation and its main strategic management tasks.

Thus, in *supplier-dominated firms*, technical change comes almost exclusively from suppliers of machinery and other production inputs. This is typically the case in both agriculture and textiles, where most new techniques originate in firms in the machinery and chemical industries. Firms' technical choices reflect input costs, and the opportunities for firm-specific technological accumulation are relatively modest, being focused on improvements and modifications in production methods and associated inputs. *The main task of innovation strategy* is therefore to use technology from elsewhere to reinforce other competitive advantages. Over the past 10 years, advances elsewhere in IT have opened up radical new applications in design, distribution, logistics and transactions, thereby making production more responsive to customer demands. But,

TABLE 5.1 Five major technological trajectories

	Supplier-dominated	Scale-intensive	Science-based	Information-intensive	Specialized suppliers
Typical core products	• Agriculture • Services • Traditional manufacture	• Bulk materials • Consumer durables • Automobiles • Civil engineering	• Electronics • Chemicals	• Finance • Retailing • Publishing • Travel	• Machinery • Instruments • Software
Main sources of technology	• Suppliers • Production learning	• Production engineering • Production learning • Suppliers • Design offices	• R&D • Basic research	• Software and systems departments • Suppliers	• Design • Advanced users

Main tasks of innovation strategy

	Supplier-dominated	Scale-intensive	Science-based	Information-intensive	Specialized suppliers
1. Positions	1. Based on non-technological advantages	1. Cost-effective and safe complex products and processes	1. Develop technically related products	1. New products and services	1. Monitor and respond to user needs
2. Paths	2. Use of IT in finance and distribution	2. Incremental integration of new knowledge (e.g. virtual prototypes, new materials, B2B*)	2. Exploit basic science (e.g. molecular biology)	2. Design and operation of complex information processing systems	2. Matching changing technologies to users' needs
3. Processes	3. Flexible response to user	3. Diffusion of best practice in design, production and distribution	3. Obtain complementary assets. Redefine divisional boundaries	3. To match IT-based opportunities with user needs	3. Strong links with lead users

* B2B = business to business.

since these revolutionary changes are available from specialized suppliers to all firms, it is still far from clear whether they can be of a lasting advantage for firms competing in supplier-dominated sectors.

In *scale-intensive firms*, technological accumulation is generated by the design, building and operation of complex production systems and/or products. Typical core sectors include the extraction and processing of bulk materials, automobiles and large-scale civil engineering projects. Given the potential economic advantages of increased scale, combined with the complexity of products and/or production systems, the risks of failure associated with radical but untested changes are potentially very costly. Process and product technologies therefore develop incrementally on the basis of earlier operating experience, and improvements in components, machinery and subsystems. The main sources of technology are in-house design and production engineering departments, operating experience, and specialized suppliers of equipment and components. In these circumstances, *the main tasks of innovation strategy* are the incremental improvement of technological improvements in complex products or production systems, and the diffusion throughout the firm of best-practice methods in design and production. Recent advances in the techniques of large-scale computer simulation and modelling now offer considerable opportunities for saving time and money in the building and testing of prototypes and pilot plant.

In *science-based firms*, technological accumulation emerges mainly from corporate R&D laboratories, and is heavily dependent on knowledge, skills and techniques emerging from academic research. Typical core sectors are chemicals and electronics; fundamental discoveries (electromagnetism, radio waves, transistor effect, synthetic chemicals, molecular biology) open up major new product markets over a wide range of potential applications. The major directions of technological accumulation in the firm are horizontal searches for new and technologically related product markets. As a consequence, the *main tasks of innovation strategy* are to monitor and exploit advances emerging from basic research, to develop technologically related products and acquire the complementary assets (e.g. production and marketing) to exploit them, and to reconfigure the operating divisions and business units in the light of changing technological and market opportunities.

Information-intensive firms have begun to emerge only in the past 10 to 15 years, particularly in the service sector: finance, retailing, publishing, telecommunications and travel. See, for example, the surveys in *The Economist*, on *retailing* (4 March 1995) and *online finance* (20 May 2000). The main sources of technology are in-house software and systems departments, and suppliers of IT hardware and of systems and applications software. The main purpose is to design and operate complex systems for processing information, particularly in distribution systems that make the provision of a service or a good more sensitive to customer demands. The *main tasks of innovation strategy* are the development and operation of complex information processing systems, and the development of related and often radically new services.

Specialized supplier firms are generally small, and provide high-performance inputs into complex systems of production, of information processing and of product development, in the form of machinery, components, instruments and (increasingly) software. Technological accumulation takes place through the design, building and operational use of these specialized inputs. Specialized supplier firms benefit from the operating experience of advanced users, in the form of information, skills and the identification of possible modifications and improvements. Specialized supplier firms accumulate the skills to match advances in technology with user requirements, which – given the cost, complexity and interdependence of production processes – put a premium on reliability and performance, rather than on price. The *main tasks of innovation strategy* are keeping up with users' needs, learning from advanced users and matching new technologies to users' needs.

Knowledge of these major technological trajectories can improve analysis of particular companies' technological strategies, by helping answer the following questions:
- Where do the company's technologies come from?
- How do they contribute to competitive advantage?
- What are the major tasks of innovation strategy?
- Where are the likely opportunities and threats, and how can they be dealt with?

Although the above taxonomy has held up reasonably well to subsequent empirical tests,[12] it inevitably simplifies.[13] For example, we can find 'supplier-dominated' firms in electronics and chemicals, but they are unlikely to be technological pacesetters. In addition, firms can belong in more than one trajectory. In particular, large firms in all sectors have capacities in *scale-intensive* (mainly mechanical and instrumentation) technologies, in order to ensure efficient production. Software technology is beginning to play a similarly pervasive role across all sectors.

5.2 Revolutionary Technologies: Biotechnology, Materials and IT

Firm-specific technological trajectories change over time as improvements in the knowledge base open up new technological opportunities. Since the beginning of the 1980s, biotechnology, new materials and IT have been widely identified by corporate R&D directors as the three fields with the greatest promise. This is confirmed by data showing

BOX 5.1
POST-WAR DEVELOPMENTS IN BIOTECHNOLOGY

'Biotechnology has been defined many times . . . The OECD definition "the application of scientific and engineering principles to the processing of materials by biological agents" has been widely adopted and encompasses the use and application of a series of skills drawn from biology, biochemistry, genetics, microbiology, biochemical engineering and separations processing.

The development of biotechnology over the past forty years has been a continuous process. The discovery of the structure of DNA, the genetic material, in the 1950s, the detailed analysis of protein synthesis in the 1960s, and the study of bacteriophage genetics in the basis of antibiotic resistance in the 1970s, laid the foundations for the scientific breakthrough of recombinant DNA (rDNA) in the 1970s. For the first time, a gene sequence could be cut and a foreign DNA sequence from another organism inserted and expressed . . .

. . . Since then the science underpinning developments in biotechnology has progressed very rapidly. Gene therapy, antisense technology, automated gene sequencing and gene discovery now present new technological opportunities to develop and apply technology. Moreover, genome analysis or genomics in particular has placed new emphasis and value on information which can be widely exploited by industry . . . biotechnology can now be regarded as an expanding series of enabling technologies . . .

. . . To date, the greatest impact of these technologies has been on the R&D programmes of companies in the pharmaceutical and agro-food sectors, where it has led to major investments by existing companies and the foundation of specialist biotechnology companies. Biotechnology is expected to continue to be of critical importance to these sectors, despite the fact that it has not yet delivered the short cuts to profitability hoped for by the pharmaceutical sector. In addition, biotechnology is currently widely used to improve the efficiency of key production processes, particularly in food processing, drinks and detergents . . .

(m)ajor new applications will emerge in industries such as textiles, leather, paper and pulp, oil refining, metals and mining, printing, environmental services and speciality chemicals . . .

Creating value through biotechnology depends upon the effective functioning of an "innovation system". This has three main components: . . . (t)he science base . . . specialist biotech companies . . . user industries.'

Source: Business Decisions and SPRU, 1996.

that the number of the world's largest firms with competencies in these fields has increased greatly since then.[14]

Third-generation biotechnology has not yet had such widespread effects, but is beginning to change methods of product development in drugs and agricultural products (see Box 5.1). Materials technology has been advancing steadily with a strengthening science base (see Box 5.2).

However, it is *information technology* that has had, so far, the most revolutionary effects, and is likely to continue to do so in the foreseeable future. As in the midst of all revolutions, the signals are incomplete, confusing and sometimes misleading, so that information and experience must be interpreted with particular care. In the 1970s, the so-called *microelectronics revolution* was associated with the spectacular achievements in semiconductor technology: in particular, the microprocessor and the capacity to store and manipulate vast quantities of information on a small and increasingly cheap electronic chip. Thereafter, the phrase *IT revolution* has come into increasing use, reflecting similar advances in the capacity to transmit information, culminating in the Internet.

Perhaps even more important, we have seen the spectacular advance in *software* technology (i.e. techniques for manipulating information), which had previously been developed and closely controlled by manufacturers of computer hardware. The steep reduction in hardware costs and the emergence of cheap standard products (such as personal computers) have resulted in the emergence of two other major sources of software technology: independent software suppliers (e.g. Microsoft) and operators of large-scale systems (e.g. banks, retail chains, airlines). As a result, the technological trajectories of firms and countries in the development of software have progressively become decoupled from their trajectories in computer hardware.

Table 5.2 compares and contrasts the characteristics of the two trajectories, which can be summarized as follows:

- The microelectronics revolution is about designing and producing electronic chips, and the IT revolution about producing software.
- The former (microelectronics) is located firmly in manufacturing, and principally involves the highly sophisticated and demanding design and manufacturing of hardware. It opens up technological opportunities mainly for firms in – or close to – the electronics industry.
- The latter (software) involves not only the design and manufacturing function, but also the administration, co-ordination and distribution functions. It opens up technological opportunities in all sectors in both manufacturing and services.
- This reflects major differences in the size of the barriers to entry into the two technologies. In chip manufacture, they are massive, with major investments required in difficult and demanding design and manufacturing activities. The development of chip technology is in fact one of the most concentrated in the world. In software,

BOX 5.2
THE IMPROVING SCIENCE BASE OF MATERIALS TECHNOLOGY

The technology of materials is of pervasive importance, across both industry and time. In a number of the *Scientific American* published in 1986, there were separate papers on materials for information and communication, for aerospace, for ground transportation and for energy utilization. There were also papers on materials that were electronic and magnetic, photonic, and that were advanced metals, advanced ceramics and advanced polymers. Underlying all these applications are the ever-closer and more productive links between materials science and materials technology.

'Until very recently the practitioners of materials engineering and materials science remained separated by a wide gulf. Craftsmanship and technology flourished. The selection, modification and processing of materials came to be fundamental elements of human cultures. Yet the science of materials – an attempt to understand their fundamental nature and why particular manipulations of them have particular effects – was slow to develop beyond the level of speculation. It was not until the 19th century that chemistry and to a lesser extent physics began to support the largely empirical efforts of artisans and engineers with applicable theories and novel analytical tools. Within the last half century the collaboration has thrived. The advent of powerful new theories and instrumentation has made investigative science an essential driving force for advances in engineering . . .

. . . The key contribution of science was the coupling of a material's external properties to its internal structure. Materials were discovered to possess an inner architecture – a hierarchy of several structural levels. The architecture was seen to be complex enough to account for the widely varying behaviour of materials. This recognition in turn implied that the behaviour of a particular material could be predicted from close study of its internal architecture.

Such study has been facilitated by a battery of new instruments and techniques that reveal increasingly fine detail . . . microscopy was followed by transmission electron microscopy, which reveals details of substructure, and then by scanning electron microscopy, which provides important three-dimensional surface information. X-ray diffraction maps the spatial arrangements of atoms or molecules in a crystal. The very identity of a material's atoms is revealed by various excitation spectroscopies, and bombardment of a material with high energy particles probes the atomic nucleus . . .

continues overleaf

> ### BOX 5.2 (*continued*)
>
> . . . Now, as more becomes known about how processing can modify a material's structure, and thus its properties and ultimately its performance, scientists are becoming more interested in processing and are having more impact on it. Their findings have been translated into improved processes in areas ranging from steel manufacture to the production of pure glass fibres. A striking example of scientific input is the development of ways to grow the very large single crystals of the semiconductor silicon from which integrated circuit chips are made.'
>
> Liedl, G. (1986) 'The science of materials', *Scientific American*, **255** (4), 104–112.

TABLE 5.2 Which technological revolution?

Description	'Microelectronic'	'Information technology'
Location	Producing electronic chips	Producing software
Key corporate functions	• Design • Production	• Design • Production • Administration • Co-ordination • Distribution
Sectors	• Electronics	• Manufacturing • Services
Pervasiveness of production	Low	High
Barriers to entry	High	Low
Visibility in available statistics	High	Low

they are much lower, with software skills and access to a workstation being all that is required for specialized applications.

- Since chip manufacture is a large-scale activity undertaken by well-established manufacturing firms, it is clearly visible in a whole range of published statistics. Software development, on the other hand, is often hidden away either in small and specialized firms in the service sector, or embedded in large organizations in sectors like retailing and finance, and is therefore hardly visible in established statistics. Perhaps the most important comprehensive statistic so far is buried in the periodic US reports *Science and Engineering Indicators* (National Science Foundation, Washington) and shows that the number of qualified scientists and engineers employed in US manufacturing was overtaken by the number employed in services in 1989, and that in finance and retailing a high proportion are software specialists.

It is particularly difficult during a period of revolutionary technological change to distinguish what is important from what is not. Fads and fashions flourish:

- *In financial markets* – see the rapid rise and fall of the so-called dot.com companies in the spring of 2000.
- *Amongst economics journalists* – see the debatable emergence of the so-called 'New Economy', when we know that revolutionary technologies – like electricity and IT – not only create the new, but transform the old. See *The Economist's Survey of the New Economy*, 23 September 2000.
- *Amongst management consultants* – see the emergence of so-called 'knowledge management', which sometimes assumes that 'knowledge' is an easily recognized, homogeneous and manipulable commodity.

We shall nonetheless risk identifying three features of the IT revolution that we think will be increasingly important for innovation strategies in future:

1. The increasingly *systemic* nature of economic and technological activities, resulting from the digitalization and interconnection of previously separate activities – for example, home electronics; logistics, sales and distribution in retailing; management information systems in large organizations. This is increasing the importance of interface technologies, and of competencies in systems integration, and is one of the factors behind the growth of external technological alliances (see Chapter 8).
2. The decreasing cost of product development through the use of simulations and virtual prototypes. For example, Paul Nightingale has shown how improvements in simulation technology have been combined with radical improvements in fundamental biomedical knowledge to begin to revolutionize the process of drug discovery.[15]
3. Given the growing importance of software technology in distribution activities, the traditional distinction between hi-tech, medium-tech and low-tech is becoming even less useful.[16] For example, does it still make sense to describe as 'traditional, low-tech' distribution-intensive activities such as banking, retailing, and the export of Dutch tulips,[17] and Australian fresh fruit?[18]

This has led to much confusion about the characteristics and implications of the 'new' or 'knowledge' economy. The more traditional notion of the knowledge economy included the broad opportunities created by developments in science and technology, and the role of intellectual capital and innovation for competitive advantage. The more recent and more narrow perspective focuses exclusively on the potential of information and communications technologies. However, these two views are based on contradictory assumptions and suggest different implications.[19] The latter ICT perspective emphasizes the low marginal costs of reproduction and near instantaneous transmission of such technologies, but too often assumes that the exchange and transfer of knowledge is almost effortless and unrestricted. The former, broader view, highlights the difficulties of capturing and transferring knowledge due to its tacit nature and context-specificity.

Roger Miller and his team at MINE (Managing Innovation in the Networked Economy) have attempted to develop a taxonomy to better characterize different business environments and the most appropriate way to manage innovation in each context. They identify four factors in the environment that influence the most effective innovation strategy and management: *velocity*, which refers to the pace of change of the relevant science, technology and markets; *institution*, which refers to the role of government, regulation and other stakeholders; *challenge*, which captures how demanding customers are in terms of product performance, customization or problem-solving; and *uncertainty*, which captures the unsolvable uncertainty and unpredictability of technology and markets. They argue that each of these four factors are higher in the new economy businesses. Table 5.3 describes some of the key findings. According to this study, in the new economy firms R&D is more closely integrated with top management and strategy formulation, but also they devote more resources to exploration, particularly involving external organizations, and cooperate more with suppliers and lead users to generate value. However, contrary to expectations, the new economy firms did not relay any more than old economy firms on innovation networks.

TABLE 5.3   Patterns of innovation in the 'new' and 'old' economies

Variable	New economy	Old economy
R&D sets strategic vision of firm	5.14	3.56
R&D active participant in making corporate strategy	5.87	4.82
R&D responsible for developing new business	5.05	3.76
Transforming academic research into products	4.64	3.09
Accelerating regulatory approval	4.62	3.02
Reliability and systems engineering	5.49	4.79
Making products *de facto* standard	3.56	2.71
Anticipating complex client needs	4.95	3.94
Exploration with potential customers and lead users	5.25	4.41
Probing user needs with preliminary designs	4.72	3.59
Using roadmaps of product generations	4.51	3.26
Planned replacement of current products	3.56	2.53
Build coalition with commercialization partners	4.18	3.38
Working with suppliers to create complementary offers	4.32	3.61

(scale 1 (low)–7 (high), only statistically significant differences shown, n = 75 firms)

Source: Derived from S. Floricel and R. Miller (2003) 'An exploratory comparison of the management of innovation in the New and Old economies', *R&D Management*, **33** (5), 501–525.

5.3 Developing Firm-specific Competencies

The ability of firms to track and exploit the technological trajectories described above depends on their specific technological and organizational competencies, and on the difficulties that competitors have in imitating them. The notion of firm-specific competencies has become increasingly influential amongst economists, trying to explain why firms are different, and how they change over time,[20] and also amongst business practitioners and consultants, trying to identify the causes of competitive success. In the 1990s, management began to shift interest from improvements in short-term operational efficiency and flexibility (through 'de-layering', 'downsizing', 'outsourcing' and 'business process re-engineering', etc.), to a concern that – if taken too far – the 'lean corporation' could become the 'anorexic corporation', without any capacities for longer-term change and survival.

Hamel and Prahalad on Competencies

The most influential business analysts promoting and developing the notion of 'core competencies' have been Gary Hamel and C. K. Prahalad.[21] Their basic ideas can be summarized as follows:

1. The sustainable competitive advantage of firms resides not in their products but in their *core competencies*: 'The real sources of advantage are to be found in management's ability to consolidate corporate-wide technologies and production skills into competencies that empower individual businesses to adapt quickly to changing opportunities' (1990, p. 81).[21]
2. Core competencies feed into more than one core products, which in turn feed into more than one business unit. They use the metaphor of the tree:

   End products   = Leaves, flowers and fruit
   Business units   = Smaller branches
   Core products   = Trunk and major limbs
   Core competencies = Root systems

   Examples of core competencies include Sony in miniaturization, Philips in optical media, 3M in coatings and adhesives and Canon in the combination of the precision mechanics, fine optics and microelectronics technologies that underlie all their products (see Box 5.3). Examples of core products include Honda in lightweight, high-compression engines and Matsushita in key components in video cassette recorders.

BOX 5.3
CORE COMPETENCIES AT CANON

Product	Competencies		
	Precision mechanics	Fine optics	Micro-electronics
Basic camera	×	×	
Compact fashion camera	×	×	
Electronic camera	×	×	
EOS autofocus camera	×	×	×
Video still camera	×	×	×
Laser beam printer	×	×	×
Colour video printer	×		×
Bubble jet printer	×		×
Basic fax	×		×
Laser fax	×		×
Calculator			×
Plain paper copier	×	×	×
Colour copier	×	×	×
Laser copier	×	×	×
Colour laser copier	×	×	×
Still video system	×	×	×
Laser imager	×	×	×
Cell analyzer	×	×	×
Mask aligners	×		×
Stepper aligners	×		×
Excimer laser aligners	×	×	×

The above list of competencies and related products in Canon are taken from Prahalad and Hamel.[21] According to Christer Oskarsson:[22]

'In the late 1950s . . . the time had come for Canon to apply its precision mechanical and optical technologies to other areas [than cameras] . . . such as business machines. By 1964 Canon had begun by developing the world's first 10-key fully electronic calculator . . . followed by entry into the coated paper copier market with the development of an electrofax copier model in 1965, and then into . . . the revolutionary Canon plain paper copier technology unveiled in 1968 . . . Following these successes of product diversification, Canon's product lines were built on a foundation of precision optics, precision engineering and electronics . . .

The main factors behind . . . increases in the numbers of products, technologies and markets . . . seem to be the rapid growth of information technology and electronics, technological transitions from analogue to digital technologies, technological fusion of audio and video technologies, and the technological fusion of electronics and physics to optronics' (pp. 24–26).

3. The importance of associated organizational competencies is also recognized: 'Core competence is communication, involvement, and a deep commitment to working across organisational boundaries' (1990, p. 82).[21]

4. Core competencies require focus: 'Few companies are likely to build world leadership in more than five or six fundamental competencies. A company that compiles a list of 20 to 30 capabilities has probably not produced a list of core competencies' (1990, p. 84).[21]

5. As Table 5.4 shows, the notion of core competencies suggests that large and multidivisional firms should be viewed not only as a collection of strategic business units (SBUs), but as bundles of competencies that do not necessarily fit tidily in one business unit.

According to Hamel and Prahalad, the concept of the corporation based on core competencies should not replace the traditional one, but a commitment to it 'will inevitably influence patterns of diversification, skill deployment, resources allocation priorities, and approaches to alliances and outsourcing' (1990, p. 86).[21] More specifically, the conventional multidivisional structure may facilitate efficient innovation within specific product markets, but may limit the scope for learning new competencies: firms with fewer divisional boundaries are associated with a strategy based on capabilities-broadening, whereas firms with many divisional boundaries are associated with a strategy based on capabilities-deepening.[23]

6. The identification and development of a firm's core competencies depend on its *strategic architecture*, defined as:

> . . . a road map of the future that identifies which core competencies to build and their constituent technologies . . . should make resource allocation priorities transparent to

TABLE 5.4 Two views of the corporation: SBUs and core competencies

	Strategic business unit	Core competencies
Basis for competition	Competitiveness of today's products	Inter-firm competition to build competencies
Corporate structure	Portfolio of businesses in related product markets	Portfolio of competencies, core products and business
Status of business unit	Autonomy: SBU 'owns' all resources other than cash	SBU is a potential reservoir of core competencies
Resource allocation	SBUs are unit of analysis	SBUs and competencies are unit of analysis
	Capital allocated to SBUs	Top management allocates capital and talent
Value added of top management	Optimizing returns through trade-offs among SBUs	Enunciating strategic architecture, and building future competencies

the whole organisation . . . Top management must add value by enunciating the strategic architecture that guides the competence acquisition process (1990, p. 89).[21]

Examples given include:

- NEC = convergence of computing and communication technologies
- Vickers, USA = being the best power and motion control company in the world
- Honda = lightweight, high-compression engines
- 3M = coatings and adhesives

Assessment of the Core Competencies Approach

The great strength of the approach proposed by Hamel and Prahalad is that it places the cumulative development of firm-specific technological competencies at the centre of the agenda of corporate strategy. Although they have done so by highlighting practice in contemporary firms, their descriptions reflects what has been happening in successful firms in science-based industries since the beginning of the twentieth century. For example, Gottfried Plumpe has shown that the world's leading company in the exploitation of the revolution in organic chemistry in the 1920s – IG Farben in Germany – had already established numerous 'technical committees' at the corporate level, in order to exploit emerging technological opportunities that cut across divisional boundaries.[24] These enabled the firm to diversify progressively out of dyestuffs into plastics, pharmaceutical and other related chemical products.

Other histories of businesses in chemicals and electrical products tell similar stories.[25] In particular, they show that the competence-based view of the corporation has major implications for the organization of R&D, for methods of resource allocation and for strategy determination, to which we shall return in Chapter 6. In the meantime, their approach does have limitations and leaves at least three key questions unanswered.

Differing potentials for technology-based diversification   It is not clear whether the corporate core competencies in all industries offer a basis for product diversification. Compare the recent historical experience of most large chemical and electronics firms, where product diversification based on technology has been the norm, with that of most steel and textile firms, where technology-related product diversification has proved very difficult (see, for example, the unsuccessful attempts to diversify by the Japanese steel industry in the 1980s).[26]

Multitechnology firms   Recommendations that firms should concentrate resources on a few fundamental (or 'distinctive') world-beating technological competencies is potentially misleading. Large firms are typically active in a wide range of technologies, in only a few of which do they achieve a 'distinctive' world-beating position.[27] In other technological fields, a *background* technological competence is necessary to enable the firm to co-ordinate and benefit from outside linkages, especially with suppliers of components, subsystems, materials and production machinery. In industries with complex products or production processes, a high proportion of a firm's technological competencies is deployed in such background competencies.[28] In addition, firms are constrained to develop competencies in an increasing range of technological fields (e.g. IT, new materials, biotechnology, see section 5.2 above) in order to remain competitive as products become even more 'multitechnological'.

Thus, as is shown in Table 5.5, a firm's innovation strategy will involve more than its distinctive core (or critical) competencies. In-house competencies in background (enabling) technologies are necessary for the effective co-ordination of changes in production and distribution systems, and in supply chains. In industries with complex product systems (like automobiles), background technologies can account for a sizeable proportion of corporate innovative activities. Background technologies can also be the sources of revolutionary and disruptive change. For example, given the major opportunities for improved performance that they offer, all businesses today have no choice but to adopt advances in IT technology, just as all factories in the past had no choice but to convert to electricity as a power source. However, in terms of innovation strategy, it is important to distinguish firms where IT is a core technology and a source of distinctive competitive advantage (e.g. Cisco, the supplier of Internet equipment) from firms where it is a background technology, requiring major changes but available

TABLE 5.5 The strategic function of corporate technological activities

Strategic functions	Definition	Typical examples
Core or critical functions	Central to corporate competitiveness Distinctive and difficult to imitate	Technologies for product design and development Key elements of process technologies
Background or enabling	Broadly available to all competitors, but essential for efficient design, manufacture and delivery of corporate products	Production machinery, instruments, materials, components (software)
Emerging or key	Rapidly developing fields of potential knowledge presenting opportunities or threats, when combined with existing core and background technologies	Materials, biotechnology, ICT-software

to all competitors from specialized suppliers, and therefore unlikely to be a source of distinctive and sustainable competitive advantage (e.g. Tesco, the UK supermarket chain).

In all industries, emerging (key) technologies can end up having pervasive and major impacts on firms' strategies and operations (e.g. software). A good example of how an emerging/key technology can transform a company is provided by the Swedish telecommunications firm Ericsson. Table 5.6 traces the accumulation of technological competencies, with successive generations of mobile cellular phones and telecommunication cables. In both cases, each new generation required competencies in a wider range of technological fields, and very few established competencies were made obsolete. The process of accumulation involved both increasing links with outside sources of knowledge, and greater expenditures on R&D, given greater product complexity. This was certainly not a process of concentration, but of diversification in both technology and product.

For these reasons, the notion of 'core competencies' should perhaps be replaced for technology by the notion of '*distributed* competencies', given that, in large firms, they are distributed:

- Over a large number of *technical fields*
- Over a variety of organizational and physical *locations* within the corporation – in the R&D, production engineering and purchasing departments of the various divisions, and in the corporate laboratory
- Amongst different *strategic objectives* of the corporation, which include not only the establishment of a distinctive advantage in existing businesses (involving both core and background technologies), but also the exploration and establishment of new ones (involving emerging technologies)

Core rigidities   As Dorothy Leonard has pointed out, 'core competencies' can also become 'core rigidities' in the firm, when established competencies become too dominant.[29] In addition to sheer habit, this can happen because established competencies are central to today's products, and because large numbers of top managers may be trained in them. As a consequence, important new competencies may be neglected or underestimated (e.g. the threat to mainframes from mini- and microcomputers by management in mainframe companies). In addition, established innovation strengths may overshoot the target. In Box 5.4, Leonard-Barton gives the fascinating example from the Japanese automobile industry: how the highly successful 'heavyweight' product managers of the 1980s (see Chapter 9) overdid it in the 1990s. Many examples show that, when 'core rigidities' become firmly entrenched, their removal often requires changes in top management.

TABLE 5.6 Technology accumulation in Ericsson's product generations

Product and generation	No. of important technologies				R&D costs (base = 100)	% of technologies acquired externally	Main technological fields (d)	No. of patent classes (e)
	Old (a)	New (b)	Total	Obsolete (c)				
Cellular phones								
1. NMT-450	n.a.	n.a.	5	n.a.	100	12	E	17
2. NMT-900	5	5	10	0	200	28	EPM	25
3. GSM	9	5	14	1	500	29	EPMC	29
Telecoms cables								
1. Coaxial	n.a.	n.a.	5	n.a.	100	30	EPM	14
2. Optical	4	6	10	1	500	47	EPCM	17

n.a. = not applicable.

Notes:
(a) No. of technologies from the previous generation.
(b) No. of new technologies, compared to previous generation.
(c) No. of technologies obsoleted from previous generation.
(d) 'Main' = >15% of total engineering stock. Categories are: E = electrical; P = physics; K = chemistry; M = mechanical; C = computers.
(e) Number of international patent classes (IPC) at four-digit level.
Source: Derived from Granstrand, O. *et al.* (1992) 'External technology acquisition in large multi-technology corporations', *R&D Management*, **22**.

Developing and Sustaining Competencies

The final question about the notion of core competencies is very practical: how can management identify and develop them?

Definition and measurement There is no widely accepted definition or method of measurement of competencies, whether technological or otherwise. One possible measure is the level of *functional performance* in a generic product, component or subsystem: in, for example, performance in the design, development, manufacture and performance of compact, high-performance combustion engines. As a strategic technological *target* for a firm like Honda, this obviously makes sense. But its achievement requires the combination of technological competencies from a wide variety of *fields* of knowledge, the composition of which changes (and increases) over time. Twenty years ago, they included mechanics (statics and dynamics), materials, heat transfer, combustion, fluid flow. Today they also include ceramics, electronics, computer-aided design, simulation techniques and software.

BOX 5.4
HEAVYWEIGHT PRODUCT MANAGERS AND FAT PRODUCT DESIGNS

'Some of the most admired features . . . identified . . . as conveying a competitive advantage [to Japanese automobile companies] were: (1) overlapping problem solving among the engineering and manufacturing functions, leading to shorter model change cycles; (2) small teams with broad task assignments, leading to high development productivity and shorter lead times; and (3) using a "heavyweight" product manager – a competent individual with extensive project influence . . . who led a cohesive team with autonomy over product design decisions. By the early 1990s, many of these features had been emulated . . . by US automobile manufacturers, and the gap between US and Japanese companies in development lead time and productivity had virtually disappeared.

However . . . there was another reason for the loss of the Japanese competitive edge – "fat product designs" . . . an excess in product variety, speed of model change, and unnecessary options . . . "overuse" of the same capability that created competitive advantages in the 1980s has been the source of the new problem in the 1990s. The formerly "lean" Japanese producers such as Toyota had overshot their targets of customer satisfaction and overspecified their products, catering to a long "laundry list" of features and carrying their quest for quality to an extreme that could not be cost-justified when the yen appreciated in 1993 . . . Moreover, the practice of using heavyweight managers to guide important projects led to excessive complexity of parts because these powerful individuals disliked sharing common parts with other car models.' (Leonard-Barton, D. *Wellsprings of Knowledge.* Harvard Business School Press, Boston, 1995, p. 33)

Thus, the functional definition of technological competencies bypasses two central tasks of corporate technology strategy: first, to identify and develop the range of disciplines or fields that must be combined into a functioning technology; second (and perhaps more important) to identify and explore the new competencies that must be added if the functional capability is not to become obsolete. This is why a definition based on the measurement of the combination of competencies in different technological fields is more useful for formulating innovation strategy, and is in fact widely practised in business.[30]

Richard Hall goes some way towards identifying and measuring core competencies.[31] He distinguishes between intangible assets and intangible competencies. Assets include intellectual property rights and reputation. Competencies include the skills and know-how of employees, suppliers and distributors, and the collective attributes which constitute organizational culture. His empirical work, based on a survey and case studies, indicates that managers believe that the most significant of these intangible resources are company reputation and employee know-how, both of which may be a function of organizational culture. Thus organizational culture, defined as the shared values and beliefs of members of an organizational unit, and the associated artefacts, becomes central to organizational learning.

Sidney Winter links the idea of competencies with his own notion of organizational 'routines', in an effort to better contrast capabilities from other generic formulas for sustainable competitive advantage or managing change.[32] A routine is an organizational behaviour that is highly patterned, is learned, derived in part from tacit knowledge and with specific goals, and is repetitious. He argues that an organizational capability is a high-level routine, or collection of routines, and distinguishes between 'zero-level' capabilities as the 'how we earn our living now', and true 'dynamic' capabilities which change the product, process, scale or markets; for example, new product development. These dynamic capabilities are not the only or even the most common way organizations can change. He uses the term '*ad hoc* problem solving' to describe these other ways to manage change. In contrast, dynamic capabilities typically involve long-term commitments to specialized resources, and consist of patterned activity to relatively specific objectives. Therefore dynamic capabilities involve both the exploitation of existing competencies and the development of new ones. For example, leveraging existing competencies through new product development can consist of de-linking existing technological or commercial competencies from one set of current products, and linking them in a different way to create new products. However, new product development can also help to develop new competencies. For example, an existing technological competence may demand new commercial competencies to reach a new market, or conversely a new technological competence might be necessary to service an existing customer.[33]

The trick is to get the right balance between exploitation of existing competencies and the exploitation and development of new competencies. Research suggests that over time some firms are more successful at this than others, and that a significant reason for this variation in performance is due to difference in the ability of managers to build, integrate and reconfigure organizational competencies and resources.[34] These 'dynamic' managerial capabilities are influenced by managerial cognition, human capital and social capital. Cognition refers to the beliefs and mental models which influence decision-making. These affect the knowledge and assumptions about future events, available alternatives and association between cause and effect. This will restrict a manager's field of vision, and influence perceptions and interpretations. Box 5.5

BOX 5.5
CAPABILITIES AND COGNITION AT POLAROID

Polaroid was a pioneer in the development of instant photography. It developed the first instant camera in 1948, first instant colour camera in 1963, and introduced sonar automatic focusing in 1978. In addition to its competencies in silver halide chemistry, it had technological competencies in optics and electronics, and mass manufacturing, marketing and distribution expertise. The company was technology-driven from its foundation in 1937, and the founder Edwin Land had 500 personal patents. When Kodak entered the instant photography market in 1976, Polaroid sued the company for patent infringement, and was awarded $924.5 m. in damages. Polaroid consistently and successfully pursued a strategy of introducing new cameras, but made almost all its profits from the sale of the film (the so-called razor-blade marketing strategy also used by Gillette), and between 1948 and 1978 the average annual sales growth was 23%, and profit growth 17% per year.

Polaroid established an electronic imaging group as early as 1981, as it recognized the potential of the technology. However, digital technology was perceived as a potential technological shift, rather than as a market or business disruption. By 1986 the group had an annual research budget of $10 m, and by 1989 42% of the R&D budget was devoted to digital imaging technologies. By 1990 28% of the firm's patents related to digital technologies. Polaroid was therefore well positioned at that time to develop a digital camera business. However, it failed to translate prototypes into a commercial digital camera until 1996, by which time there were 40 other companies in the market, including many strong Japanese camera and electronics firms. Part of the problem was adapting the product development and marketing channels to the new product needs. However, other more fundamental problems related to long-held cognitions: a continued commitment to the razor-blade business model, and pursuit of image quality. Profits from the new market for digital cameras were derived from the cameras rather than the consumables (film). Ironically, Polaroid had rejected the development of ink-jet printers, which rely on consumables for profits, because of the relatively low quality of their (early) outputs. Polaroid had a long tradition of improving its print quality to compete with conventional 35 mm film.

Source: Tripsas, M. and G. Gavetti (2000) 'Capabilities, cognition, and inertia: evidence from digital imaging', *Strategic Management Journal*, **21**, 1147–1161.

> **BOX 5.6**
> **ON VISIONS**
>
> In recent years the notion of 'visions' has become popular to describe informal expectations of the future, as distinct from the formally derived forecasting of conventional strategic planning. Jonathan Sapsed's study of vision statements and strategic change argues that visions are essentially tactical devices: internally they serve to announce new directions and close debate, whilst externally they allure investors or signal to competitors.
>
> Examples are the media wars over video-on-demand in the late 1990s. BT (British Telecom) announced plans for full-motion-video to be transmitted through the existing copperwire telephone network, using the then unproven technology ADSL (asymmetric digital subscriber line). This was heavily publicized as BT conducted a high-profile technical and market trial of the interactive service. The vision of consumers dialling up premium movies for instant transmission on their telephone handsets was widely interpreted as a spoiler to investment in the nascent UK cable television industry. It had the effect of lowering market valuations of the struggling cable franchisees. They, in response, revealed their alternative visions and trials of a rival cable-based technical solution, ATM (asynchronous transfer mode). The press quoted both camps expressing scepticism about the viability of the other's technology.
>
> In fact, neither service was eventually introduced to the market, although the technological learning of the trials was redeployed on new applications and product markets. Sapsed shows that visions and strategies are ephemeral and much more volatile than the underlying technological paths. Announcements of visions serve various purposes, but should not be taken to signify genuine strategic intention. They can, however, be extremely important, particularly for start-up firms in new, uncertain technologies who often have nothing more tangible to sell to prospective investors. The stock market valuations of high-tech start-ups are often based on the 'buying' of visions.
>
> Source: Sapsed, J. (2005) *Restricted Vision: Strategizing under uncertainty*. Imperial College Press, London.

discusses the role of (limited) cognition in the case of Polaroid and digital imaging. Human capital refers to the learned skills that require some investment in education, training experience and socialization, and these can be generic, industry- or firm-specific. It is the firm-specific factors that appear to be the most significant in dynamic

BOX 5.7

THE OVERVALUATION OF TECHNOLOGICAL WONDERS

In 1986, Schnaars and Berenson published an assessment of the accuracy of forecasts of future growth markets since the 1960s, with the benefit of 20 or more years of hindsight.[36] The list of failures is as long as the list of successes. Below are some of the failures.

'The 1960s were a time of great economic prosperity and technological advancement in the United States . . .

One of the most extensive and widely publicized studies of future growth markets was TRW Inc.'s "Probe of the Future". The results . . . appeared in many business publications in the late 1960s . . . Not all . . . were released. Of the ones that were released, nearly all were wrong! Nuclear-powered underwater recreation centres, a 500-kilowatt nuclear power plant on the moon, 3-D colour TV, robot soldiers, automatic vehicle control on the interstate system, and plastic germproof houses were amongst some of the growth markets identified by this study.

. . . In 1966, industry experts predicted, "The shipping industry appears ready to enter the jet age." By 1968, large cargo ships powered by gas turbine engines were expected to penetrate the commercial market. The benefits of this innovation were greater reliability, quicker engine starts and shorter docking times.

. . . Even dentistry foresaw technological wonders . . . in 1968, the Director of the National Institute of Dental Research, a division of the US Public Health Service, predicted that "in the next decade, both tooth decay and the most prevalent form of gum disease will come to a virtual end". According to experts at this agency, by the late 1970s false teeth and dentures would be "anachronisms" replaced by plastic teeth implant technology. A vaccine against tooth decay would also be widely available and there would be little need for dental drilling.'

managerial capability, which can lead to different decisions when faced with the same environment. Social capital refers to the internal and external relationships which effect a managers' access to information, their influence, control and power.

Top management and 'strategic architecture' for the future   The importance given by Hamel and Prahalad to top management in determining the 'strategic architecture' for the development of future technological competencies is debatable. As *The Economist* has argued:[35]

It is hardly surprising that companies which predict the future accurately make more money than those who do not. In fact, what firms want to know is what Mr. Hamel and

---

BOX 5.8

LEARNING ABOUT OPTOELECTRONICS IN JAPANESE COMPANIES

Using a mixture of bibliometric and interview data, Kumiko Miyazaki traced the development and exploitation of opto-electronics technologies in Japanese firms. Her main conclusions were as follows:

'. . . Competence building is strongly related to a firm's past accomplishments. The notions of path dependency and cumulativeness have a strong foundation. Competence building centres in key areas to enhance a firm's core capabilities.

. . . by examining the different types of papers related to semiconductor lasers over a 13-year period, it was found that in most firms there was a decrease in experimental type papers accompanied by a rise in papers marking "new developments" or "practical applications". The existence of a wedge pattern for most firms confirmed . . . that competence building is a cumulative and long process resulting from trial and error and experimentation, which may eventually lead to fruitful outcomes. The notion of search trajectories was tested using . . . INSPEC and patent data. Firms search over a broad range in basic and applied research and a narrower range in technology development . . . In other words, in the early phases of competence building, firms explore a broad range of technical possibilities, since they are not sure how the technology might be useful for them. As they gradually learn and accumulate their knowledge bases, firms are able to narrow the search process to find fruitful applications.'

Source: Miyazaki, K. (1994) 'Search, learning and accumulation of technological competencies: the case of optoelectronics', *Industrial and Corporate Change*, **3**, 653.

Mr. Prahalad steadfastly fail to tell them: how to guess correctly. As if to compound their worries, the authors are oddly reticent about those who have gambled and lost.

The evidence in fact suggests that the successful development and exploitation of core competencies does not depend on management's ability to forecast accurately long-term technological and product developments: as Boxes 5.6 and 5.7 illustrate, the record here is not at all impressive.[33] Instead, the importance of new technological opportunities and their commercial potential emerge not through a flash of genius (or a throw of the dice) from senior management, but gradually through an incremental corporate-wide process of learning in knowledge-building and strategic positioning. New core competencies cannot be identified immediately and without trial and error: thus

BOX 5.9
MARKET VISIONS AND TECHNOLOGICAL INNOVATION AT CORNING

Corning has a long tradition of developing radical technologies to help to create emerging markets. It was one of the first companies in the USA to establish a corporate research laboratory in 1908. The facility was originally set up to help to solve some fundamental process problems in the manufacture of glass, and resulted in improved glass for railroad lanterns. This led to the development of Pyrex in 1912, which was Corning's version of the German-invented borosilicate glass. This led to new markets in medical supplies and consumer products.

In the 1940s the company began to develop television tubes for the emerging market for colour televisions sets, drawing upon its technology competencies developed for radar during the war. Corning did not have a strong position in black-and-white television tubes, but the tubes for colour television followed a different and more challenging technological trajectory, demanding a deep understanding of the fundamental phenomena to achieve the alignment of millions of photorescent dots to a similar pattern of holes.

In 1966, in response from a joint enquiry from the British Post Office and British Ministry of Defence, Corning supplied a sample of high-quality glass rods to determine the performance in transmitting light. Based on the current performance of copper wire, a maximum loss of 20 db/km was the goal. However, at that time the loss of the optical fibre (waveguide) was ten times this: 200 db/km. The target was theoretically possible given the properties of silica, and Corning began research on optical fibre. Corning pursued a different approach to others, using pure silica which demanded very high temperatures making it difficult to work with. The company had developed this tacit knowledge in earlier projects, and this would take time for others to acquire. In 1970 the research group developed a composition and fibre design that exceeded the target performance. Excluded from the USA market by an agreement with AT&T, Corning formed a five-year joint development agreement with five companies from the UK, Germany, France, Italy and Japan. Corning subsequently developed key technologies for waveguides, filed the 12 key patents in the field, and after a number of high-profile but successful patent infringement actions against European, Japanese and Canadian firms, it came to dominate what would become $10m annual sales by 1982.

Corning also had close relationships with the main automobile manufacturers as a supplier of headlights, but it had failed to convince these companies to adopt its safety glass for windscreens (windshields) due to the high cost and low importance of safety at that time. Corning had also developed a ceramic heat exchanger for petrol (gasoline) turbine engines, but the automobile manufactures were not willing to reverse their huge investments for the production of internal combustion engine. However, discussion with GM, Ford and Chrysler indicated that future legislation would demand reduced vehicle emissions, and therefore some form of catalytic converter would become standard for all cars in the USA. However, no one knew how to make these at that time. The passing of the Clean Air Act in 1970 required reductions in emissions by 1975, and accelerated development. Competitors included 3M and GM. However, Corning had the advantage of having already developed the new ceramic for its (failed) heat exchanger project, and its competencies in R&D organization and production processes. Unlike its competitors which organized development along divisional lines, Corning was able to apply as many researchers as it had to tackle the project, what became known as 'flexible critical mass'. In 1974 it filed a patent for its new extrusion production technology, and in 1975 for a new development of its ceramic material. The competitors' technologies proved unable to match the increasing reduction in emissions needed, and by 1994 catalytic converters generated annual sales of $1bn for Corning.

Source: Graham, M. and A. Shuldiner (2001) *Corning and the Craft of Innovation*. Oxford University Press, Oxford.

Box 5.3 above neglects to show that Canon failed in electronic calculators and in recording products.[38]

It was through a long process of trial and error that Ericsson's new competence in mobile telephones first emerged.[39] As Box 5.8 shows, it is also how Japanese firms developed and exploited their competencies in optoelectronics.

A study of radical technological innovations found how visions can influence the development or acquisition of competencies, and identified three related mechanisms through which firms link emerging technologies to markets that do not yet exist: motivation, insight and elaboration.[40] Motivation serves to focus attention and to direct energy, and encourages the concentration of resources. It requires senior management to communicate the importance of radical innovation, and to establish and enforce challenging goals to influence the *direction* of innovative efforts. Insight represents the critical connection between technology and potential application. For radical technological innovations, such insight is rarely from the marketing function, customers or competitors, but is driven by those with extensive technical knowledge and expertise with a sense of both market needs and opportunities. Elaboration involves the demonstration of technical feasibility, validating the idea within the organization, prototyping and the building and testing of different business models. At this point the concept is sufficiently well elaborated to work with the marketing function and potential customers. Market visioning for radical technologies not necessarily the result of individual or technological leadership. 'There were multiple ways for a vision to take hold of an organization . . . our expectation was that a single individual would create a vision of the future and drive it across the organization. But just as we discovered that breakthrough innovations don't necessarily arise simply because of a critical scientific discovery, neither do we find that visions are necessarily born of singular prophetic individuals' (ref. 40, pp. 239–244). Box 5.9 illustrates how Corning developed its ceramic competencies to develop products for the emerging demand for catalytic converters in the car industry, and for glass fibre for telecommunications.

5.4 Technological Paths in Small Firms

Unlike large firms, small firms tend to be specialized rather than diversified in their technological competencies and product range. However, as with large firms, it is impossible to make robust generalizations about their technological trajectories and innovation strategies. Kurt Hoffman and his colleagues have recently pointed out that relatively little research has been undertaken on innovation in small firms; what research has been done tends to concentrate on the small group of spectacular high-tech successes (or failures) rather than the much more numerous run-of-the-mill small firms coping (say) with the introduction of IT into their distribution systems.[41]

Table 5.7 tries to categorize these differences. Until recently, attention has been focused on the left-hand side of the table – the spectacular and visible successes amongst small innovating firms – in particular, the 'superstars' that became big, and those of the new technology-based firms (NTBFs) that often want to become big. As we have seen earlier in this chapter and in Chapter 4, recent more systematic surveys of innovative activities and of small firms show two other classes of small firm with less spectacular innovation strategies, but of far greater importance to the overall economy: specialized suppliers of production inputs, and firms whose sources of innovation are mainly their suppliers.

Superstars are large firms that have emerged from small beginnings, through high rates of growth based on the exploitation of a major invention (e.g. instant photography, reprography), or a rich technological trajectory (e.g. semiconductors, software), enabling small firms to exploit first-mover advantages like patent protection and

TABLE 5.7 Categories of innovating small firms

	Superstars: small firms into big since 1950	New technology-based firms (NTBFs)	Specialized suppliers	Supplier-dominated
Examples	Polaroid, DEC, TI, Xerox, Intel, Microsoft, Compaq, Sony, Casio, Benetton	Start-ups in electronics, biotechnology and software	Producer goods (machines, components, instruments, software)	Traditional products (e.g. textiles, wood products, food products) and many services
Sources of competitive advantage	Successful exploitation of major invention or technological trajectory	1. Product or process development in fast-moving and specialized area 2. Privatizing academic research	Combining technologies to meet users' needs	Integration and adaptation of innovations by suppliers
Main tasks of innovation strategy	Preparing replacements for the original invention (or inventor)	1. 'Superstar' or 'specialized supplier'? 2. Knowledge or money?	Links to advanced users and pervasive technologies	Exploiting new IT-based opportunities in design, distribution and co-ordination

learning curves (see Chapter 3). Successful innovators often either accumulated their technological knowledge in large firms before leaving to start their own, or they offered their invention to large firms but were refused (examples: Polaroid, Xerox). Few superstars have emerged either in the chemical industry over the past 50 years, or – contrary to expectations – out of biotechnology firms over the past 15 years, probably because the barriers to entry (in R&D, production, or marketing) remain high.

The examples in Table 5.7 show that many superstars are from the USA, although we can find European and Japanese examples. Experience suggests that one of the main challenges facing the management of superstars is their transition from the original innovator and the original innovation to new management and a new line of products. Beyond the period of spectacular growth, the characteristics behind the original success can become sources of 'core rigidities' (see section 5.3). Successful innovators are often strong characters who do not necessarily encourage diversity in ideas and approaches within the firm. Successful innovations are often well protected by patents and other first-comer advantages, which can blunt the drive for improvement and change. These difficulties have beset companies like DEC, Polaroid and Xerox. One of the most successful in maintaining its innovative performance has been Sony. An interesting exercise is to speculate about the future of today's superstars: what will happen to Microsoft after Bill Gates?

New technology-based firms (NTBFs) are small firms that have emerged recently from large firms and large laboratories in such fields as electronics, software and biotechnology. They are usually specialized in the supply of a key component, subsystem, service or technique to larger firms, who may often be their former employers. Contrary to a widespread belief, most of the NTBFs in electronics and software have emerged from corporate or government laboratories involved in development and testing activities. It is only with the advent of biotechnology (and, more recently software), that university laboratories have become regular sources of NTBFs, thereby strengthening the strong direct links that have always existed between university-based research and the pharmaceutical industry. However, some observers criticize this trend, and fear the 'privatization' of university research in biotechnology will in the long term reduce the rate of scientific progress and innovation and their contribution to economic and social welfare.

The management of NTBFs faces two sets of strategic problems:

1. The first relates to long-term prospects for growth. Very few technology-based small firms can become superstars, since they provide mainly specialized 'niche' products with no obvious or spectacular synergies with other markets. How far the firm will grow, or how long it will survive, will often depend on its ability to negotiate the

transition from the first to the second (improved) generation of products, and to develop the supporting managerial competencies.

2. How far the NTBF will grow depends on the second strategic choice: whether the management is aiming to maximize long-term value of the business, or merely seeking an increase in income and independence. Thus, owners of small firms often sell their firms after a few years and live off their investments. And university researchers set up consultancy firms, either to increase their personal income (the BMW effect), or to find supplementary income for their university-based research and teaching activities in times of increasing financial stringency.

These will be discussed in more detail in Chapter 12.

Specialized supplier firms have already been described earlier in this chapter. They design, develop and build specialized inputs into production, in the form of machinery, instruments and (increasingly) software, and interact closely with their (often large) technically progressive customers. They perform relatively little formal R&D, but are nonetheless a major source of the active development of significant innovations, with major contributions being made by design and production staff.[42]

Finally, most small firms fall into the *supplier-dominated* category, with their suppliers of production inputs as their main sources of new technology. As we saw in Chapter 3, these firms depend heavily on their suppliers for their innovations, and therefore are often unable to appropriate firm-specific technology as a source of competitive advantage. Technology will become more important in future, with the growing range of potential IT applications offered by suppliers, especially in service activities like distribution and co-ordination. An increasing range of small firms will therefore need to obtain the technological competencies to be able to specify, purchase, install and maintain software systems that help increase their competitiveness. Whether these competencies will become distinctive, *core* competencies is less clear, given that they can be adopted by all small firms. Distinctive advantage will emerge only where the software competencies are difficult to imitate, namely in developing and operating complex systems. Amongst small firms, such competencies are less likely to emerge in those *using* software, than amongst those *supplying* software services.

5.5 Summary and Further Reading

In this chapter we have shown how firms are inevitably constrained in their choice of innovation strategies by their accumulated skills, and by the opportunities that they are capable of exploiting. In other words, they are on technological trajectories. We identified five broad technological trajectories, each of which has distinct sources and directions of technical change, and which defines key tasks for innovation strategy. We also identified key technologies that are expected to create major technological opportunities in the future. In particular, we stress the importance of distinguishing between the nature and strategic implications of the microelectronics revolution (i.e. the costly business of making electronic chips) from the IT revolution (i.e. the pervasive possibilities of applying software technology to the corporate functions of R&D, logistics, distribution, transactions and co-ordination, as well as production).

We reviewed the concepts developed by Prahalad and Hamel in strategic management, which are (at last!) giving technological competencies a centrally important role. They show the potential tensions between corporate structures built around into strategic business units, and the requirements to reconfigure organizations and associated competencies to exploit emerging technological opportunities. However, the emphasis on *distinctive* core competencies neglects the wider function of technological competencies in establishing linkages with the outside world, and in exploring future opportunities. Similarly, the emphasis on *strategic architecture* (or *vision*) neglects the inevitable uncertainties in the development of future competencies, and the consequent need for experiment and experience.

Given these uncertainties, management concepts and tools for identifying and developing technological competencies should be kept simple, robust and capable of constant revision of assumptions and inputs in the light of new evidence or insights. Probably the most useful remains the technology–product matrix, where existing competencies can be mapped against products, and technical judgment can explore systematically (see Box 5.10).

In addition the matrix in Box 5.10 can be used to explore the:

- Likely effects of emerging new technical fields on the firm's existing competencies and products.
- Possibilities of product diversification based on existing competencies.
- Additional competencies that need to be acquired to allow product diversification.

There has been a plethora of papers and books on competencies since Prahalad and Hamel published their original article in the *Harvard Business Review* in 1990, ranging from statistical analysis of patents and other technology indicators (see e.g. N. Argyres, 1996, 'Capabilities, technological diversification and divisionalization', *Strategic Management Journal*, **17**, 395–410), through to conceptual papers based on case studies in specific sectors (see e.g. R. Henderson and I. Cockburn, 1995, 'Measuring competence? Exploring firm effects in pharmaceutical research', *Strategic Management Journal*, **15**, Special Issue, 63–84). Despite this, most studies have not been sufficiently critical or robust to provide theoretical or practical insights for managers. The best introduction

BOX 5.10

THE TECHNOLOGY–PRODUCT MATRIX

	Strategic technical area (STA)		
	Product line 1	Product line 2	Product line 3
Integrated circuit fabrication	×		
Integrated circuit design	×	×	×
System architecture		×	
Software engineering	×	×	×

When he was Director of Planning for GTE Laboratories Incorporated, Graham Mitchell wrote the following:

'. . . the issue which is . . . most frustrating and difficult to resolve concerns the degree to which technical programs should be targeted merely to the support of the existing strategies of the business, as opposed to providing opportunities for significant and sometimes revolutionary change. . . .

. . . For business management, it requires acceptance of an enlarged role for the technologist in the formulation of business goals and strategy. For technical management, it implies that the traditional emphasis on the *management of projects* be extended to include a greater emphasis on the strategically more important issue of the *management of technology* . . . The underlying situation which gives particular value to the STA (strategic technical area) is that, in many large industrial corporations, similar technical skills and expertise are widely applicable in different parts of the organisation. . . .

The STA/product line, service or business matrix . . . [see above] provides the framework needed to integrate the conflicting approaches to technical strategy.

A review of the distribution of technical effort *vertically* – that is, by product line or organisation – deals directly with the general management question: "Is the technical strategy viable, and to what extent is the distribution of technical resources consistent with the stated strategy of the business?"

A *horizontal* view through the matrix begins to answer the technical managers' question: "Where are the greatest technical opportunities for the corporation?" The horizontal view defines those technical areas which are most important to the corporation in aggregate and, at the same time, identifies the specific present and future business lines which will be the most affected by advances in the underlying technology.'

Source: Mitchell, G. (1986) 'New approaches for the strategic management of technology', in Horwich, M. (ed.), *Technology and the Modern Corporation: A strategic perspective*, Pergamon, New York, 134.

is still *Competing for the Future* by Prahalad and Hamel (Harvard Business School Press, 1994), followed by *Competence-Based Competition* edited by Gary Hamel (John Wiley & Sons Ltd/Inc., Chichester, 1994). Dorothy Leonard-Barton provides a more critical assessment in her book *Wellsprings of Knowledge* (Harvard Business School Press, 1995), and a range of more practical tools and techniques are provided by *Measuring Strategic Competencies: Technological, market and organizational indicators of innovation*, edited by Joe Tidd (Imperial College Press, London, 2000). A special issue of the *Strategic Management Journal*, 2003 (volume 24), is devoted to a discussion of the resource-based view of the firm.

The implications of radical new technologies (and especially IT) for corporate strategy are immense, ever-changing and difficult to define with certainty. Good papers on the subject are emerging in management journals, but some of the best descriptive – and analytical – material can be found in business journals like *The Economist*.

On *retailing*, see *The Economist*, 26 Feb., 2001: 42; C.H. Cristensen and R.S. Tedlow, 'Patterns of disruption in retailing', *Harvard Business Review*, Jan.–Feb. 2000, 42–45; N.C. Kim and R. Hamborgne, 'Creating new market space', *Harvard Business Review*, Jan.–Feb. 1999, 83–93; M. Harvey, 'Innovation and competition in UK supermarkets', CRIC Briefing Paper No. 3, June 1999; http://les1.man.ac.uk/cric/Pdfs/BP3.pdf

On *bookselling*, see *The Economist*, 20 November 1999, 108–111; 22 August 1998, 24–25; 10 May 1997, 'Survey of electronic commerce'. On the *music industry*, see *The Economist*, 8 May 1999, 91–92.

Intermediate users: *pharmaceuticals, biotech and new materials*, see *The Economist*, 26 June 1999, 17–29; 21 Feb. 1998, 14–16; 14 Mar. 1998, 123; 3 Jan. 1998, 78–79; 20 Apr. 1996, 96; G.P. Pisano, 'Learning before doing in the development of new process technology', *Research Policy*, **25**, 1996, 1097–1119.

The suppliers, see *The Economist*, 8 Apr. 2000; 6 June 1998, 97–100; 28 Mar. 1998, 98; 12 July 1997, 71–72; 29 Mar. 1997, 14–20.

E-commerce: see N. Venkatraman, 'Five steps to a dot-com strategy: how to find your footing on the Web', *Strategic Management Review*, Spring 2000, 15–28; *The Economist*, 26 Feb. 2000; 6 Nov. 1999, 97–98.

References

1 Teece, D. and G. Pisano (1994) 'The dynamic capabilities of firms: an introduction', *Industrial and Corporate Change*, **3**, 537–556.
2 Prof. J. Meisel, cited in Fluss, J. (1996) 'The electric vehicle: development and potential', MBA dissertation, The Management School, Imperial College, London.
3 Fulkerson, W., R. Judkins and M. Sanghvi (1990) 'Energy from fossil fuels', *Scientific American*, September, 83–89.

4 Patel, P. and K. Pavitt (2000) 'How technological competencies help define the core (not the boundaries) of the Firm', in Dosi, G., Nelson, R. and Winter, S. (eds), *The Nature and Dynamics of Organizational Capabilities*, Oxford University Press, Oxford, 313–333.

5 Mowery, D. and N. Rosenberg (1989) *Technology and the Pursuit of Economic Growth*. Cambridge University Press, Cambridge.

6 Patel, P. and K. Pavitt (1998) 'The wide (and increasing) spread of technological competencies in the world's largest firms: a challenge to conventional wisdom', in Chandler, A., Hagstrom, P. and Solvell, O. (eds), *The Dynamic Firm*. Oxford University Press, Oxford.

7 Nelson, R. and S. Winter (1977) 'In search of a useful theory of innovation', *Research Policy*, **6**, 36–76.

8 Dosi, G. (1982) 'Technological paradigms and technological trajectories', *Research Policy*, **11**, 147–162.

9 Pavitt, K. (1984) 'Sectoral patterns of technical change: towards a taxonomy and a theory', *Research Policy*, **13**, 343–373; Pavitt, K. (1990) 'What we know about the strategic management of technology', *California Management Review*, **32**, 17–26.

10 Freeman, C., J. Clark and L. Soete (1982) *Unemployment and Technical Innovation: A study of long waves and economic development*. Frances Pinter, London.

11 Mowery, D. and N. Rosenberg (1989) *Technology and the Pursuit of Economic Growth*. Cambridge University Press, Cambridge.

12 Arundel, A., G. van de Paal and L. Soete (1995) *Innovation Strategies of Europe's Largest Industrial Firms*. PACE Report, MERIT, University of Limbourg, Maastricht; Cesaretto, S. and S. Mangano (1992) 'Technological profiles and economic performance in the Italian manufacturing sector, *Economics of Innovation and New Technology*, **2**, 237–256.

13 Coombs, R. and A. Richards (1991) 'Technologies, products and firms' strategies', *Technology Analysis and Strategic Management*, **3**, 77–86, 157–175.

14 Patel, P. and K. Pavitt (1998) 'The wide (and increasing) spread of technological competencies in the world's largest firms: a challenge to conventional wisdom', in Chandler, A., Hagstrom, P. and Solvell, O. (eds), *The Dynamic Firm*. Oxford University Press, Oxford.

15 Nightingale, P. (2000) 'Economies of scale in experimentation: knowledge and technology in pharmaceutical R&D', *Industrial and Corporate Change*, **9**, 315–359.

16 Belussi, F. (1989) 'Benetton – a case study of corporate strategy for innovation in traditional sectors', in Dodgson, M. (ed.), *Technology Strategy and the Firm: Management and public policy*. Longman, London; Barras, R. (1990) 'Interactive innovation in financial and business services: the vanguard of the service revolution', *Research Policy*, **19**, 215–238.

17 Schneider, A. (1986) 'High tech comes to the tulip bed', *Business Week* (Industrial-Technology Edition) Issue 2946, May 12, 72D, 72H.

18 Kennedy, A. (1993) 'Up and comers: all wrapped up in a great gift idea', *BRW* (International Edition), **3** (5), 32–33.

19 Floricel, S. and R. Miller, (2003) 'An exploratory comparison of the management of innovation in the New and Old economies', *R&D Management*, **33** (5), 501–525.

20 Dosi, G., D. Teece and S. Winter (1992) 'Towards a theory of corporate coherence: preliminary remarks', in Dosi, G., Giannetti, R. and Toninelli, P. (eds), *Technology and Enterprise in a Historical Perspective*. Clarendon Press, Oxford.

21 Prahalad, C. and G. Hamel (1990) 'The core competencies of the corporation', *Harvard Business Review*, May–June, 79–91; Prahalad, C. and Hamel, G. (1994) *Competing for the Future*. Harvard Business School Press, Cambridge, Mass.

22 Oskarsson, C. (1993) *Technology Diversification: The phenomenon, its causes and effects*. Department of Industrial Management and Economics, Chalmers University, Gothenburg.

23 Argyres, N. (1996) 'Capabilities, technological diversification and divisionalization', *Strategic Management Journal*, **17**, 395–410.

24 Plumpe, G. (1995) 'Innovation and the structure of IG Farben', in Caron, F., Erker, P. and Fischer, W. (eds), *Innovations in the European Economy between the Wars.* De Gruyter, Berlin.

25 Graham, M. (1986) *RCA and the Videodisc: The business of research.* Cambridge University Press, Cambridge; Hounshell, D. and J. Smith (1988) *Science and Corporate Strategy: Du Pont R&D, 1902–1980.* Cambridge University Press, New York; Reader, W. (1975) *Imperial Chemical Industries, a History.* Oxford University Press, Oxford; Reich, L. (1985) *The Making of American Industrial Research: Science and business at GE and Bell.* Cambridge University, Cambridge. For a discussion of the implications for innovation strategy of these and related studies, see Pavitt, K. and W. Steinmueller (2001) 'Technology in corporate strategy: change, continuity and the information revolution', in Pettigrew, A., Thomas, H. and Whittington, R. (eds), *Handbook of Strategy and Management*, Sage.

26 *The Economist* (1989) 'Japan's smokestack fire-sale', 19 August, 63–64.

27 Granstrand, O., P. Patel and K. Pavitt (1997) 'Multi-technology corporations: why they have "distributed" rather than "distinctive core" competencies', *California Management Review*, **39**, 8–25; Patel, P. and K. Pavitt (1998) 'The wide (and increasing) spread of technological competencies in the world's largest firms: a challenge to conventional wisdom', in Chandler, A., Hagstrom, P. and Solvell, O. (eds), *The Dynamic Firm.* Oxford University Press, Oxford.

28 Prencipe, A. (1997) 'Technological competencies and product's evolutionary dynamics: a case study from the aero-engine industry,' *Research Policy*, **25**, 1261.

29 Leonard-Barton, D. (1995) *Wellsprings of Knowledge.* Harvard Business School Press, Boston.

30 Capon, N. and R. Glazer (1987) 'Marketing and technology: a strategic coalignment', *Journal of Marketing*, **51**, 1–14.

31 Hall, R. (1994) 'A framework for identifying the intangible sources of sustainable competitive advantage', in Hamel, G. and Heene, A. (eds), *Competence-Based Competition.* John Wiley & Sons, Ltd, Chichester, pp. 149–169.

32 Winter, S.G. (2003) 'Understanding dynamic capabilities', *Strategic Management Journal*, **24**, 991–995

33 Danneels, E. (2002) 'The dynamic effects of product innovation and firm competencies', *Strategic Management Journal*, **23**, 1095–1121.

34 Adner, R. and Helfat, C. (2003) 'Corporate effects and dynamic managerial capabilities', *Strategic Management Journal*, 1011–1025.

35 *The Economist* (1994) 'The vision thing', 3 September, 77.

36 Schnaars, S. and C. Berenson (1986) 'Growth market forecasting revisited: a look back at a look forward', *California Management Review*, **28**, 71–88.

37 For more detail, see Schnaars, S. (1989) *Megamistakes: Forecasting and the myth of rapid technological change.* Free Press, New York.

38 Sandoz, P. (1997) *Canon.* Penguin, London.

39 Granstrand, O. *et al.* (1992) 'External technology acquisition in large multi-technology corporations', *R & D Management*, **22** (2), 111–133.

40 O'Connor, G. and R. Veryzer, (2001) 'The nature of market visioning for the technology-based radical innovation', *Journal of Product Innovation Management*, **18**, 231–246.

41 Hoffman, K., M. Parejo and J. Bessant (1998) 'Small firms, R&D, technology and innovation in the UK: a literature review', *Technovation*, **18**, 39–56.

42 Pavitt, K., M. Robson and J. Townsend (1989) 'Technological accumulation, diversification and organisation in UK companies, 1945–83', *Management Science*, **35**, 81–99.

Processes: Integration for Strategic Learning

Amongst the important dynamic capabilities of the firm, the most critical are the processes that ensure effective *integration* and *learning*.[1] Integration has long been recognized as a major task of management, especially in R&D departments with scientists and engineers from specialized disciplines[2] and in large firms with specialized functions and divisions.[3] And as we have seen many times already, continuous learning is central to the survival and success of firms operating in changing and complex environments.

The trick in innovation strategy is to be able to do both together. Effective learning in innovation requires strong feedback between decisions and their implementation (in other words, between analysis and action), and this often requires the effective integration of information and knowledge across functional and divisional boundaries. In the literature on innovation management, probably the best known and the most studied of these interfaces is between the R&D and marketing functions.[4] This and others closely related to the implementation of innovations will be discussed in Chapter 9.

Here we shall concentrate on three areas where integration and learning are essential for the success of innovation strategy: the location of R&D (and other technological activities) within the corporation; the role of the R&D and related functions in determining the allocation of corporate financial resources; and the links between innovation strategy and corporate strategy. Throughout this chapter we will use the term R&D, but in many organizations there is no such group or function. For example, in the service sector the locus of such activities is often in groups called 'business development' or 'technology', or in the marketing function. Similarly, in smaller firms such activities often occur in 'design' or 'technical support'. Whatever the term used, the issues are similar. We shall emphasize the nature and the importance of the tasks to be addressed, rather than the specific procedures designed to implement them. At the very least, these will probably require the exchange of information across organizational boundaries within the firm. Given the importance of transferring accumulated tacit knowledge, they will be all the more effective if they involve the mobility of practitioners across organizational boundaries. This is consistent with more recent notions of 'open innovation', rather than 'closed innovation' which relies on internal development. For a study of Japanese practice in this respect, see Kenney and Florida.[5]

6.1 Locating R&D Activities – Corporate versus Divisional

Building organizations that are responsive to change, i.e. are capable of continuous learning, is one of the major tasks of innovation strategy. The R&D and related technical functions within the firm are central features of this capacity to learn. However, as can be seen in Table 6.1, the nature and purpose of the large firms' technological activities vary greatly. Contrast R&D activities of interest to different parts of the firm:

- *Corporate* level: time horizons are long, learning feedback loops slow, internal linkages weak, linkages to external knowledge sources strong, and projects relatively cheap.
- *Business unit* level: time horizons are short, learning feedback loops fast, internal linkages (with production and marketing) strong, and projects expensive.

Balancing these various activities is a demanding activity involving choices between (i) R&D (and other technological activities) performed in the operating divisions, and in the corporate laboratory; (ii) R&D (and other technological activities) performed in the home country, and in foreign countries. Until recently, a useful rule of thumb for deciding where R&D should be performed was the following:

- *R&D supporting existing businesses* (i.e. products, processes, divisions) should be located in established divisions.

TABLE 6.1   The heterogeneity of large firms' technological activities[6]

Locus of interest	Time horizon (years)	Focus
Corporate-wide	~10	• Monitoring major scientific and technical developments • Knowledge-building • Creating new options • Technology positioning • Technical and human resource development
Group/division	~5	• Exploit synergies across business units
Business unit	~2–3	Implementing business objectives in product: • Cost • Quality • Development time, etc.

- *R&D supporting new businesses* (i.e. products, processes, divisions) should initially be located in central laboratories, then transferred to divisions (established or newly created) for exploitation.
- *R&D supporting foreign production* should be located close to that foreign production, and concerned mainly with adapting products and processes to local conditions.

Corporate versus Divisional Laboratories

As we shall now see, the decisions about the location of R&D have become more complicated.[7] Recent experience shows that there are in fact two dimensions in the organizational location of a firm's R&D activities:

- *Physical location*, determined mainly by the importance of the main organizational interface: the corporate laboratory towards the general development of fundamental fields of science and technology, and the divisional laboratories towards present-day businesses
- Its *funding*, determined by where the potential benefits will be captured: by some of the established divisions or by the corporation as a whole

As Table 6.2 shows, this leads to four possible categories of R&D activities in the firm. The two already described are in quadrants 1 and 4: activities funded and performed

TABLE 6.2   The location and funding of a company's R&D

	Corporate-level performance: Where important interfaces are with general advances in generic science and technologies	Divisional-level performance: Where important interfaces are with production, customers and suppliers
	Quadrant 1	Quadrant 2
Corporate-level funding:	Scanning external research threats and opportunities	Commercializing radical new technologies
When potential benefits are corporate-wide	Assimilating and assessing radical new technologies	Exploiting interdivisional synergies (e.g. production and materials technologies)
	Quadrant 3	Quadrant 4
Divisional-level funding:	Exploratory development of radical new technologies	Mainstream product and process development
When potential benefits are division-specific	Contract research for specific problem-solving for established divisions	Incremental improvements

by corporate-level laboratories (quadrant 1), and those funded and performed by division-level laboratories (quadrant 4). The growth of interest in those in quadrants 2 and 3 reflects a major problem that has emerged in the last 20 years, namely the gap within large firms between the corporate-level and division-level laboratories, at a time when competitive success often depends on rapid product development (see Box 6.1).

Activities in quadrant 3 reflect the attempt to ensure stronger linkages between the central and divisional laboratories by strengthening the financial contribution of the divisions to the corporate laboratory, thereby encouraging the interest of the former in the latter, and the sensitivity of the latter to the former. Activities in quadrant 2 recognize that the full-scale commercial exploitation of radically new technologies does not always fit tidily within established divisional structures, so that central funding and initiative may be necessary.

The dynamic capabilities approach has placed most emphasis on capability-building at the business unit level, and has somewhat neglected the potential contribution of the corporate centre to the development of new capabilities. The potential roles of the corporate centre can be grouped into four areas: *leveraging, integration, reconfiguration* and *learning*.[8] The centre may leverage existing capabilities by identifying resources within business units, recognizing where these might be exploited by other business units, and implementing the necessary organizational changes to execute the transfer. For example, the drug Viagra was originally developed for cardiac markets, but trials revealed side effects which could be beneficial in the treatment of erectile dysfunction. The centre may also co-ordinate and integrate resources across business units, by supporting cross-business development groups. For example, the development of products and services for home automation demands the integration of capabilities in electronics, software, telecommunications, and the manufacture of both brown and white goods. The most successful firms in this emerging market have been those that have successfully integrated these capabilities, which reside in different business units.[9] In extreme cases a potential innovation may not have an obvious home in any existing business unit, because the market does not yet exist or is not served by a unit, or the business model is fundamentally different. This is a common problem with more radical innovations.[10] In these cases the centre may need to create a new business unit to develop and exploit the innovation. This has been the strategy adopted successfully by 3M. The reconfiguration of resources is different to integration, and consists of the recombination of capabilities to produce economies of scale or scope. The centre is best placed to identify the potential for consolidation. For example, the need for customer support might be common across a number of business units, and a solution using a call centre might only be practical if these needs are pooled. Finally, the centre can influence learning within the business units by its funding and assessment of R&D at the business unit level, and by encouraging interaction across business units.

BOX 6.1
PROBLEMS IN LINKING CORPORATE R&D TO THE REST OF THE FIRM

Corporate R&D became unpopular in US firms in the 1980s, when two specific examples were very influential. First, at RCA, failure in the early 1980s of the *videodisc* – in competition with the then emerging video-cassette recorder – resulted in part from the dominant influence and mistaken technical choices made by the Corporate Laboratories, in opposition to the Consumer Electronics Division.[11] Second, Xerox's Palo Alto Research Centre (PARC) in the 1970s made major inventions that were subsequently exploited by other companies. Examples in personal computing include 'bit mapped' displays, windows, the mouse and word processing programmes. According to Bro Uttal, Xerox had difficulty translating first-rate research into money-making products for the following reasons:

> the process takes time at any large company . . . Sheer size slows decision-making, and the need to concentrate on existing businesses impairs management's ability to move deftly into small, fast-changing markets. This is a special problem for Xerox, still overwhelmingly a one-product company . . .
> Serious organizational flaws, acknowledged by high Xerox executives, have proved a handicap. PARC had weak ties with the rest of Xerox, and the rest of Xerox had no channel for marketing products based on the researchers' efforts . . . The company has revamped marketing five times in the last six years . . . disgruntled researchers have left in frustration. These Xeroids, as they call themselves, have showered PARC's concepts – for designing personal computers, office equipment, and other products – on competing companies.[12]

Similar cutbacks in central corporate R&D in the UK have been analysed by Whittington.[13] Writing in the mid-1980s, Margaret Graham concluded the following:

> Whether or not corporate R&D organizations continue to generate long-term research in-house, the ability to translate research performed elsewhere into usable technology for a specific firm will almost certainly remain a responsibility internal to the firm. For this purpose, if corporate R&D did not exist in some form, sooner or later it would have to be invented.[14]

The growing interest in the 1990s in 'core competencies' suggests that she was right.

Corporate R&D: Centralization, Decentralization or External Collaboration?

Within this framework, there are no simple rules that tell managers how to strike the right balance between corporate and divisional initiatives in R&D – a situation that is further complicated by the growth of so-called 'strategic alliances' in R&D with organizations external to the firm. The histories of major firms in science-based industries reveals a cyclical pattern, suggesting there is no right answer, and that finding and maintaining the proper balance is not easy (see, for example, the long history and experience of Du Pont[15]). Nonetheless, we can identify four sets of factors that will influence the proper balance:

1. *The firm's main technological trajectory*. This gives strong guidance on the appropriate balance. At one extreme, the corporate initiatives are very important in the chemically based – and particularly the pharmaceutical – industry, where fundamental discoveries at the molecular level are often directly applicable in technological development. At the other extreme, corporate-level laboratories are less important in sectors – like aircraft and automobiles – that are based on complex products and production systems, where the benefits of basic advances are more indirect (e.g. the use of simulation technologies), and the critical interface is between R&D and design, on the one hand, and production, on the other.

2. *The degree of maturity of the technology*. The examples of optoelectronics and biotechnology[16] show that, after the emergence of a fundamental technological breakthrough, extended periods of trial, error and learning are necessary before specific technological opportunities begin to emerge. During the early 'incubation' stage, there are advantages in isolating such learning processes from immediate commercial pressure by locating them in the corporate laboratory, before transfer to a more market-oriented framework in an established division or internal venture group.

3. *Corporate strategic style*. The corporate R&D laboratory will have low importance in firms whose strategies are entirely driven by short-term financial performance in existing products (see section 6.4). Such 'market-led' strategies will concentrate on the bottom row of division-level funding in Table 6.2, and miss the opportunities emerging from the development and exploitation of radical new technologies. Such financially driven strategies became increasingly apparent in US and UK firms in the 1980s,[17] although Rod Coombs and Albert Richards reported a reversal of the trend in some British firms in the 1990s.[18]

4. *Links to 'new science'-based technologies*. New forms of corporate linkages with basic and academic research are emerging in the 'new sciences' that have grown out of

recent advances in molecular biology, nanotechnology and IT.[19] Advances in these fields are the basis of the growth of firms spun off from universities, since they have reduced the costs of technical experimentation to a level where university-type laboratories and research methods can make significant technical advances. This has also had the effect of increasing both the range of technological opportunities that large firms can exploit, and the uncertainties surrounding their eventual usefulness. Large firms therefore prefer to explore these opportunities through collaborations until the uncertainties are reduced.[20]

6.2 Locating R&D Activities – Global versus Local

Since the 1980s, some analysts and practitioners have argued that, following the 'globalization' of product markets, financial transactions and direct investment, large firms' R&D activities should also be globalized – not only in their traditional role of supporting local production – but also in order to create interfaces with specialized skills and innovative opportunities at a world level.[21] However, although striking examples of the internationalization of R&D can be found (e.g. the large Dutch firms, particularly Philips[22]), more comprehensive evidence casts doubt on the strength of such a trend. This evidence is based on the countries of origin of the inventors cited on the front page of patents granted in the USA, to nearly 359 of the world's largest, technologically active firms (and which account for about half of all patenting in the USA). This information turns out to be an accurate guide to the international spread of large firms' R&D activities.

Table 6.3 summarizes the available evidence. Column 1 gives the country of origin and number of large firms; columns 2 and 3 the percentage shares of corporate patenting in the USA from inside and outside the home country (2 + 3 = 100); column 4 the percentage share of R&D outside the home country (should be similar to column 3); columns 5–8 break down the foreign patenting in column 3 by region (5 + 6 + 7 + 8 = 3); and column 9 the rate of increase of foreign patenting from the 1980s to the 1990s.

Taken together, this evidence and some further analysis[23] shows that:

- The world's large firms perform about 12% of their innovative activities outside their home country. The equivalent share of production is about 25%.
- The most important factor explaining each firm's share of foreign innovative activities is its share of foreign production. Firms from smaller countries in general have

TABLE 6.3 Indicators of the geographic location of large firms' innovative activities in the 1990s

Nationality of large firms (no.) (1)	% share of origin of US patents in 1992–6 (2–3)		% share of foreign-performed R&D expenditure (year) (4)	% share of foreign origin of US patents in 1992–6 (5–8)				% change in foreign origin of US patents, since 1980–4 (9)
	Home	Foreign		USA	Europe	Japan	Other	
Japan (95)	97.4	2.6	2.1 (1993)	1.9	0.6	0.0	0.1	−0.7
USA (128)	92.0	8.0	11.9 (1994)	0.0	5.3	1.1	1.6	2.2
Europe (136)	77.3	22.7		21.1	0.0	0.6	0.9	3.3
Belgium	33.2	66.8		14.0	52.6	0.0	0.2	4.9
Finland	71.2	28.8	24.0 (1992)	5.2	23.5	0.0	0.2	6.0
France	65.4	34.6		18.9	14.2	0.4	1.2	12.9
Germany	78.2	21.8	18.0 (1995)	14.1	6.5	0.7	0.5	6.4
Italy	77.9	22.1		12.0	9.5	0.0	0.6	7.4
Netherlands	40.1	59.9		30.9	27.4	0.9	0.6	6.6
Sweden	64.0	36.0	21.8 (1995)	19.4	14.2	0.2	2.2	−5.7
Switzerland	42.0	58.0		31.2	25.0	0.9	0.8	8.2
UK	47.6	52.4		38.1	12.0	0.5	1.9	7.6
All firms (359)	87.4	12.6	11.0 (1997)	5.5	5.5	0.6	0.9	2.4

Sources: Derived from Patel, P. and K. Pavitt (2000) 'National systems of innovation under strain: the internationalization of corporate R&D' in Barrell, R., G. Mason and M. O'Mahoney (eds), *Productivity, Innovation and Economic Performance*. Cambridge UP; and Patel, P. and M. Vega (1998) *Technology Strategies of Large European Firms*, P.R.Patel@sussex.ac.uk.

higher shares of foreign innovative activities. On average, foreign production is less innovation-intensive than home production.

- Most of the foreign innovative activities are performed in the USA and Europe (in fact, Germany). They are *not* 'globalized'.
- Since the late 1980s, European firms – and especially those from France, Germany and Switzerland – have been performing an increasing share of their innovative activities in the USA, in large part in order to tap into local skills and knowledge in such fields as biotechnology and IT.

What Should Management Make of 'Techno-globalism'?

Controversy remains both in the interpretation of this general picture, and in the identification of implications for the future. Our own views are as follows:

1. There are major efficiency advantages in the geographic concentration in one place of strategic R&D for *launching major new products and processes* (first model and production line). These are:
   (a) dealing with unforeseen problems, since proximity allows quick, adaptive decisions;
   (b) integrating R&D, production and marketing, since proximity allows integration of tacit knowledge through close personal contacts (see Box 6.2).
2. The nature and degree of international dispersion of R&D will also depend on the company's major technological trajectory, and the strategically important points for integration and learning that relate to it. Thus, whereas automobile firms find it difficult to separate their R&D geographically from production when launching a major new product, drug firms can do so, and instead locate their R&D close to strategically important basic research and testing procedures.
3. In deciding about the internationalization of their R&D, managers must distinguish between:
   (a) becoming part of global *knowledge networks* – in other words, being aware of, and able to absorb – the results of R&D being carried out globally. Practising scientists and engineers have always done this, and it is now easier with modern IT. However, business firms are finding it increasingly useful to establish relatively small laboratories in foreign countries in order to become strong members of local research networks and thereby benefit from the person-embodied knowledge behind the published papers;
   (b) the *launching of major innovations*, which remains complex, costly, and depends crucially on the integration of tacit knowledge. This remains difficult to achieve

BOX 6.2

GLOBALIZATION OF PRODUCT DEVELOPMENT? LESSONS FROM FORD AND CHRYSLER[24]

'April 20th [1994] . . . Ford's chairman . . . announced a sweeping management reorganization, removing power from regional fiefs and giving it to global teams. Five so-called "platform teams" will work on the basic Ford models to be sold around the world . . .

Ford's problem is no longer learning to make cars cheaply . . . but also designing them quickly and cheaply enough. Ford's most ambitious attempt at global design, the simultaneous development of the Contour (as it is known in America) and the Mondeo (Europe) is a case in point. Both cars were based on the same basic chassis. Design teams in Michigan and Cologne were wired into the same computer. Yet the Contour/Mondeo project ran a year late and cost $6 billion. Too many regional managers managed, as before, to get in the way. Now Mr Trotman is giving a small cadre of managers global authority, while, lower down the ladder, local managers will have more freedom to take smaller decisions without referring them to head office. GM's reorganization is likely to follow similar lines.

. . . both Ford and GM managers admit they have learnt something from Chrysler. For the past three years, Chrysler has been concentrating on developing models faster – and stripping out bureaucracy. On April 19th Chrysler unveiled its best-ever quarterly net profit . . .'

Chrysler is reaping the profits from having already reorganized its designers and engineers into 'platform teams' – as Ford is now doing. Anyone involved in a particular vehicle – from marketing to manufacturing – works out of the same office, eliminating potential design snags, speeding up product development and sharply reducing costs. The Chrysler Cirrus and Dodge Stratus models will come to market this autumn, less than three years after the project won approval (and two years faster than the models they replace). Their development cost of $900m. is less than a sixth of that for Ford's Mondeo.

across national boundaries. Firms therefore still tend to concentrate major product or process developments in one country. They will sometimes choose a foreign country when it offers identifiable advantages in the skills and resources required for such developments, and/or access to a lead market.[25]

4. Matching global knowledge networks with the localized launching of major innovations will require increasing international mobility amongst technical personnel, and the increasing use of multi-national teams in launching innovations.

5. Advances in IT will enable spectacular increases in the international flow of codified knowledge in the form of operating instructions, manuals and software. They may also have some positive impact on international exchanges of tacit knowledge through teleconferencing, but not anywhere near to the same extent. The main impact will therefore be at the second stage of the 'product cycle',[26] when product design has stabilized, and production methods are standardized and documented, thereby facilitating the internationalization of production. Product development and the first stage of the product cycle will still require frequent and intense personal exchanges, and be facilitated by physical proximity. Advances in IT are therefore more likely to favour the internationalization of production than of the process of innovation.

The two polar extremes of organizing innovation globally are the specialization-based and integration-based, or network structure.[27] In the specialization-based structure the firm develops global centres of excellence in different fields, which are responsible globally for the development of a specific technology or product or process capability. The advantage of such global specialization is that it helps to achieve a critical mass of resources and makes coordination easier. As one R&D director notes:

> ... the centre of excellence structure is the most preferable. Competencies related to a certain field are concentrated, coordination is easier, and economies of scale can be achieved. Any R&D director has the dream to structure R&D in such a way. However, the appropriate conditions seldom occur.[27]

In addition, it may allow location close to a global innovation cluster. The main disadvantages of global specialization are the potential isolation of the centre of excellence from global needs, and the subsequent transfer of technologies to subsidiaries worldwide. In contrast, in the integration-based structure different units around the world each contribute to the development of technology projects. The advantage of this approach is that it draws upon a more diverse range of capabilities and international perspectives. In addition, it can encourage competition amongst different units. However, the integrated approach suffers from very high costs of coordination, and commonly suffers from duplication of efforts and inefficient use of scarce resources. In practice, hybrids of these two extreme structures are common, often as a result of practical compromises and trade-offs necessary to accommodate history, acquisitions and politics. For example, specialization by centre of excellence may include contributions from other units, and integrated structures may include the contribution of specialized units. The main factors influencing the decision where to locate R&D globally are, in order of importance:[28]

1. The availability of critical competencies for the project;
2. The international credibility (within the organization) of the R&D manager responsible for the project;
3. The importance of external sources of technical and market knowledge, e.g. sources of technology, suppliers, and customers;
4. The importance and costs of internal transactions, e.g. between engineering and production;
5. Cost and disruption of relocating key personnel to the chosen site.

6.3 Allocating Resources for Innovation

Given their mathematical skills, one might have expected R&D managers to be enthusiastic users of quantitative methods for allocating resources to innovative activities. The evidence suggests otherwise: practising R&D managers have been sceptical for a long time (see Box 6.3). An exhaustive report by practising European managers on R&D project evaluation classifies and assesses more than 100 methods of evaluation and presents 21 case studies on their use.[29] However, it concludes that no method can guarantee success, that no single approach to pre-evaluation meets all circumstances, and that – whichever method is used – the most important outcome of a properly structured evaluation is *improved communication*. These conclusions reflect three of the characteristics of corporate investments in innovative activities:

1. They are uncertain, so that success cannot be assured.
2. They involve different stages that have different outputs that require different methods of evaluation.
3. Many of the variables in an evaluation cannot be reduced to a reliable set of figures to be plugged into a formula, but depend on expert judgments: hence the importance of communication, especially between the corporate functions concerned with R&D and related innovative activities, on the one hand, and with the allocation of financial resources, on the other.

Uncertainty

Given the complexities involved, the outcomes of investments in innovation are uncertain, so that the forecasts (of costs, prices, sales volume, etc.) that underlie project and programme evaluations can be unreliable. According to Joseph Bower, management finds it easier, when appraising investment proposals, to make more accurate forecasts of reductions in production cost than of expansion in sales, whilst their ability to forecast the financial consequences of new product introductions is very limited indeed.[30] This last conclusion is confirmed by the study by Edwin Mansfield and his colleagues of project selection in large US firms.[31] By comparing project forecasts with outcomes, he showed that managers find it difficult to pick technological and commercial winners:

- Probability of *technical* success of projects (P_t) = 0.80.
- Subsequent probability of *commercial* success (P_c) = 0.20.
- Combined probability for all stages: $0.8 \times 0.2 = 0.16$.

He also found that managers and R&D workers cannot predict accurately the *development costs, time periods, markets and profits* of R&D projects. On average, costs were greatly *underestimated*, and time periods *overestimated* by 140–280% in incremental product improvements, and by 350–600% in major new products. Other studies have found that:

- About half business R&D expenditures are on *failed* R&D projects. The higher rate of success in *expenditures* than in *projects* reflects the weeding out of unsuccessful projects at their early stages and before large-scale commercial commitments are made to them.[32]
- R&D scientists and engineers are often deliberately over-optimistic in their estimates, in order to give the illusion of a high rate of return to accountants and managers.[33]

How to Evaluate Learning?

The potential benefits of innovative activities are twofold. First, *extra profits* derived from increased sales and/or higher prices for superior products, and from lower costs and/or increased sales from superior production processes. Conventional project appraisal methods can be used to compare the value of these benefits against their cost. Second, *accumulated firm-specific knowledge* ('learning', 'intangible assets') that may be useful for the development of *future* innovations (e.g. new uses for solar batteries, carbon fibres, robots, word processing). This type of benefit is relatively more important in R&D projects that are more long term, fundamental and speculative.

Conventional techniques cannot be used to assess this second type of benefit, because it is an 'option' – in other words, it creates the *opportunity* for the firm to invest in a potentially profitable investment, but the realization of the benefits still depends on a decision to commit further resources. Conventional project appraisal techniques cannot evaluate options (see Box 6.4).

The inherent uncertainty in most R&D projects limits the ability of managers to predict the outcomes and benefits of projects. Research suggests that changes to R&D plans and goals are common, being driven by external factors, such as technological breakthroughs, as well as internal factors, such as changes in the project goals. Together the impact of changes to project plans and goals overwhelm the effects of the quality of formal project planning and management.[34] This reality is consistent with the real options approach to investing in R&D, because investments are sequential and managers have some influence on the timing, resourcing and continuation or abandonment of projects at different stages. By investing relatively small amounts in a wide range of projects, a greater range of technological opportunities can be explored. Once uncertainty has been reduced, only the most promising projects are allowed to continue.

For a given level of R&D investment this real option approach should increase the value of the project portfolio. However, because options interact, a decision regarding one project can affect the option value of another project (unlike NPV calculations,

BOX 6.4

WHY CONVENTIONAL FINANCIAL EVALUATION METHODS DO NOT WORK WITH INVESTMENTS IN TECHNOLOGY

The following text was written by the Professor of Finance at the Sloan School of Management at MIT.[35]

'Suppose a firm invests in a negative NPV (net present value) project in order to establish a foothold in an attractive market. Thus a valuable second-stage investment is used to justify the immediate project. The second stage must depend on the first: if the firm could take the second project without having taken the first, then the future opportunity should have no impact on the immediate decision . . .

At first glance, this may appear to be just another forecasting problem. Why not estimate cash flows for both stages, and use discounted cash flow to calculate the NPV for the two stages taken together?

You would not get the right answer. The second stage is an option, and conventional discounted cash flow does not value options properly. The second stage is an option because the firm is not committed to undertaking it. It will go ahead if the first stage works and the market is still attractive. If the first stage fails, or if the market sours, the firm can stop after stage 1 and cut its losses. Investing in stage 1 purchases an intangible asset: a call option on stage 2. If the option's present value offsets the first stage's negative NPV, the first stage is justified . . .

. . . DCF (discounted cash flow) is readily applied to "cash cows" – relatively safe businesses held for the cash they generate . . . It also works for "engineering investments", such as machine replacements, where the main benefit is reduced cost in a defined activity.

. . . DCF is less helpful in valuing businesses with substantial growth opportunities or intangible assets. In other words, it is not the whole answer when options account for a large fraction of a business's value.

. . . DCF is no help at all for pure research and development. The value of R&D is almost all option value. Intangible assets' value is usually options value.'

which rarely include interaction effects). Therefore the creation of further options through R&D projects may not increase the overall option value of the R&D portfolio, and conversely the interaction of options arising from different projects can give rise to a non-linear increase in the combined option value.[36] However, in almost all cases it is impossible to calculate the value of R&D using real options, because unlike financial options it is difficult to predict technological breakthroughs, estimate future sales from products flowing from the R&D (or project pay-off), or to identify and model project-specific risks, and the time varying volatilities of the processes and eventual values.[37] Nonetheless, the real options perspective remains a useful way of conceptualizing R&D investment, particularly at the portfolio level. It can help to make more explicit and to identify future growth options created by R&D, even when these are not related to the (current) goals of the R&D. Combined with decision trees, a real options approach can help to identify risks and pay-offs, key uncertainties, decision points and future branches (options).[38] It is particularly effective where high volatility demands flexibility, placing a premium on the certainty of information and timing of decisions.

How Practising Managers Cope

These two sets of difficulties – in evaluating the potential contributions of technological investments to firm-specific intangible assets, and in dealing with uncertainty – are reflected in how successful managers allocate resources to technological activities. In particular, they:

- Encourage *incrementalism* – step-by-step modification of objectives and resources, in the light of new evidence.
- Use *simple rules* models for allocating resources, so that the implications of changes can be easily understood.
- Make explicit from the outset criteria for *stopping* the project or programme.
- Use *sensitivity analysis* to explore if the outcome of the project is 'robust' (unchanging) to a range of different assumptions (e.g. 'What if the project costs twice as much, and takes twice as long, as the present estimates?').
- Seek the reduction of *key uncertainties* (technical and – if possible – market) before any irreversible commitment to full-scale – and costly – commercialization.
- Recognize that *different types* of R&D should be evaluated by *different criteria*.

Organizing Resource Allocation to Innovative Activities

In other words, the corporate R&D community recognizes that the successful allocation of resources to innovation depends less on robustness of decision-making techniques than on the organizational processes in which they are embedded. According to Mitchell and Hamilton,[39] there are three (overlapping) categories of R&D that large firms must finance. Each category has different objectives and criteria for selection, the implications of which are set out in Table 6.4.

TABLE 6.4 Criteria and procedures for evaluating different types of R&D

Objective	Technical activity	Evaluation criteria (% of all R&D)	Decision-takers	Market analysis	Nature of risk	Higher volatility	Longer time horizons	Nature of external alliances
Knowledge building	Basic research, monitoring	Overhead cost allocation (2–10%)	R&D	None	Small = cost of R&D	Reflects wide potential	Increases search potential	Research grant
Strategic positioning	Focused applied research, exploratory development	'Options' evaluation (10–25%)	Chief executive R&D division	Broad	Small = cost of R&D	Reflects wide potential	Increases search potential	R&D contract Equity
Business investment	Development and production engineering	'Net present value' analysis (70–99%)	Division	Specific	Large = total cost of launching	Uncertainty reduces net present value	Reduces present value	Joint venture Majority control

Knowledge-building   This is the early-stage and relatively inexpensive research for nurturing and maintaining expertise in fields that could lead to future opportunities or threats. It is often treated as a necessary overhead expense, and sometimes viewed with suspicion (and even incomprehension) by senior management obsessed with short-term financial returns and exploiting existing markets, rather than creating new ones.

With knowledge-building programmes, the central question for the company is: 'What are the potential costs and risks of *not* mastering or entering the field?' Thus, no successful large firm in manufacture can neglect to explore the implications of development in IT, even if IT is not a potential core competence. And no successful firm in pharmaceuticals could avoid exploring recent developments in biotechnology. Decisions about such programmes should be taken solely by R&D staff on the basis of technical judgments, and especially those staff concerned with the longer term. Market analysis should not play any role. Outside financial linkages are likely to be with academic and other specialist groups, and to take the form of a grant.

Strategic positioning   These activities are in between knowledge-building and business investment, and are an important – and often neglected – link between them. They involve applied R&D and feasibility demonstration, in order to reduce technical uncertainties, and to build in-house competence, so that the company is capable of transforming technical competence into profitable investment. For this type of R&D, the appropriate question is: 'Is the programme likely to create *an option* for a profitable investment at a later date?' Comparisons are sometimes made with financial 'stock options', where (for a relatively small sum) a firm can purchase the *option* to buy a stock at a specified price, before a specified date – in anticipation of increase in its value in future (see Box 6.4).

Decisions about this category of R&D programme should involve divisions, R&D directors and the chief executive, precisely because – as their description implies – these programmes will help determine the strategic options open to the company at a later date. At this stage, market analysis should be broad (e.g. where could genetic engineering create new markets for vegetables in a food company?). A variety of evaluation methods may be used (e.g. the product–technology matrix), but they will be more judgmental than rigorously quantitative. Costs will be higher than those of knowledge-building, but much lower than those of full-scale business investment. As with knowledge-building programmes, both high volatility in predictions and expectations, and long time horizons, are *not* unwelcome signs of unacceptably high risk, but welcome signs of rich possibilities and sufficient time to explore them. Outside linkages require tighter management than those related to knowledge-building, probably through a contract or equity participation.

BOX 6.5

NET PRESENT VALUE

Net present value (NPV) = $\sum_{0}^{T} p_t / (1+i)t - C$

where p_t = forecast cash flow in time period t;
T = project life;
i = expected rate of return on securities equivalent in risk to project being evaluated;
C = cost of project at time $t = 0$.

Business investment   This is the development, production and marketing of new and better products, processes and services. It involves relatively large-scale expenditures, evaluated with conventional financial tools such as *net present value* (see Box 6.5).

In such projects, the appropriate question is: 'What are the potential costs and benefits in continuing with the project?' Decisions should be taken at the level of the division bearing the costs and expecting the benefits. Success depends on meeting the precise requirements of specific groups of users, and therefore depends on careful and targeted marketing. Financial commitments are high, so that volatility in technological and market conditions is unwelcome, since it increases risk. Long time horizons are also financially unwelcome, since they increase the financial burden. Given the size and complexity of development and commercialization, external linkages need to be tightly controlled through majority ownership or a joint venture. Given the scale of resources involved, careful and close monitoring of progress against expectations is essential. For such projects most firms rely on financial methods to evaluate their project portfolio, around 77% of firms according to a recent survey. However, the same survey revealed that only 36% of the best performing firms rely on financial methods, compared to 39% which use strategic methods.[40] An explanation for the relatively poor performance of financial methods is that the sophistication of the models often far exceeds the quality of the data inputs, particularly at the early stages of a project's life.

6.4 Technology and Corporate Strategy

In linking technology to corporate strategy, it must be remembered that the links run both ways. Not only does corporate strategy define objectives for technology.

Technology defines opportunities and constraints for corporate strategy. We shall discuss each of these links in turn.

How Technology Contributes to Corporate Strategy

According to a report prepared by practising managers in large European firms, the R&D function should play a central role in the formulation of an innovation strategy, as part of overall corporate strategy:[41]

> It is essential that the R&D function is totally integrated with the company's activities and strategic thinking. This is the most effective way of judging the relevance of technology past, present and future, to the company's fortunes – its strengths, weaknesses, opportunities and threats. R&D management needs to take the initiative in playing a proactive part in strategy formulation and in executing the plan . . .
>
> For R&D to participate fully, a person with an overall technical awareness of the company's activities, such as the Technical Director, at senior management level is essential. His/her primary functions are:
> - to provide a technical awareness within the company and a 'window' on the external world of technology;
> - to ensure the appropriate level of technology for maintaining or regenerating the company's existing business;
> - to provide a technical input into reviews of new business opportunities;
> - to determine the overall technical strategy consistent with corporate requirements.

The relevance of this recommendation is supported by systematic statistical studies in the USA and the UK,[42] showing that a company's commitment to, and performance in, innovation depends in part on the share of top management with training in science and engineering. The EIRMA report also showed that the technical director had influence over corporate strategy in about 60% of the firms.

R&D strategies – defining the contribution of technology to corporate objectives – are developed formally in most European large firms at the level of the division and the corporation. Table 6.5 shows that formal R&D strategies had their greatest impact in the areas of direct responsibility of the R&D director (namely, the volume, composition, and balance of R&D activities). They had less influence on the organizational location and methods of funding R&D – organization, new venture areas, balance between divisions, methods of funding, and on positioning – role in market, technological position, patent and licensing policies.

Compatibility between Corporate Strategy and the Nature of Technological Opportunities

The EIRMA report makes no mention of the compatibility (or otherwise) between the technological opportunities open to the firm, on the one hand, and its organizational

TABLE 6.5 The areas of impact of R&D strategy in large European firms

Tasks of R&D strategy	% of companies observing impact
Focal R&D areas	91
Volume of R&D expenditures	85
R&D balance – short, medium and long term	83
R&D organization	71
Internal versus external R&D	69
New venture areas	60
R&D balance between divisions	60
Role in the market	59
Technological position	54
Personnel policy	51
Methods of funding	48
Patent policy	37
Licence policy	33

Source: Derived from EIRMA (1986) *Developing R&D Strategies*, Paris.

structure and strategic style, on the other. One view is that overall corporate strategy should determine both organizational *structure* and technology *strategy*. Another view (our own) is that different kinds of technological opportunities require different kinds of strategies and structures if they are to be exploited effectively.

According to Chandler, large firms' corporate headquarters have two functions: entrepreneurial promotion and administrative control (see Box 6.6).

Gould and Campbell identify three generic corporate strategic styles that each have a different balance between the entrepreneurial and administrative functions.[43] As Table 6.6 shows, three 'strategic styles' are appropriate for different types of technology and market:

1. *Financial control* strategies are reflected in a strong administrative monitoring function in corporate HQ, and an expectation of high, short-term financial returns. Technological investments in knowledge-building and strategic positioning will be neither understood nor encouraged, but will concentrate instead on low-risk incremental improvements in established businesses. As such, this strategic style is suited to conglomerates in low-technology industries. There will therefore be relatively little organic growth, and in high-tech industries many technology-based opportunities that are missed rather than exploited.

2. *Strategic planning* strategies are reflected in strong central entrepreneurial direction with corporate HQ giving strong encouragement to investments in knowledge-building and strategic positioning, and taking a central role in deciding technological priorities and – more generally – of showing what Leonard-Barton calls 'strategic

---

BOX 6.6

THE ENTREPRENEURIAL AND ADMINISTRATIVE FUNCTIONS OF CORPORATE HQ[44]

'One was *entrepreneurial*, that is to determine strategies to utilise for the long term the firm's organisational skills, facilities and capital to allocate resources – capital and product-specific technical and managerial skills – to pursue these strategies. The second was more *administrative*. It was to monitor the performance of the operating divisions; to check on the use of the resources allocated; and, when necessary, redefine the product lines of the divisions so as to continue to use effectively the firm's organisational capabilities.

. . . there are differences in the ways the two basic economic functions of the corporate office are carried out – the *entrepreneurial* function of planning for the future health and growth of the enterprise and the *administrative* function of controlling the operation of its many divisions.' (Our italics)

---

intent'.[45] As such, this strategic style is most appropriate for high-tech, focused and lumpy businesses, such as automobiles, drugs and petroleum, where experimentation is costly, and customer markets clearly defined.

3. *Strategic control* strategies also give high priorities to entrepreneurial technological investments, but devolve the formulation and execution of strategies much more to the divisions and business units. Instead of exercising 'strategic intent', the HQ shows 'strategic recognition' in recognizing and reinforcing successful entrepreneurial ventures emerging from the divisions, and which become separate divisions themselves. As such, this style of strategy is best suited to high-tech businesses with pervasive technologies, varied markets and relatively low costs of experimentation. Examples include 3M Corporation, with its widespread applications of adhesives and coatings technologies, and consumer electronics firms.

Mismatches between a firm's strategic style and its core technology will inevitably cause instability, and can have two causes. First, the imposition of a strong style of financial control in a sector where high technological investments – especially in knowledge-building and strategic positioning – are necessary for long-term survival. Examples include GEC in the UK, which has progressively reduced its competence in electronic computers and components, and moved out of high-tech sectors with expanding opportunities; and ITT in the USA (see Box 6.7). Neither GEC or ITT exist in name today, but the lessons remain relevant.

TABLE 6.6 Strategic styles and technology strategies

Strategic style	Functions of HQ	Emphasis in HQ	Firm type	Examples	Mismatch
Financial control	Financial monitoring and control of divisions	Administrative: Tight control and quick profits	Low tech: Low capital intensity	Conglomerates like Hanson	GEC (UK) ITT (USA)
Strategic planning	Heavy influence in divisions' strategic plans	Entrepreneurial: Decides new business and costly projects	High tech: Lumpy projects, focused markets	Drugs, oil and automobile firms	IBM (1980s) ICI
Strategic control	Strategic plans devolved to divisional HQ	Entrepreneurial: But leaves choices to divisions	High tech: Pervasive technology, varied markets	3M Consumer electronics firms	Digitalization in consumer electronics firms

BOX 6.7
ITT AND GEC[46]

'... ITT was the quintessential conglomerate: a collection of hundreds of businesses held together not by the logic of their activities but by the force of their management. Now, ... the company has decided to split itself into three easier-to-understand firms ... The biggest of the three, *ITT Hartford*, will take over *ITT's* sprawling insurance operations ... *ITT Industries* will take in the old *ITT's* car, defence, electronics and fluid technology businesses ... Finally, hotels, entertainment and information-services will go into a "new" ITT Corporation.

In the 1960s and 1970s they [conglomerates] were stock analysts' darlings. They were thought to be immune from the vagaries of the business cycle ... They were also thought to demonstrate that good management techniques can be applied to any business, however remote from the company's core concerns ... At ITT, these management techniques were synonymous with Harold Geenen ... Under his ... stare, ITT, which started out as a Caribbean telephone company, acquired a rag-bag of companies, including Avis, Continental Baking, Rayonier, Sheraton and Hartford Fire Insurance. The ostensible aim ... of this spending spree was to insure against the uncertain future of the telephone business, but the immediate effect was to expand Mr. Geenen's empire. By 1970, he had control of 400 separate companies operating in 70 different countries ... To all these businesses, Mr. Geenen applied the same ruthless, financially driven approach.

Today, focus, not size, is the essence of managerial correctness ... they must decide what they are good at, and get out of non-related businesses ... Conglomerates are not on the way out entirely, however. Even Anglo-Saxon capitalism has a place for combative "antique-dealer" conglomerates such as Britain's Hanson, which specialises in buying under-valued companies, polishes them up and then flogging them off.'

'GEC is the creation of one man, Arnold Weinstock – the most impressive British industrialist of his generation. In the 1960s, he mopped up most of Britain's electronics and electrical engineering industry and cornered a fair chunk of its defence budget. A little monopolistic perhaps, but then many of the best businessmen try to be so. Mr. Weinstock was a management pioneer, forcing the companies he acquired to follow strict financial disciplines ...

Unfortunately ... the times have changed and GEC has not. True, Lord Weinstock, as he is now, has continued to add British defence businesses, but he is better known for sitting atop a pile of cash that many shareholders would rather invest for themselves . . . every boss worth his MBA nowadays looks at the books, but the more modern ones also give their managers freedom . . . The young technology entrepreneurs whom Britain is beginning to produce see GEC as a horrible relic of the past. "We arranged a $1m. contract with them: it went to head office and never came out," complains one.'

GEC subsequently renamed itself 'Marconi', and focused on expanding its telecommunications business through a series of expensive acquisitions. However, its timing was poor, and it suffered from the collapse of technology shares in 2000, and subsequently had to re-invent itself to avoid financial collapse.

Second, the changing nature of technological opportunities, which require a changed strategic style for their effective exploitation. Contemporary examples include:

- The *chemical industry*, where the focus of technical change over the past 30 years has shifted from large-scale process innovations in bulk plastics and synthetic fibres, to product innovation in drugs and other fine chemicals, and smaller-scale developments in speciality chemicals. This raises two questions for corporate strategy: (1) Are there still *technological* synergies between bulk and fine chemicals? If not, should the large chemical companies demerge? (2) Are there *organizational* incompatibilities between bulk chemicals requiring centralized strategic planning, and speciality chemicals requiring decentralized strategic control? (see Box 6.8).
- The *computer industry*, where technological change over the past 20 years has resulted in a revolutionary change from a market requiring the centralized style of strategic planning (selling large and lumpy mainframe computers to large organizations) to one requiring the decentralized style of strategic control (selling a wide range of relatively cheap hardware and software products in a wide range of market applications). Thus, although IBM and other large firms had the *technological* capacity to make the necessary transformations, they did not succeed in making the matching changes in strategic style and organization.

These examples shed important new light on the changing nature of the threats posed to established firms by radically new technology. In the past the main threat was considered to be the inability of the established firms to master the new technology (e.g. makers of horse-drawn carriages could not make automobiles). However, large firms today all typically have R&D laboratories that enable them to monitor, assess and (if necessary) master most new technology. The more difficult challenge now is to deal

BOX 6.8
WHY ICI CHOSE TO DEMERGE[47]

'. . . Imperial Chemical Industries (ICI) sought new sources of growth during the 1980s to offset the sluggish sales of its older products but, in the end, only increased the complexity of its already complicated and hard-to-manage portfolio of businesses . . . ICI's response . . . was unusual: a merger that split the organization into two separate companies: new ICI and Zeneca. The rationale underlying the structure composition and expansion of the old ICI . . . had been technology and vertical integration. Investment in research and development traditionally ensured a flow of new products . . . Huge chemical complexes grew up around very large plants, such as ethylene crackers, and ICI exploited each by-product stream to develop a new business . . . in the 1970s . . . the chemical industry began to mature. [In the 1980s] ICI pushed strongly into high-value specialty chemicals, using both in-house development and acquisitions . . . Such moves did little to reduce the complexity of ICI's businesses.

. . . the key to successful restructuring was recognising a technological fault line with ICI. Pharmaceuticals and other bioscience-related activities fell on one side of the fault line; the traditional chemical businesses fell to the other. The fault line divided two coherent groups of businesses that could be managed as separate companies. Within each group – but not across them – there was the potential for mutual support and interdependence . . . the demerger . . . preserved the advantages of retaining under a single ownership businesses that shared technical competencies and benefited from common services and informed central direction. Each of the two companies has had to evolve a balance between the centre and the businesses that suits its particular circumstances. The narrower focus and greater homogeneity of the problems that the new head offices handle shorten and sharpen the lines of communication between headquarters and the operating businesses.'

In 1993 ICI (the chemicals business) and Zeneca (the pharmaceuticals and biotechnology businesses) demerged. Subsequently ICI disposed of its bulk chemicals businesses, and began to focus on the higher growth and less cyclical speciality chemicals business. In 1997 it acquired the speciality chemicals business of Unilever for $4.9bn. The group now consists of the 'traditional' paints division (Dulux), flavours and foods (Quest) and food starches and adhesives (National Starch). In the meantime, Zeneca outperformed most other pharmaceutical firms in terms of sales growth, and in 2000 merged with Astra to become AstraZeneca. By means of such organizational evolution, ICI has continued to develop new businesses based on development in chemistry, pharmaceuticals and the biosciences.

with the *organizational* implications of the new technology, which may require radical and disruptive changes in, for example, products, markets, degree of centralization, the boundaries of corporate divisions, the key internal interfaces and external networks, and the relative power and influence of various professional groups.[48]

6.5 Organizational Processes in Small Firms

In small firms, deliberate organizational processes to integrate the technical function with production, marketing, strategy and resource allocation are of less central importance than in large firms. In general, these functions are less specialized, and less likely to be separated by physical and organizational distance. Table 6.7 tries to contrast the differences between large and small firms in how certain key tasks of innovation strategy are accomplished. In large firms, deliberate organizational design and formal procedures are essential as a means of integrating knowledge, of supporting professional judgments and of getting things done. In small firms, the characteristics of senior managers – their training, experience, responsibilities and external linkages – play a central role. In particular, their level of technical and organizational skills will determine

TABLE 6.7 How tasks of innovation strategy are accomplished in large and small firms

Strategic tasks	Large firms	Small firms
Integrating technology with production and marketing	• Organizational design • Organizational processes for knowledge flows across boundaries	• Responsibilities of senior managers
Monitoring and assimilating new technical knowledge	• Own R&D and external networks	• Trade and technical journals • Training and advisory services • Consultants • Suppliers and customers
Judging the learning benefits of investments in technology	• Judgements based on formal criteria and procedures	• Judgements based on qualifications and experience of senior management
Matching strategic style with technological opportunities	• Deliberate organizational design	• Qualifications of managers and staff

whether or not they will be able to develop and commercially exploit a firm-specific technological advantage.

6.6 Summary and Further Reading

There are no correct recipes for locating R&D and related innovative activities in the firm. Tensions are inevitable between *organizational decentralization* for rapid implementation, and sensitivity to production and customers, on the one hand, and *organizational centralization* for the exploration of radical and long-term opportunities, not linked to current businesses, on the other. There are also inevitable tensions between *geographic dispersion* for adapting to local markets and integrating local skills, on the one hand, and *geographic concentration* for the effective launching of major innovations, on the other. The important management challenges are therefore:

- Reconciling rapid product development and quick profit returns, with long-term search for major new technological opportunities.
- Reconciling effective new product launches, with the assimilation of competencies located in foreign countries.

Formal inputs from the corporate technical function into corporate strategy have been strongest in the determination of R&D priorities, but less strong in R&D funding and organization, and in market positioning. In addition:

- Corporate strategic style must be compatible with *the nature of technological opportunities*, if these are to be effectively exploited. Complete emphasis on financial control will discourage innovation. And the appropriate degree of centralization of entrepreneurial initiative will depend on the size of each innovative investment, and on the degree of similarity of user markets.
- Failures in established firms to benefit from radical new technologies now arise less from the inability to master them, and more from the inability or unwillingness to deal with their consequences for *organizational change*.
- Conventional project appraisal techniques are of *only limited usefulness* for R&D project evaluation and resource allocation, given high uncertainties in outcomes and difficulties in putting a financial value on the technological learning associated with R&D. In particular, in the early stages of R&D, uncertainties are high, costs (and therefore risks) are low, and choices essentially *judgmental*.
- Of great importance are the *processes* of resource allocation that recognize the essentially incremental nature of progress in R&D and related innovative activities, and that *integrate* the skills and methods (technical, financial and other) appropriate to

the purpose and nature of three types of innovative activity: knowledge-building, strategic positioning and business investment.

For further reading Professor Rubenstein's text *Managing Technology in the Decentralised Firm* (John Wiley & Sons, Ltd/Inc., Chichester, 1989) covers most of the ground very thoroughly, or for a more up-to-date account see Vittorio Chiesa's *R&D Strategy and Organization* (Imperial College Press, 2001). The journal *R&D Management* is a major source of the latest research results on R&D project evaluation and related subjects, a recent special issue being on the application of real options to the evaluation of R&D projects – *R&D Management* (2001) **31** (2). In addition, there are the papers by R. Cooper, S. Edgett and A. Kleinschmidt, 'Best practices for managing R&D portfolios', *Research-Technology-Management*, **41** (1), 1998; and by A. Henricksen, 'A technology assessment primer for management of technology' (*International Journal of Technology Management*, **13** (5), 1997). Also, the European Industrial Research Management Association (EIRMA) continues to produce excellent reports by practitioners on R&D strategy and asssessment (the most recent being in 2004 – *Assessing R&D Effectiveness*). Professor Chandler's books (*The Visible Hand: The managerial revolution in American business*, The Belknap Press of Harvard University Press, 1977; and *Scale and Scope: The dynamics of industrial competition*, Harvard University Press, 1990) are the classic texts on the relationship between corporate strategy and structure. More recently, the implications of radical technical change for established organizational practices have been explored by Clayton Christensen in *The Innovator's Dilemma* (Harvard University Press, 1997), and his more recent text which focuses more on the organization implications of disruptive innovation, *The Innovator's Solution* (Harvard, 2003, with Michael Raynor). Research results on the internationalization of corporate R&D can be found in O. Granstrand, L. Hakanson and S. Sjolander, *Technology Management and International Business* (John Wiley & Sons, Ltd, Chichester, 1992), the special numbers of *Research Policy* on 'The internationalization of industrial R&D' (28/2/1999), and of the *Journal of Product Innovation Management* on the 'Internationalization of innovation' (17/5/2000). A major series of case studies on the problems of managing innovation in an international framework is presented in the book by R. Boutellier, O. Gassman and M. von Zedwitz, *Managing Global Innovation* (Springer, 1999). Finally, John Cantwell has edited two recent texts on the globalization of innovation: *Multinational Enterprises, Innovation Strategies and Systems of Innovation* (2004, edited with Jose Molero); and *Globalization and the Location of Firms* (2004, both published by Edward Elgar).

The organizational implications of advances in IT are not obvious, and are still being discovered through experience and experiment. Recent contributions include: K. Werback, 'Syndication: the emerging model for business in the Internet era', *Harvard Business Review*, May–June 2000, 85–93; D. B. Yoffie and M. A. Cusumano, 'Building a company on Internet time: lessons from Netscape', *California Management Review*, **41** (3), Spring 1999, 8–28; *The Economist*, 18 November 2000, 'Survey of E-management'.

References

1 Teece, D. and G. Pisano (1994) 'The dynamic capabilities of firms: an introduction', *Industrial and Corporate Change*, **3**, 537–556.
2 Allen, T. (1977) *Managing the Flow of Technology*. MIT Press, Cambridge, Mass.
3 Lawrence, W. and J. Lorsch (1967) *Organization and Environment: Managing differentiation and integration*. Graduate School of Business Administration, Harvard University, Boston, Mass.
4 Cooper, R. (1983) 'A process model for industrial new product development', *IEEE Transactions on Engineering Management*, EM-30; Rothwell, R. (1977) 'The characteristics of successful innovators and technically progressive firms', *R&D Management*, **7**, 191–206.
5 Kenney, M. and R. Florida (1994) 'The organization and geography of Japanese R&D: results from a survey of Japanese electronics and biotechnology firms', *Research Policy*, **23**, 305–323.
6 Rieck, R. and K. Dickson (1993) 'A model of technology strategy', *Technology Analysis and Strategic Management*, **5**, 397–412; Mitchell, G. (1986) 'New approaches for the strategic management of technology,' in Horwitch, M. (ed.), *Technology in the Modern Corporation: A strategic perspective'*. Pergamon, New York.
7 Rousell, P., K. Saad and T. Erickson (1991) *Third Generation R&D*. Harvard Business School Press, Boston, Mass.
8 Bowman, C. and V. Ambrosini (2003) 'How the resource-based and the dynamic capability views of the firm inform corporate level strategy', *British Journal of Management*, **14**, 289–303.
9 Tidd, J. (1995) 'The development of novel products through intra- and inter-organizational networks: The case of home automation', *Journal of Product Innovation Management*, **12** (4), 307–322.
10 McDermott, C. and G. O'Connor (2002) 'Managing radical innovation: An overview of emergent strategy issues', *Journal of Product Innovation Management*, **19**, 424–438.
11 Graham, M. (1986) *RCA and the Videodisc: The business of research*. Cambridge University Press, Cambridge.
12 Uttal, B. (1983) 'The lab that ran away from Xerox', *Fortune*, 5 September.
13 Whittington, R. (1990) 'The changing structures of R&D: from centralisation to fragmentation', in Loveridge, R. and Pitt, M. (eds), *The Strategic Management of Technological Innovation*. John Wiley & Sons, Ltd, Chichester.
14 Graham, M. (1986) 'Corporate research and development: The latest transformation', in Horwitch, M. (ed.), *Technology in the Modern Corporation: A strategic perspective*. Pergamon, New York.
15 Hounshell, D. and J. Smith (1988) *Science and Corporate Strategy: Du Pont R&D, 1902–1980*. Cambridge University Press, New York; Rubenstein, A. (1989). *Managing Technology in the Decentralised Firm*. John Wiley & Sons, Ltd, Chichester.
16 Miyazaki, K. (1994) 'Search, learning and accumulation of technological competencies: the case of optoelectronics', *Industrial and Corporate Change*, **3**, 631–654. For a more detailed analysis, see Miyazaki, K. (1995) *Building Competencies in the Firm*. Macmillan, Basingstoke; Sharp, M. (1991) 'Technological trajectories and corporate strategies in the diffusion of biotechnology', in Deiaco, E., Hornell, E. and Vickery, G. (eds), *Technology and Investment*. Pinter, London.

17 Graham, M. (1986) 'Corporate research and development: the latest transformation', in Horwitch, M. (ed.), *Technology in the Modern Corporation: A strategic perspective'*. Pergamon, New York; Whittington, R. (1990) 'The changing structures of R&D: from centralisation to fragmentation', in Loveridge, R. and Pitt, M. (eds), *The Strategic Management of Technological Innovation*. John Wiley & Sons, Ltd, Chichester.

18 Coombs, R. and A. Richards (1993) 'Strategic control of technology in diversified companies with decentralised R&D', *Technology Analysis and Strategic Management*, **5**, 385–396.

19 See, for example, Koumpis, K. and K. Pavitt (1999) 'Corporate activities in speech recognition and natural language: another "New Science"-based Technology', *International Journal of Innovation Management*, **3**, 335–366; Mahdi, S. and K. Pavitt (1997) 'Key national factors in the emergence of computational chemistry firms', *International Journal of Innovation Management*, **1**, 355–386.

20 In biotechnology, see Pisano, G. (1991) 'The governance of innovation: vertical integration and collaborative relationships in the biotechnology industry', *Research Policy*, **20**, 237–249.

21 Ohmae, K. (1990) *The Borderless World: Power and strategy in the interlinked economy*. Collins, London.

22 Ghoshal, S. and C. Bartlett (1987) 'Innovation processes in multinational corporations', *Strategic Management Journal*, **8**, 425–439.

23 Cantwell, J. (1992) 'The internationalisation of technological activity and its implications for competitiveness', in Granstrand, O., Hakanson, L. and Sjolander, S. (eds), *Technology Management and International Business*, John Wiley & Sons, Ltd, Chichester; Patel, P. (1996) 'Are large firms internationalising the generation of technology? Some new evidence', *IEEE Transactions on Engineering Management*, **43**, 41–47; Ariffin, L. and M. Bell (1999) 'Firms, politics and political economy: patterns of subsidiary–parent linkages and technological capability-building in electronics TNC subsidiaries in Malaysia', in Jomo, K., Felker, G. and Rasiah, R. (eds), *Industrial Technology Development in Malaysia: Industry and firm studies*, Routledge, London and New York; Hu, Y-S. (1995) 'The international transferability of competitive advantage', *California Management Review*, **37**, 73–88; Senker, J. (1995) 'Tacit knowledge and models of innovation', *Industrial and Corporate Change*, **4**, 425–447; Senker, J., P. Benoit-Joly and M. Reinhard (1996) *Overseas Biotechnology Research by Europe's Chemical-Pharmaceuticals Multinationals: Rationale and implications*. STEEP Discussion Paper No. 33, Science Policy Research Unit, University of Sussex, Brighton; Niosi, J. (1999) 'The internationalization of industrial R&D', *Research Policy*, **29**, 107.

24 *The Economist* (1994) 'Following Chrysler', 23 April, 86–87.

25 See Gerybadze, A. and G. Reger (1999) 'Globalisation of R&D: recent changes in the management of innovation in transnational corporations', *Research Policy*, **28**, 251–274.

26 See Vernon, R. (1966) 'International investment and international trade in the product cycle', *Quarterly Journal of Economics*, **80**, 190–207.

27 Chiesa, V. (2001) *R&D Strategy and Organization*. Imperial College Press, London.

28 Chiesa, V. (2000) 'Global R&D project management and organization: a taxonomy', *Journal of Product Innovation Management*, **17**, 341–359.

29 EIRMA (European Industrial Research Management Association) (1995) *Evaluation of R&D Projects*. Paris.

30 Bower, J. (1986) *Managing the Resource Allocation Process*. Harvard Business School, Division of Research, Boston, Mass.

31 Mansfield, E. *et al.* (1972) *Research and Innovation in the Modern Corporation*. Macmillan, London.

32 Booz Allen and Hamilton (1982) *New Product Management in the 1980s*. New York.

33 Freeman, C. and L. Soete (1997) *The Economics of Industrial Innovation*, 3rd edn. Pinter, London.

34 Dvir, D. and T. Lechler (2004) 'Plans are nothing, changing plans is everything: the impact of changes on project success', *Research Policy*, **33**, 1–15.

35 Myers, S. (1984) 'Finance theory and financial strategy', *Interfaces*, **14**, 126–137.

36 McGrath, R. and A. Nerkar (2004) 'Real options reasoning and a new look at the R&D investment strategies of pharmaceutical firms', *Strategic Management Journal*, **25**, 1–21.

37 Paxon, D. (2001) 'Introduction to real R&D options', *R&D Management*, **31** (2), 109–113.

38 Loch, C. and K. Bode-Greual (2001) 'Evaluating growth options as sources of value for pharmaceutical research projects', *R&D Management*, **31** (2), 231–245.

39 Mitchell, G. and W. Hamilton (1988) 'Managing R&D as a strategic option', *Research-Technology Management*, **31**, 15–22.

40 Cooper, R., S. Edgett and E. Kleinschmidt (2001) 'Portfolio management for new product development: results of an industry practices study', *R&D Management*, **31** (4), 361–380.

41 EIRMA (European Industrial Research Management Association) (1986) *Developing R&D Strategies*. Paris; (2004) *Assessing R&D Effectiveness*.

42 Scherer, F. and K. Huh (1992) 'Top management education and R&D investment', *Research Policy*, **21**, 507; Bosworth, D. and R. Wilson (1992) *Technological Change: The role of scientists and engineers*. Avebury, Aldershot.

43 Gould, M. and A. Campbell (1987) *Strategies and Styles: The role of the centre in managing diversified corporations*. Blackwell, Oxford.

44 Chandler, A. (1991) 'The functions of the HQ unit in the multibusiness firm', *Strategic Management Journal*, **12**, 31–50.

45 Leonard-Barton, D. (1995) *Wellsprings of Knowledge: Building and sustaining the sources of innovation*. Harvard Business School Press, Boston.

46 *The Economist* (1995) 'The death of the Gene machine', 17 June, 86–92; *The Economist* (1996) 'Changing of the guard', 7 September, 72; *The Economist* (2000) 'Reinventing Marconi', 2 December, 42–47.

47 Owen, G. and T. Harrison (1995) 'Why ICI chose to demerge', *Harvard Business Review*, March–April, 133–142.

48 Pavitt, K. (2000) 'Innovating routines in the business firm: what matters, what's staying the same, and what's changing?' Keynote speech at Annual Meeting of the Schumpeter Society, Manchester, 1 July. SPRU Electronic Working Paper No. 45, University of Sussex.

Part III

ESTABLISHING EFFECTIVE EXTERNAL LINKAGES

In Part I we identified a number of generic processes for the management of innovation, which include scanning the environment, resourcing development and implementing innovation. In Part II we discovered how market position, technological paths and organizational processes influence and constrain these generic processes in the development and implementation of an innovation strategy. In this section we are concerned with the enabling routines for building effective linkages outside the organization in order to identify, resource and implement innovations. This is the essence of what has been called 'open innovation'.

Chapter 7 focuses on linkages with customers and markets. As we saw in Chapter 2, much of the early innovation research suggests that 'understanding user needs' and the involvement of 'lead users' improves the likelihood of new product success: involving users in the development process helps firms to acquire knowledge from the users, and encourages the subsequent acceptance of the innovation and commitment to its use. However, there is increasing evidence that the nature of different technologies and markets affects the process of selection and involvement of customers. In short, there is no 'one best way', but rather a range of alternatives. For example, where new products or services are very novel or complex, potential users may not be aware of, or able to articulate, their needs. In such cases traditional methods of market research are of little use, and there will be a greater burden on developers of radical new products and services to 'educate' potential users. Therefore we examine how the novelty and complexity of technologies and markets influence the identification, development and adoption of innovations.

Chapter 8 focuses on linkages with suppliers, competitors and other external sources of knowledge. Organizations collaborate for many reasons, including efficiency and flexibility, but here we are concerned with gaining access to technological and market knowledge. Such relationships may take many forms, ranging from simple licensing agreements, loose coalitions or so-called strategic alliances, to more formal joint ventures. Therefore the technological and market competencies of a specific firm may be a less reliable indicator of innovative potential than its position in a network.

Increasingly, such networks of relationships are the most appropriate unit of analysis for understanding the innovation process. A network is as much a process as a structure, which both constrains firms, and in turn is shaped by firms. In these terms collaboration can be understood as an attempt to cope with the increasing complexity and interrelatedness of different technologies and markets. We examine the technological and market motives for collaboration, and identify the organizational processes necessary to exploit it as an opportunity for knowledge acquisition and learning.

Learning from Markets

In Chapters 5 and 6 we examined how firms identify their technological competencies and develop the appropriate organizational structures and processes to support these. In this chapter we focus on the identification and development of market innovations. Market innovation includes the identification of market trends and opportunities, the translation of these requirements into new products and services, and the promotion and diffusion of these products and services. In this chapter we examine three issues:

1. How do the *characteristics of an innovation* constrain the options for development and marketing? Most marketing texts focus on relatively mature, simple low-technology products or services, but different factors will be relevant in the case of novel, complex high-technology products or services.
2. In what way do the *characteristics of potential users* affect the development and adoption of innovations? Most research on marketing examines the behaviour of consumers, but industrial and business users demand different relationships.
3. Which *commercialization or diffusion processes* are most effective in promoting the awareness and use of new products and services? The traditional distinction between 'early adopters' and 'laggards' in the take-up of innovations is unhelpful, and we need to understand what factors affect the adoption of novel products and services.

7.1 How Do Technology and Markets Affect Commercialization?

Marketing focuses on the needs of the customer, and therefore should begin with an analysis of customer requirements, and attempt to create value by providing products and services that satisfy those requirements.

The marketing mix is the set of variables that are to a large extent controllable by the company, normally referred to as the 'four Ps': product, price, place and promotion. All four factors allow some scope for innovation: product innovation results in new or improved products and services, and may change the basis of competition; product innovation allows some scope for premium pricing, and process innovation may result in price leadership; innovations in logistics may affect how a product or service is made available to customers, including distribution channels and nature of sales points; innovations in media provide new opportunities for promotion.

However, we need to distinguish between strategic marketing – that is whether or not to enter a new market – and tactical marketing, which is concerned mainly with the problem of differentiating existing products and services, and extensions to such products. There is a growing body of research that suggests that factors which contribute to new product success are not universal, but are contingent upon a range of technological and market characteristics. A recent study of 110 development projects found that complexity, novelty and whether the project was for hardware or software development affected the factors that contributed to success.[1] Our own research confirms that different managerial processes, structures and tools are appropriate for routine and novel development projects.[2] For example, in terms of frequency of use, the most common methods used for high novelty projects are segmentation, prototyping, market experimentation and industry experts, whereas for the less novel projects the most common methods are partnering customers, trend extrapolation and segmentation. The use of market experimentation and industry experts might be expected where market requirements or technologies are uncertain, but the common use of segmentation for such projects is harder to justify. However, in terms of usefulness, there are statistically significant differences in the ratings for segmentation, prototyping, industry experts, market surveys and latent needs analysis. Segmentation is the only method more effective for routine development projects, and prototyping, industry experts, focus groups and latent needs analysis are all more effective for novel development projects (Table 7.1). For example, IDEO, the global design and development consultancy, finds conventional market research methods insufficient and sometimes misleading for new products and services, and instead favours the use of direct observation and prototyping (see Box 7.1).

Clearly then, many of the standard marketing tools and techniques are of limited utility for the development and commercialization of novel or complex new products or services. A number of weaknesses can be identified:

- *Identifying and evaluating novel product characteristics.* Marketing tools such as conjoint analysis have been developed for variations of existing products or product extensions, and therefore are of little use for identifying and developing novel products or applications.
- *Identifying and evaluating new markets or businesses.* Marketing techniques such as segmentation are most applicable to relatively mature, well-understood products and markets, and are of limited use in emerging, ill-defined markets.

TABLE 7.1 The effect of product novelty on the tools used for new product and service and development

	High novelty		Low novelty	
	Usage (%)	Usefulness	Usage (%)	Usefulness
Segmentation*	89	3.42	42	4.50
Prototyping*	79	4.33	63	4.08
Market experimentation	63	4.00	53	3.70
Industry experts*	63	3.83	37	3.71
Surveys/focus groups*	52	4.50	37	4.00
Trend extrapolation	47	4.00	47	3.44
Latent needs analysis*	47	3.89	32	3.67
User-practice observation	47	3.67	42	3.50
Partnering customers	37	4.43	58	3.67
User-developers	32	4.33	37	3.57
Scenario development	21	3.75	26	2.80
Role-playing	5	4.00	11	1.00

* Denotes difference in usefulness rating is statistically significant at 5% level ($n = 50$).
Source: Adapted from Tidd, J. and K. Bodley (2002) 'Effect of project novelty on the effectiveness of tools used to support new product development', *R&D Management*, **32**, 2, 127–138.

- *Promoting the purchase and use of novel products and services.* The traditional distinction between consumer and business marketing is based on the characteristics of the customers or users, but the characteristics of the innovation and the relationship between developers and users is more important in the case of novel and complex products and services.

Therefore before applying the standard marketing techniques, we must have a clear idea of the maturity of the technologies and markets. Figure 7.1 presents a simple two-by-two matrix, with technological maturity as one dimension, and market maturity as the other. Each quadrant raises different issues and will demand different techniques for development and commercialization:

- *Differentiated.* Both the technologies and markets are mature, and most innovations consist of the improved use of existing technologies to meet a known customer need. Products and services are differentiated on the basis of packaging, pricing and support.
- *Architectural.* Existing technologies are applied or combined to create novel products or services, or new applications. Competition is based on serving specific market niches and on close relations with customers. Innovation typically originates or is in collaboration with potential users.
- *Technological.* Novel technologies are developed which satisfy known customer needs. Such products and services compete on the basis of performance, rather than price or quality. Innovation is mainly driven by developers.

BOX 7.1
LEARNING FROM USERS AT IDEO

IDEO is one of the most successful design consultancies in the world, based in Palo Alto, California and London, UK, it helps large consumer and industrial companies worldwide to design and develop innovative new products and services. Behind its rather typical Californian wackiness lies a tried and tested process for successful design and development:

1. Understand the market, client and technology.
2. Observe users and potential users in real life situations.
3. Visualize new concepts and the customers who might use them, using prototyping, models and simulations.
4. Evaluate and refine the prototypes in a series of quick iterations.
5. Implement the new concept for commercialization.

The first critical step is achieved through close *observation* of potential users in context. As Tom Kelly of IDEO argues, 'We're not big fans of focus groups. We don't much care for traditional market research either. We go to the source. Not the "experts" inside a (client) company, but the actual people who use the product or something similar to what we're hoping to create . . . we believe you have to go beyond putting yourself in your customers' shoes. Indeed we believe it's not even enough to *ask* people what they think about a product or idea . . . customers may lack the vocabulary or the palate to explain what's wrong, and especially what's *missing*.'

The next step is to develop prototypes to help evaluate and refine the ideas captured from users. 'An iterative approach to problems is one of the foundations of our culture of prototyping . . . you can prototype just about anything – a new product or service, or a special promotion. What counts is moving the ball forward, achieving some part of your goal.'

Source: Kelly, T. *The Art of Innovation: Lessons in creativity from IDEO.* HarperCollinsBusiness, 2002.

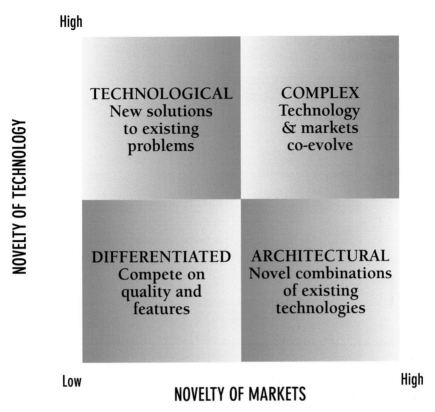

High

NOVELTY OF TECHNOLOGY

TECHNOLOGICAL
New solutions
to existing
problems

COMPLEX
Technology
& markets
co-evolve

DIFFERENTIATED
Compete on
quality and
features

ARCHITECTURAL
Novel combinations
of existing
technologies

Low   **High**

NOVELTY OF MARKETS

FIGURE 7.1   Technological and market maturity determine the marketing process

- *Complex.* Both technologies and markets are novel, and co-evolve. In this case there is no clearly defined use of a new technology, but over time developers work with lead users to create new applications. The development of multimedia products and services is a recent example of such a co-evolution of technologies and markets.

Assessing the maturity of a market is particularly difficult, mainly due to the problem of defining the boundaries of a market. The real rate of growth of a market provides a good estimate of the stage in the product life cycle and, by inference, the maturity of the market. In general high rates of market growth are associated with high R&D costs, high marketing costs, rising investment in capacity and high product margins (Figure 7.2). At the firm level there is a significant correlation between expenditure on R&D, number of new product launches and financial measures of performance such as value-added and market to book value.[3] Generally, profitability declines as a market matures as the scope for product and service differentiation reduces, and competition shifts towards price.

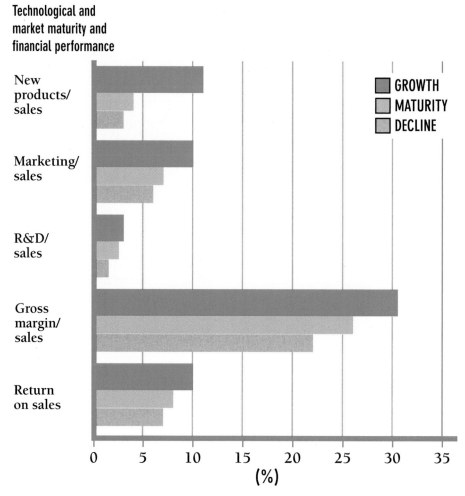

FIGURE 7.2 Market maturity affects the innovation process

Source: Derived from Buzzell, R. D. and B. T. Gale (1987) *The PIMS Principle*. Free Press, New York.

7.2 Differentiating Products

In Chapter 3 we discussed generic corporate strategies based on price leadership or differentiation. Here we are concerned with the specific issue of how to differentiate a product from competing offerings where technologies and markets are relatively stable. It is in these circumstances that the standard tools and techniques of marketing are most useful. We assume the reader is familiar with the basics of marketing, so here we shall focus on product differentiation by quality and other attributes.

Differentiation measures the degree to which competitors differ from one another in a specific market. Markets in which there is little differentiation and no significant difference in the relative quality of competitors are characterized by low profitability, whereas differentiation on the basis of relative quality or other product characteristics is a strong predictor of high profitability in any market conditions. Where a firm achieves a combination of high differentiation and high perceived relative quality, the return on investment is typically twice that of non-differentiated products. Analysis of the Strategic Planning Institute's database of more than 3000 business units helps us to identify the profit impact of market strategy (PIMS):[4]

- *High relative quality is associated with a high return on sales.* One reason for this is that businesses with higher relative quality are able to demand higher prices than their competitors. Moreover, higher quality may also help reduce costs by limiting waste and improving processes. As a result companies may benefit from both higher prices and lower costs than competitors, thereby increasing profit margins.
- *Good value is associated with increased market share.* Plotting relative quality against relative price provides a measure of relative value: high quality at a high price represents average value, but high quality at a low price represents good value. Products representing poor value tend to lose market share, but those offering good value gain market share.
- *Product differentiation is associated with profitability.* Differentiation is defined in terms of how competitors differ from each other within a particular product segment. It can be measured by asking customers to rank the individual attributes of competing products, and to weight the attributes. Customer weighting of attributes is likely to differ from that of the technical or marketing functions.

Analysis of the PIMS data reveals a more detailed picture of the relationships between innovation, value and market performance (Figure 7.3). Process innovation helps to improve relative quality and to reduce costs, thereby improving the relative value of the product. Product innovation also affects product quality, but has a greater effect on reputation and value. Together, innovation, relative value and reputation drive growth in market share. For example, there is an almost linear relationship between product innovation and market growth: businesses with low levels of product innovation – that is having less than 1% of products introduced in the last three years – experience an average real annual market growth of less than 1%; whereas businesses with high levels – that is having around 8% of products introduced in the past three years – experience real annual market growth of around 8%.[5] The compound effect of such differences in real growth can have a significant impact on relative market share over a relatively short period of time. However, in consumer markets maintaining high levels of new product introduction is necessary, but not sufficient. In addition, reputation, or brand image,

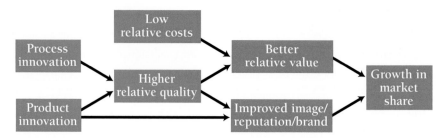

FIGURE 7.3   Relationship between innovation and market performance

Source: Adapted from Clayton, T. and G. Turner (2000) 'Brands, innovation and growth', in Tidd, J. (ed.), *From Knowledge Management to Strategic Competence: Measuring technological, market and organizational innovation.* Imperial College Press, London.

must be established and maintained, as without it consumers are less likely to sample new product offerings whatever the value or innovativeness. Witness the rapid and consistent growth of Nokia in the mobile phone market (see Box 7.2).

Quality function deployment (QFD) is a useful technique for translating customer requirements into development needs, and encourages communication between engineering, production and marketing. Unlike most other tools of quality management, QFD is used to identify opportunities for product improvement or differentiation, rather than to solve problems. Customer-required characteristics are translated or 'deployed' by means of a matrix into language which engineers can understand (Figure 7.4). The construction of a relationship matrix – also known as 'the house of quality' – requires a significant amount of technical and market research. Great emphasis must be made on gathering market and user data in order to identify potential design trade-offs, and to achieve the most appropriate balance between cost, quality and performance. The construction of a QFD matrix involves the following steps:[6]

1. Identify customer requirements, primary and secondary, and any major dislikes.
2. Rank requirements according to importance.
3. Translate requirements into measurable characteristics.
4. Establish the relationship between the customer requirements and technical product characteristics, and estimate the strength of the relationship.
5. Choose appropriate units of measurement and determine target values based on customer requirements and competitor benchmarks.

Symbols are used to show the relationship between customer requirements and technical specifications, and weights attached to illustrate the strength of the relationship. Horizontal rows with no relationship symbol indicate that the existing design is

BOX 7.2
NOKIA: DIFFERENTIATION BY DESIGN AND INNOVATION

Founded in 1865, Nokia began as a forestry company, and for almost 100 years remained in the pulp and paper industry until a series of unrelated acquisitions in the 1960s. However, in the past decade Nokia has transformed itself from a sprawling conglomerate, with a wide range of mature, low-margin products, to the world's largest manufacturer of mobile phones, with around 30% of the global market for handsets, and 60% share of the industry profits.

When Jorma Ollila, a graduate of the LSE in London, became chief executive in 1992, his strategy consisted of four phrases: 'global', 'telecom-orientated', 'focus' and 'value-added'. In 1993 Ollila told the European Commission that he thought that Europe had lost the computer market to the USA, and consumer electronics to Japan, but could still dominate the emerging telecom market. By 1992 Nokia had become a conglomerate with businesses in aluminium, cables, paper, rubber, televisions, tyres, power generation and real estate. In the mid-1980s Nokia acquired its first interest in the electronics sector, the company Teleste, and in 1991 acquired the British telecoms company, Techophone. In 1985 the share of Nokia's revenues derived from telecoms was just 14%, but following disposal of almost all non-telecoms businesses, this share had grown to almost 90% by 1995. Its success has been based on a combination of technological innovation and product design. It has benefited from its early focus on digital and data traffic, rather than analogue and voice, and the co-development of the GSM (global system of mobile communications) standard, but also has developed a strong brand and carefully segmented the market by means of product design.

Annual growth has regularly been some 40%, and in 2000 the company sold more than 400 million phones. Despite the trend towards a high-volume mass market, Nokia achieves product margins of up to 25%, compared to its rivals' 1 to 3%. The company is now the fifth largest in Europe, employs 44000 people in 11 countries, almost half of which are Finns. Due to the high rate of growth, half of Nokia's staff has worked for the company for less than three years, and the average age is only 32. It spends around 9% of revenues on R&D, and around a third of its staff work in design or R&D.

In 2004 Nokia began to lose sales and market share, faced with increased competition from SonyEricsson and Samsung. One reason was a delay in developing newer clam-shell style handsets pioneered by competitors such as Samsung and Motorola; another reason is it being late in developing higher-end camera phones. As a result margins dropped to below 20%, and market share to below 30%, from a peak of 35% in 2003. In 2004 Nokia planned to introduce between 15 to 20 new phones to remedy this, and remains the most profitable producer of mobile phones.

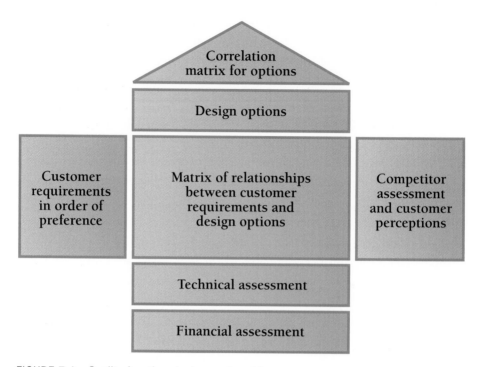

FIGURE 7.4   Quality function deployment matrix

incomplete. Conversely, vertical columns with no relationship symbol indicate that an existing design feature is redundant as it is not valued by the customer. In addition, comparisons with competing products, or benchmarks, can be included. This is important because relative quality is more relevant than absolute quality: customer expectations are likely to be shaped by what else is available, rather than some ideal.

In some cases potential users may have latent needs or requirements which they cannot articulate. In such cases three types of user need can be identified: 'must be's', 'one-dimensionals' and attractive features or 'delighters'.[7] Must be's are those features which must exist before a potential customer will consider a product or service. For example, in the case of an executive car it must be relatively large and expensive. One-dimensionals are the more quantifiable features which allow direct comparison between competing products. For example, in the case of an executive car, the acceleration and braking performance. Finally, the delighters, which are the most subtle means of differentiation. The inclusion of such features delights the target customers, even if they do not explicitly demand them. For example, delighters in the case of an executive car include ultrasonic parking aids, rain-sensitive windscreen wipers and photochromatic mirrors. Such features are rarely demanded by customers or identified by regular market research. However, indirect questioning can be used to help identify latent requirements.

QFD was originally developed in Japan, and is claimed to have helped Toyota to reduce its development time and costs by 40%. More recently many leading American firms have adopted QFD, including AT&T, Digital and Ford, but results have been mixed: only around a quarter of projects have resulted in any quantifiable benefit.[8] In contrast, there has been relatively little application of QFD by European firms. This is not the result of ignorance, but rather a recognition of the practical problems of implementing QFD.

Clearly, QFD requires the compilation of a lot of marketing and technical data, and more importantly the close co-operation of the development and marketing functions. Indeed, the process of constructing the relationship matrix provides a structured way of getting people from development and marketing to communicate, and therefore is as valuable as any more quantifiable outputs. However, where relations between the technical and marketing groups are a problem, which is too often the case, the use of QFD may be premature. The problem of cross-functional communication in product development will be discussed in more detail in Chapter 9.

7.3 Creating Architectural Products

Architectural products consist of novel combinations of existing technologies that serve new markets or applications. In such cases the critical issue is to identify or create new market segments.

Market share is associated with profitability: on average, market leaders earn three times the rate of return of businesses ranked fifth or less.[9] Therefore the goal is to segment a market into a sufficiently small and isolated segment which can be dominated and defended. This allows the product and distribution channels to be closely matched to the needs of a specific group of customers.

Market or buyer segmentation is simply the process of identifying groups of customers with sufficiently similar purchasing behaviour so that they can be targeted and treated in a similar way. This is important because different groups are likely to have different needs. By definition the needs of customers in the same segment will be highly homogeneous. In formal statistical terms the objective of segmentation is to maximize across-group variance and to minimize within-group variance.

In practice segmentation is conducted by analysing customers' buying behaviour and then using factor analysis to identify the most significant variables influencing behaviour – descriptive segmentation – and then using cluster analysis to create distinct segments which help identify unmet customer needs – prescriptive segmentation. The principle of segmentation applies to both consumer and business markets, but the process and basis of segmentation is different in each case.

Segmenting Consumer Markets

Much of the research on the buying behaviour of consumers is based on theories adapted from the social and behavioural sciences. Utilitarian theories assume that consumers are rational and make purchasing decisions by comparing product utility with their requirements. This model suggests a sequence of phases in the purchasing decision: problem recognition, information search, evaluation of alternatives and finally the purchase. However, such rational processes do not appear to have much influence on actual buying behaviour. For example, in the UK the Consumers' Association routinely tests a wide range of competing products, and makes buying recommendations based on largely objective criteria. If the majority of buyers were rational, and the Consumers' Association successfully identified all relevant criteria, these recommendations would become best-sellers, but this is not the case.

Behavioural approaches have greater explanatory power. These emphasize the effect of attitude, and argue that the buying decision follows a sequence of changing attitudes to a product – awareness, interest, desire and finally action. The goal of advertising is to stimulate this sequence of events. However, research suggests that attitude alone explains only 10% of decisions, and can rarely predict buyer behaviour.

In practice the balance between rational and behavioural influences will depend on the level of customer involvement. Clearly, the decision-making process for buying an aircraft or machine tool is different from the process of buying a toothpaste or shampoo. Many purchasing decisions involve little cost or risk, and therefore low involvement. In such cases consumers try to minimize the financial, mental and physical effort involved in purchasing. Advertising is most effective in such cases. In contrast, in high-involvement situations, in which there is a high cost or potential risk to customers, buyers are willing to search for information and make a more informed decision. Advertising is less effective in such circumstances, and is typically confined to presenting comparative information between rival products.

There are many bases of segmenting consumer markets, including by socio-economic class, life-cycle groupings and by lifestyle or psychographic (psychological–demographic) factors. An example of psychographic segmentation is the Taylor–Nelson classification that consists of self-explorers, social registers, experimentalists, achievers, belongers, survivors and the aimless. Better-known examples include the *Yuppy* (young upwardly mobile professional) and *Dinky* (dual income, no kids), and the more recent *Yappy* (young affluent parent), *Sitcoms* (single income, two children, oppressive mortgage), and *Skiers* (spending the kids' inheritance). There is often a strong association between a segment and particular products and services. For example, the yuppy of the 1980s was defined by a striped shirt and braces, personal organizer, brick-sized mobile phone and, of course, a BMW. Annual sales of Filofax were just £100 000 in the UK in 1980, but the deregulation of the City of London in 1986 created 50 000 new, highly paid jobs. As a result annual sales of Filofax reached a peak of £6 m. in 1986, the year before the City crashed.

Such segmentation is commonly used for product development and marketing in fast-moving consumer goods such as foods or toiletries and consumer durables such as consumer electronics or cars (see Box 7.3). It is of particular relevance in the case of product variation or extension, but can also be used to identify opportunities for new products, such as functional foods for the health-conscious, and emerging requirements such as new pharmaceuticals and health-care services for the wealthy elderly.

Segmenting Business Markets

Business customers tend to be better informed than consumers and, in theory at least, make more rational purchasing decisions. Business customers can be segmented on the basis of common buying factors or purchasing processes. The basis of segmentation should have clear operational implications, such as differences in preferences, pricing, distribution or sales strategy. For example, customers could be segmented on the basis of how experienced, sophisticated or price-sensitive they are. However, the process is complicated by the number of people involved in the buying process:

- The actual customer or buyer, who typically has the formal authority to choose a supplier and agree terms of purchase.
- The ultimate users of the product or service, who are normally, but not always, involved in the initiation and specification of the purchase.
- Gatekeepers, who control the flow of information to the buyers and users.
- Influencers, who may provide some technical support to the specification and comparison of products.

Therefore it is critical to identify all relevant parties in an organization, and determine the main influences on each. For example, technical personnel used to determine the specification may favour performance, whereas the actual buyer may stress value for money.

The most common basis of business segmentation is by the benefits customers derive from the product, process or service. Customers may buy the same product for very different reasons, and attach different weightings to different product features. For example, in the case of a new numerically controlled machine tool, one group of

BOX 7.3
THE MARKETING OF PERSIL POWER

In 1994 the Anglo-Dutch firm Unilever launched its revolutionary new washing powder 'Persil Power' across Europe ('Omo Power' in some European markets). It was heralded as the first major technological breakthrough in detergents for 15 years. Development had taken 10 years and more than £100m. The product contained a manganese catalyst, the so-called 'accelerator', which Unilever claimed washed whiter at lower temperatures. The properties of manganese were well known in the industry, but in the past no firm had been able to produce a catalyst which did not also damage clothes. Unilever believed that it had developed a suitable manganese catalyst, and protected its development with 35 patents. The company had test marketed the new product in some 60 000 households and more than 3 million washes, and was sufficiently confident to launch the product in April 1994. However, reports by Procter & Gamble, Unilever's main rival, and subsequent tests by the British Consumers' Association found that under certain conditions Persil Power significantly damaged clothes. After a fierce public relations battle Unilever was forced to withdraw the product, and wrote off some £300m. in development and marketing costs. What went wrong?

There were many reasons for this, but with the benefit of hindsight two stand out. First, the nature of the test marketing and segmentation. Unilever had conducted most of its tests in Dutch households. Typically, northern Europeans separate their whites from their coloured wash, and tend to read product instructions. In contrast, consumers in the South are more likely to wash whites and dyed fabrics together, and to wash everything on a hot wash irrespective of any instructions to the contrary. The manganese catalyst was fine at low temperatures for whites only, but reacted with certain dyes at higher temperatures. Second, the nature of the product positioning. Persil Power was launched as a broad-base detergent suitable for all fabrics, but in practice was only a niche product effective for whites at low temperatures. Unilever learned a great deal from this product launch, and has since radically reorganized its product development process to improve communication between the research, development and marketing functions. Now product development is concentrated in a small number of innovation centres, rather than being split between central R&D and the product divisions, and the whole company uses the formal new product development process based on the development funnel discussed in Chapter 9.

customers may place the greatest value on the reduction in unit costs it provides, whereas another group may place greater emphasis on potential improvements in precision or quality of the output (see Box 7.4).

It is difficult in practice to identify distinct segments by benefit because these are not strongly related to more traditional and easily identifiable characteristics such as firm size or industry classification.[10] Therefore benefit segmentation is only practical where such preferences can be related to more easily observable and measurable customer characteristics. For example, in the case of the machine tool, analysis of production volumes, batch sizes, operating margins and value added might help differentiate between those firms which value higher efficiency from those which seek improvements in quality.

This suggests a three-stage segmentation process for identifying new business markets:

1. First, a segmentation based on the functionality of the technology, mapping functions against potential applications.
2. Next, a behavioural segmentation to identify potential customers with similar buying behaviour, for example regarding price or service.
3. Finally, combine the functional and behavioural segmentations in a single matrix to help identify potential customers with relevant applications and buying behaviour.

In addition, analysis of competitors' products and customers may reveal segments not adequately served, or alternatively an opportunity to redefine the basis of segmentation. For example, existing customers may be segmented on the basis of size of company, rather than the needs of specific sectors or particular applications. However, in the final analysis segmentation only provides a guide to behaviour as each customer will have unique characteristics.

There is likely to be a continuum of customer requirements, ranging from existing needs, to emerging requirements and latent expectations, and these must be mapped on to existing and emerging technologies.[11] Whereas much of conventional market research is concerned with identifying the existing needs of customers and matching these to existing technological solutions, in this case the search has to be extended to include emerging and new customer requirements. There are three distinct phases of analysis:

1. Cross-functional teams including customers are used to generate new product concepts by means of brainstorming, morphology and other structured techniques.
2. These concepts are refined and evaluated, using techniques such as QFD.

BOX 7.4
THE MARKETING OF MONDEX

Mondex is a smart card which can be used to store cash credits – in other words, an electronic purse. The card incorporates a chip which allows cash-free transfers of monetary value from consumer to retailer, and from retailer to bank. NatWest bank first conceived of Mondex in 1990. The rationale for development of the system was the huge costs involved in handling small amounts of cash, estimated to be some £4.5 bn in the UK each year, and therefore the banks and retailers are the main potential beneficiaries. The benefits to consumers are less clear.

In 1991 NatWest created a venture to franchise the system worldwide, and in the UK entered alliances with Midland Bank and BT. Interviews with customer focus groups were conducted in the UK, USA, France, Germany and Japan to determine the likely demand for the service. The results of this initial market research suggested that up to 80% of potential customers would use Mondex, if available. Therefore internal technical trials went ahead in 1992, based on 6000 staff of NatWest. As a result, minor improvements were made, such as a key fob to read the balance remaining on a card, and a locking facility. Market trials began in Swindon in 1995, chosen for its demographic representativeness. Almost 70% of the town's retailers were recruited to the pilot, although several large multiple retailers declined to participate as they were planning their own cards. Some 14000 customers of NatWest and Midland applied for a free card, but this represented just 25% of their combined customer base in the town. The main barrier to adoption appeared to be the lack of clear benefits to users, whereas the banks and retailers clearly benefited from reduced handling and security costs.

Nevertheless, in 1996 it was announced that Mondex would be offered to all students of Essex University, and cards were to include a broader range of functions including student identification and library access, as well as being accepted by all the banks, shops and bars on campus. University students are ideal consumers of such innovative services, and the campus environment represents a controllable environment in which to test the attractiveness of the service where universal acceptance is guaranteed. Five other universities were subsequently recruited to the three-year trial.

In 1996 Mondex was spun off from NatWest Bank, and is now owned by a consortium headed by Mastercard International. The main competing products are Visa Cash and Belgium's Proton technology. Only 2 million Mondex cards were in use in 2000, but many millions more are to be used by large credit card companies such as JCB of Japan which plans to replace 15 million credit, debit and loyalty cards over the next few years. In addition, Mondex technology, in particular its well-regarded operating system MultOS, has since successfully licensed its technology in more than 50 countries. In 2000 it was announced that Mondex technology was to be used in the Norwegian national lottery, and Mondex was part of a bid consortium for the UK national lottery. Thus the technology and associated business have evolved from a narrow focus on electronic cash, to the broader issue of smart card applications.

3. Parallel prototype development and market research activities are conducted. Prototypes are used not as 'master models' for production, but as experiments for internal and external customers to evaluate.

Where potential customers are unable to define or evaluate product design features, in-depth interview clinics must be carried out with target focus groups or via antenna shops. In antenna shops, market researchers and engineers conduct interactive customer interviews, and use marketing research tools and techniques to identify and quantify perceptions about product attributes.

Product mapping can be used to expose the technological and market drivers of product development, and allows managers to explore the implications of product extensions. It helps to focus development efforts and limit the scope of projects by identifying target markets and technologies. This helps to generate more detailed functional maps for design, production and marketing. An initial product introduction, or 'core' product, can be extended in a number of ways:

- An enhanced product, which includes additional distinctive features designed for an identified market segment.
- An 'upmarket' extension. This can be difficult because customers may associate the company with a lower quality segment. Also, sales and support staff may not be sufficiently trained or skilled for the new segments.
- A 'downmarket' extension. This runs the risk of cannibalizing sales from the higher end, and may alienate existing customers and dealers.
- Custom products with additional features required by a specific customer or distribution channel.
- A hybrid product, produced by merging two core designs to produce a new product.

In his detailed analysis of the disk drive industry, Clayton Christensen distinguishes between two types of architectural innovation.[12] The first, *sustaining* innovation, which continues to improve existing product functionality for existing customers and markets. The second, *disruptive* innovation, provides a different set of functions which are likely to appeal to a very different segment of the market. As a result, existing firms and their customers are likely to undervalue or ignore disruptive innovations, as these are likely to underperform existing technologies in terms of existing functions in established markets. This illustrates the danger of simplistic advice such as 'listening to customers', and the limitations of traditional management and marketing approaches. Therefore established firms tend to be blind to the potential of disruptive innovation, which is more likely to be exploited by new entrants. Segmentation of current markets and close relations with existing customers will tend to reinforce sustaining innovation, but will fail to identify or wrongly reject potential disruptive innovations. Instead firms must develop and maintain a detailed understanding of potential applications and changing users' needs.

A fundamental issue in architectural innovation is to identify the need to change the architecture itself, rather than just the components within an existing architecture. New product introduction is, up to a point, associated with higher sales and profitability, but very high rates of product introduction become counterproductive as increases in development costs exceed additional sales revenue. This was the case in the car industry, when Japanese manufacturers reduced the life cycle to just four years in the 1990s, but then had to extend it again. Alternatively, expectations of new product introductions can result in users skipping a generation of products in anticipation of the next generation. This has happened in both the PC and cell phone markets, which has had knock-on effects in the chip industry. Put another way, there is often a trade-off between high rates of new product introduction and product life. The development of common product platforms and increased modularity is one way to try to tackle this trade-off in new product development. Incremental product innovation within an existing platform can either introduce benefits to *existing* customers, such as lower price or improved performance, or additionally attract *new* users and enter new market niches. A study of 56 firms and over 240 new products over a period of 22 years found that a critical issue in managing architectural innovation is the precise balance between the frequency of radical change of product platform, and incremental innovation within these platforms.[13] This suggests that a strategy of ever-faster new product development and introduction is not sustainable, but rather the aim should be to achieve an optimum balance between platform change and new product based on existing platforms. This logic appears to apply to both manufactured products and services (see Box 7.5).

BOX 7.5
PRODUCT STRATEGIES IN SERVICES

Services differ from manufactured goods in many ways, but the two characteristics that most influence innovation management are their intangibility and the interaction between production and consumption. The intangibility of most services makes differentiation more difficult as it is harder to identify and control attributes. The near simultaneous production and consumption of many service offerings blurs the distinction between process (how) and product (what) innovation, and demands the integration of back and front end operations.

For example, in our study of 108 service firms in the UK and USA, we found that a strategy of rapid, reiterative redevelopment ('RRR') was associated with higher levels of new service development success and higher service quality. This approach to new service development combines many of the benefits of the polar extremes of radical and incremental innovation, but with lower costs and risks. This strategy is less disruptive to internal functional relationships than infrequent but more radical service innovations, and encourages knowledge reuse through the accumulation of numerous incremental innovations. For example, in 1995 the American Express Travel Service Group implemented a strategy of RRR. In the previous decade, the group had introduced only two new service products. In 1995 a vice-president of product development was created, cross-functional teams were established, a formal development process adopted, and computer tools, including prototyping and simulation, were deployed. Since then the group has developed and launched more than 80 new service offerings, and has become the market leader.

Source: Tidd, J. and F. Hull (2003) *Service Innovation: Organizational responses to technological opportunities and market imperatives*. Imperial College Press, London.

7.4 Marketing Technological Products

Technological products are characterized by the application of new technologies in existing products or relatively mature markets. In this case the key issue is to identify existing applications where the technology has a cost or performance advantage.

The traditional literature on industrial marketing has a bias towards relatively low-technology products, and has failed largely to take into account the nature of high-technology products and their markets.

The first and most critical distinction to make is between a technology and a product.[14] Technologists are typically concerned with developing devices, whereas potential customers buy products, which marketing must create from the devices. Developing a product is much more costly and difficult than developing a device. Devices that do not function or are difficult to manufacture are relatively easy to identify and correct compared to an incomplete product offering. A product may fail or be difficult to sell due to poor logistics and branding, or difficult to use because insufficient attention has been paid to customer training or support. Therefore attempting to differentiate a product on the basis of its functionality or the performance of component devices can be expensive and futile.

For example, a personal computer (PC) is a product consisting of a large number of devices or sub-systems, including the basic hardware and accessories, operating system, application programs, languages, documentation, customer training, maintenance and support, advertising and brand development. Therefore a development in microprocessor technology, such as RISC (reduced instruction set computing) may improve product performance in certain circumstances, but may be undermined by more significant factors such as lack of support for developers of software and therefore a shortage of suitable application software.

Therefore in the case of high-technology products it is not sufficient to carry out a simple technical comparison of the performance of technological alternatives, and conventional market segmentation is unlikely to reveal opportunities for substituting a new technology in existing applications. It is necessary to identify why a potential customer might look for an alternative to the existing solution. It may be because of lower costs, superior performance, greater reliability, or simply fashion. In such cases there are two stages to identify potential applications and target customers: technical and behavioural.[15]

Statistical analysis of existing customers is unlikely to be of much use because of the level of detail required. Typically technical segmentation begins with a small group of potential users being interviewed to identify differences and similarities in their requirements. The aim is to identify a range of specific potential uses or applications. Next, a behavioural segmentation is carried out to find three or four groups of customers with similar situations and behaviour. Finally, the technical and behavioural segments are combined to define specific groups of target customer and markets that can then be evaluated commercially (Figure 7.5).

Several features are unique to the marketing of high-technology products, and affect buying behaviour:[16]

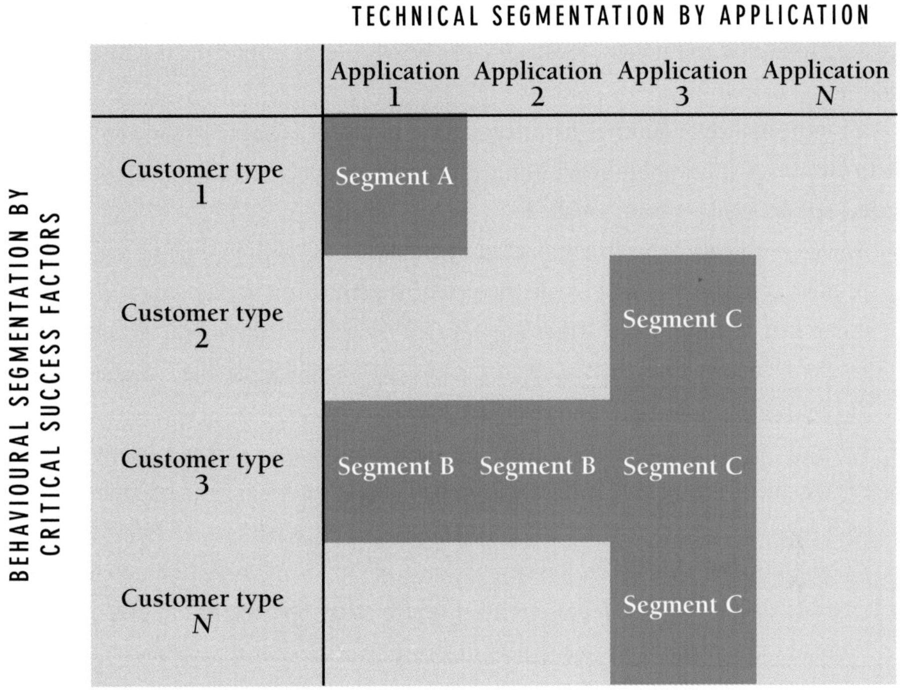

TECHNICAL SEGMENTATION BY APPLICATION

	Application 1	Application 2	Application 3	Application N
Customer type 1	Segment A			
Customer type 2			Segment C	
Customer type 3	Segment B	Segment B	Segment C	
Customer type N			Segment C	

BEHAVIOURAL SEGMENTATION BY CRITICAL SUCCESS FACTORS

FIGURE 7.5 Technical and behavioural segmentation for marketing high-technology products

- *Buyers' perceptions of differences in technology affect buying behaviour.* In general, where buyers believe technologies to be similar, they are likely to search for longer than when they believe there to be significant differences between technologies.
- *Buyers' perceptions of the rate of change of the technology affects buying behaviour.* In general, where buyers believe the rate of technological change is high, they put a lot of effort in the search for alternatives, but search for a shorter time. In non-critical areas a buyer may postpone a purchase.
- *Organizational buyers may have strong relationships with their suppliers, which increases switching costs.* In general, the higher the supplier-related switching costs, the lower the search effort, but the higher the compatibility-related switching costs, the greater the search effort.

Exploiting Intellectual Property

In addition, in some cases technology can be commercialized by licensing or selling the intellectual property rights (IPR), rather than developing products or processes. For example, in 1998 IBM reported licence income of US$1bn, and in the USA the total royalty income from licensing was around US$100bn. There is a range of IPR that can be used to exploit technology, the main types being patents, copyright and design rights and registration.

All industrialized countries have some form of patent legislation, the aim of which is to encourage innovation by allowing a limited monopoly, usually for 20 years, provided certain legal tests are satisfied:

- *Novelty* – no part of 'prior art', including publications, written, oral or anticipation. In most countries the first to file the patent is granted the rights, but in the USA it is the first to invent. The American approach may have the moral advantage, but results in many legal challenges to patents, and requires detailed documentation during R&D.

- *Inventive step* – 'not obvious to a person skilled in the art'. This is a relative test, as the assumed level of skill is higher in some fields than others. For example, Genentech was granted a patent for the plasminogen activator t-PA which helps to reduce blood clots, but despite its novelty, a Court of Appeal revoked the patent on the grounds that it did not represent an inventive step because its development was deemed to be obvious to researchers in field.

- *Industrial application* – utility test requires the invention to be capable of being applied to a machine, product or process. In practice a patent must specify an application for the technology, and additional patents sought for any additional application. For example, Unilever developed Ceramides and patented their use in a wide range of applications. However, it did not apply for a patent for application of the technology to shampoos, which was subsequently granted to a competitor.

- *Patentable subject* – for example, discoveries and formula cannot be patented, and in Europe neither can software (the subject of copyright) nor new organisms, although both these are patentable in the USA. For example, contrast the mapping of the human genome in the USA and Europe: in the USA the research is being conducted by a commercial laboratory which is patenting the outcomes, and in Europe by a group of public laboratories which is publishing the outcomes on the Internet.

- *Clear and complete disclosure*. Note that a patent provides only certain legal property rights, and in the case of infringement the patent holder needs to take the appropriate legal action. In some cases secrecy may be a preferable strategy. Conversely, national patent databases represent a large and detailed reservoir of technological innovations which can be interrogated for ideas.

Copyright is concerned with the expression of ideas, and not the ideas themselves. Therefore the copyright exists only if the idea is made concrete – for example, in a book or recording. There is no requirement for registration, and the test of originality is low compared to patent law, requiring only that 'the author of the work must have used his own skill and effort to create the work'. Like patents, copyright provides limited legal rights for certain types of material for a specific term. For literary, dramatic, musical and artistic works copyright is normally for 70 years after the death of the author, 50 in the USA, and for recordings, film, broadcast and cable programmes 50 years from their creation. Typographical works have 25 years' copyright. The type of materials covered by copyright include:

- 'Original' literary, dramatic, musical and artistic works, including software and in some cases databases.
- Recordings, films, broadcasts and cable programmes.
- Typographical arrangement or layout of a published edition.

Design rights are similar to copyright protection, but mainly apply to three-dimensional articles, covering any aspect of the 'shape' or 'configuration', internal or external, whole or part, but specifically excluding integral and functional features, such as spare parts. Design rights exist for 15 years, and 10 years if commercially exploited. Design registration is a cross between patent and copyright protection, is cheaper and easier than patent protection, but more limited in scope. It provides protection for up to 25 years, but covers only visual appearance – shape, configuration, pattern and ornament. It is used for designs that have aesthetic appeal – for example, consumer electronics and toys. For example, the knobs on top of Lego bricks are functional, and would therefore not qualify for design registration, but were also considered to have 'eye appeal', and therefore granted design rights.[17]

Licensing IPR can have a number of benefits:

- Reduce or eliminate production and distribution costs and risks.
- Reach a larger market.
- Exploit in other applications.
- Establish standards.
- Gain access to complementary technology.
- Block competing developments.
- Convert competitor into defender.

Considerations when drafting a licensing agreement include degree of exclusivity, territory and type of end use, period of licence and type and level of payments – royalty, lump sum or cross-licence. Pricing a licence is as much an art as a science, and depends on a number of factors such as the balance of power and negotiating skills. Common methods of pricing licences are:[18]

- Going market rate – based on industry norms, e.g. 6% of sales in electronics and mechanical engineering.
- 25% rule – based on licensee's gross profit earned through use of the technology.

- Return on investment – based on licensor's costs.
- Profit-sharing – based on relative investment and risk. First, estimate total life-cycle profit. Next, calculate relative investment and weight according to share of risk. Finally, compare results to alternatives, e.g return to licensee, imitation, litigation.

There is no 'best' licensing strategy, as it depends on the strategy of the organization and the nature of the technology and markets (see Box 7.6). For example, Celltech licensed its asthma treatment to Merck for a single payment of $50 m., based on sales projections. This isolated Celltech from the risk of clinical trials and commercialization,

---

BOX 7.6
ARM HOLDINGS

ARM Holdings designs and licenses high-performance, low-energy-consumption 16- and 32-bit RISC (reduced instruction set computing) chips, which are used extensively in mobile devices such as cell phones, cameras, electronic organizers and smart cards. ARM was established in 1990 as a joint venture between Acorn Computers in the UK and Apple Computer. Acorn did not pioneer the RISC architecture, but it was the first to market a commercial RISC processor in the mid-1980s. Perhaps ironically, the first application of ARM technology was in the relatively unsuccessful Apple Newton PDA (personal digital assistant). One of the most recent successful applications has been in the Apple i-Pod. ARM designs but does not manufacture chips, and receives royalties of between 5 cents and US$2.50 for every chip produced under licence. Licensees include Apple, Ericsson, Fujitsu, HP, NEC, Nintendo, Sega, Sharp, Sony, Toshiba and 3Com. In 1999 it announced joint ventures with leading chip manufacturers such as Intel and Texas Instruments to design and build chips for the next generation of hand-held devices. It is estimated that ARM-designed processors were used in 10 million devices in 1996, 50 million in 1998, 120 million devices sold in 1999, and a billion sold in 2004, representing around 80% of all mobile devices. In 1998 the company was floated in London and on the Nasdaq in New York, and it achieved a market capitalization of £3bn in December 1999, with an annual revenue growth of 40% to £15.7m. The company employs around 400 staff, 250 of which are based in Cambridge in the UK, with an average age of 27. It spends almost 30% of revenues on R&D. The company has created 30 millionaires amongst its staff.

and provided a much-needed cash injection. Toshiba, Sony and Matsushita license DVD technology for royalties of only 1.5% to encourage its adoption as the industry standard. Until the recent legal proceedings, Microsoft applied a 'per processor' royalty its OEM (original equipment manufacturer) customers for Windows to discourage its customers from using competing operating systems. The successful exploitation of IPR also incurs costs and risks:

- Cost of search, registration and renewal.
- Need to register in various national markets.
- Full and public disclosure of your idea.
- Need to be able to enforce.

In most countries the basic registration fee for a patent is relatively modest, but in addition applying for a patent includes the cost of professional agents, such as patent agents, translation for foreign patents, official registration fees in all relevant countries and renewal fees. For example, the lifetime cost for a single non-pharmaceutical patent in the main European markets would be around £80 000, and the addition of the USA and Japan some £40 000 more. Patents in the other Asian markets are cheaper, at up to £5000 per country, but the cumulative cost becomes prohibitive, particularly for lone inventors or small firms. Pharmaceutical patents are much more expensive, up to five times more, due to the complexity and length of the documentation. In addition to these costs, firms must consider the competitive risk of public disclosure, and the potential cost of legal action should the patent be infringed. Some of these issues and most appropriate strategies for dealing with them were discussed in Chapter 4.

7.5 Commercializing Complex Products

Complex products or systems are a special case in marketing because neither the technology nor markets are well defined or understood. Therefore technology and markets co-evolve over time, as developers and potential users interact. Note that technological complexity does not necessarily imply market complexity, or vice versa. For example, the development of a passenger aircraft is complex in a technological sense, but the market is well defined and potential customers are easy to identify. We are concerned here with cases where both technologies and markets are complex – for example, telecommunications, multimedia and pharmaceuticals.

The traditional distinction between consumer and industrial marketing in terms of the nature of users, rather than the products and services themselves, is therefore unhelpful. For example, a new industrial product or process may be relatively simple, whereas a new consumer product may be complex. The commercialization process for complex products has certain characteristics common to consumer and business markets:[19]

- Products are likely to consist of a large number of interacting components and sub-systems, which complicates development and marketing.
- The technical knowledge of customers is likely to be greater, but there is a burden on developers to educate potential users. This requires close links between developers and users.
- Adoption is likely to involve a long-term commitment, and therefore the cost of failure to perform is likely to be high.
- The buying process is often lengthy, and adoption may lag years behind availability and receipt of the initial information.

The Nature of Complex Products

Complex products typically consist of a number of components, or sub-systems. Depending on how open the standards are for interfaces between the various components, products may be offered as bundled systems, or as sub-systems or components. For bundled systems, customers evaluate purchases at the system level, rather than at the component level. For example, many pharmaceutical firms are now operating managed healthcare services, rather than simply developing and selling specific drugs. Similarly, robot manufacturers offer 'manufacturing solutions', rather than stand-alone robot manipulators. Bundled systems can offer customers enhanced performance by allowing a package of optimized components using proprietary interfaces of 'firmware', and in addition may provide the convenience of a single point of purchase and after-sales support. However, bundled systems may not appeal to customers with idiosyncratic needs, or knowledgeable customers able to configure their own systems.

The growth of systems integrators and 'turnkey' solutions suggests that there is additional value to be gained by developing and marketing systems rather than components: typically, the value added at the system level is greater than the sum of the value added by the components. There is, however, an important exception to this rule. In cases where a particular component or sub-system is significantly superior to competing offerings, unbundling is likely to result in a larger market.[20] The increased market is due to additional customers who would not be willing to purchase the bundled system, but would like to incorporate one of the components or sub-systems into their own systems. For example, Intel and Microsoft have captured the dominant market shares of microprocessors and operating systems respectively, by selling components rather than by incorporating these into their own PCs.

Links between Developers and Users

The development and adoption process for complex products, processes and services is particularly difficult. The benefits to potential users may be difficult to identify and value, and because there are likely to be few direct substitutes available the market may not be able to provide any benchmarks. The choice of suppliers is likely to be limited, more an oligopolistic market than a truly competitive one. In the absence of direct competition, price is less important than other factors such as reputation, performance and service and support.

Innovation research has long emphasized the importance of 'understanding user needs' when developing new products,[21] but in the special case of complex products and services potential users may not be aware of, or may be unable to articulate, their needs. In such cases it is not sufficient simply to understand or even to satisfy existing customers, but rather it is necessary to lead existing customers and identify potential new customers. Conventional market research techniques are of little use, and there will be a greater burden on developers to 'educate' potential users. Hamel and Prahalad refer to this process as *expeditionary marketing*.[22] The main issue is how to learn as quickly as possible through experimentation with real products and customers, and thereby anticipate future requirements and pre-empt potential competitors.

The relationship between developers and users will change throughout the development and adoption process (Figure 7.6). Three distinct processes need to be managed, each demanding different linkages: development, adoption and interfacing. Numerous frameworks have been formulated to help structure and manage the development process, and these will be discussed in the next chapter. The process of diffusion and adoption is examined in section 7.6. However, relatively little guidance is available for managing the interface between the developers and adopters of an innovation.

The interface process can be thought of as consisting of two flows: information flows and resource flows.[23] Developers and adopters will negotiate the inflows and outflows of both information and resources. Therefore developers should recognize that resources committed to development and resources committed to aiding adoption should not be viewed as independent or 'ring-fenced'. Both contribute to the successful commercialization of complex products, processes and services. Developers should also identify and manage the balance and direction of information and resource flows at different stages of the process of development and adoption. For example, at early stages managing information inflows may be most important, but at later stages managing outflows of information and resources may be critical. In addition, learning will require the management of knowledge flows, involving the exchange or secondment of appropriate staff.

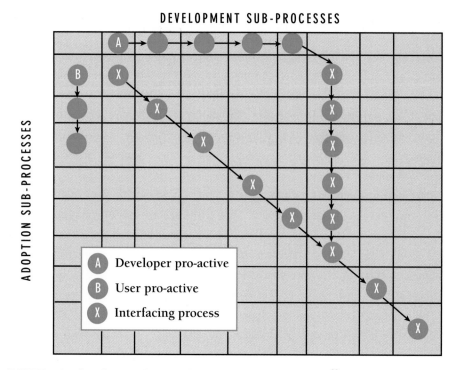

FIGURE 7.6 Developer–adopter relationships for development[23]

Two dimensions help determine the most appropriate relationship between developers and users: the range of different applications for an innovation; and the number of potential users of each application:[24]

- *Few applications and few users.* In this case direct face-to-face negotiation regarding the technology design and use is possible.
- *Few applications, but many users.* This is the classic marketing case, which demands careful segmentation, but little interaction with users.
- *Many applications, but few users.* In this case there are multiple stakeholders amongst the user groups, with separate and possibly conflicting needs. This requires skills to avoid optimization of the technology for one group at the expense of others. The core functionality of the technology must be separated and protected, and custom interfaces developed for the different user groups.
- *Many applications and different users.* In this case developers must work with multiple archetypes of users and therefore aim for the most generic market possible, customized for no one group.

In general, where there are relatively few potential users, as is usually the case with complex products for business customers, customers are likely to demand that developers have the capability to solve their problems, and be able to transfer the solution to them. However, customer expectations vary by sector and nationality. For example, firms in the paper and pulp industry do not expect suppliers to have strong problem-solving capabilities, but do require solutions to be adapted to their specific needs. Conversely, firms in the speciality steel industry demand suppliers to possess strong problem-solving capabilities. Overall, German and Swedish customers expect suppliers to have problem-solving and adaptation capabilities, but British, French and Italian customers appear to be less demanding.[25]

Role of Lead Users

Lead users are critical to the development and adoption of complex products. As the title suggests, lead users demand new requirements ahead of the general market of other users, but are also positioned in the market to significantly benefit from the meeting of those requirements.[26] Where potential users have high levels of sophistication, for example in business-to-business markets such as scientific instruments, capital equipment and IT systems, lead users can help to co-develop innovations, and are therefore often early adopters of such innovations. The initial research by Von Hippel suggests lead users adopt an average of seven years before typical users, but the precise lead time will depend on a number of factors, including the technology life cycle (see section 7.6). A recent empirical study identified a number of characteristics of lead users:[27]

- *Recognize requirements early* – are ahead of the market in identifying and planning for new requirements.
- *Expect high level of benefits* – due to their market position and complementary assets.
- *Develop their own innovations and applications* – have sufficient sophistication to identify and capabilities to contribute to development of the innovation.
- *Perceived to be pioneering and innovative* – by themselves and their peer group.

This has two important implications. First, those seeking to develop innovative complex products and services should identify potential lead users with such characteristics to contribute to the co-development and early adoption of the innovation. Second, that lead users, as early adopters, can provide insights to forecasting the diffusion of innovations. For example, a study of 55 development projects in telecommunications computer infrastructure found that the importance of customer inputs increased with technological newness and, moreover, the relationship shifted from customer surveys and focus groups to co-development because 'conventional marketing techniques proved to be of limited utility, were often ignored, and in hindsight were sometimes strikingly inaccurate'.[28] Clayton Christensen and Michael Raynor make a similar point in their book *The Innovator's Solution*, and argue that conventional segmentation of

---

BOX 7.7

IDENTIFYING POTENTIALLY DISRUPTIVE INNOVATIONS

In their book *The Innovator's Solution: Creating and sustaining successful growth* (Harvard Business School Press, 2003), Clayton Christensen and Michael Raynor argue that segmentation of markets by product attributes or type of customer will fail to identify potentially disruptive innovations. Building on the seminal marketing work of Theodore Levitt, they recommend *circumstance*-based segmentation, which focuses on the 'job to be done' by an innovation, rather than product attributes or type of users. This perspective is likely to result in very different new products and services than traditional ways of segmenting markets. One of the insights this approach provides is the idea of innovations from *non-consumption*. So instead of comparing product attributes with competing products, identify target customers who are trying to get a job done, but due to circumstances – wealth, skill, location, etc. – do not have access to existing solutions. These potential customers are more likely to compare the disruptive innovation with the alternative of having nothing at all, rather than existing offerings. This can lead to the creation of whole new markets – for example, the low-cost airlines in the USA and UK, such as Southwest and Ryanair, or Intuit's QuickBooks. Similarly, in the MBA market, distance learning programmes were once considered inferior to conventional programmes, and instead leading business schools competed (and many still do) for funds for larger and ever-more expensive buildings in prestigious locations. However, improvements to technology, combined with other forms of learning to create 'blended' learning environments, have created whole new markets for MBA programmes, for those who are unable or unwilling to pursue more conventional programmes.

---

markets by product attributes or user types cannot identify potentially disruptive innovations (Box 7.7)

Adoption of Complex Products

The buying process for complex products is likely to be lengthy due to the difficulty of evaluating risk and subsequent implementation. Perceived risk is a function of a buyer's level of uncertainty and the seriousness of the consequences of the decision to purchase. There are two types of risk; the performance risk, that is the extent to which

BOX 7.8
THE EMI CAT SCANNER

In 1972 the British firm EMI launched the first computer-assisted tomography (CAT) scanner for use in medical diagnosis. The CAT scanner converted conventional X-ray information into three-dimensional pictures which could be examined using a monitor. EMI had invented and patented all the key technologies of the CAT scanner. The initial slow scanning speed of early machines meant that they were only suitable for organs with minimal movement, such as the brain. In 1976 EMI introduced a faster machine which had a scan time of only 20 seconds, and therefore could be used for whole body scans. It was generally acknowledged that at that time the EMI CAT scanner provided a scanned image superior to that of competing machines, therefore allowing more detailed diagnosis.

Established suppliers of conventional X-ray equipment such as Siemens in Europe and General Electric in the USA responded by differentiating their CAT scanners from those offered by EMI. They competed with the technically superior machines of EMI by emphasizing the faster scan speed of their machines, which they claimed improved patient throughput times. EMI argued that there was a trade-off between scan time and image quality, and that in any case scan time was insignificant relative to the total consultation time required for a patient. However, in North American hospitals, which were the largest market for such machines, patient throughput was of critical importance. Worse still, early machines provided by EMI were highly complex and proved unreliable, and the company was unable to provide worldwide service and support until much later. Early users unfairly compared the reliability of the CAT scanners to more mature and less complex X-ray machines. As a result, the EMI scanner gained a reputation for being unreliable and slow. The machines supplied by its competitors were technically inferior in terms of scanning quality, but gained market share through clever marketing and better customer support. By 1977 the Medical Division of EMI was making a loss, and in 1979 the company was purchased by the Thorn Group.

EMI had invented the CAT scanner, but failed to identify the requirements of its key customers, and underestimated the technical and marketing response of established firms.

the purchase meets expectations, and the psychological risk associated with how other people in the organization react to the decision. Low-risk decisions are likely to be made autonomously, and therefore it is easier to target decision-makers and identify buying criteria. For complex products there is greater uncertainty, and the consequences of the purchase are more significant, and therefore some form of joint or group decision-making is likely.

If there is general agreement concerning the buying criteria, a process of information-gathering and deliberation can take place in order to identify and evaluate potential suppliers. However, if there is disagreement concerning the buying criteria, a process of persuasion and bargaining is likely to be necessary before any decision can be made.

In the case of organizational purchases, the expectations, perceptions, roles and perception of risk of the main decision-makers may vary. Therefore we should expect and identify the different buying criteria used by various decision-makers in an organization. For example, a production engineer may favour the reliability or performance of a piece of equipment, whereas the finance manager is likely to focus on life-cycle costs and value for money (see Box 7.8). Three factors are likely to affect the purchase decision in an organization:[29]

1. *Political and legal environment.* This may affect the availability of, and information concerning, competing products. For example, government legislation might specify the tender process for the development and purchase of new equipment.
2. *Organizational structure and tasks.* Structure includes the degree of centralization of decision-making and purchasing; tasks include the organizational purpose served by the purchase, the nature of demand derived from the purchaser's own business, and how routine the purchase is.
3. *Personal roles and responsibilities.* Different roles need to be identified and satisfied. Gatekeepers control the flow of information to the organization, influencers add information or change buying criteria, deciders choose the specific supplier or brand, and the buyers are responsible for the actual purchase. Therefore the ultimate users may not be the primary target.

7.6 Forecasting the Diffusion of Innovations

A great deal of research has been conducted to try to identify what factors affect the rate and extent of adoption of an innovation. In this section we examine three issues relevant to the marketing of innovations:

1. How the characteristics of an innovation affect adoption.
2. How the process of commercialization and diffusion affects adoption.
3. What techniques are available for forecasting future patterns of adoption.

Characteristics of an Innovation Affecting Diffusion

A number of characteristics of an innovation have been found to affect diffusion:[30]

- relative advantage;
- compatibility;
- complexity;
- trialability;
- observability.

Relative advantage   Relative advantage is the degree to which an innovation is perceived as better than the product it supersedes, or competing products. Relative advantage is typically measured in narrow economic terms – for example, cost or financial payback – but non-economic factors such as convenience, satisfaction and social prestige may be equally important. In theory, the greater the perceived advantage, the faster the rate of adoption.

It is useful to distinguish between the primary and secondary attributes of an innovation. Primary attributes, such as size and cost, are invariant and inherent to a specific innovation irrespective of the adopter. Secondary attributes, such as relative advantage and compatibility, may vary from adopter to adopter, being contingent upon the perceptions and context of adopters.

Incentives may be used to promote the adoption of an innovation, by increasing the perceived relative advantage of the innovation, subsidizing trials or reducing the cost of incompatibilities.

Compatibility   Compatibility is the degree to which an innovation is perceived to be consistent with the existing values, experience and needs of potential adopters. There are two distinct aspects of compatibility: existing skills and practices; and values and norms. The extent to which the innovation fits the existing skills, equipment, procedures and performance criteria of the potential adopter is important, and relatively easy to assess.

So-called 'network externalities' can affect the adoption process. For example, the cost of adoption and use, as distinct from the cost of purchase, may be influenced by: the availability of information about the technology from other users, of trained skilled users, technical assistance and maintenance, and of complementary innovations, both technical and organizational.

However, compatibility with existing practices may be less important than the fit with existing values and norms.[31] Significant misalignments between an innovation and an adopting organization will require changes in the innovation or organization, or both. In the most successful cases of implementation, mutual adaptation of the innovation and organization occurs.[32]

Complexity   Complexity is the degree to which an innovation is perceived as being difficult to understand or use. In general, innovations which are simpler for potential users to understand will be adopted more rapidly than those which require the adopter to develop new skills and knowledge.

Trialability   Trialability is the degree to which an innovation can be experimented with on a limited basis. An innovation that is trialable represents less uncertainty to potential adopters, and allows learning by doing. Innovations which can be trialed will generally be adopted more quickly than those which cannot. The exception is where the undesirable consequences of an innovation appear to outweigh the desirable characteristics. In general, adopters wish to benefit from the functional effects of an innovation, but avoid any dysfunctional effects. However, where it is difficult or impossible to separate the desirable from the undesirable consequences trialability may reduce the rate of adoption.

Observability   Observability is the degree to which the results of an innovation are visible to others. The easier it is for others to see the benefits of an innovation, the more likely it will be adopted. The simple epidemic model of diffusion assumes that innovations spread as potential adopters come into contact with existing users of an innovation.

Processes of Diffusion

Research on diffusion attempts to identify what influences the rate of adoption of an innovation. The diffusion of an innovation is typically described by an S-shaped (logistic) curve (Figure 7.7). Initially, the rate of adoption is low, and adoption is confined to so-called 'innovators'. Next to adopt are the 'early adopters', then the 'late majority', and finally the curve tails off as only the 'laggards' remain. Such taxonomies are fine with the benefit of hindsight, but provide little guidance for future patterns of adoption.

Hundreds of marketing studies have attempted to fit the adoption of specific products to the S-curve, ranging from television sets to new drugs. In most cases mathematical techniques can provide a relatively good fit with historical data, but research

DIFFUSION OF COLOUR TELEVISIONS IN THE UK

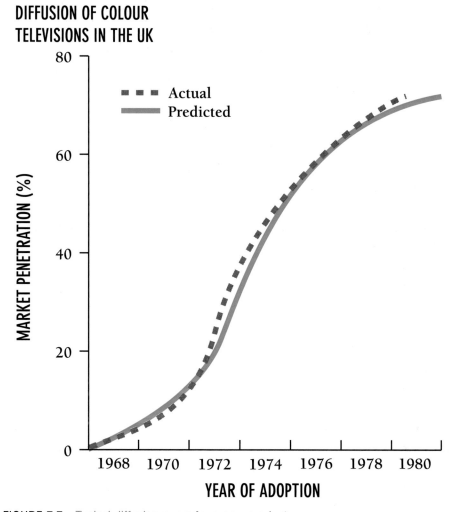

FIGURE 7.7 Typical diffusion curve for a new product

Source: Data from Meade, N. (1984) 'The use of growth curves in forecasting market development – a review and appraisal', *Journal of Forecasting*, **3**, 429–451.

has so far failed to identify robust generic models of adoption. In practice the precise pattern of adoption of an innovation will depend on the interaction of demand-side and supply-side factors:

1. Demand-side models, mainly statistical:
   (a) epidemic, based on direct contact with or imitation of prior adopters;
   (b) Bass, based on adopters consisting of innovators and imitators;
   (c) Probit, based on adopters with different benefit thresholds;
   (d) Bayesian, based on adopters with different perceptions of benefits and risk.

2. Supply-side models, mainly sociological:
   (a) appropriability, which emphasizes relative advantage of an innovation;
   (b) dissemination, which emphasizes the availability of information;
   (c) utilization, which emphasizes the reduction of barriers to use;
   (d) communication, which emphasizes feedback between developers and users.

The epidemic model was the earliest and is still the most commonly used. It assumes a homogeneous population of potential adopters, and that innovations spread by information transmitted by personal contact and the geographical proximity of existing and potential adopters. This model suggests that the emphasis should be on communication, and the provision of clear technical and economic information. However, the epidemic model has been criticized because it assumes that all potential adopters are similar and have the same needs.

As a result, the Bass model of diffusion is modified to include two different groups of potential adopters: innovators, who are not subject to social emulation; and imitators, for whom the diffusion process takes the epidemic form. This produces a skewed S-curve because of the early adoption by innovators, and suggests that different marketing processes are needed for the innovators and subsequent imitators. The Bass model is highly influential in economics and marketing research.

The Probit model takes a more sophisticated approach to the population of potential adopters. It assumes that potential adopters have different threshold values for costs or benefits, and will only adopt beyond some critical or threshold value. In this case differences in threshold values are used to explain different rates of adoption. This suggests that the more similar potential adopters are, the faster the diffusion.

In the Probit model, potential adopters know the value of adoption, but delay adoption until the benefits are sufficient. However, it is unrealistic to assume that adopters will have perfect knowledge of the value of an innovation. Therefore Bayesian models of diffusion introduce lack of information as a constraint to diffusion. Potential adopters are allowed to hold different beliefs regarding the value of the innovation, which they may revise according to the results of trials to test the innovation. Because these trials are private, imitation cannot take place and other potential adopters cannot learn from the trials. This suggests better-informed potential adopters may not necessarily adopt an innovation earlier than the less well informed, which was an assumption of earlier models.[33]

The choice between the four models will depend on the characteristics of the innovation and nature of potential adopters. The simple epidemic model appears to provide a good fit to the diffusion of new processes, techniques and procedures, whereas the Bass model appears to best fit the diffusion of consumer products. However, the mathematical structure of the epidemic and Bass models tends to overstate the importance of differences in adopter characteristics, but tends to underestimate the effect of macro-economic and supply-side factors. In general, both these models of diffusion work best where the total potential market is known, that is for derivatives of existing products and services, rather than totally new innovations. All demand-side models have limitations:

- Adopters are assumed to be relatively homogeneous, apart from some difference in progressiveness or threshold values. They do not consider the possibility that the rationality and the profitability of adopting a particular innovation might be different for different adopters. For example, local 'network externalities' such as the availability of trained skilled users, technical assistance and maintenance, or complementary technical or organizational innovations are likely to affect the cost of adoption and use, as distinct from the cost of purchase.
- The population of potential adopters and the innovation are assumed to be the same at the beginning and at the end of the diffusion period. However, research confirms that many innovations change over the course of diffusion, and that this may change the potential population of adopters, who in turn may lead to subsequent modifications to the innovation.
- They focus almost exclusively on the adopters' or demand side of the diffusion process, and ignore supply-side factors. In reality both demand- and supply-side factors must be taken into account (see Box 7.9).

Sociological models place greater emphasis on the relationship between demand- and supply-side factors.[34] The early appropriability models focus almost exclusively on the supply side, and assume that innovations of sufficient value will be adopted. This suggests that the most important issues are the relative advantage of an innovation. The subsequent dissemination model assumes that the availability of information and communication channels are the most critical issues in diffusion. The utilization model incorporates demand-side issues, in particular problems of adoption and application, both structural and perceptual. Finally, there are most recent communication models of diffusion, which are based on feedback between developers and potential adopters.

The communication perspective considers how individual psychological characteristics such as attitude and perception affect adoption. Individual motivations, perceptions, likes and dislikes determine what information is reacted to and how it is processed. Potential adopters will be guided and prejudiced by experience, and will have 'cognitive maps' which filter information and guide behaviour. Social context will also influence individual behaviour. Social structures and meaning systems are locally constructed, and therefore highly context-specific. These can distort the way in which information is interpreted and acted upon. Therefore the perceived value of an innovation, and therefore its subsequent adoption, is not some objective fact, but instead

BOX 7.9
DIFFUSION OF ROBOTICS

The first commercial industrial robots were developed in the 1960s, and in the 1980s many believed that robots would replace most forms of manual work. However, in 1995 the density of robots in use – the number of robots per 10 000 workers – was only 21 in the UK, 33 in the United States, 69 in Germany and 338 in Japan. This raises two questions: why has the diffusion of robots been so slow, and why are there such large differences in the rates of diffusion in different countries? Economists tend to argue that the level of investment in industrial robots, like all capital equipment, is a function of labour costs: the higher labour costs, the more likely firms are to substitute robots for labour. This is clearly part of the explanation, as rankings of labour costs are similar to those for robot density: Japan, Sweden, Germany, United States, France and the UK. However, more detailed studies reveal other factors affecting the diffusion of robots, such as industry structure and work organization.

The automobile industry has traditionally been the largest customer for robots. The most common applications in this sector are in spot welding, machine loading and surface treatment. The main requirement is high accuracy, or rather repeatability. Therefore those countries with the largest automobile sector electronics industry have begun to employ large numbers of robots in assembly, changing the rate and pattern of diffusion.

Work organization has also affected the number and type of robots used in different countries. Expensive, sophisticated robots are common in the UK and USA in firms characterized by low levels of training. Therefore robots have to be made 'idiot proof' in order to minimize operator involvement. In contrast, cheaper, less complex robots are more common in Japan where operators are trained to work with the robots and to perform routine programming and maintenance. Moreover, many firms which have poor control of the quality of materials and components have been forced to adopt sophisticated component feeding and sensor systems to allow for component variability, whereas firms which have made investments in quality management have been able to use cheaper, simpler equipment.

Thus the diffusion of industrial robots has not followed a simple logistics curve based on some cost function. Rather, patterns of adoption have varied as industrial robots and users' needs have co-evolved.[35] Robots were originally conceived as being direct replacements for workers, so-called 'steel collar' workers, or 'universal automation', but have become increasingly specialized such that there are now robots designed specifically for assembly, spraying or welding applications. At the same time users have grown sophisticated and their needs have fragmented, some demanding 'turnkey' integrated solutions, others preferring to build their own systems from various components. Therefore it is meaningless to think of the adoption of industrial robots as a process of simple diffusion.

depends on individual psychology and social context. These factors are particularly important in the later stages of diffusion.

Initially, the needs of early adopters or innovators dominate, and therefore the characteristics of an innovation are most important. Innovations tend to evolve over time through improvements required by these early users, which may reduce the relative cost to later adopters. However, early adopters are almost by definition 'atypical'; for example, they tend to have superior technical skills. As a result the preferences of early adopters can have a disproportionate impact on the subsequent development of an innovation, and result in the establishment of inferior technologies or abandonment of superior alternatives.

Bandwagons may occur where an innovation is adopted because of pressure caused by the sheer number of those who have already adopted an innovation, rather than by individual assessments of the benefits of an innovation. In general, as soon as the number of adopters has reached a certain threshold level, the greater the level of ambiguity of the innovations benefits, the greater the subsequent number of adopters. This process allows technically inefficient innovations to be widely adopted, or technically efficient innovations to be rejected. Examples include the QWERTY keyboard, originally designed to prevent professional typists from typing too fast and jamming typewriters; and the DOS operating system for personal computers, designed by and for computer enthusiasts.

Bandwagons occur due to a combination of competitive and institutional pressures. Where competitors adopt an innovation, a firm may adopt because of the threat of lost competitiveness, rather than as a result of any rational evaluation of benefits. For example, many firms adopted flexible manufacturing systems (FMS) in the 1980s in response to increased competition, but most failed to achieve significant benefits. The main institutional pressure is the threat of lost legitimacy; for example, being considered by peers or customers as being less progressive or competent. For example, in the early 1990s most leading firms established websites on the World Wide Web (WWW) because it was perceived to be progressive, rather than because of any immediate commercial benefits.

The critical difference between bandwagons and other types of diffusion is that they require only limited information to flow from early to later adopters. Indeed, the more ambiguous the benefits of an innovation, the more significant bandwagons are on rates of adoption.[36] Therefore the process of diffusion must be managed with as much care as the process of development. In short, better products do not necessarily result in more sales. Not everybody requires a better mousetrap.

Forecasting Patterns of Adoption

Forecasting can help to identify what might be required in the future, and to estimate how many are likely to be required in a given time period. However, in the case of innovative products forecasting is difficult, as the products and markets may not be well defined. For example, when Apple launched its first personal digital assistant (PDA), the Newton, it forecast first-year sales of 400 000, but in fact sold just 70 000. This was partly due to the poor performance of the software for recognizing handwriting, but also because Apple failed to position the product clearly between traditional paper organizers and sub-notebook computers.

The most appropriate choice of forecasting method will depend on:
- what we are trying to forecast;
- rate of technological and market change;
- availability and accuracy of information;
- the company's planning horizon;
- the resources available for forecasting.

In practice there will be a trade-off between the cost and robustness of a forecast. The more common methods of forecasting such as trend extrapolation and time series are of limited use for new products, because of the lack of past data. However, regression analysis can be used to identify the main factors driving demand for a given product, and therefore provide some estimate of future demand, given data on the underlying drivers.

For example, a regression might express the likely demand for the next generation of digital mobile phones in terms of rate of economic growth, price relative to competing systems, rate of new business formation, and so on. Data are collected for each of the chosen variables and coefficients for each derived from the curve which best describes the past data. Thus the reliability of the forecast depends a great deal on selecting the right variables in the first place. The advantage of regression is that, unlike simple extrapolation or time series analysis, the forecast is based on cause and effect relations. Econometric models are simply bundles of regression equations, including their interrelationship. However, regression analysis is of little use where future values of an explanatory value are unknown, or where the relationship between the explanatory and forecast variables may change.

Leading indicators and analogues can improve the reliability of forecasts, and are useful guideposts to future trends in some sectors. In both cases there is a historical relationship between two trends. For example, new business start-ups might be a leading indicator of the demand for fax machines in six months' time. Similarly, business users of mobile telephones may be an analogue for subsequent patterns of domestic use.

Such 'normative' techniques are useful for estimating the future demand for existing products, or extensions of existing products, but are of limited utility in the case of radical new products. Exploratory forecasting, in contrast, attempts to explore the range of future possibilities. The most common methods are:

- customer and market surveys;
- brainstorming;
- delphi or expert opinion;
- scenario development.

Most companies conduct customer surveys of some sort. In consumer markets this can be problematic simply because customers are unable to articulate their future needs. For example, market research would not have been very helpful to Sony in the development of the first domestic video cassette recorder or personal stereo. In industrial markets, customers tend to be better equipped to communicate their future requirements. Consequently, in industrial markets innovations often originate from customers. Companies can also consult their direct sales force, but these may not always be the best guide to future customer requirements. Information is often filtered in terms of existing products and services, and biased in terms of current sales performance rather than long-term development potential.

Structured idea generation, or brainstorming, aims to solve specific problems or to identify new products or services. Typically, a small group of experts is gathered together and allowed to interact. A chairman records all suggestions without comment or criticism. The aim is to identify, but not evaluate, as many opportunities or solutions as possible. Finally, members of the group vote on the different suggestions. The best results are obtained when representatives from different functions are present, but this can be difficult to manage. Brainstorming does not produce a forecast as such, but can provide useful input to other types of forecasting.

The opinion of outside experts, or Delphi method, is useful where there is a great deal of uncertainty or for long time horizons. The relevant experts may include suppliers, dealers, customers, consultants and academics. The Delphi method begins with a postal survey of expert opinion on what the future key issues will be, and the likelihood of the developments. The response is then analysed, and the same sample of experts resurveyed with a new, more focused questionnaire. This procedure is repeated until some convergence of opinion is observed, or conversely if no consensus is reached. In Europe, governments and transnational agencies use Delphi studies to help formulate policy. In Japan, large companies and the government routinely survey expert opinion in order to reach some consensus in those areas with the greatest potential for long-term development. Used in this way, the Delphi method can to a large extent become a self-fulfilling prophecy.

Scenario development may involve many different forecasting techniques, including computer-based simulation. Typically, it begins with the identification of the critical indicators, which might include use of brainstorming and Delphi techniques. Next, the reasons for the behaviour of these indicators is examined, perhaps using regression techniques. The future events which are likely to affect these indicators are identified. These are used to construct the best, worst and most likely future scenarios. Finally, the company assesses the impact of each scenario on its business. The goal is to plan for the outcome with the greatest impact, or better still, retain sufficient flexibility to respond to several different scenarios. Scenario development is a key part of the long-term planning process in those sectors characterized by high capital investment, long lead times and significant environmental uncertainty, such as energy, aerospace and telecommunications.

7.7 Summary and Further Reading

In this chapter we have examined how the maturity of technologies and markets affects the process of marketing an innovation. Where both technologies and markets are relatively mature, the key issue is how to differentiate a product or service for competing offerings. In this case many of the standard marketing techniques can be applied, but other tools such as quality function deployment (QFD) are useful. Where existing technologies are applied to new markets, what we call architectural innovation, the key issue is the resegmentation of markets to identify potential new applications. Where new technologies are applied to existing markets, the key issue is to assess the advantage the technology may have over existing solutions in specific applications, and then identify target users based on behavioural characteristics. Finally, where both technologies and markets are complex, the key issue is the relationship between developers and potential users.

The diffusion of an innovation depends on the characteristics of the innovation, the nature of potential adopters, and the process of communication. The relative advantage, compatibility, complexity, trialability and observability of an innovation all affect the rate of diffusion. The skills, psychology, social context and infrastructure of adopters also affect adoption. Epidemic models assume that innovations spread by communication between adopters, but bandwagons do not require this. Instead, early adopters influence the development of an innovation, but subsequent adopters may be more influenced by competitive and peer pressures.

The standard marketing texts for management students is Phillip Kotler's *Principles of Marketing* (Prentice-Hall, European and International Editions, 2002), which focuses on consumer products, and for business and industrial markets, *Business Marketing Management*, by Michael Hutt and Thomas Speh (South Western College Publishing, 2nd edition, 2003). However, neither adequately deals with the particular case of innovative new products and services. A number of American texts cover the related but more narrow issue of marketing high-technology products, including William Davidow's *Marketing High Technology* (Free Press, 1986) and *Essentials of Marketing High Technology* by William Shanklin and John Ryans, Jr (Lexington, 1987). The former is written by a practising engineer/marketing manager, and therefore is strong on practical advice, and the latter is written by two academics, and provides a more coherent framework for analysis. Vijay Jolly's *Commercializing New Technologies* (Harvard Business School Press, 1998) provides a process model based on the experiences of leading firms such as 3M and Sony, which consists of five sub-processes or stages, but the framework is biased towards mass consumer markets. Everett Roger's classic text the *Diffusion of Innovations*, first published in 1962, remains the best overview of this subject, the most recent and updated edition being published in 2003 (Simon & Schuster).

There are few texts which focus exclusively on the more generic problems of applying conventional marketing tools and techniques to innovative new products and processes, but the best attempts to date are the chapter on 'Securing the future' in Gary Hamel and C. K. Prahalad's *Competing for the Future* (Harvard Business School Press, 1994) and the chapter on 'Learning from the market' in Dorothy Leonard's *Wellsprings of Knowledge* (Harvard Business School Press, 1995). Dawn Iacobucci has edited an excellent compilation of current theory and practice of business-to-business and other relationship-based marketing in *Networks in Marketing* (Sage, 1996), much of which is relevant to the development and marketing of complex products and services. It also provides a sound introduction to the more general subject of networks, which will be discussed in the next chapter. We discuss the special case of complex product systems in a special issue of *Research Policy*, **29**, 2000, and in *The Business of Systems Integration*, edited by Andrea Prencipe, Andy Davies and Mike Hobday (Oxford University Press, 2003).

References

1 Dvir, D. *et al.* (1998) 'In search of project classification: a non-universal approach to project success factors', *Research Policy*, **27**, 915–935.

2 Tidd, J. and K. Bodley (2002) 'Effect of novelty on new product development processes and tools', *R&D Management*, **32** (2), 127–138.

3 Tidd, J. and C. Driver (2000) 'Technological and market competencies and financial performance', in Tidd, J. (ed.), *From Knowledge Management to Strategic Competence: Measuring technological, market and organizational innovation*. Imperial College Press, London, 94–125.

4 Luchs, B. (1990) 'Quality as a strategic weapon', *European Business Journal*, **2** (4), 34–47.

5 Clayton, T. and G. Turner (2000) 'Brands, innovation and growth', in Tidd, J. (ed.), *From Knowledge Management to Strategic Competence: Measuring technological, market and organizational innovation*. Imperial College Press, London, 77–93.

6 Burn, G. (1990) 'Quality function deployment', in Dale, B. and Plunkett, J. (eds), *Managing Quality*. Philip Allan, London, 66–88.

7 Dimancescu, D. and K. Dwenger (1995) *World-Class New Product Development*. American Management Association, New York.

8 Griffin, A. (1992) 'Evaluating QFD's use in US firms as a process for developing products', *Journal of Product Innovation Management*, **9**, 171–187.

9 Buzzell, P. and B. Gale (1987) *The PIMS Principle*. Free Press, New York.

10 Moriarty, P. and D. Reibstein (1986) 'Benefit segmentation in industrial markets', *Journal of Business Research*, **14**, 463–486.

11 Lauglaug, A. (1993) 'Technical-market research – get customers to collaborate in developing products', *Long Range Planning*, **26** (2), 78–82.

12 Christensen, C. (2000) *The Innovator's Dilemma*. HarperCollins, New York.

13 Jones, N. (2003) 'Competing after radical technological change: the significance of product line management strategy', *Strategic Management Journal*, **24**, 1265–1287.

14 Davidow, W. (1986) *Marketing High Technology*. Free Press, New York.

15 Millier, P. (1989) *The Marketing of High-Tech Products: Methods of analysis*. Editions d'Organisation, Paris (in French).

16 Weiss, A. and J. Heide (1993) 'The nature of organizational search in high technology markets', *Journal of Marketing Research*, **30**, 220–233.

17 Anselm, M., A. Bleakley and K. Watson (1999) *Intellectual Property and Media Law*. Blackstone Press, London.

18 Byrne, N. (1998) *Licensing Technology*. Jordan, Bristol.

19 Hobday, M., H. Rush and J. Tidd (2000) 'Complex product systems', *Research Policy*, **29**, 793–804.

20 Wilson, L., A. Weiss and G. John (1990) 'Unbundling of industrial systems', *Journal of Marketing*, **27**, 123–138.

21 Cooper, R. and E. Kleinschmidt (1993) 'Screening new products for potential winners', *Long Range Planning*, **26** (6), 74–81.

22 Hamel, G. and C. Prahalad, (1994) *Competing for the Future*. Harvard Business School Press, Boston, Mass.

23 More, P. (1986) 'Developer/adopter relationships in new industrial product situations', *Journal of Business Research*, **14**, 501–517.

24 Leonard-Barton, D. and D. Sinha (1993) 'Developer–user interaction and user satisfaction in internal technology transfer', *Academy of Management Journal*, **36** (5), 1125–1139.

25 Hakansson, H. (1995) 'The Swedish approach to Europe', in Ford, D. (ed.), *Understanding Business Markets*. The Dryden Press, London, 232–261.

26 Von Hippel, E. (1986) 'Lead users: A source of novel product concepts', *Management Science*, **32** (7), 791–805; (1988) *The Sources of Innovation*. Oxford University Press, Oxford.

27 Morrison, P., J. Roberts and D. Midgley (2004) 'The nature of lead users and measurement of leading edge status', *Research Policy*, **33**, 351–362.

28 Callahan, J. and E. Lasry (2004) 'The importance of customer input in the development of very new products', *R&D Management*, **34** (2), 107–117.

29 Webster, F. Jr (1991) *Industrial Marketing Strategy*, 3rd edn. John Wiley & Sons, Inc., New York.

30 Rogers, E. (2003) *Diffusion of Innovations*. Free Press, New York.

31 Leonard-Barton, D. and D. Sinha (1993) 'Developer–user interaction and user satisfaction in internal technology transfer', *Academy of Management Journal*, **36** (5), 1125–1139.

32 Leonard-Barton, D. (1990) 'Implementing new production technologies: exercises in corporate learning', in Von Glinow, M. and Mohmian, S. (eds), *Managing Complexity in High Technology Organizations*. Cambridge University Press, 160–187.

33 Lissom, F. and J. Metcalfe (1994) 'Diffusion of innovation ancient and modern: a review of the main themes', in Dodgson, M. and Rothwell, P. (eds), *The Handbook of Industrial Innovation*. Edward Elgar, Cheltenham, 106–141.

34 Williams, F. and D. Gibson (1990) *Technology Transfer: A communications perspective*. Sage, London.

35 Tidd, J. (1991) *Flexible Manufacturing Technologies and International Competitiveness*. Pinter, London.

36 Abrahamson, E. and L. Plosenkopf (1993) 'Institutional and competitive band-wagons: using mathematical modelling as a tool to explore innovation diffusion', *Academy of Management Journal*, **18** (3), 487–517.

Chapter 8

Learning Through Alliances

Almost all innovations demand some form of collaborative arrangement, for development or commercialization, but the failure rate of such alliances remains high. In this chapter we examine the role of collaboration in the development of new technologies, products and businesses. Specifically, we address the following issues:

- Why do firms collaborate?
- What types of collaboration are most appropriate in different circumstances?
- How do technological and market factors affect the structure of an alliance?
- What organizational and managerial factors affect the success of an alliance?
- How can a firm best exploit alliances for learning new technological and market competencies?

We begin with a discussion of the main theoretical arguments for collaboration, and review some of the more practical benefits. Next we link the rationale for collaboration with different forms and structures of alliances, focusing on the specific cases of supplier relations, strategic alliances and innovation networks. Next we identify recent trends and patterns in collaborative activity, and finally, we discuss how to better manage alliances, including the potential to acquire new market or technological knowledge.

8.1 Why Collaborate?

In Chapter 2 we examined the role of technological and market competencies in developing an innovation strategy. We discussed the advantages and disadvantages of a strategy of technological leadership, and examined the links between different strategies and the sources of innovation. In this chapter we review strategies for exploiting external sources of innovation. Firms collaborate for a number of reasons:

- To reduce the cost of technological development or market entry.
- To reduce the risk of development or market entry.
- To achieve scale economies in production.
- To reduce the time taken to develop and commercialize new products.
- To promote shared learning (see Chapter 11).

In any specific case, a firm is likely to have multiple motives for an alliance (see Box 8.1). However, for the sake of analysis it is useful to group the rationale for collaboration into technological, market and organizational motives (Figure 8.1). Technological reasons include the cost, time and complexity of development. In the current highly competitive business environment, the R&D function, like all other aspects of business, is forced to achieve greater financial efficiency, and to examine critically whether in-house development is the most efficient approach. In addition, there is an increasing recognition that one company's peripheral technologies are usually another's core activities, and that it often makes sense to source such technologies externally, rather than to incur the risks, costs and most importantly of all, timescale associated with in-house development. The rate of technological change, together with the increasingly complex nature of many technologies, means that few organizations can now afford to maintain in-house expertise in every potentially relevant technical area. Many products incorporate an increasing range of technologies as they evolve; for example, automobiles now include much computing hardware and software to monitor and control the engine, transmission, brakes and in some cases suspension. Therefore most R&D and product managers now recognize that no company, however large, can continue to survive as a technological island. For example, when developing the Jaguar XK8 Ford collaborated with Nippondenso in Japan to develop the engine management system and ZF in Germany to develop the transmission system and controls. In addition, there is a greater appreciation of the important role that external technology sources can play in providing a window on emerging or rapidly advancing areas of science. This is particularly true when developments arise from outside a company's traditional areas of business, or from overseas.

Two factors need to be taken into account when making the decision whether to 'make or buy' a technology: the transaction costs, and strategic implications.[1] Transaction cost analysis focuses on organizational efficiency, specifically where market transactions involve significant uncertainty. Risk can be estimated, and is defined in terms of a probability distribution, whereas uncertainty refers to an unknown outcome. Projects involving technological innovation will feature uncertainties associated with completion, performance and pre-emption by rivals. Projects involving market entry will feature uncertainties due to lack of geographical or product market knowledge. In such cases firms are often prepared to trade potentially high financial returns for a reduction in uncertainty.

However, sellers of technological or market know-how may engage in opportunistic behaviour. By opportunistic behaviour we mean high pricing or poor performance. Generally, the fewer potential sources of technology, the lower the bargaining power of the purchaser, and the higher the transaction costs. In addition, where the technology is complex it can be difficult to assess its performance. Therefore transaction costs are

BOX 8.1
PHILIPS AND SONY

The alliance between Philips and Sony to develop, produce and commercialize the compact disc (CD) is a good example of the multiple objectives of strategic alliances.

Philips had developed the prototype for the CD by 1978, after six years of development, but recognized that it would be difficult for the company to turn the concept into a world standard. Philips had previously experienced the commercial failure of its video laser disc system. Therefore in 1979 Philips approached Sony to form a strategic alliance. Sony was chosen because it had the requisite development and manufacturing capability, and provided access to the Japanese market. Also, like Philips, Sony had recently suffered commercial defeat with its Betamax video format.

Philips had developed the basic prototypes of the recording technologies, but the two firms jointly developed the commercial chips necessary for the modulation, control and correction of the digital signal. Sony also developed three integrated circuits which eliminated 500 components, making the CD player smaller, cheaper to produce and more reliable.

Philips and Sony quickly moved to establish their technology as the international standard, both by official and de facto means. Their format was adopted by the influential Electronic Association of Japan, which effectively blocked competing standards from Japanese manufacturers. Moreover, both firms used their in-house recording and pressing facilities to produce CD recordings, CBS/Sony in Japan and Philips/PolyGram in Germany, and thus ensure a supply of music titles. In 1982 the CD was launched in the Japanese market, and in Europe and the USA in 1983. Sales of CD players and recording exceeded all forecasts: 3 million players in 1985, 9 million in 1986; a cumulative total of 59 million CD recordings by 1985, and 136 million by 1986.

In short, the alliance between Philips and Sony had many motives, including access to complementary technologies, economies of scale in production, establishment of international standards and access to international markets. It was successful because in each case the motives of the respective partners were complementary, rather than competitive.

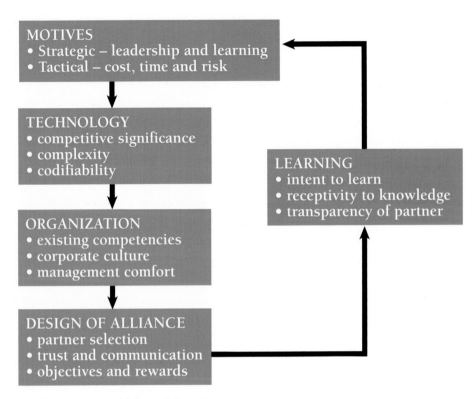

FIGURE 8.1   A model for collaboration

increased where a potential purchaser of technology has little knowledge of the technology. In this respect the acquisition of technology differs from subcontracting more routine tasks such as production or maintenance work, as it is difficult to specify contractually what must be delivered.[2]

As a result, the acquisition of technology tends to require a closer relationship between buyers and sellers than traditional market transactions, resulting in a range of possible acquisition strategies and mechanisms. The optimal technology acquisition strategy in any specific case will depend on the maturity of the technology, the firm's technological position relative to competitors and the strategic significance of the technology.[3] Some form of collaboration is normally necessary where the technology is novel, complex or scarce. Conversely, where the technology is mature, simple or widely available, market transactions such as subcontracting or licensing are more appropriate. However, the cumulative effect of outsourcing various technologies on the basis of comparative transaction costs may limit future technological options and reduce competitiveness in the long term.[4]

Therefore in practice, transaction costs are not the most significant factors affecting the decision to acquire external technology. Factors such as competitive advantage, market expansion and extending product portfolios are more important.[5] Adopting a more strategic perspective focuses attention on long-term organizational effectiveness, rather than short-term efficiency. The early normative strategy literature emphasized the need for technology development to support corporate and business strategies, and therefore technology acquisition decisions began with an evaluation of company strengths and weaknesses. The more recent resource-based approach emphasizes the process of resource accumulation or learning.[6] Competency development requires a firm to have an explicit policy or intent to use collaboration as an opportunity to learn rather than minimize costs. This suggests that the acquisition of external technology should be used to complement internal R&D, rather than being a substitute for it. In fact, a strategy of technology acquisition is associated with diversification into increasingly complex technologies.[7]

Thus neither transaction costs nor strategic behaviour fully explain actual behaviour, and to some extent the approaches are complementary. For example, a survey of top executives found that the two most significant issues considered when evaluating technological collaboration were the strategic importance of the technology and the potential for decreasing development risk.[8] Thus both strategic and transaction cost factors appear to be significant. Strategic considerations suggest *which* technologies should be developed internally, and transaction costs influence *how* the remaining technologies should be acquired. Firms attempt to reduce transaction costs when purchasing external technology by favouring existing trading partners to other sources of technology.[9] In short, for successful technology acquisition the choice of partner may be as important as the search for the best technology. For both partners, the transaction costs will be lower when dealing with a firm with which they are familiar: they are likely to have some degree of mutual trust, shared technical and business information and existing personal social links.

There is also a growing realization that exposure to external sources of technology can bring about other important organizational benefits, such as providing an element of 'peer review' for the internal R&D function, reducing the 'not invented here' syndrome, and challenging in-house researchers with new ideas and different perspectives. In addition, many managers realize the tactical value of certain types of externally developed technology. Some of these are increasingly viewed as a means of gaining the goodwill of customers or governments, of providing a united front for the promotion of uniform industry-wide standards, and to influence future legislation.

The UMIST survey of more than 100 UK-based alliances confirms the relative importance of market-induced motives for collaboration (Table 8.1). Specifically, the most common reasons for collaboration for product development are in response to changing customer or market needs. However, these data provide only the motives for collaboration, not the outcomes. The same survey found that although many firms formed

TABLE 8.1   Reasons for collaboration ($n = 106$)

	Mean score (1 = low, 5 = high)
In response to key customer needs	4.1
In response to a market need	4.1
In response to technology changes	3.8
To reduce risk of R&D	3.8
To broaden product range	3.7
To reduce R&D costs	3.7
To improve time to market	3.6
In response to competitors	3.5
In response to a management initiative	3.3
To be more innovative in product development	3.3

Source: From Littler, D. (1993) *Risks and Rewards of Collaboration*, UMIST.

alliances in an effort to reduce the time, cost or risk of R&D, they did not necessarily realize these benefits from the relationship. In fact, the study concluded that around half of the respondents believed that collaboration made development more complicated and costly. However, it is important to relate benefits to the objectives of collaboration. For example, firms that entered into alliances specifically to reduce the cost or time of development often achieved this, whereas firms that formed alliances for other reasons were more likely to complain that the cost and time of development increased. The study also identified a number of potential risks associated with collaboration:

- leakage of information;
- loss of control or ownership;
- divergent aims and objectives, resulting in conflict.

Around a third of respondents claimed to have experienced such problems. The problem of leakage is greatest when collaborating with potential competitors, as it is difficult to isolate the joint venture from the rest of the business and therefore it is inevitable that partners will gain access to additional knowledge and skills. This additional information may take the form of market intelligence, or more tacit skills or knowledge. Consequently a firm may lose control of the venture, resulting in conflict between partners.

A study of the 'make or buy' decisions for sourcing technology in almost 200 firms concluded that product and process technology from external sources often provides immediate advantages, such as lower cost or a shorter time to market, but in the longer term can make it harder for firms to differentiate their offerings, and difficult to achieve or maintain any positional advantage in the market.[10] Instead, successful strategies of cost leadership or differentiation (the two polar extremes of Porter's model, see Chapter 3) are associated with internal development of process and product technologies. However, in highly dynamic environments, characterized by market uncertainty and technological change, sourcing technology externally is a superior strategy to relying entirely on internal capabilities. This pattern of collaboration is observed in a range of sector studies, for example, high levels of collaboration in the information and communications technology and biotechnology industries, but lower levels in more mature sectors. In the more high-technology sectors, organizations generally seek *complementary* resources – for example, the many relationships between biotechnology firms (for basic research), and pharmaceutical firms (for clinical trials, production and marketing and distribution channels). In the pharmaceutical sector the number of *exploration* alliances with biotechnology firms is predictive of the number of products in development, which in turn is predictive of the number of *exploitation* alliances for sales and distribution.[11] In more mature sectors, more often partners pool *similar* resources to share costs or risk, or to achieve critical mass or economies of scale. There are also differences in the choice of partner. Firms in higher technology sectors tend to favour *horizontal* relationships with their peers and competitors, whereas those in more mature sectors more commonly have *vertical* relations with suppliers and customers.[12] A sector is usually defined as 'high technology' on the basis of the industry average R&D intensity (R&D expenditure/turnover). However, this represents an industry average of a measure of only one input to innovation. Using measures of new product introduction and novelty reveal significant variance in collaborative strategies within sectors. At the firm level, R&D intensity is still associated with the propensity to collaborate, but firms developing products 'new to the market' are much more likely to collaborate than those developing products only 'new to the firm'.[13] This is because the more novel innovations demand more inputs or novelty of inputs, and are associated with greater market uncertainty. We examine these strategies and patterns of collaboration in more detail in sections 8.3 and 8.4. Before that, we will identify and discuss the many different forms of collaboration available.

8.2 Forms of Collaboration

No single form of collaboration is optimal in any generic sense. Some firms, such as Philips, favour joint ventures, whereas others, such as Asea Brown Boveri (ABB), favour acquisition. However, in practice technological and market characteristics will constrain options, and company culture and strategic considerations will determine what is

TABLE 8.2   Forms of collaboration

Type of collaboration	Typical duration	Advantages (rationale)	Disadvantages (transaction costs)
Subcontract/ supplier relations	Short term	Cost and risk reduction Reduced lead time	Search costs, product performance and quality
Licensing	Fixed term	Technology acquisition	Contract cost and constraints
Consortia	Medium term	Expertise, standards, share funding	Knowledge leakage Subsequent differentiation
Strategic alliance	Flexible	Low commitment Market access	Potential lock-in Knowledge leakage
Joint venture	Long term	Complementary know-how Dedicated management	Strategic drift Cultural mismatch
Network	Long term	Dynamic, learning potential	Static inefficiencies

possible and what is desirable. For example, in the case of cross-border company acquisition, the potential for synergy and likelihood of success is greatest where there is some overlap in technologies, products or markets as this creates the potential for consolidation of R&D, production or marketing. In contrast, such overlaps are a major cause of failure of alliances because they create the potential for conflict and competition. Therefore firms should consider alliance partners with complementary technology, products or markets.[14]

Alliances can be characterized in a number of different ways. For example, whether they are horizontal or vertical. Horizontal relationships include cross-licensing, consortia and collaboration with potential competitors of sources of complementary technological or market know-how. Vertical relationships include subcontracting, and alliances with suppliers or customers. The primary motive of horizontal alliances tends to be access to complementary technological or market know-how, whereas the primary motive for vertical alliances is cost reduction. An alternative way of viewing alliances is in terms of their strategic significance or duration (Table 8.2). In these terms, contracting and licensing are tactical, whereas strategic alliances, formal joint ventures and innovation networks are strategic and more appropriate structures for learning. We discuss each of these in turn, focusing on the three most distinct cases of supplier relations, strategic alliances and innovation networks.

Supplier Relations and Subcontracting

The subcontracting or 'outsourcing' of non-core activities has become popular in recent times. Typically, arguments for subcontracting are framed in terms of strategic focus, or 'sticking to the knitting', but in practice most subcontracting or outsourcing arrangements are based on the potential to save costs; suppliers are likely to have lower overheads and variable costs, and may benefit from economies of scale if serving other firms.

Resource dependence and agency theory are more commonly used to explain vertical relationships, and are concerned with the need to control key technologies in the value chain. The seminal work of Von Hippel[15] and subsequent work by others have encouraged firms to identify and form relationships with 'lead' users, and more recently perceptions of the practices of Japanese manufacturers have led many firms to form closer relationships with suppliers.[16] Indeed, closer links between firms, their suppliers and customers may help to reduce the cost of components, through specialization and sharing information on costs. However, factors such as the selection of suppliers and users, timing and mode of their involvement, and the novelty and complexity of the system being developed may reduce or negate the benefit of close supplier–user links.[17]

The quality of the relationship with suppliers and the timing of their involvement in development are critical factors. Traditionally such relationships have been short-term, contractual, arm's-length agreements focusing on the issue of the cost, with little supplier input into design or engineering. In contrast the 'Japanese' or 'partnership' model is based on long-term relationships, and suppliers make a significant contribution to the development of new products. The latter approach increases the visibility of cost–performance trade-offs, reduces the time to market and improves the integration of component technologies. In certain sectors, particularly machine tools and scientific equipment, there is a long tradition of collaboration between manufacturers and lead users in the development of new products. Figure 8.2 presents a range of potential relationships with suppliers. Note that in this diagram we are not suggesting any trend from left to right, but rather that different types of relationship are appropriate in different circumstances, in essence an argument for carefully segmenting supply needs and suppliers, instead of the wholesale adoption of simplistic fashions such as 'partnerships' or business-to-business (so-called 'B2B') supply intranets.

On the vertical axis we have objectives ranging from cost reduction, quality improvement, lead-time reduction through to product and process innovation. On the horizontal axis we distinguish between three types of supply market:
- *Homogeneous* – all potential suppliers have very similar performance
- *Differentiated* – suppliers differ greatly and one clearly superior
- *Indeterminate* – suppliers differ greatly under different conditions

In the case of homogeneous supply conditions and a primary objective to reduce costs, we would argue that a traditional market/contractual relationship is the ideal

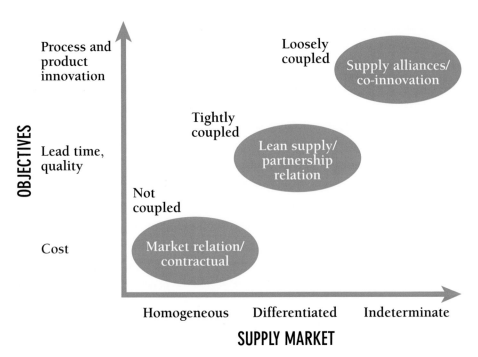

FIGURE 8.2 How objectives and nature of supply market influence supplier relationships

arrangement. In its most recent form this might be achieved by means of a business-to-business intranet exchange or club, whereby potential suppliers to a specific customer or sector pool their price and other data, or bid for specific contracts. Examples include Covisint in the automobile industry, established by Ford, General Motors and DaimlerChrysler, and MetalSite formed by a group of the largest steel producers in the USA. Such developments are not confined to manufacturing, and British Airways, American, United, Delta and Continental have established an electronic procurement hub for routine supplies with an annual turnover of $32bn. In the UK the retailers Kingfisher, Tesco and Marks & Spencer have joined the Worldwide Retail Exchange (WWRX) in an effort to reduce the cost of purchases by up to 20%. Savings of 5 to 10% are more typical of such exchanges, but as with other applications of Internet technology the most significant savings are in transaction costs rather than the goods purchased. Estimates and efficiencies vary, but reports suggest transactions costs can be just 10% of conventional supply chains. Such developments attempt to exploit buyer power, and make supplier prices more transparent. They are the closest thing in the real world to the market of 'perfect information' found in economics textbooks. Nonetheless, there are still some concerns that these might evolve into cartels controlled by the existing dominant companies, and thereby restrict new entrants and potential competition. However, where the supply market is more differentiated, other types of relationship are likely be more appropriate. In this case some form of 'partnership' or 'lean' relationship is often advocated, based on the quality and development lead-time benefits experienced by Japanese manufacturers of consumer durables, specifically cars and electronics. Lamming identifies a number of defining characteristics of such partnership or 'lean' supply relations:[18]

- fewer suppliers, longer-term relations;
- greater equity – real 'cost transparency';
- focus on value flows – the relationship, not the contract;
- vendor assessment, plus development;
- two-way or third-party assessment;
- mutual learning – share experience, expertise, knowledge and investment.

These principles are based on a distillation of the features of the best Japanese manufacturers in the automobile and electronics sectors, and more recent experiments in other contexts, such as aerospace in the UK and USA,[19] and as such may represent best practice under certain conditions. Nishiguchi compared supplier relations in Japan and the UK, and found that lean or partnership approaches had significant advantages over market relations, including more supportive customers and less erratic trade.[20] This resulted in measurable differences in operational performance, such as a reduction in inventory held by customers of 90%, and tool development time reduction by some 70%. However, trade-offs existed. In the lean relationships customers were rated by suppliers as being significantly more demanding than in the market relationships, and involved a much higher degree of monitoring by customers. Perhaps of greater strategic significance, in the lean relationships the suppliers' sales were dominated by a few key customers, and asset specificity – a measure of how much a suppliers' plant and equipment are dedicated to a particular customer – was much higher. These two factors make suppliers in lean relations very vulnerable to the fortunes of their key customers. For example, in the UK the retail chain Marks & Spencer was often presented as the model of supplier relations, but following its poor market and financial performance in the late 1990s, many of its long-term supply 'partners' have been abandoned or ordered to cut costs or be deselected. Nevertheless, 'partnership' models have fast become the norm in both the private and public sectors, irrespective of the supply market conditions or objectives of the relationship. For example, one study found that the main explanation for the adoption of lean supply practices was managerial choice, rather than any rationale based on external factors such as industry structure or supply needs.[21]

However, in the case of indeterminate supply markets a partnership or lean supply strategy may be sub-optimal or even dysfunctional. We shall revisit the case of Japanese business groups later in this chapter, but in anticipation of that discussion there is evidence that such rigid supply structures may offer static efficiencies in terms of cost savings, quality improvement and reduction in development lead time, but may suffer dynamic inefficiencies when it comes to developing novel technologies, products and processes. On the one hand, the increase in the global sourcing of technology has reduced the chance that an existing 'partner' will be the most appropriate supplier, and on the other hand the tacit nature or 'stickiness' of technological knowledge suggests that a market transaction would be inadequate.[22] Therefore where innovation is the primary objective of the supply relationship, and the supply market is neither homogenous nor clearly differentiated, a temporary, ad hoc relationship with a supplier may be more appropriate. These have some features common to horizontal strategic alliances, in that they are clearly focused, project-based forms of collaboration. In such cases the relationship is neither market nor partnership, but a hybrid (see Box 8.2). Loose coupling is appropriate where multi-technology products are characterized by uneven rates of advance in the underlying technologies, and in such cases technology consultants or systems integrators act as a buffer between the suppliers and users of the technology.[23] For suppliers, technological competencies and problem-solving capabilities are associated with high gross margins and a larger share of overseas business.[24] A survey of companies offering specialist services to support new product development found that the most common service offered was industrial design (58% of firms), but 30% offered a complete range of services, including R&D, market research, design, development and development of production processes.[25] The USA accounts for almost half of such firms, and within Europe the UK accounts for more than half.

Table 8.3 lists some of the management practices found to contribute to a supplier relationship for successful new product development. This list suggests a number of good practices common to partnership or lean approaches, but unbundles these practices from the need for long-term, stable co-dependent relationships. The low rating given to co-location and shared equipment suggests a more arm's-length relation, albeit highly integrated for the purposes of the project. Note the relatively high ranking of the need for consensus that the right supplier has been chosen.

Technology Licensing

The issue of intellectual property and the basis of licensing was discussed in the previous chapter. Licensing offers a firm the opportunity to exploit the intellectual property of another firm, normally in return for payment of a fee and royalty based on sales. Typically a technology licence will specify the applications and markets in which the technology may be used, and often will require the buyer to give the seller access to any subsequent improvements in the technology.

BOX 8.2
GLOBETRONICS: EVOLUTION OF SUPPLY RELATIONSHIPS

Globetronics Bhd. was formed in 1990 by two Malaysians formerly employed by Intel. The Malaysian Technology Development Corporation (MTDC) provided 30% of the venture capital, and the company was subsequently floated in 1997 to raise additional capital for growth. The company's primary activities are similar to the majority of transnational semiconductor firms based in Malaysia, and involve post-fabrication manufacture of semiconductors, including assembly and packaging. Indeed, the company's main customers are American and Japanese transnationals. The significant difference is that domestic ownership and management have allowed Globetronics to more easily capture value-added activities such as development and marketing.

The company now has seven business divisions and a new plant in the Philippines. Two of the businesses are joint ventures with the Japanese firm Sumitomo. The relationship with Sumitomo began as a simple subcontracting agreement, but over the years a high level of trust has been achieved and two joint ventures have been established. The first, SGT, was created in 1994, and is 49% owned by Globetronics. It is the largest manufacturer in the world and the only company outside of Japan to produce ceramic substrate semiconductor packages. The second joint venture, SGTI, was created in 1996, and is 30% owned by Globetronics. In both cases the Japanese partner has maintained majority ownership, but it is clear that the Malaysian partner has made some progress in assimilating the technological and design capabilities. This provides a promising model for companies in developing countries, to escape dependent subcontracting relationships by using joint ventures to upgrade their technological and market competencies.

Source: Extract from Tidd, J. and M. Brocklehurst (1999) 'Routes to technological learning and development: an assessment of Malaysia's innovation policy and performance', *Technological Forecasting and Social Change*, **63** (2).

In theory, licensing-in a technology has a number of advantages over internal development, in particular lower development costs, less technological and market risk, and faster product development and market entry. Potential drawbacks to licensing-in include restrictive clauses imposed by the licenser, loss of control of operational issues such as pricing, production volume and product quality, and the potential transaction costs of search, negotiation and adaptation.

TABLE 8.3 Supplier relationship factors contributing to successful new product development

Factor	Most successful	Least successful	Difference*
Strength of supplier's top management commitment	6.14	5.22	0.91
Direct cross-functional, inter-company communication	6.05	4.87	1.18
Strength of customer's top management commitment	5.70	4.95	0.75
Familiarity with supplier's capability prior to project	5.64	4.58	1.07
Customer requirements information-sharing	5.12	4.22	0.90
Joint agreement on performance measures	5.07	4.20	0.88
Supplier membership/participation on customer's project team	5.02	3.73	1.29
Technology-sharing	4.84	3.77	1.07
Strength of consensus that right supplier was selected	4.83	3.88	0.95
Formal trust development practices	4.14	3.07	1.07
Common and linked information systems	4.07	2.96	1.11
Shared education and training	3.44	2.29	1.15
Risk/reward-sharing schemes	3.13	2.47	0.65
Co-location of customer/supplier personnel	2.95	1.84	1.11
Technology information-sharing	2.44	1.62	0.82
Shared plant and equipment	2.44	1.62	0.82

* All differences statistically significant at 5% level.
1 = no use, 7 = significant/extensive. $N = 83$.
Source: Derived from Ragatz, G., R. Handfield and T. Scannell (1997) 'Success factors for integrating suppliers into new product development', *Journal of Product Innovation Management*, **14**, 190–202.

In practice, the relative costs and benefits of licensing-in will depend on the nature of the technologies and markets and strategy and capability of the firm. A survey of more than 200 firms in the chemical, engineering and pharmaceutical industries found that the most important reasons for licensing were related to the speed of access, rather than cost. Factors such as quickly acquiring knowledge required for product development, keeping pace with competitors and increasing sales were found to be most important, and factors such as the cost of development least important.[26] The study found that the most significant problems associated with licensing-in are entry costs such as the choice of suitable technology and licenser, and the loss of control of decision-making. Differences in emphasis exist across sectors; for example, pharmaceutical firms experience higher search costs than engineering firms, and engineering firms place greater emphasis on the potential for reducing the cost and improving the speed of market entry. For example, Eli Lilly licensed-in basic cephalosporin technology from the National Research and Development Corporation. Using its in-house skills, it was able to produce a wide range of these antibiotics, hence adding value to the licensed technology.

In some cases, however, there is a reluctance to license-in technology which may adversely affect the differentiation of end products, if customers became aware of the fact. Many firms express concerns regarding the constraints imposed by international licensing agreements, specifically the common requirement to 'grant-back' any improvements made to the technology. For example, ICI claims increasing globalization and concentration within the chemical industry as reducing the scope for licensing technology. For these reasons an increasing number of firms are careful to license only components of any process or product in order to allow scope for subsequent improvement and differentiation. For example, Mitsubishi Chemical licensed a well-established process technology from a US competitor, but chose not to license the catalyst or polymer design. This allowed the company to avoid having to grant-back its subsequent improvements to the catalyst and polymer design to the American competitor. Nippondenso adopted a similar strategy when licensing technology from potential competitors.

However, this approach to licensing is only viable where the technology can be easily 'unbundled'. For example, Kirin, the Japanese food and drink manufacturer, finds that in most cases it is able to negotiate and exploit simple licences for its brewery and food businesses, but not for its pharmaceutical products. The company prefers formal joint ventures to develop new pharmaceutical products because of the complex interrelated technologies, patents and skills required.

Research Consortia

Research consortia consist of a number of organizations working together on a relatively well-specified project. The rationale for joining a research consortium includes sharing the cost and risk of research, pooling scarce expertise and equipment, performing pre-competitive research and setting of standards. They may take many different forms, the most centralized being pooled investment in a common research facility or new venture, and the least centralized being co-ordinated research co-located in the various member firms. Typically, European firms have favoured the former, whereas American firms have tended to adopt the latter type. Japanese firms appear to favour a hybrid form in which shared research facilities are used in parallel to co-ordinated in-house research. These differences in structure are due to technical, competitive and legal reasons.

The original research consortia were the industry-based research associations formed in the UK during the First World War. The idea was to encourage small firms to fund research and the research associations were largely a response to the competitive threat of German manufacturers. They were funded by a combination of government funds and contributions from member firms. Many of the associations have survived and are now commercial contract research organizations. The concept of research associations transferred to Japan in the 1950s, and a number of research consortia were formed to support the development of the automobile component industry. In 1961 the Engineering Association Act was passed in Japan, allowing firms to form consortia without being prosecuted under anti-trust law, and also provided tax benefits for consortia. However, the goals and structure of the Japanese consortia are very different from those in the UK.

In Japan, the Technology Research Association system provides a structure to bring together firms from a wide range of different industries. Unlike their British counterparts, the Japanese associations tend to be temporary organizations, and are disbanded when the project is completed, whereas the British associations have tended to become permanent organizations. Members of the Japanese associations tend to be large firms with extensive in-house R&D capabilities, whereas members of British associations tend to be much smaller companies with little in-house R&D. Finally, the Japanese associations tend to be in high-technology areas, whereas the British associations originally concentrated on support for mature industries.

The USA was relatively late to adopt research consortia because of a strong belief in the efficiency of a free market economy resulting in severe anti-trust (anti-monopoly) legislation, which made it difficult to organize consortia. Also the early dominance of large firms in many of the leading sectors made it easier for firms to rely on their own resources to conduct research. However, the National Cooperative Research Act was passed in the USA in 1984 in an effort to emulate the apparent success of the Japanese consortia. As a result, the Microelectronic and Computer Technology Corporation (MCC) (see Box 8.3) and SEMATECH (semiconductor manufacturing technology initiative) were able to be established.[27]

BOX 8.3
MICROELECTRONIC AND COMPUTER TECHNOLOGY CORPORATION (MCC)

MCC was formed by a number of US firms in 1982, including computer manufacturers Sperry, NCR, DEC, Honeywell, and producers of semiconductor devices such as AMD and Motorola. MCC was formed in response to the perceived loss of competitiveness of the US computing industry, and fear of the Japanese fifth-generation computer project. Indeed no foreign-owned firms were allowed to join the MCC. The formation of the MCC challenged the existing anti-trust (anti-monopoly) laws in the USA, and subsequently the National Cooperative Research Act came into effect in 1984 to allow firms to collaborate for basic research under certain circumstances. Since the Act more than 400 R&D consortia have been formed in the USA.

A state-of-the-art research facility was built in Austin, Texas, and in its first 10 years of operation was funded for more than $500m. by its member firms. Four broad areas were targeted for research: computer-aided design, packaging inter-connect, software and advanced computer architecture. The success of MCC has been mixed. On the one hand it was awarded 117 US patents, licensed 182 tech-nologies and published more than 2400 technical reports. However, there is little evidence of any commercial outputs or any significant technology transfer to its member firms. Thus in the 1990s the research emphasis of the MCC has shifted from strategic research for its corporate members, to lower-level and shorter-term projects, many funded by the US and foreign governments. Two modes of tech-nology transfer were evaluated in the case of MCC. First, transfer to member firms. Second, the establishment of new-technology spin-off firms. Every effort was made to promote technology transfer to member firms, and a formal standard technology package (STP) was introduced whereby technology release dates were announced and shareholders invited to MCC to receive the technology package. However, in a survey of member firms, only a fifth of managers believed that the MCC would create useful technology, but almost half predicted problems of trans-ferring these technologies to the member firms. Researchers at the MCC shared these concerns:

> 'we were caught in the paradox that faces many R&D consortia sooner or later. Our success depended on how successfully our shareholders adopted MCC technology. Yet we had virtually no control once MCC technology was in the hands of the shareholders.'

> 'clearly the technology transfer problem must be viewed from both the researcher and customer viewpoints . . . most of the participants have been literally wasting their research investment in MCC by failing to put in place effective mechanisms and people for marketing, selling, developing, and supporting MCC technology once it reached the successful prototype stage.'

(quoted in Gibson and Rogers, 1994, 339–340)[27]

There are significant differences between the Japanese and American research consortia.[28] Almost all members of the Japanese consortia conduct research in member firms, compared to less than a half of the American consortia members. The US consortia favour separate joint facilities and research in universities. Therefore the Japanese and American consortia face different organizational problems. The biggest problem the Japanese must tackle is how to co-ordinate research in member firms, whereas the biggest problem for the Americans is how to manage technology transfer from the centre to the member firms. In addition, the Japanese consortia are more focused on applied product development and pilot production, whereas the American consortia tend to concentrate on idea generation and technical feasibility studies. Therefore the success of research consortia depends on their motives, structure and membership.

Consortia, defined as multi-firm collaborations, take two main forms, between competitors or between non-competing firms. Firms commonly collaborate with competitors in the development of pre-competitive technologies. This form of collaboration is particularly attractive when supported by government or EU funds, as in the case of the Framework programme for European research and technological development. For example, CRL, part of the Thorn group, participated some years ago in a government-funded consortium working on ferroelectric liquid crystal (FLC) technology, bringing its expertise in colour and grey scale to the collaboration. Since the end of this consortium, the company has continued to develop the technology itself, and recently launched a low-power, low-cost monochrome FLC display. In addition, CRL's printer shuttle arrays were developed with support from the EU-funded ESPRIT programme.

Collaboration between firms in different industries appears to raise much less concern about proprietary positions. In most cases, they are viewed as an attractive means of leveraging in-house skills by working with organizations possessing complementary technical capabilities. Intra-industry collaborations are more important in non-competitive areas, such as in the areas of health, safety and the environment, and in setting new standards or influencing legislation. For example, the 1990 US Clean Air Act placed the onus on automobile manufacturers and oil companies to provide the basic science which will act as a realistic foundation for future legislation. Competing oil companies and auto manufacturers set up the Auto-oil consortium to provide this science.

Nevertheless, even in consortia involving non-competing firms, it appears that vested interests can sometimes lead to difficulties. For example, in the collaboration between auto manufacturers and oil companies aimed at reducing toxic emissions from car exhausts, there exist serious differences of opinion between these two groups of firms about whether the main thrust of the research should be directed towards improved engine efficiencies, or towards better gasoline formulations.

Therefore both industry and firm level factors influence the formation of consortia. Industry level factors that increase the formation of and participation in consortia are weak competition and or weak appropriability conditions, including intellectual property. The main firm level factor that influences the formation and participation in consortia is R&D capability, although previous experience of consortia also has an effect. The technological capability of a firm increases both the opportunity and incentive to participate. Greater technological capability (or perceptions of) makes a firm a more attractive potential member, and increase the opportunity to participate. In addition, technological capability (and market) position should allow it to learn more easily or absorb more knowledge. Empirical studies confirm that increased technological capability is associated with increased participation in consortia, controlling for other effects, and that industries with weaker competition and appropriability tend to have form more consortia.[29]

Strategic Alliances and Joint Ventures

Strategic alliances, whether formal or informal, typically take the form of an agreement between two or more firms to co-develop a new technology or product. Whereas research consortia tend to focus on more basic research issues, strategic alliances involve near-market development projects. However, unlike more formal joint ventures, a strategic alliance typically has a specific end goal and timetable, and does not normally take the form of a separate company. There are two basic types of formal joint venture: a new company formed by two or more separate organizations, which typically allocate ownership based on shares of stock controlled; or a more simple contractual basis for collaboration. The critical distinction between the two types of joint venture is that an equity arrangement requires the formation of a separate legal entity. In such cases management is delegated to the joint venture, which is not the case for other forms of collaboration. A range of strategic alliances and joint ventures are discussed in detail in section 8.4.

Doz and Hamel identify a range of motives for strategic alliances and suggest strategies to exploit each:[30]

- To build critical mass through co-option.
- To reach new markets by leveraging co-specialized resources.
- To gain new competencies through organizational learning.

In a co-option alliance, critical mass is achieved through temporary alliances with competitors, customers or companies with complementary technology, products or services. Through co-option a company seeks to group together other relatively weak companies to challenge a dominant competitor. Co-option is common where scale or network size is important, such as mobile telephony and airlines. For example, Airbus (see Box 8.4) was originally created in response to the dominance of Boeing, and Symbian in response to Microsoft's dominance. Greater international reach is a common related motive for co-option alliances. Fujitsu initially used its alliance with ICL to develop a market presence in Europe, as did Honda with Rover. However, co-option alliances may be inherently unstable and transitory. Once the market position has been achieved, one partner may seek to take control through acquisition, as in the case of Fujitsu and ICL, or to go unilateral, as in the case of Honda and Rover.[31]

In a co-option alliance, partners are normally drawn from the same industry, whereas in co-specialization partners are usually from different sectors. In a co-specialized alliance, partners bring together unique competencies to create the opportunity to enter

---

BOX 8.4
AIRBUS INDUSTRIE

Airbus Industrie was formed in France in 1969 as a joint venture between the German firm MBB (now DASA) and French firm Aérospatiale, to be joined by CASA of Spain in 1970 and British Aerospace (now BAe Systems) in 1979. Airbus is not a company, but a Groupment d'Intérêt Economique (GIE), which is a French legal entity which is not required to publish its own accounts. Instead, all costs and any profits or losses are absorbed by the member companies. The partners make components in proportion to their share of Airbus Industrie: Aérospatiale and DASA each have 37.9%, BAe 20% and CASA 4.2%.

At that time the international market for civil aircraft was dominated by the US firm Boeing, which in 1984 accounted for 40% of the airframe market in the non-communist world. The growing cost and commercial risk of airframe development had resulted in consolidation of the industry and a number of joint ventures. In addition, product life cycles had shortened due to more rapid improvements in engine technology. The partners identified an unfilled market niche for a high-capacity/short medium-range passenger aircraft, as more than 70% of the traffic was then on routes of less than 4600 km. Thus the Airbus A300 was conceived in 1969. The A300 was essentially the result of the French and German partners, the former insisting on final assembly in France, and the latter gaining access to French technology. The first A300 flew in 1974, followed by a series of successful derivatives such as the A310 and the A320. The British partner played a leading role in the subsequent projects, bringing both capital and technological expertise to the venture. Airbus has since proved to be highly innovative with the introduction of fly-by-wire technology, and common platforms and control systems for all its aircraft to reduce the cost of crew training and aircraft maintenance. In 2000 the group announced plans to develop a double-decker 'super' jumbo, the A380, with seats for 555 passengers and costing an estimated US$12 bn to develop. Airbus estimates a global market of 1163 very large passenger aircraft and an additional 372 freighters, but needs to sell only 250 A380s to achieve breakeven. This would challenge Boeing in the only market it continues to dominate. (However, Boeing predicts a market of just 320 very large aircraft, as it assumes a future dominance of point-to-point air travel by smaller aircraft, whereas Airbus assume a growth in the hub-and-spoke model, which demands large aircraft for travel between hubs.)

In 1998 Airbus outsold Boeing for the first time in history. In 1999 Daimler-Chrysler (DASA), Aérospatiale and CASA merged to form the European Aeronautic Defence and Space Company (EADS), making BAe Systems, formerly British Aerospace, the only non-EADS member of Airbus. The group plans to move from the unwieldy GIE structure to become a company. This would allow streamlining of its manufacturing operations, which are currently geographically dispersed across the UK, France, Germany and Spain, and more importantly help create financial transparency to help identify and implement cost savings. Also, some customers have reported poor service and support as Airbus has to refer such work to the relevant member company.

Airbus demonstrates the complexity of joint ventures. The primary motive was to share the high cost and commercial risk of development. On the one hand, the French and German participation was underwritten by their respective governments. This fact has not escaped the attention of Boeing and the US Government, which provides subsidies indirectly via defence contracts. On the other hand, all partners had to some extent captive markets in the form of national airlines, although almost three-quarters of all Airbus sales were ultimately outside the member countries. Finally, there were also technology motives for the joint venture. For example, BAe specializes in development of the wings, Aérospatiale the avionics, DASA the fuselages and CASA the tails. However, as suggested above, there are now strong financial, manufacturing and marketing reasons for combining the operations within a single company.

new markets, develop new products or build new businesses. Such co-specialization is common in systems or complex products and services. However, there is a risk associated with co-specialization. Partners are required to commit to partners' technology and standards. Where technologies are emerging and uncertain and standards are yet to be established, there is a high risk that a partner's technology may become redundant. This has a number of implications for co-specialization alliances. First, that at the early stages of an emerging market where the dominant technologies are still uncertain, flexible forms of collaboration such as alliances are preferable, and at later stages when market needs are clearer and the relevant technological configuration better defined, more formal joint ventures become appropriate.[32] Second, to restrict the use of alliances to instances where the technology is tacit, expensive and time-consuming to develop. If the technology is not tacit, a licence is likely to be cheaper and less risky, and if the technology is not expensive or time-consuming to develop, in-house development is preferable.[33] If resources and relationships allow, some form of network hedging strategy might be advisable, as suggested later in the section on innovation networks. We will discuss the acquisition of competencies through alliances in section 8.5.

There has been a spectacular growth in strategic alliances, and at the same time more formal joint ventures have declined as a means of collaboration. In the mid-1980s less than 1000 new alliances were announced each year, but by the year 2000 this had grown to almost 10 000 per year (based on data from Thomson Financial). There are a number of reasons for the increase in alliances overall, and more specifically the switch from formal joint ventures to more transitory alliances:[34]

- *Speed: transitory alliances versus careful planning.* Under turbulent environmental conditions, speed of response and learning and lead time are more critical than careful planning, selection and development of partnerships.
- *Partner fit: network versus dyadic fit.* Due to the need for speed, partners are often selected from existing members of a network, or alternatively reputation in the broader market.
- *Partner type: complementarity versus familiarity.* Transitory alliances increasingly occur across traditional sectors, markets and technologies, rather than from within. Microsoft and Lego to develop an Internet-based computer game, Deutsche Bank and Nokia to create mobile financial services.
- *Commitment: aligned objectives versus trust.* The transitory nature of relationships make the development of commitment and trust more difficult, and alliances rely more on aligned objectives and mutual goals.
- *Focus: few, specific tasks versus multiple roles.* To reduce the complexity of managing the relationships, the scope of the interaction is more narrowly defined, and focused more on the task than the relationship.

Innovation Networks

The concept of innovation networks has become popular in recent years, as it appears to offer many of the benefits of internal development, but with few of the drawbacks of collaboration. Networks have been claimed by some to be a new hybrid form of organization that has the potential to replace both firms (hierarchies) and markets, in essence the 'virtual corporation', whereas others believe them to be simply a transitory form of organization, positioned somewhere between internal hierarchies and external market mechanisms. Whatever the case, there is little agreement on what constitutes a network, and the term and alternatives such as 'web' and 'cluster' have been criticized for being too vague and all-inclusive.[35] A recent review of the field concluded that:

> little is known about innovation networks . . . there is no clear definition of what an inno-
> vation network is. Rather there are numerous models, each emphasizing different aspects
> depending on the research questions. It is also not clear whether there are common char-
> acteristics applicable to all spheres of innovation, or disparate phenomena with little or
> no commonality. Not very much can be found in the literature about the dynamics of
> innovation networks: how they arise, the growth processes they undergo, and the way
> they die or merge into other networks . . . Open questions are: What characterizes an inno-
> vation network in comparison with classical firms of organization? What is the structure
> of an innovation network, what are its elements, what are their basic interactions, what
> co-ordination mechanisms are important and what dynamics emerge from internal inter-
> actions? What is the relationship between an innovation network and its environment and
> how does the environment influence the dynamics.[36]

Different authors adopt different meanings, levels of analysis and attribute networks with different characteristics. For example, academics on the Continent have focused on social, geographical and institutional aspects of networks, and the opportunities and constraints these present for innovation.[37] In contrast, Anglo-Saxon studies have tended to take a systems perspective, and have attempted to identify how best to design, manage and exploit networks for innovation.[38] Figure 8.3 presents a framework for the analysis of different network perspectives in innovation studies.

Whilst there is little consensus in aims or means, there appears to be some agree-ment that a network is more than an aggregation of bilateral relationships or dyads, and therefore the configuration, nature and content of a network impose additional constraints and present additional opportunities. A network can be thought of as con-sisting of a number of positions or nodes, occupied by individuals, firms, business units, universities, governments, customers or other actors, and links or interactions between these nodes. A network perspective is concerned with how these economic actors are influenced by the social context in which they are embedded and how actions can be influenced by the position of actors. For example, national systems of innovation, discussed in Chapter 4, are an example of an innovation network at a high level of

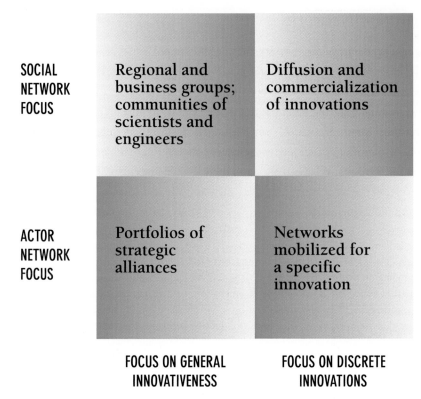

SOCIAL NETWORK FOCUS	Regional and business groups; communities of scientists and engineers	Diffusion and commercialization of innovations
ACTOR NETWORK FOCUS	Portfolios of strategic alliances	Networks mobilized for a specific innovation
	FOCUS ON GENERAL INNOVATIVENESS	**FOCUS ON DISCRETE INNOVATIONS**

FIGURE 8.3   Different network perspectives in innovation research

Source: Derived from Conway, S. and F. Steward (1998) 'Mapping innovation networks', *International Journal of Innovation Management*, **2** (2), 223–254.

aggregation. Innovations networks can exist at any level: global, national, regional, sector, organizational, or individual. Whatever the level of analysis, the most interesting attribute of an innovation network is the degree and type of interaction between actors, which results in a dynamic but inherently unstable set of relationships. Innovation networks are an organizational response to the complexity or uncertainty of technology and markets, and as such innovations are not the result of any linear process. This makes it very difficult, if not impossible, to predict the path or nature of innovation resulting from network interactions (see Box 8.5 for an example of this unpredictability of innovation in networks). The generation, application and regulation of an innovation within a network is unlike the trial-and-error process within a single firm or venture, or variation and selection within a market. Instead, actors in an innovation network attempt to reduce the uncertainty associated with complexity through a process of recursive learning and testing.

BOX 8.5
AN ENVIRONMENTAL INNOVATION NETWORK FOR IKEA

The catalogue of IKEA has one of the world's highest circulations, with a print run of more than 100 million per year, needing 50 000 tonnes of high-quality paper each year. However, in the 1990s there were growing environmental concerns about the discharge of chlorinated compounds from the processes used to create the relatively high-quality paper used in such promotional materials, as well as the more general issue of paper recycling. In response to these concerns, in 1992 IKEA introduced two new goals for the production of its catalogue: be printed on paper that was totally chlorine-free (TCF), and to include a high proportion of recycled paper.

However, these goals demanded significant innovation. No such paper product existed at the time, and the dominant industry suppliers believed that the combination of no chlorine and high levels of recycled pulp to be impossible. To achieve the necessary paper brightness for catalogue printing, a minimum of 50% chlorine-dioxide-bleached pulp had been used. Chlorine had been used for 50 years as the bleaching agent for high-quality paper. Moreover, the high-quality paper used for such catalogues consisted of a very thin paper base, which is coated with clay, which makes the insertion of recycled fibre very difficult. The manager of R&D at Svenska Cellulosa Aktiebolaget (SCA), one of Europe's largest producers of high-quality paper, argued that 'the high quality demands and the large volume of filling substances is the main reason that it is neither realistic nor necessary to use recycled fibre'. SCA reinforced this view with the decision to build a new SEK 2.4 billion plant to produce conventional high-quality coated paper. At that time SCA was not a supplier to IKEA.

In Sweden, the paper manufacturer Aspa worked with the chemical firm Eka Nobel to develop an environmentally acceptable bleaching process with less damaging discharges, but this was still based on chlorine dioxide and failed to achieve the necessary brightness for use in high-quality paper, and was marketed as 'semi-bleached'. Following customer demand for a true TCF product, including a request from Greenpeace for TCF paper for production of its newsletter, Aspa was forced to develop a stable product with secure supplies. At this stage the pulp and fibre company Sodra Cell became involved, and identified the need to reach full brightness to create a broader market for TCF paper. Sodra worked with the German company Kvaerner to develop an alternative but equally effective bleaching

continues overleaf

BOX 8.5   *(continued)*

---

process, and Kvaerner established a research project on ozone bleaching with Lentzing and STORA Billerud. The ozone bleaching process was adapted from an established process for water purification with the help of AGA Gas. However, the use of ozone in place of chlorine for bleaching required the quality of the pulp-wood to be improved, so the harvesting system had to be changed to ensure that wood was better sorted and available within weeks of harvesting. To improve the brightness and strength of the paper, the impurities in the pulp from de-inked recycled paper had to be reduced, which required a new washing process. The changes in the chemistry of the pulp subsequently reduced the strength of the paper, which required changes in the paper production process. The printing processes had to be adapted to the characteristics of the new paper. Initially Sodra Cell supplied the new product to SCA through its relationship with Aspa, but also to the Italian paper producer Burgo, which provided the paper for the IKEA catalogue.

Thus the organization evolved beyond a simple industrial supply relationship, to an innovation network including customers, printers, paper manufacturers, pulp and fibre producers, forestry companies, research institutes, environmental and a lobby group, across many different countries. At the same time, the intended innovation shifted from a high-quality TCF clay-coated paper, to a TCF uncoated fresh pulp and 10% de-inked recycled pulp product.

Source: Hakansson, H. and A. Waluszewski (2003) *Managing Technological Development: IKEA, the environment and technology.* Routledge, London.

A network can influence the actions of its members in two ways.[39] First, through the flow and sharing of information within the network. Second, through differences in the position of actors in the network, which causes power and control imbalances. Therefore the position an organization occupies in a network is a matter of great strategic importance, and reflects its power and influence in that network. Sources of power include technology, expertise, trust, economic strength and legitimacy. Networks can be tight or loose, depending on the quantity (number), quality (intensity) and type (closeness to core activities) of the interactions or links. Such links are more than individual transactions, and require significant investment in resources over time. Hakansson identifies a number of important types of interaction within innovation networks:[40]

- *Product interactions* – products and groups of products and services interact, are adapted and evolve.
- *Process interactions* – the interdependencies between product and process, and between different processes and production facilities are another interaction within a network, together with their use and utilization.
- *Social interaction within the organization* – for example, business units are more than a combination of product and process facilities. They consist also of social interactions, with knowledge of and an ability to work with, other business units within the organization.
- *Social interaction between organizations* – business relationships both restrict and provide opportunities for innovation, particularly for systemic innovations.

Networks are appropriate where the benefits of co-specialization, sharing of joint infrastructure and standards and other network externalities outweigh the costs of network governance and maintenance. Where there are high transaction costs involved in purchasing technology, a network approach may be more appropriate than a market model, and where uncertainty exists a network may be superior to full integration or acquisition. Historically, networks have evolved from long-standing business relationships. Any firm will have a group of partners that it does regular business with – universities, suppliers, distributors, customers and competitors. Over time mutual knowledge and social bonds develop through repeated dealings, increasing trust and reducing transaction costs. Therefore a firm is more likely to buy or sell technology from members of its network.[41] Firms may be able to access the resources of a wide range of other organizations through direct and indirect relationships, involving different channels of communication and degrees of formalization. Typically, this begins with stronger relationship between a firm and a small number of primary suppliers, which share knowledge at the concept development stage. The role of the technology gatekeeper, or heavyweight project manager, is critical in this respect. In many cases organizational linkages can be traced to strong personal relationships between key individuals in each organization. These linkages may subsequently evolve into a full network of secondary and tertiary suppliers, each contributing to the development of a sub-system or component technology, but links with these organizations are weaker and filtered by the primary suppliers. However, links amongst the primary, secondary and tertiary supplier groups may be stronger to facilitate the exchange of information.

This process is path-dependent in the sense that past relationships between actors increase the likelihood of future relationships, which can lead to inertia and constrain innovation. Indeed much of the early research on networks concentrated on the constraints networks impose on members – for example, preventing the introduction of 'superior' technologies or products by controlling supply and distribution networks.

Organizational networks have two characteristics that affect the innovation process: activity cycles and instability.[42] The existence of activity cycles and transaction chains creates constraints within a network. Different activities are systematically related to each other and through repetition are combined to form transaction chains. This repetition of transactions is the basis of efficiency, but systemic interdependencies create constraints to change. For example, the Swiss watch industry was based on long-established networks of small firms with expertise in precision mechanical movements, but as a result was slow to respond to the threat of electronic watches from Japan.

Similarly, Japan has a long tradition of formal business groups, originally the family-based *zaibatsu*, and more recently the more loosely connected *keiretsu*. The best-known groups are the three ex-*zaibatsu* – Mitsui, Mitsubishi and Sumitomo, and the three newer groups based around commercial banks – Fuji, Sanwa and Dal Ichi Kangyo (DKB). There are two types of *keiretsu*, although the two overlap. The vertical type organizes suppliers and distribution outlets hierarchically beneath a large, industry-specific manufacturer, for example Toyota Motor. These manufacturers are in turn members of *keiretsu* which consist of a large bank, insurance company, trading company and representatives of all major industrial groups. These inter-industry *keiretsu* provide a significant internal market for intermediate products. In theory, benefits of membership of a *keiretsu* include access to low-cost, long-term capital, and access to the expertise of firms in related industries. This is particularly important for high-technology firms. In practice, research suggests that membership of *keiretsu* is associated with below-average profitability and growth,[43] and independent firms like Honda and Sony are often cited as being more innovative than established members of *keiretsu*. However, the *keiretsu* may not be the most appropriate unit of analysis, as many newer, less formal clusters of companies have emerged in modern Japan.

However, as the role of a network is different for all its members, there will always be reasons to change the network and possibilities to do so. A network can never be optimal in any generic sense, as there is no single reference point, but it is inherently adaptable. This inherent instability and imperfection mean that networks can evolve over time. For example, Belussi and Arcangeli discuss the evolution of innovation networks in a range of traditional industries in Italy.[44]

More recent research has examined the opportunities networks might provide for innovation, and the potential to explicitly design or selectively participate in networks for the purpose of innovation, that is a path-creating rather than path-dependent process.[45] A study of 53 research networks found two distinct dynamics of formation and growth. The first type of network emerges and develops as a result of environmental interdependence, and through common interests – an *emergent* network. However, the other type of network requires some triggering entity to form and develop – an *engineered* network.[46] In an engineered network a nodal firm actively recruits other

TABLE 8.4 Competitive dynamics in network industries

	Type of network	
	Unconnected, closed	Connected, open
System attributes	Incompatible technologies	Compatible across vendors and products
	Custom components and interfaces	Standard components
Firm strategies	Control standards by protecting proprietary knowledge	Shape standards by sharing knowledge with rivals and complementary markets
Source of advantage	Economies of scale, customer lock-in	Economies of scope, multiple segments

Source: Adapted from Garud, R. and A. Kumaraswamy (1993) 'Changing competitive dynamics in network industries', *Strategic Management Journal*, **14**, 351–369.

members to form a network, without the rationale of environmental interdependence or similar interests. Different types of network may present different opportunities for learning (Table 8.4). In a closed network, a company seeks to develop proprietary standards through scale economies and other actions, and thereby lock customers and other related companies into its network.[47] In such cases established companies are able to reinforce their positional advantage by adopting new technologies which have implications for compatibility, whereas new entrants or existing firms at the periphery of the network will find it extremely difficult to gain a positional advantage through innovation.[48] Obvious examples include Microsoft in operating systems and Intel in microprocessors for PCs. In the case of open networks, complex products, services and businesses have to interface with others and it is in everyone's interest to share information and to ensure compatibility (see Box 8.6). Open networks or systems often involve multiple hierarchical levels or sub-systems, each controlled by a different technical community. Therefore innovations in one technical sub-field may influence some relationships within the network, but not the whole network. Therefore innovation by established firms at the periphery of the network or by new entrants is more common. Examples include telephony and power generation and distribution.

Virtual innovation networks are beginning to emerge, based on firms that are connected via intranet/extranet/Internet and exchange information within a business relationship to create value. To date such virtual networks are most common in supply chain and customer order automation, but recent examples include product development. For example, in supply chain management Herve Thermique, a French manufacturer of heating and air conditioners, uses an extranet to coordinate its 23 offices and 8000 suppliers, and General Electric has an extranet bidding and trading system

BOX 8.6
THE DEVELOPMENT OF LINUX: A CASE OF NETWORK INNOVATION?

The computer operating system Linux has been largely developed by a network of voluntary programmers, often referred to as the 'Linux community'. This is probably one of the few true examples of a 'cyber' or virtual organization. Linus Torvalds first suggested the development of a free operating system to compete with the DOS/Windows monopoly in 1991, and quickly attracted the support of a group of volunteer programmers: 'having those 100 part-time users was really great for all the feedback I got. They found bugs that I hadn't because I hadn't been using it the way they were . . . after a while they started sending me fixes or improvements . . . this wasn't planned, it just happened'. Thus Linux grew from 10 000 lines of code in 1991 to 1.5 million lines by 1998. Its development coincided with and fully exploited the growth of Internet and later Web forms of collaborative working. The provision of the source code to all potential developers promotes continuous incremental innovation, and the close and sometimes indistinguishable developer and user groups promotes concurrent development and debugging. The weaknesses are potential lack of support for users and new hardware, availability of compatible software and forking in development.

By 1998 there were estimated to be more than 7.5 million users and almost 300 user groups across 40 countries. Linux has achieved a 25% share of the market for server operating systems, although its share of the PC operating system market was much lower, and Apache, a Linux application Web server program, accounted for half the market. Although Linux is available free of charge, a number of businesses have been spawned by its development. These range from branding and distribution of Linux, development of complementary software and user support and consultancy services. For example, although Linux can be downloaded free of charge, RedHat Software provides an easier installation program and better documentation for around US$50, and in 1998 achieved annual revenues of more than US$10m. RedHat was floated in 1999. In China, the lack of legacy systems, low costs and government support have made Linux-based systems popular on servers and desktop applications. In 2004 Linux began to enter consumer markets, when Hewlett-Packard launched its first Linux-based notebook computer, which helped to reduce the units' cost by US$60.

to manage its 1400 suppliers; Boeing exploited a virtual network in the development of the 777, and has a Web-based order system for its 700 customers worldwide which features 410 000 spare parts; and in product development, Caterpillar's customers can amend designs during assembly, and Adaptec co-ordinates design and production of microchips in Hong Kong, Taiwan and Japan.[49]

8.3 Patterns of Collaboration

In this section we examine how technology and markets affect how firms collaborate. Research on collaborative activity has been plagued by differences in definition and methodology. Essentially there have been two approaches to studying collaboration. The approach favoured by economists and strategists is based on aggregate data and examines patterns within and across different sectors. This type of research provides useful insights into how technological and market characteristics affect the level, type and success of collaborative activities. The other type of research is based on structured case studies of specific alliances, usually within a specific sector, but sometimes across national boundaries, and provides richer insights into the problems and management of collaboration.

Industry structure and technological and market characteristics result in different opportunities for joint ventures across sectors, but other factors determine the strategy of specific firms within a given sector. At the industry level, high levels of R&D intensity are associated with high levels of technologically oriented joint ventures, probably as a result of increasing technological rivalry. This suggests that technologically oriented joint ventures are perceived to be a viable strategy in industries characterized by high barriers to entry, rapid market growth and large expenditures on R&D. However, within a specific sector, joint venture activity is not associated with differences in capital expenditure or R&D intensity. A study of joint ventures in the USA found that technologically oriented alliances tend to increase with the size of firm, capital expenditure and R&D intensity.[50] Similarly, the number of marketing and distribution-oriented joint ventures increases with firm size and capital expenditure, but is not affected by R&D intensity. At the level of the firm, different factors are more important. For example, there are significant differences in the motives of small and large firms. In general, large firms use joint ventures to acquire technology, whilst smaller firms place greater emphasis on the acquisition of market knowledge and financial support.

Joint venture activity is high in the chemical, mechanical and electrical machinery sectors, as firms seek to acquire external technological know-how in order to reduce the inherent technological uncertainty in those sectors. In contrast, joint ventures are much less common in consumer goods industries, where market position is the result of product differentiation, distribution and support. If obtaining complementary assets or resources are a primary motive for collaboration, we would expect alliances to be concentrated in those sectors in which mutual ignorance of the partner's technology or markets is likely to be high.[51] Similarly, joint ventures would occur more frequently between partners who are in industries relatively unrelated to one another, and that such alliances are likely to be short-lived as firms learn from each other. Surveys of alliances in so-called high-technology sectors such as software and automation appear to confirm that access to technology is the most common motive. Market access appears to be a more common motive for collaboration in the computer, microelectronics, consumer electronics and telecommunications sectors (Table 8.5).

However, these data need to be treated with some caution as in many cases partners exchange market access for technology access or vice versa. For example, Japanese firms rarely sell technology, but are often prepared to exchange technology for access to markets. Conversely, European firms commonly trade market access for technology.[52] In this way firms limit the potential for paying high price premiums for market or technologies because of their lack of knowledge.

A breakdown of alliances by region provides some further explanation. Patterns within and between triad regions are very different. Alliances between US firms appear to be common in all fields. Alliances between European firms are concentrated in software development and telecommunications, but there is relatively little collaborative activity within the European automation, microelectronics and computing industries. Alliances between Japanese firms appear to be much less common than expected. This

TABLE 8.5 Technology strategic alliances by type and sector

	Number	Technology (%)	Market (%)	Technology/ market ratio*	Primary motive
Aerospace	228	34	13	2.6	Technology
Automation	278	41	31	1.3	Technology
Automotive	205	27	52	0.5	Market
Chemicals	410	16	51	0.3	Market
Computers	198	28	51	0.6	Market
Consumer electronics	58	19	53	0.4	Market
Energy	141	31	23	1.4	Technology
Microelectronics	383	33	52	0.6	Market
Software	344	38	24	1.6	Technology
Telecommunications	366	28	35	0.8	Market

* Technology/market ratio >1 implies technology-intensive; <1 implies market-intensive.
Source: Data derived from Hagedoorn, J. (1993) 'Understanding the rationale of strategic technology partnering', *Strategic Management Journal*, **14**, 371–385.

may reflect the weakness of the database, but is more likely to reflect the rationale for strategic alliances. The most common reason for international alliances is market access, whereas the most common reason for intra-regional alliances is technology acquisition.

The patterns of collaboration between the different triad regions provide some support for this argument. The data do not provide any indication of the direction of technology transfer, but knowledge of national strengths and weaknesses allows some analysis. Alliances between American and European firms are significant in all fields. Alliances between American and Japanese firms are only significant in computers and microelectronics, presumably the former dominated by the US partners, and the latter by the Japanese. There appears to be relatively little collaboration between Japanese and European companies, perhaps reflecting the weakness of the European electronics industry.

Given the problems of management and organization, potential for opportunistic behaviour and the limited success of alliances it might be expected that the popularity of alliances might decline as firms gain experience of such problems. However, according to the Co-operative Agreements and Technology Indicators (CATI) database the number of technology alliances increased from less than 300 in 1990 to more than 500 by 2000. It is possible to identify a number of significant trends in recent years (Figure 8.4).

Overall, the number of alliances has increased over time, and networks of collaboration appear to have become more stable, being based around a number of nodal firms in different sectors. These networks are not necessarily closed, but rather represent the dynamic partnering behaviour of large, leading firms in each of the sectors. The nodal firms are relatively stable, but their partners change over time. Contrary to the claims of globalization, the number of domestic alliances has increased faster than international ones. As a result, international partnerships fell from around 80% of all new agreements in 1976, to below 50% by 2000. This trend is particularly strong in the USA. Distinct sectoral patterns exist. In the more high-technology sectors such as pharmaceuticals, biotechnology and information and communications technologies, most of the collaborative activity is confined within each of the triad regions: Europe, Japan and North America, the exceptions being aerospace and defence. In contrast, most of the activity in the chemical and automotive sectors is across the triad regions. This suggests that the primary motive for collaborating with domestic firms is access to technology, but market access is more important in the case of cross-border alliances. This concentration of high-technology collaboration within regions appears to be more problematic for some regions than others. For example, a study of European electronics firms found that intra-European R&D agreements had no effect on firm patenting, even when sponsored by the EU. However, R&D collaboration with extra-European firms had a positive effect, which in this case means with US partners.[53]

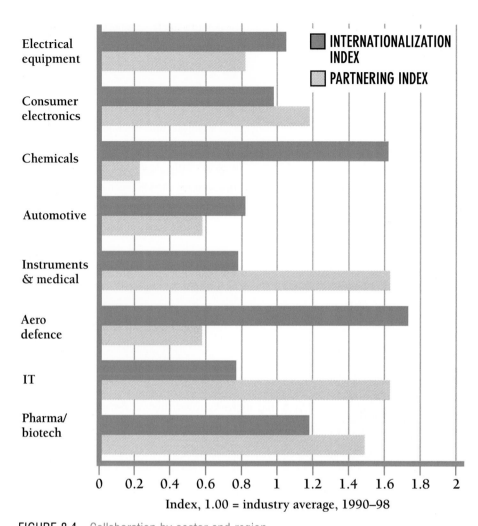

FIGURE 8.4   Collaboration by sector and region

Source: Derived from Hagedoorn, J. (2002) 'Inter-firm R&D partnerships', *Research Policy*, **31**, 477–492.

The most recent data from the MERIT-CATI database indicate that flexible forms of collaboration such as strategic alliances have become more popular than the more formal arrangements such as joint ventures. In 1970 more than 90% of the relationships were formal equity joint ventures, but this had fallen to 50% by the mid-1980s and is currently only 10%, the balance being contractual joint ventures and more transitory alliances of some type.[54] This trend has been most marked in high-technology sectors where firms seek to retain the flexibility to switch technology. Together, the pharmaceutical (including biotechnology) and information and communications technology sectors account for almost all 80% of the growth in technology collaboration since the mid-1980s. The other most common sectors are aerospace and instrumentation and medical equipment, but collaboration in the aerospace and defence industries has declined. Collaboration in 'mid-technology' sectors such as chemicals, automotive and electronics has shown little or no increase over the same period.

8.4 Effect of Technology and Organization

Our study of how 23 UK and 15 Japanese firms acquired technology externally identified the conditions under which each particular method is favoured.[55] The relative importance of different external sources to each of the UK-based companies is summarized in Figure 8.5. Each of these sources is discussed in turn, and examples provided.

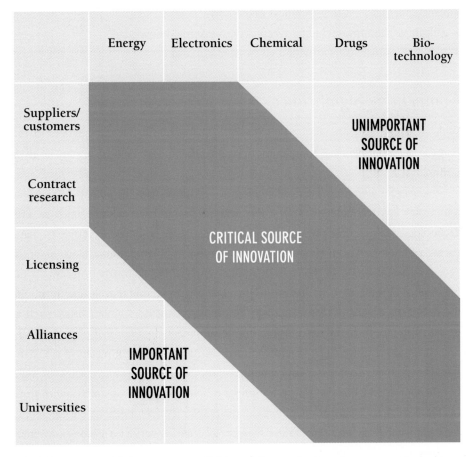

FIGURE 8.5 Relative importance to UK-based firms of external sources of technology

Source: Adapted from Tidd, J. and M. Trewhella (1997) 'Organizational and technological antecedents for knowledge acquisition', *R&D Management*, **27** (4), 359–375.

It is possible to identify two dimensions which affect companies' attitudes towards technology acquisition: the characteristics of the technology and the organization's 'inheritance' (Table 8.6). Together, the eight factors in Table 8.6 determine the knowledge acquisition strategy of a firm. The relevant characteristics of the technology include:

- competitive significance of the technology;
- complexity of the technology;
- codifiability, or how easily the technology is encoded;
- credibility potential, or political profile of the technology.

An organization's inheritance encompasses those characteristics which, at least in the short run, are fixed and therefore represent constraints within which the R&D function develops its strategies for acquiring technology. These include:

- corporate strategy, for example, a leadership versus follower position;
- capabilities and existing technical know-how;
- culture of the firm, including receptivity to external knowledge;
- 'comfort' of management with a given technical area.

Competitive Significance

Without doubt, the competitive significance of the technology is the single most important factor influencing companies' decisions about how best to acquire a given technology.

Strategies for acquiring pacing technologies – i.e. those with the potential to become tomorrow's key technologies – vary. For example, some organizations, such as AEA Technology, seek to develop and maintain at least some in-house expertise in many pacing technologies, so they will not be 'wrong-footed' if conditions change or unexpected advances occur. In the past, this policy enabled the company to recognize the importance of finite element analysis to its modelling core competence, and to acquire the necessary aspects of this technology before its competitors. Other firms, such as Kodak, also recognize the need to monitor developments in a number of pacing technologies, but see universities or joint ventures as the most efficient means of achieving this. The company sponsors a large amount of research in leading universities throughout the world, and has also set up a number of joint venture programmes with firms in complementary industries. Guinness, for example, identified genetic engineering as a pacing technology and seconded a member of staff to work at a leading university for three years. The outcome of this initiative was a new biological product, protected by a confidentiality agreement with the university. Although this genetically engineered species cannot yet be used in food and drink products, the company has successfully

TABLE 8.6 Links between technology acquisition strategy, organizational factors and characteristics of technology

Organizational and technological factors	Acquisition mechanism (most favoured/alternative)	Rationale for decision
I. Characteristics of the organization		
Corporate strategy:		
Leadership	In-house R&D/equity acquisition	Differentiation, first-mover, proprietary technology
Follower	Licence/customers and suppliers/contract	Low-cost imitation
Fit with competencies:		
Strong	In-house R&D	Options to leverage competencies
Weak	Contract/licence/consortia	Access to external technology
Company culture:		
External focus	Various	Cost-effectiveness of source
Internal focus	In-house/joint venture	Learning experience
Comfort with new technology:		
High	In-house corporate/university	High risk and potential high reward
Low	Licence/customers and suppliers/consortia	Lowest risk option
II. Characteristics of the technology		
Base	Licence/contract/customers/suppliers	Cost-effective/secure source
Key	In-house R&D/joint venture	Maximize competitive advantage
Pacing	In-house corporate/university	Future position/learning
Emerging	University/in-house corporate	Watching brief
Complexity:		
High	Consortia/universities/suppliers	Specialization of know-how
Low	In-house R&D/contract/suppliers	Division of labour
Codifiability:		
High	Licence/contract/university	Cost-effectiveness of source
Low	In-house R&D/joint venture	Learning/tacit know-how
Credibility potential:		
High	Consortia/customer/government	High profile source
Low	University/contract/licence	Cost-effectiveness of source

Source: Adapted from Tidd, J. and M. Trewhella (1997) 'Organizational and technological antecedents for knowledge acquisition', *R&D Management*, **27** (4), 359–375.

internalized the technology, understands its potential, and is well placed to evaluate new developments in the area, or to take advantage of any changes in legislation and public attitudes.

In the UK universities are a widely used external source of technology. These relationships range from support for Ph.D. candidates, extramural research awards for post-doctoral staff to carry out research in a specified area, to more formal contract research and collaborative schemes such as the LINK scheme jointly funded by the DTI and a number of companies to conduct precompetitive research in a specified area. Firms use university research for a number of reasons: to access specialist technical support; to extend in-house research; and to provide a window on emerging technologies.

Extensions to existing in-house research typically involve using universities to conduct either fundamental research, aimed at gaining a better understanding of an underlying area of science, or more speculative extensions to existing in-house programmes which cannot be justified internally because of their high risk, or because of limited in-house resources. For example, Zeneca has made extensive use of universities to undertake fundamental studies into the molecular biology of plants and the cloning of genes. Although not key technologies, access to state-of-the-art knowledge in these areas is vital to support a number of the organization's core agricultural activities. Similarly, CRL, the research organization of Thorn, utilized technology originally developed at Edinburgh and Cambridge universities in its spatial light modulators and ultra-low power consumption gas sensors, respectively.

University-funded research can also be used as windows on emerging or rapidly advancing fields of science and technology. Companies view access to such information as being critical in making good decisions about if or when to internalize a new technology. For example, Azko launched a series of university-funded research programmes in the USA during the late 1980s. During its first three years, these programmes yielded 40 patent applications.

Most companies look to acquire base technologies externally or, in the case of non-competitive technologies, by co-operative efforts. Companies recognize that their base technologies are often the core competencies of other firms. In such cases, the policy is to acquire specific pieces of base technology from these firms, who can almost always provide better technology, at less cost, than could have been obtained from in-house sources. Materials testing, routine analysis and computing services are common examples of technical services now acquired externally.

Complexity of the Technology

The increasingly interdisciplinary nature of many of today's technologies and products means that, in many technical fields, it is not practical for any firm to maintain all necessary skills in-house. This increased complexity is leading many organizations to conclude that, in order to stay at the forefront of their key technologies, they must somehow leverage their in-house competencies with those available externally. For example, the need to acquire external technologies appears to increase as the number of component technologies increases. In extreme cases of complexity, networks of specialist developers may emerge which serve companies which specialize in systems integration and customization for end-users.

Zeneca's core skills centre on genetic engineering/molecular biology (which allows new characteristics to be inserted into plants), cell biology (which enables the regeneration of plants) and rapid screening and testing techniques for plants and seeds. These key technologies have been developed, and are retained, internally. Such is their complexity, however, that the company seeks to leverage these proprietary skills with complementary technologies accessed from other sources. Examples include techniques based on genetic fingerprinting, to speed up plant breeding by enabling desirable characteristics to be identified without having to grow the plant to maturity; and the use of anti-sense technology to isolate and clone agronomically important genes.

In recent years, acquisition has again become a popular means of acquiring technology. Normally, the rationale is to establish a position quickly in a particular technical area. However, feelings about the effectiveness of this route are mixed; many acquisitions have suffered from the loss or demotivation of key staff, or have failed to realize their expected potential for other reasons. Nevertheless, a few companies claim spectacular successes amongst their acquisitions. The common factors here appear to be prior experience of the markets in which the new technology would be used, and a compatible culture between the two organizations. For example, Eli Lilly identified a small US company, Agouron Pharmaceuticals, working on an enzyme believed to be a catalyst in tumour growth. The company considered this technology sufficiently promising to invest $4.5 m. to acquire an 8% stake in the company and first claim to any resulting drugs.

Alliances between large pharmaceutical firms and smaller biotechnology firms have received a great deal of management and academic attention over the past few years. On the one hand, pharmaceutical firms have sought to extend their technological capabilities through alliances with and the acquisition of specialist biotechnology firms. Each of the leading drug firms will at any time have about 200 collaborative projects, around half of which are for drug discovery. On the other hand, small biotechnology firms have sought relationships with pharmaceutical firms to seek funding, development, marketing and distribution. In general, pharmaceutical and biotechnology firms each use alliances to acquire complementary assets, and such alliances are found to contribute significantly to new product development and firm performance.[56] For the pharmaceutical firms, there is a strong positive correlation between the number of alliances and markets sales. For the biotechnology firms the benefits of such relationships is less clear. Two trajectories co-exist. The first is based on increasing specification of biological hypotheses. The second is based on platform technologies related to the generation and screening of compounds and molecules, such as combinational chemistry, genomic libraries, bio-informatics and proteomics. The former type of biotechnology firm remains dependent upon the complementary assets of the pharmaceutical firms, whereas the latter type appears to have the capacity to benefit from a broader range of network relationships.[57] A biotechnology firm's *exploration* alliances with pharmaceutical firms is a significant predictor of products in development (along with technological diversity), and in turn products in development are a predictor of *exploitation* alliances with pharmaceutical firms, and these exploitation alliances predict a firm's products in the market.[58] However, different forms of alliance yield different benefits. Research contracts and licenses with biotechnology firms are associated with an increase in biotechnology-based *patents* by pharmaceutical firms, whereas the acquisition of biotechnology firms is associated with an increase in biotechnology-related *products* from pharmaceutical firms. This increase in biotechnology-related products includes only those products developed subsequent to the acquisition, and does not include those products directly acquired with the biotechnology firms. Interestingly, minority equity interests in biotechnology firms and joint ventures between pharmaceutical and biotechnology firms are associated with a reduction in biotechnology-related patents and products.[59] This may be due to the very high organizational costs of joint ventures (see section 8.5), or the fact that joint ventures tend to tackle more complex and risky projects than simpler licensing or research contracts.

Codifiability of the Technology

The more that knowledge about a particular technology can be codified, i.e. described in terms of formulae, blueprints and rules, the easier it is to transfer, and the more speedily and extensively such technologies can be diffused. Knowledge that cannot easily be codified – often termed 'tacit' – is, by contrast, much more difficult to acquire, since it can only be transferred effectively by experience and face-to-face interactions. All else being equal, it appears preferable to develop tacit technologies in-house. In the absence of strong intellectual property rights (IPR) or patent protection, tacit technologies provide a more durable source of competitive advantage than those which can easily be codified.

For example, Kodak maintains all of its strategic technologies in-house, even if these are considered mature, because of the large amount of tacit know-how embodied in even these technologies. The existence of difficult-to-codify tacit knowledge is one of the factors that has allowed it to maintain a competitive advantage in one particular core technology, even though the basic features of this technology have been in use for over 100 years. Similarly, the design skills of many Italian firms have allowed them to remain internationally competitive despite significant weaknesses in other dimensions. The difficulty of maintaining a competitive advantage when technology is easily codifiable is highlighted by Guinness, which developed a small, plastic, gas-filled device that gives canned beer the same creamy head as keg beer. This 'widget' initially provided the company with a source of competitive advantage and extra sales, but the innovation was soon copied widely throughout the industry, to the extent that widgets are now almost a requirement for any premium canned beer.

Credibility Potential

The credibility given to the company by a technology, or by the source of the technology, is a significant factor influencing the way companies decide to acquire a technology. Particular value is placed on gaining credibility or goodwill from governments, customers, market analysts, and even from the company's own top management, academic institutions and potential recruits. For example, Celltech's collaboration with a large US chemical firm appears to have enhanced the former's market credibility. Not only did the collaboration demonstrate the organization's ability to manage a multimillion-dollar R&D project, but the numerous patents and academic publications that arose from it were also felt to have improved the company's scientific standing. Similarly, in Japan the mobile telecoms services provider DoCoMo worked closely with the national telephone services provider NTT, although it had the depth and range of technologies required to develop telephony equipment and products. The rationale for the relationship was to influence future standards and to increase the credibility of its consumer telephone products in a market in which it was increasingly difficult to differentiate by means of product or service (see Box 8.7).

Corporate Strategy

One of the most important factors affecting the balance between in-house generated, and externally acquired, technology is the degree to which company strategy dictates that it should pursue a policy of technological differentiation or leadership (see Chapter 3). For example, Kodak distinguishes between two types of technical core competencies: strategic, i.e. those activities in which the company must be a world leader because they represent such an important source of competitive advantage, and enabling, i.e. skills required for success, but which do not have to be controlled internally. Although

---

BOX 8.7
SYMBIAN

Symbian is a joint venture between Psion, Nokia, Ericsson and Motorola which develops and licenses operating systems for wireless devices such as cell phones and electronic organizers. The members, all electronics manufacturers, formed Symbian in 1998 to help reduce the chance that they would become low-margin 'box-shifters', as has happened in the PC market. The venture was initially established to exploit the Epoc operating system developed by Psion, which owned 40% of Symbian. The two Scandinavian telecoms each invested £57.5m. for their original 30% share. Motorola joined Symbian later. It is estimated that in 2000 Psion invested £8m. in Symbian, but accounted for as much as 90% of Psion's stock market value. Psion sold its share in 2003, and Nokia now owns 48%, and Panasonic, Siemens and Sony Ericsson have all increased their shares of Symbian. Licensees are required to pay a royalty per device. By 2003 Symbian's operating systems were in 6.7 million handsets globally, and in 2004 were in 41% of all personal organizers and smart phones sold, compared to 23% with PalmSource, and just 5% with Microsoft software. Symbian, based in London, planned to increase its staff from 900 to 1200 in 2004. However, two doubts remain. First, the competitive response of companies such as Microsoft, with Windows CE, and second, strategic conflicts between members of Symbian, in particular the growing dominance of Nokia.

---

all strategic activities are retained in-house, the company is prepared to access enabling technologies externally, if the overall technology is sufficiently complex.

Some companies adopt a policy of intervention in the technology supply market, until the market becomes sufficiently competitive to ensure reliable sources of technology continue to be available at reasonable prices. For example, the extent to which BP is prepared to rely on external sources of technology depends, amongst other things, on the nature of the supply market. When only a few suppliers exist, BP will develop key items of technology itself, and pass these on to its suppliers in order to ensure their availability. However, once sufficient suppliers have entered the market to make it competitive, its policy is to conduct no further in-house development in that area. Indeed, one of the declared aims of BP's in-house R&D activities is to 'force the pace' at which the industry as a whole innovates.

Firm Competencies

An organization's internal technical capabilities are another factor influencing the way in which it decides to acquire a given technology. Where these are weak, a firm normally has little choice but to acquire from outside, at least in the short run, whereas strong in-house capabilities often favour the internal development of related technologies, because of the greater degree of control afforded by this route. In such cases, the main driving force behind the acquisition strategy is speed to market. For example, speed to market is a critical success factor for many firms in consumer markets. Such firms select the technology acquisition method that provides the fastest means of commercialization. When the required expertise is available in-house, this route is normally favoured because it allows greater control of the development process, and is therefore usually quicker. However, where suitable in-house capabilities are lacking, external sourcing is almost always faster than building the required skills internally. Gillette, for example, found that one of its new products required laser spot-welding competencies that the company lacked and, given the limited market window, was forced to go outside to acquire this technology.

Company Culture

Every company has its own culture, that is 'the way we do things around here'. We will discuss culture in more detail in the next chapter, but here we are concerned with the underlying values and beliefs that play an important role in technology acquisition policies. A culture of 'we are the best' is likely to contribute towards to a rather myopic view of external technology developments, and limit the potential for learning from external partners. Some organizations, however, consistently reinforce the philosophy that important technical developments can occur almost anywhere in the world. Consequently, staff in these companies are encouraged to identify external developments, and to internalize potentially important technologies before the competition. However, in practice few firms have formal 'technology scouting' personnel or functions.

For example, Glaxo emphasizes that companies need to guard against becoming captives of their own in-house expertise, since this limits the scope of its activities to what can be achieved through internal resources. With this in mind, the company has expanded its research effort by placing many of its more specialized R&D activities overseas. This, it is claimed, allows its research to benefit from different cultural and scientific approaches, and from being brought into intimate contact with the many different markets it serves. Local perspectives are particularly important for product development, but international networks can also be used to acquire access to basic research.

Kodak's philosophy is that world class organizations must access technology wherever it resides, and that a culture of 'not invented here' is a prescription for second-class citizenship in the global marketplace. Like Glaxo, this firm also has a number of foreign research laboratories. For example, Japan is now the centre of this organization's worldwide efforts in molecular beam epitaxy, a method of growing crystals for making gallium arsenide chips. A key role for overseas laboratories is to monitor technology developments in host countries. Local champions from around the world are closely networked so that technical advances made in one geographic location are rapidly disseminated around the organization as a whole. Such is this company's determination to maintain a 'window' on potential sources of technology that it has set up joint ventures with many large and small companies worldwide, including links with Matsushita, Canon, Nikon, Minolta, Fuji and Apple. In the past, the company has also worked with biotechnology companies such as Cetus on human diagnostics kits, and with Amersham International to develop immunoassay kits and biosensors.

Management Comfort

The degree of comfort that management has with a given technology manifests itself at the level of the individual R&D manager or management team, rather than at the level of the organization as a whole. Management comfort is multifaceted. One aspect is related to a management team's familiarity with the technology. Another reflects the degree of confidence that the team can succeed in a new technical area, perhaps because of a research group's track record of success in related fields. Attitude to risk is also a factor.[60]

All else being equal, the more comfortable a company's managers feel with a given technology, the more likely that technology is to be developed in-house. For example, the current business of Zeneca Seeds was built on the basis of providing an outlet for the parent organization's in-house biotechnology expertise. This means that the concept of world-leading in-house technology sits very comfortably with the organization's top management, which is reflected by the fact that annual in-house R&D expenditure exceeds the firm's annual capital expenditure by a considerable margin. Similarly, AEA Technology's core technologies of plant life extension, environmental sciences, modelling and land remediation treatment all derive from its nuclear industry background. Top management's comfort with these technologies has led them to encourage staff to build on these skills, and to use these as a springboard for diversification into new scientific areas.

8.5 Managing Alliances for Learning

So far we have discussed collaboration as a means of accessing market or technological know-how, or acquiring assets. However, alliances can also be used as an opportunity to learn new market and technological competencies, in other words to internalize a partner's know-how. Seen in this light, the success of an alliance becomes difficult to measure.

Collaboration is an inherently risky activity, and less than half achieve their goals. A study of almost 900 joint ventures found that only 45% were mutually agreed to have been successful by all partners.[61] Other studies confirm that the success rate is less than 50%.[62,63]

It is difficult to assess the success of a collaborative venture, and in particular termination of a partnership does not necessarily indicate failure if the objectives have been met. For example, around half of all alliances are terminated within seven years, but in some cases this is because the partners have subsequently merged. It is common for a collaborative arrangement to evolve over time, and objectives may change. For example, a licensing agreement may evolve into a joint venture. Finally, an apparent failure may result in knowledge or experience that may be of future benefit. An alliance is likely to have a number of different objectives – some explicit, others implicit – and outcomes may be planned or unplanned. Therefore any measure of success must be multidimensional and dynamic in order to capture the different objectives as they evolve over time. Reasons for failure include strategic divergence, procedural problems and cultural mismatch. Table 8.7 presents the most common reasons for the failure of alliances, based on a meta-analysis of the 16 studies. The studies reviewed differ in their samples and methodologies, but 11 factors appear in a quarter of the studies, which provides some level of confidence.

Firms have different expectations of alliances and these affect their evaluation of success. Those firms which view product development collaboration as discrete events with specific aims and objectives are more likely to evaluate the success of the relationship in terms of the project cost and time and ultimate product performance. However, a small proportion of firms view collaboration as an opportunity to learn new skills and knowledge and to develop longer-term relationships. In such cases measures of success need to be broader. If learning is a major goal, it is necessary for partners to have complementary skills and capabilities, but an even balance of strength is also important. The more equal the partners, the more likely an alliance will be successful. Both partners must be strong financially and in the technological, product or market contribution they make to the venture. A study of 49 international alliances by management consultants McKinsey found that two-thirds of the alliances between equally

TABLE 8.7 Common reasons for the failure of alliances based on a review of 16 studies

Reason for failure	% studies reporting factor (n = 16)
Strategic/goal divergence	50
Partner problems	38
Strong–weak relation	38
Cultural mismatch	25
Insufficient trust	25
Operational/geographic overlap	25
Personnel clashes	25
Lack of commitment	25
Unrealistic expectations/time	25
Asymmetric incentives	13

Source: Derived from Duysters, G., G. Kok and M. Vaandrager (1999) 'Crafting successful strategic technology partnerships', *R&D Management*, **29** (4), 343–351.

matched partners were successful, but where there was a significant imbalance of power almost 60% of alliances failed.[64] Consequently in the case of a formal joint venture equal ownership is the most successful structure, 50–50 ownership being twice as likely to succeed as other ownership structures. This appears to be because such a structure demands continuous consultation and communication between partners, which helps anticipate and resolve potential conflicts, and problems of strategic divergence. Our own study of Anglo-Japanese joint ventures identified three sources of strategic conflict between parent firms: product strategy; market strategy; and pricing policy. These were primarily the result of coupling complementary resources with divergent strategies, what we refer to as the 'trap of complementarity'. In essence, parents with complementary resources almost inevitably have different long-term strategic objectives. Too many joint ventures are established to bridge gaps in short-term resources, rather than for long-term strategic fit.[65]

This suggests that firms must learn to design alliances with other firms, rather than pursue ad hoc relationships. By design we do not mean the legal and financial details of the agreement, but rather the need to select a partner which can contribute what is needed, and needs what is offered, of which there is sufficient prior knowledge or experience to encourage trust and communication, to allow areas of potential conflict such as overlapping products or markets to be designed out. Partners must specify mutual expectations of respective contributions and benefits. They should agree on a business plan, including contingencies for possible dissolution, but allow sufficient flexibility for the goals and structure of the alliance to evolve. It is important that partners communicate on a routine basis, so that any problems are shared. Without such explicit design, collaboration may make product development more costly, complex and difficult to

TABLE 8.8 The effect of collaboration on the product development process ($n = 106$)

	Agree/strongly agree	Disagree/strongly disagree
Makes product development more costly	51	22
Complicates product development	41	35
Makes development more difficult to control	41	38
Makes development more responsive to supplier needs	36	26
Allows development to adapt better to uncertainty	27	43
Accelerates product development	25	58
Makes development more responsive to customer needs	22	50
Allows development to respond better to market opportunities	15	63
Enhances competitive benefits arising through development	12	65
Facilitates the incorporation of new technology in development	7	70

Source: Adapted from Bruce *et al.* (1995, p. 542), with kind permission from Elsevier Science Ltd, The Boulevard, Langford Lane, Kidlington OX5 1GB, UK.[66]

control (Table 8.8). Thus whilst the *failure* of an alliance is most likely to be the result of strategic divergence, the *success* of an alliance depends to a large extent on what can be described as operational and people-related factors, rather than strategic factors such as technological, market or product fit (Table 8.9). The most important operational factors are agreement on clearly stated aims and responsibilities, and the most important people factors are high levels of commitment, communication and trust. A survey of 135 German firms gives us a better idea of the relative importance of these different factors.[66] The study found that firms take people-related, economic and technological factors into consideration, but that these three groups of variables are largely independent of each other. Factor analysis confirms that the people-related factors are more significant than either the economic or technological considerations, specifically creation of trust, informal networking and learning. However, managers often put greater effort into the 'harder' technical and operational issues, than the 'softer' but more important people issues, and focus more on 'deal making' to form alliances, than the processes necessary to sustain them. One study of alliances between high-technology firms found that more than half of the problems in the first year of an alliance relate to the relationship, rather than the strategic or operational factors. The most common problems were poor communication – quality and frequency – and conflicts due to differences in national or corporate cultures.[67] The study identified three strategies for minimizing these cultural mismatches. First, for one partner to adopt the culture of the other

TABLE 8.9   Factors affecting outcomes of collaborative product development ($n = 106$)

Factor	Respondents freely mentioning factor
Establishing ground rules	67
Clearly defined objectives agreed by all parties	41
Clearly defined responsibilities agreed by all parties	19
Realistic aims	10
Defined project milestones	11
People factors	54
Collaboration champion	22
Commitment at all levels	11
Top management commitment	10
Personal relationships	10
Staffing levels	3
Process factors	45
Frequent communication	20
Mutual trust/openness/honesty	17
Regular progress reviews	13
Deliver as promised	9
Flexibility	3
Ensuring equality	42
Mutual benefit	22
Equality in power/dependency	11
Equality of contribution	9
Choice of partner	39
Culture/mode of operation	13
Mutual understanding	12
Complementary strengths	12
Past collaboration experience	2

Source: Adapted from Bruce, M., F. Leverick and D. Littler (1995) 'A management framework for collaborative product development', in Bruce, M. and Biemans, W. (eds), *Product Development: Meeting the challenge of the design–marketing interface*, John Wiley & Sons, Ltd, Chichester, 171.

(unlikely outside an acquisition). Second, to limit the degree of cultural contact necessary through the operational design of the project. Finally, to appoint cultural translators or liaisons to help identify, interpret and communicate different cultural norms.

Other factors which contribute to the success of an alliance include:[68]

- The alliance is perceived as important by all partners.
- A collaboration 'champion' exists.
- A substantial degree of trust between partners exists.
- Clear project planning and defined task milestones are established.
- Frequent communication between partners, in particular between marketing and technical staff.

- The collaborating parties contribute as expected.
- Benefits are perceived to be equally distributed.

Mutual trust is clearly a significant factor, when faced with the potential opportunistic behaviour of the partners; for example, failure to perform or the leakage of information. Trust may exist at the personal and organizational levels, and researchers have attempted to distinguish different levels, qualities and sources of trust.[69] For example, the following bases of trust in alliances have been identified:

- Contractual – honouring the accepted or legal rules of exchange, but can also indicate the absence of other forms of trust.
- Goodwill – mutual expectations of commitment beyond contractual requirements.
- Institutional – trust based on formal structures.
- Network – because of personal, family or ethnic/religious ties.
- Competence – trust based on reputation for skills and know-how.
- Commitment – mutual self-interest, committed to the same goals.

These types of trust are not necessarily mutually exclusive, although over-reliance on contractual and institutional forms may indicate the absence of the types of trust. Goodwill is normally a second-order effect based on network, competence or commitment. In the case of innovation, problems may occur where trust is based on the network, rather than competence or commitment, as discussed earlier. Clearly, high levels of interpersonal trust are necessary to facilitate communication and learning in collaboration, but inter-organizational trust is a more subtle issue. Organizational trust may be defined in terms of organizational routines, norms and values which are able to survive changes in individual personnel.[70] In this way organizational learning can take place, including new ways of doing things (operational or lower-level learning) and doing new things through diversification (strategic or higher-level learning). Organizational trust requires a longer time horizon to ensure that reciprocity can occur, as for any particular collaborative project one partner is likely to benefit disproportionately. In this way organizational trust may mitigate against opportunistic behaviour. However, in practice this may be difficult where partners have different motives for an alliance or differential rates of learning.

In Chapter 5 we examined the nature of core competencies. Conceiving of the firm as a bundle of competencies, rather than technology or products, suggests that the primary purpose of collaboration is the acquisition of new skills or competencies, rather than the acquisition of technology or products. Therefore a crucial distinction must be made between acquiring the skills of a partner, and simply gaining access to such skills. The latter is the focus of contracting, licensing and the like, whereas the internalization of a partner's skills demands closer and longer contact, such as formal joint ventures or strategic alliances. An example would be Kodak which for many years dominated the photographic market exploiting its competencies based on 'wet chemistry in the dark'. However, the advent of digital photography threatened many (but not all) these competencies, and through a combination of corporate ventures, alliances and acquisitions Kodak successfully managed the transition from chemistry to digital photography, unlike its rival Polaroid (see Box 8.8).

BOX 8.8
KODAK DEVELOPS DIGITAL COMPETENCIES THROUGH ALLIANCES AND ACQUISITIONS

Faced with developments in digital imaging technology, Kodak redefined its business as 'pictures, not technology', stressing that the market competencies were still relevant to the digital photographic markets, but it lacked the relevant technological competencies. The board hired George Fisher from Motorola to be the new CEO. Fisher pursued a two-tier strategy for new business development. For the medical imaging business, Kodak acquired a number of specialist digital technology firms, including Imation Corporation, which had developed a hybrid dry laser imaging technology. It combined these new competencies with its existing market knowledge, as it accounted for around 30% of the global medical imaging market at that time.

For the consumer imaging market, Kodak established a new Digital and Applied Imaging division, but this suffered from the parent company's organizational routines, which had evolved to monitor relatively stable mass markets and slow-moving technology, and were therefore inappropriate for digital imaging at that time. As a result of organizational problems, the division was made organizationally independent in 1997, and in 1998 formed a joint venture with Intel to develop the 'Picture CD' project. Similarly, initial attempts to develop digital cameras in the existing Consumer Imaging division resulted in cameras that failed to meet the technological and market demands. These developments were also later moved to the new Digital and Applied Imaging division, which had routines more suited to the needs of emerging technologies and markets. A series of successful products followed, and by 2004 Kodak had 20% of the global market share in digital cameras.

Source: Derived from Jeffrey T. Macher and Barak D. Richman (2004) 'Organizational responses to discontinuous innovation', *International Journal of Innovation Management*, **8** (1), 87–114.

TABLE 8.10 Determinants of learning through alliances

	Factors which promote learning
A. Intent to learn	
1. Competitive posture	Co-operate now, compete later
2. Strategic significance	High, to build competencies, rather than to fix a problem
3. Resource position	Scarcity
4. Relative power balance	Balance creates instability, rather than harmony
B. Transparency or potential for learning	
5. Social context	Language and cultural barriers
6. Attitude towards outsiders	Exclusivity, but absence of 'not invented here'
7. Nature of skills	Tacit and systemic, rather than explicit
C. Receptivity or absorptive capacity	
8. Confidence in abilities	Realistic, not too high or too low
9. Skills gap	Small, not too substantial
10. Institutionalization of learning	High, transfer of individual learning to organization

Source: Adapted from Hamel, G. (1991) 'Learning in international alliances', *Strategic Management Journal*, **12**, 91.

It is possible to identify three factors that affect learning through alliances: intent, transparency and receptivity (Table 8.10). Intent refers to a firm's propensity to view collaboration as an opportunity to learn new skills, rather than to gain access to a partner's assets. Thus where there is intent, learning takes place by design rather than by default, which is much more significant than mere leakage of information. Transparency refers to the openness or 'knowability' of each partner, and therefore the potential for learning. Receptivity, or absorptiveness, refers to a partner's capacity to learn. Clearly, there is much a firm can do to maximize its own intent and receptivity, and minimize its transparency. Intent to learn will influence the choice of partner and form of collaboration. Transparency will depend on the penetrability of the social context, attitudes towards outsiders, i.e. clannishness, and the extent to which the skills are discrete and encodable. Explicit knowledge, such as designs and patents, are more easily encoded than tacit knowledge. This suggests that a harmonious alliance may not necessarily represent a win–win situation. On the contrary, where two partners attempt to extract value from their alliance in the same form, whether in terms of short-term economic benefits or longer-term skills acquisition, managers are likely to frequently engage in arguments over value sharing. Where partners have different goals, for example one partner seeks short-term benefits whereas the other seeks the acquisition of new skills, the relationship tends to be more harmonious, at least until one partner is no longer dependent on the other. For example, where a firm works with a university or commercial research organization, the goals of the alliance are likely to be very

TABLE 8.11 Factors influencing satisfaction of firms and research organizations in collaboration

Significant factor	For firm	For research organization
Previous links	significant	significant
Commitment	significant	significant
Partner's reputation	not significant	significant
Definition of objectives	significant	not significant
Communication	not significant	significant
Conflict	significant	not significant
Organizational design	not significant	not significant
Geographical proximity	not significant	not significant

Source: Derived from Mora-Valentin, E., A. Montoro-Sanchez and L. Guerras-Martin (2004) 'Determining factors in the success of R&D cooperative agreements between firms and research organizations', *Research Policy*, **33**, 17–40.

different, and therefore the factors influencing a successful outcome may differ (Table 8.11).

Therefore the preferred structure for an alliance will depend on the nature of the knowledge to be acquired, whereas the outcome will be determined largely by a partner's ability to learn, which is a function of skills and culture. Tactical alliances are most appropriate to obtain migratory or explicit knowledge, but more strategic relationships are necessary to acquire embedded or tacit knowledge.[71] Alliances for explicit knowledge focus on trades in designs, technologies or products, but by the very nature of such knowledge this provides only temporary advantages because of its ease of codification and movement. Alliances for embedded knowledge present a more subtle management challenge. This involves the transfer of skills and capabilities, rather than discrete packages of know-how. This requires personnel to have direct, intimate and extensive exposure to the staff, equipment, systems and culture of the partnering organization. However, the absorptive capacity of an organization is not a constant, and depends on the fit with the partner's knowledge base, organizational structures and processes, such as the degree of management formalization and centralization of decision-making and research.[72] Studies suggest that knowledge creation in an alliance is more likely to occur where there is a clear intent and specific goals exist, but conversely individual autonomy within a joint project is associated with a reduction in knowledge creation. One of the most significant factors influencing knowledge creation and learning in an alliance is the use of formal environmental scanning, and this effect increases with the complexity of projects.[73] There appear to be two reasons for the importance of scanning in such alliances. First, the need to identify relevant knowledge in the environment, and second, to ensure that the developments continue to be relevant to the changing environment.

The conversion of tacit to explicit knowledge is a critical mechanism underlying the link between individual and organizational learning.[74] Through a process of dialogue, discussion, experience-sharing and observation, individual knowledge is amplified at the group and organizational levels. This creates an expanding community of interaction, or 'knowledge network', which crosses intra- and inter-organizational levels and boundaries. These knowledge networks are a means to accumulate knowledge from outside the organization, share it widely within the organization, and store it for future use. Therefore the interaction of groups with different cultures, whether within or beyond the boundaries of the organization, is a potential source of learning and innovation.

Organizational structure and culture will determine absorptive capacity in inter-organizational learning. Culture is a difficult concept to grasp and measure, but it helps to distinguish between national, organizational, functional and group cultures.[75] Differences in national culture have received a great deal of attention in studies of cross-border alliances and acquisitions, and the consensus is that national differences do exist and that these affect both the intent and ability to learn. In general, British and American firms focus more on the legal and financial aspects of alliances, but rarely have either the intent or ability to learn through alliances. In contrast, French, German and Japanese firms are more likely to exploit opportunities for learning.[76] The issue of national stereotypes aside, there may be structural reasons for these differences in the propensity to learn.

For example, Japanese firms have good historical reasons for exploiting alliances as opportunities for learning. Initially, Western firms typically entered Japan through alliances in which they provided technology in return for access to Japanese sales and distribution channels. This exchange of technology for market access appeared to offer value to both sides. However, while the Western partner often remained dependent on the Japanese partner for distribution and sales, the Japanese partner typically built up its technological skills and became less reliant on the Western partner. As a result, in the 1980s European and American partners began to lose technological leadership in many fields, and were forced to trade distribution and sales channels at home for access to the Japanese market. Therefore collaboration has shifted from relatively simple and well-defined licensing agreements or joint ventures, to more complex and informal relationships which are much more difficult to manage.

Most recently, firms from the USA and Europe have begun to use alliances for operational learning. Operational learning provides close exposure to what competitors are doing in Japan and how they are doing it. For example, to learn how Japanese partners manage their production facilities, supplier base or product development process. This is not possible from a distance, and requires close alliances with potential competitors. However, fewer firms in the West have exploited fully the potential of alliances for strategic learning, that is the acquisition of new technological and market competencies (see Box 8.9).

In contrast, many American and British firms find it difficult to learn through alliances. This appears to be because firms focus on financial control and short-term financial benefits, rather than the longer-term potential for learning. For example, firms will attempt to minimize the number and quality of people they contribute to a Japanese joint venture, and the time committed. As a result, little learning takes place and little or no corporate memory is built up.

At the lower level of analysis, different functional groups and project teams may have different cultures. For example, the differences between technical and marketing cultures are well documented, and are a major barrier to communication within an organization.[77] When such groups are required to communicate across organizations the potential for problems is even greater. There is some evidence that employees attempt to trade information based on the perceived economic interests of their firms, but that these perceptions differ. A study of 39 managers involved in alliances in the steel industry identified three clusters of behaviour regarding information trading: value-oriented, competition-oriented and complex decision-makers.[78] Value-oriented employees base their behaviour on the importance of the information to their own firm, independent of its potential value to the partner. Competition-oriented employees base their behaviour solely on the value of the information to competitors. The complex decision makers include both considerations, and also the potential for trading information. Some firms develop reputations for being very secretive, while others are seen as more open. No doubt this contrasting approach to knowledge-sharing will interest

---

BOX 8.9
ROVER, HONDA AND BMW

In Christmas 1978 BL, now Rover, signed a deal with Honda to build under licence the Ballade, as the new Triumph Acclaim. The project highlighted the vast quality and cost differentials between BL and Honda products, and exposed BL to Japanese engineering and work practices.

Since then Rover has relied on customizing Honda designs to develop its own derivatives: the Honda Civic and Rover 200/400, Honda Legend and Rover 800, and Honda Accord and Rover 600. The Rover 200/Honda Civic, launched in 1995, were the last product of the partnership. The alliance between Rover and Honda, and the subsequent acquisition of Rover by BMW, illustrates how alliances evolve, and how different firms choose to acquire technological, market and organizational competencies. Initially, Rover sought a relationship with Honda in order to introduce a new product, which it had neither the financial nor technological resources to develop itself. For Honda, this represented a low-cost and low-risk way of entering the European market and a means of developing local suppliers and dealers. Subsequently, Rover and Honda co-developed a number of cars, although Honda was clearly the dominant partner. As a result, Rover made some improvements to its work organization, product engineering and quality, and Honda shared components costs and learned a great deal about customizing its products for European markets, and how to manage British workers, suppliers and dealers. However, the learning was somewhat asymmetric, as Honda successfully established its own manufacturing facilities in the UK, whereas Rover failed to improve significantly its process capability or product quality. In 1994 some 16 years of collaboration ended when BMW purchased Rover for £920m. Honda was forced to terminate its relationship, and to sell its 20% share in Rover. The following year Rover attempted to recruit an additional 500 engineers, which is indicative of Honda's engineering support.

BMW had different motives for purchasing Rover. First and foremost, it hoped to gain access to the Land Rover division, with its strong brand, four-wheel-drive technology and profitable products. Second, it gained additional production capacity, with a relatively inexpensive workforce, which was a major consideration for a relatively small, independent car manufacturer. Finally, it completed its product portfolio: despite its previous aspirations to be a prestige car manufacturer, Rover's expertise is in small, front-wheel-drive cars, whereas BMW specializes in larger rear-wheel-drive cars. However, BMW significantly underestimated the problems at Rover, and its own ability to become a volume manufacturer. After five years of investment and substantial losses, in 2000 BMW disposed of almost all of Rover, selling it for a nominal £10 to a consortium optimistically called 'Phoenix', which was established by a group of ex-Rover managers.

Sales of Rover cars continued to fall due to the absence of sufficient resources to develop any new models, but in 2004 MG Rover (as it was now called) launched a new small car, the CityRover, which was a rebadged car developed and manufactured in India by Tata. MG Rover also announced a strategic alliance with Shanghai Automotive Industry Corporation (SAIC) in China to develop a new medium-sized car for launch in Europe by 2006.

enthusiasts of game theory, but the empirical evidence suggests that firms that share their knowledge with their peers and competitors – for example, through conferences and journals – have a higher innovative performance than those that do not share, controlling for the level of R&D spending and number of patents.[79] The reasons for this apparent reward for generosity include the need to motivate and recruit researchers, and a strategy to be perceived as a technology leader to influence technological trajectories and attract alliance partners.

8.6 Summary and Further Reading

In this chapter we have discussed the arguments for and against collaboration for the development of new technology, products and processes, and have identified the different forms and patterns of collaboration which exist. Essentially, firms collaborate to reduce the cost, time or risk of access to unfamiliar technologies or markets. Transaction costs analysis focuses on the static, short-term trade-offs between developing an innovation in-house versus external mechanisms, whereas a strategic learning framework focuses on the dynamic, longer-term potential for acquiring new technological, market or organizational competencies. The precise form of collaboration will be determined by the motives and preferences of the partners, but their choice will be constrained by the nature of the technologies and markets, specifically the degree of complexity and tacitness. The success of an alliance depends on a number of factors, but organizational issues dominate, such as the degree of mutual trust and level of communication. The transaction costs approach better explains the relationship between the reason for collaboration, and the preferred form and structure of an alliance. The strategic learning approach better explains the relationship between the management and organization of an alliance and the subsequent outcomes.

A good review of the theoretical issues is provided by a compilation of papers edited by Rod Coombs *et al.*, *Technological Collaboration* (Edward Elgar, 1996). The book includes a discussion of both economic and sociological analyses of collaboration, including transaction costs and evolutionary theories, as well as a review of the more recent network approaches. The literature on innovation networks is large and growing, but the following provide a good introduction: O. Jones, S. Conway and F. Steward, *Social Interaction and Organizational Change: Aston perspectives on innovation networks* (Imperial College Press, London, 2001); *International Journal of Innovation Management*, Special Issue on Networks, **2** (2) (1998); R. Gulati, 'Alliances and networks', *Strategic Management Journal*, **19**, 293–317 (1998); and F. Belussi and F. Arcangeli, 'A typology of networks' in *Research Policy*, **27**, 415–428 (1998).

For a less academic treatment of alliances, Bleeke and Ernst provide a practical guide, albeit a little dated, for managers of collaborative projects in *Collaborating to Compete* (John Wiley & Sons, Inc., 1993), written by two management consultants at McKinsey & Co., and based on a survey of international alliances and acquisitions. In *Alliance Advantage* (Harvard Business School Press, 1998) Yves Doz and Gary Hamel develop a framework to help understand and better manage alliances, drawing on their earlier work on learning through alliances. On the more specific subject of customer–supplier alliances, Jordan Lewis provides a practical guide based on studies of a number of American and British present and past exemplars such as Motorola and Marks & Spencer in *The Connected Corporation* (Free Press, 1995). The example of Marks & Spencer is now even more interesting as it demonstrates some of the limitations of supplier partnerships. More academic and rigorous treatments of customer–supplier alliances are provided by Richard Lamming in *Beyond Partnership* (Prentice-Hall, 1993) and Toshihiro Nishiguchi in *Strategic Industrial Sourcing: The Japanese advantage* (Oxford University Press, 1994), both based mainly on the experience of the automobile industry.

References

1 McGee, J. and M. Dowling (1994) 'Using R&D cooperative arrangements to leverage managerial experience', *Journal of Business Venturing*, **9**, 33–48.
2 Hauschildt, J. (1992) 'External acquisition of knowledge for innovations – a research agenda', *R&D Management*, **22** (2), 105–110.
3 Welch, J. and P. Nayak (1992) 'Strategic sourcing: a progressive approach to the make or buy decision', *Academy of Management Executive*, **6** (1), 23–31.
4 Bettis, R., S. Bradley and G. Hamel (1992) 'Outsourcing and industrial decline', *Academy of Management Executive*, **6** (1), 7–21.
5 Atuaheme-Gima, K. and P. Patterson (1993) 'Managerial perceptions of technology licensing as an alternative to internal R&D in new product development: an empirical investigation', *R&D Management*, **23** (4), 327–336.
6 Robins, J. and M. Wiersema (1995) 'A resource-based approach to the multibusiness firm', *Strategic Management Journal*, **16** (4), 277–300.
7 Granstrand, O. *et al.* (1992) 'External technology acquisition in large multi-technology corporation', *R&D Management*, **22** (2), 111–133.
8 Tyler, B. and H. Steensma (1995) 'Evaluating technological collaborative opportunities: a cognitive modeling perspective', *Strategic Management Journal*, **16**, 43–70.
9 Bidault, F. and T. Cummings (1994) 'Innovating through alliances: expectations and limitations', *R&D Management*, **24** (2), 33–45.
10 Scott Swan, K. and B. Allred (2003) 'A product and process model of the technology-sourcing decision', *Journal of Product Innovation Management*, **20**, 485–496.
11 Rothaermel, F. and D. Deeds (2004) 'Exploration and exploitation alliances in biotechnology: a system of new product development', *Strategic Management Journal*, **25**, 201–221.
12 Miotti, L. and F. Sachwald (2003) 'Co-operative R&D; why and with whom? An integrated framework of analysis', *Research Policy*, **32**, 1481–1499.

13 Teher, B. (2002) 'Who co-operates for innovation, and why? An empirical analysis', *Research Policy*, **31**, 947–967.

14 Bleeke, J. and D. Ernst (1993) *Collaborating to Compete*. John Wiley & Sons, Inc., New York.

15 Von Hippel, E. (1988) *The Sources of Innovation*. Oxford University Press, Oxford.

16 Nishiguchi, T. (1994) *Strategic Industrial Sourcing: The Japanese advantage*. Oxford University Press, Oxford.

17 Leonard-Barton, D. and D. Sinha (1993) 'Developer–user interaction and user satisfaction in internal technology transfer', *Academy of Management Journal*, **36** (5), 1125–1139.

18 Lamming, R. (2000) 'Assessing supplier performance', in Tidd, J. (ed.), *From Knowledge Management to Strategic Competence*. Imperial College Press, London, 229–253.

19 Bozdogan, K. *et al.* (1998) 'Architectural innovation in product development through early supplier integration', *R&D Management*, **28** (3), 163–173.

20 Nishiguchi, T. (1994) *Strategic Industrial Sourcing: The Japanese advantage*. Oxford University Press, Oxford.

21 Bidault, F., C. Despres and C. Butler (1998) 'The drivers of cooperation between buyers and suppliers for product innovation', *Research Policy*, **26**, 719–732; Ragatz, G., R. Handfield and T. Scannell (1997) 'Success factors for integrating suppliers into new product development', *Journal of Product Innovation Management*, **14**, 190–202.

22 Andersen, P. (1999) 'Organizing international technological collaboration in subcontractor relationships', *Research Policy*, **28**, 625–642.

23 Brusconi, S., A. Prencipe and K. Pavitt (2002) 'Knowledge specialisation and the boundaries of the firm', *Administrative Science Quarterly*, **46** (4), 597–621.

24 Kaufman, A., C. Wood and G. Theyel (2000) 'Collaboration and technology linkages: a strategic supplier typology', *Strategic Management Journal*, **21**, 649–663.

25 Chiesa, V., R. Manzini, and E. Pizzurno (2004) 'The externalization of R&D activities and the growing market of product development services', *R&D Management*, **34**, 65–75.

26 Atuahene-Gima, K. and P. Patterson (1993) 'Managerial perceptions of technology licensing as an alternative to internal R&D in new product development: an empirical investigation', *R&D Management*, **23** (4), 327–336.

27 Gibson, D. and E. Rogers (1994) *R&D Collaboration on Trial*. Harvard Business School Press, Boston, Mass.

28 Aldrich, H. and T. Sasaki (1995) 'R&D consortia in the US and Japan', *Research Policy*, **24** (2), 301–316.

29 Sakakibara, M. (2002) 'Formation of R&D consortia: industry and company effects', *Strategic Management Journal*, **23**, 1033–1050.

30 Doz, Y. and G. Hamel (1998) *Alliance Advantage: The art of creating value through partnering*. Harvard Business School Press, Boston, Mass.

31 Carr, C. (1999) 'Globalisation, strategic alliances, acquisitions and technology transfer: Lessons from ICL/Fujitsu and Rover/Honda and BMW', *R&D Management*, **29** (4), 405–421.

32 Mauri, A. and G. McMillan (1999) 'The influence of technology on strategic alliances', *International Journal of Innovation Management*, **3** (4), 367–378.

33 Jay Lambe, C. and R. Spekman (1997) 'Alliances, external technology acquisition, and discontinuous technological change', *Journal of Product Innovation Management*, **14**, 102–116.

34 Duysters, G. and A. de Man (2003) 'Transitionary alliances: an instrument for surviving turbulent industries?', *R&D Management*, **33**, 49–58.

35 DeBresson, C. and F. Amesse (1991) 'Networks of innovators: a review and introduction to the issues', *Research Policy*, **20**, 363–379.

36 Pyka, A. and G. Kuppers (2002) *Innovation Networks: Theory and Practice.* Edward Elgar, Cheltenham.

37 Camagni, R. (1991) *Innovation Networks: Spatial perspectives.* Belhaven Press, London.

38 Nohria, N. and R. Eccles (1991) *Networks and Organizations.* Harvard Business School Press, Boston, Mass.

39 Gulati, R. (1998) 'Alliances and networks', *Strategic Management Journal,* **19**, 293–317.

40 Hakansson, H. and A. Waluszewski (2003) *Managing Technological Development.* Routledge, London.

41 Bidault, F. and W. Fischer (1994) 'Technology transactions: networks over markets', *R&D Management,* **24** (4), 373–386.

42 Hakansson, H. (1995) 'Product development in networks', in Ford, D. (ed.), *Understanding Business Markets: Interaction, relationships and networks.* The Dryden Press, New York, 487–507.

43 Nakateni, I. (1984) 'The economic role of financial corporate groupings', in Masahiko Aoki (ed.), *The Economic Analysis of the Japanese Firm.* North-Holland, Amsterdam, 227–258.

44 Belussi, F. and F. Arcangeli (1998) 'A typology of networks: flexible and evolutionary firms', *Research Policy,* **27**, 415–428.

45 Galaskiewicz, J. (1996) 'The "new network analysis" and its application to organizational theory and behaviour', and Aranjo, L. and G. Easton, 'Networks in socioeconomic systems' in Iacobucci, D. (ed.), *Networks in Marketing.* Sage, London, 19–31.

46 Doz, Y., P. Olk and P. Ring (2000) 'Formation processes of R&D consortia: Which path to take? Where does it lead?', *Strategic Management Journal,* **21**, 239–266.

47 Tidd, J. (1997) 'Complexity, networks and learning: integrative themes for research on innovation management', *International Journal of Innovation Management,* **1** (1), 1–22.

48 Hooi-Soh, P. and E. Roberts (2003) 'Networks of innovators: a longitudinal perspective', *Research Policy,* **32**, 1569–1588.

49 Passiante, G. and P. Andriani (2000) 'Modelling the learning environment of virtual knowledge networks', *International Journal of Innovation Management,* **4** (1), 1–31.

50 Berg, S., J. Duncan and P. Friedman (1982) *Joint Venture Strategies and Corporate Innovation.* Gunn & Ham, Cambridge, Mass.

51 Balakrishnan, S. and M. Koza (1995) 'An information theory of joint ventures', in Gomez-Mejia, L. and Lawless, M. (eds), *Advances in Global High Technology Management: Strategic alliances in high technology.* JAI Press, Greenwich, Conn., Vol. 5, Part B, 59–72.

52 Krubasik, E. and H. Lautenschlager (1993) 'Forming successful strategic alliances in high-tech businesses', in Bleeke, J. and Ernst, D. (eds), *Collaborating to Compete.* John Wiley & Sons, Inc., New York, 55–65.

53 Duysters, G., G. Kok and M. Vaandrager (1999) 'Crafting successful strategic technology partnerships', *R&D Management,* **29** (4), 343–351; Hagedoorn, J. (2002) 'Inter-firm R&D partnerships: an overview of major trends and patterns since 1960', *Research Policy,* **31**, 477–492; Giarrantana, M. and S. Torrisi (2002) 'Competence accumulation and collaborative ventures: evidence form the largest European electronics firms and implications for EU technological policies', in Lundan, S. (ed.), *Network Knowledge in International Business.* Edward Elgar, Cheltenham, 196–215.

54 Tidd, J. and M. Trewhella (1997) 'Organizational and technological antecedents for knowledge acquisition and learning', *R&D Management,* **27** (4), 359–375.

55 Tyler, B. and P. Nayak (1995) 'Evaluating technological collaborative opportunities: a cognitive modeling perspective', *Strategic Management Journal,* **16**, 43–70.

56 Rothaermel, F. (2001) 'Complementary assets, strategic alliances, and the incumbent's advantage: an empirical study of industry and firms effects in the biopharmaceutical industry', *Research Policy,* **30**, 1235–1251.

57 Orsenigo, L., F. Pammolli and M. Riccaboni (2001) 'Technological change and network dynamics: lessons from the pharmaceutical industry', *Research Policy*, **30**, 485–508.

58 Nicholls-Nixon, C. and C. Woo (2003) 'Technology sourcing and the output of established firms in a regime of encompassing technological change', *Strategic Management Journal*, **24**, 651–666.

59 Rothaermel, F. and D. Deeds (2004) 'Exploration and exploitation alliances in biotechnology: a system of new product development', *Strategic Management Journal*, **25**, 201–221.

60 Harrigan, K. (1986) *Managing for Joint Venture Success*. Lexington Books, Lexington, Mass.

61 Dacin, M., M. Hitt and E. Levitas (1997) 'Selecting partners for successful international alliances', *Journal of World Business*, **32** (1), 321–345.

62 Spekmen, R. *et al.* (1996) 'Creating strategic alliances which endure', *Long Range Planning*, **29** (3), 122–147.

63 Bleeke, J. and D. Ernst (1993) *Collaborating to Compete*. John Wiley & Sons, Inc., New York.

64 Tidd, J. and Y. Izumimoto (2001) 'Knowledge exchange and learning through international joint ventures: an Anglo-Japanese experience', *Technovation*, **21** (2).

65 Brockhoff, K. and T. Teichert (1995) 'Cooperative R&D partners' measures of success', *International Journal of Technology Management*, **10** (1), 111–123.

66 Bruce, M., F. Leverick and D. Littler (1995) 'Complexities of collaborative product development', *Technovation*, **15** (9), 535–552.

67 Kelly, M., J. Schaan and H. Joncas (2002) 'Managing alliance relationships: key challenges in the early stages of collaboration', *R&D Management*, **32** (1), 11–22.

68 Hoecht, A. and P. Trott (1999) 'Trust, risk and control in the management of collaborative technology development', *International Journal of Innovation Management*, **3** (3), 257–270.

69 Dodgson, M. (1993) 'Learning, trust, and technological collaboration', *Human Relations*, **46** (1), 77–95.

70 Badaracco, J. (1991) *The Knowledge Link: How firms compete through strategic alliances*. Harvard Business School Press, Boston, Mass.

71 Lane, P. and M. Lubatkin (1998) 'Relative absorptive capacity and inter-organizational learning', *Strategic Management Journal*, **19**, 461–477.

72 Nonaka, I. and H. Takeuchi (1995) *The Knowledge-Creating Company*. Oxford University Press, Oxford.

73 Johnson, W. (2002) 'Assessing organizational knowledge creation theory in collaborative R&D projects', *International Journal of Innovation Management*, **6** (4), 387–418.

74 Levinson, N. and M. Asahi (1995) 'Cross-national alliances and interorganizational learning', *Organizational Dynamics*, Autumn, 50–63.

75 Hamel, G. (1991) 'Competition for competence and inter-partner learning within international strategic alliances', *Strategic Management Journal*, **12**, 83–103.

76 Jones, K. and W. Shill (1993) 'Japan: allying for advantage', in Bleeke, J. and Ernst, D. (eds), *Collaborating to Compete*. John Wiley & Sons, Inc., New York, 115–144; Sasaki, T. (1993) 'What the Japanese have learned from strategic alliances', *Long Range Planning*, **26** (6), 41–53.

77 Biemans, W. (1995) 'Internal and external networks in product development', in Bruce, M. and Biemans, W. (eds), *Product Development: Meeting the challenge of the design-marketing interface*. John Wiley & Sons, Ltd, Chichester, 137–159.

78 Schrader, S. (1995) 'Informal alliances: information trading between firms', in Gomez-Mejia, L. and Lawless, M. (eds), *Advances in Global High-Technology Management: Strategic alliances in high technology*. JAI Press, Greenwich, Conn., Vol. 5, Part B, 31–55.

79 Spencer, J. (2003) 'Firms' knowledge-sharing strategies in the global innovation system: evidence from the flat panel display industry', *Strategic Management Journal*, **24**, 217–233.

Part IV

BUILDING EFFECTIVE IMPLEMENTATION MECHANISMS

So far we have been looking at how an organization identifies and positions itself strategically, and how it establishes effective linkages with its marketplace and with other organizations. We now move to look in more detail at how it mobilizes and organizes the internal and external processes which enable innovation to happen.

Innovation success depends not only on clear strategic direction and effective external positioning, but also on being able to manage projects from initial idea or opportunity to a successful commercial product or service, or an effective new internal process. This involves a sequence of problem-solving activities and needs a staged framework for decision-making about whether or not to continue with development, allocation of resources, and so on. And this in turn requires skills in managing projects, linking different functional resources together, knowledge-sharing, managing both technical and market development, managing the change process itself, and ensuring that learning is captured from the experience of the project. Increasingly these activities take place in distributed form across organizational boundaries and in various kinds of formal (and sometimes informal) networks. Chapter 9 explores these themes and identifies emerging routines for their effective management.

Implementation also raises the question of whether the innovation can be managed within the existing organizational framework, or whether new arrangements are needed. One option for more radical innovations – where either the technologies or markets are unfamiliar – is to establish an internal corporate venture. The aim is to encourage entrepreneurial activity and experimentation, but at the same time to draw on the resources of the parent organization. Therefore internal ventures can be a means to leverage existing competencies in new markets or to acquire new competencies, perhaps in response to disruptive innovation. Chapter 10 looks in detail at this.

Managing the Internal Processes

Throughout the book we have been looking at the innovation process in terms of phases in time, moving from searching for trigger signals, through selecting strategic options, to implementation. It is worth revisiting the process model which we developed in Chapter 2 and in particular to note the funnel shape, moving in from a wide set of possibilities through to increasing certainty but with increasing commitment of resources.

Implicit in this is the presence of *routines* which help enable management of activities through these phases. For example:

- Picking up a wide range of signals and processing them for relevance to the organization.
- Selecting projects which have a good strategic fit.
- Monitoring and managing projects through the various stages of development.
- Deciding where and when to stop projects, and where and when to accelerate them.
- Preparing the ground for effective launch, often through early involvement of key stakeholders.
- Reviewing and capturing learning from completed projects.

For many enterprises these activities are often carried out on an ad hoc basis, and not in systematic fashion. As we pointed out in Chapter 2 the process is by no means linear – there is considerable overlap, backtracking, false starts and recycling. So dealing with it is unlikely to involve standard operating procedures or procedural rules, but this doesn't mean that the process is unmanageable. It is possible to create a structured framework within which the process can operate in something approaching a repeatable fashion – it can be built around behavioural routines and these can be learned and refined over time. Research evidence confirms the view that a degree of structure and discipline is an important component of success.[1]

A key theme throughout the book has been the need to adapt and configure generic routines to deal with particular circumstances. In particular organizations face the challenge that much innovative activity is increasingly distributed across organizational boundaries. Working within various kinds of networks offers enhanced access to knowledge and resources but it also poses difficulties in terms of managing at arm's length where the levers of structures and authority may not apply – and where key issues of

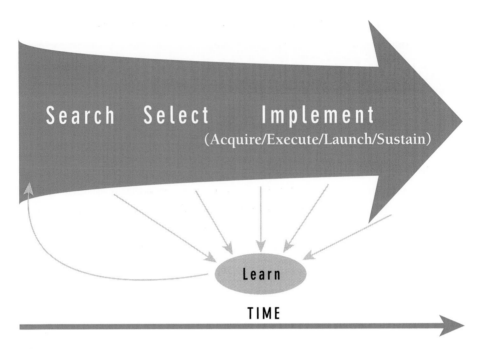

FIGURE 9.1   Innovation process model

trust, information-sharing and conflict resolution become important elements in the innovation process.

In similar fashion there are challenges which emerge when the organization has to deal with innovative activities outside the normal 'steady state'. Where external conditions lead to discontinuities in the innovation environment – for example, the emergence of radically new technology or of completely new markets – the old 'rules of the game' no longer apply. Under these conditions organizations need to deploy alternative routines to help deal with high levels of uncertainty and to manage a process of co-evolution of new rules of the game.

This chapter looks in more detail at how organizations manage these elements within the innovation process, and reports on some of the routines which appear to have been associated with effective performance at each of the stages. Innovation management is essentially about building and embedding such routines within the organization – but it is also about reviewing, improving and on occasions replacing them with new and more appropriate ones to cope with what is a constantly changing environment. In other words, it is about building dynamic capability.

Many of the routines described in this chapter involve the use of different innovation tools – essentially structured aids to help analyse and act in managing the innovation process. In the website accompanying the book there is an extensive 'innovation toolbox' which describes these in more detail and provides signposts to further information about their use.

9.1 Enabling Effective Search

The process of innovation begins with picking up various kinds of trigger signals. These might be about technology, markets, competitor behaviour, shifts in the political or regulatory environment, new social trends, etc. – and they could come from inside or far outside the organization. How can an effective searching and scanning process for picking these up in timely fashion be organized and managed? What lessons have organizations learned about effective strategies to improve the range and quality of signals being picked up? And how can we ensure that the sheer volume of all this information coming in does not swamp the important signals with all the 'noise'?

Organizations pick up signals about innovation possibilities through exploring a particular 'selection environment' – essentially a search space made up of knowledge about technologies, markets, competitors and other sources. So we need effective routines to make sure that space is thoroughly explored – and to stretch its boundaries to create

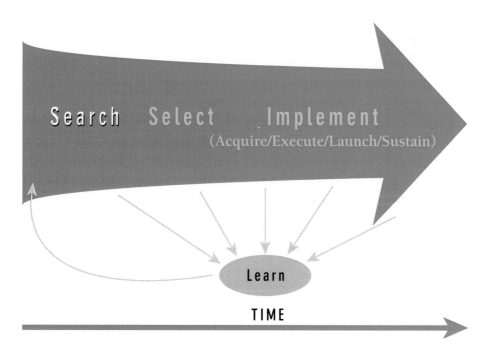

FIGURE 9.2

new space. For large organizations significant resources can be devoted to this activity, but for many others it will raise the issue of developing and leveraging networks and connections.

There are a number of approaches which can be used to explore and extend this search space, and we will look at some of these in the following section.

Defining the Boundaries of the Marketplace

Essentially this involves asking the deceptively simple question, 'What business are we in?' This kind of question prompts discussion of current and potential markets and assists in looking for new opportunities. Sometimes innovation can take the form of repositioning – offering the same basic product or service but addressed in a new way to different markets. For example, Amazon.com is seen as an on-line retailer but is trying to broaden its business by positioning itself also as a software developer and supplier.[2]

Understanding Market Dynamics

Closely linked to the above is understanding where potential markets may arise as a consequence of various kinds of change. For example, the cellular phone business has moved from a specialist, high-price business tool into the general marketplace as a result of both technological and cultural change. Similarly low cholesterol and other healthy foods are increasingly becoming relevant to a large segment of the population as a result of changing social attitudes and education. Building up such understanding of the changing marketplace requires various forms of communication and interaction, from monitoring through to customer panels and surveys.[3,4]

Knowledge of this kind is a powerful input to planning both product and process innovation – for example, the shift to widespread use of wireless technology is a powerful trigger for process innovation around increasing capacity and flexibility within the factory. Firms like Zara and Benetton have sophisticated IT systems installed in each of their shops such that they can quickly identify which lines are selling well on a daily basis – and tailor production to this. The same information also guides the development of their internal business and manufacturing systems.[4,5]

'Trend-spotting'

One difficulty in exploring a market space arises when the market does not exist or where it suddenly takes a turn in a new direction. Developing antennae to pick up on the early warnings of trends is important, particularly in consumer-related innovation. For example, much of the development of the mobile phone industry has been on the back of the different uses to which schoolchildren put their phones and delivering innovations which support this. Examples include text messaging, image/video exchange and downloadable personal ring tones where the clues to the emergence of these innovation trajectories were picked up by monitoring what such children were doing or aspiring towards.[6] Chapter 8 explores some of these themes in greater detail.

Monitoring Technological Trends

Related to this is the identification of emerging trends in technologies, where existing trajectories may be challenged or redirected as a result of new knowledge. Picking up on these requires active search and scanning at the periphery – for example, through monitoring websites and chat rooms, visiting conferences, seminars and exhibitions,

---

BOX 9.1
OUT OF THE MOUTHS OF BABES . . .

The first text message sent via the Short Message Service (SMS) protocol was sent in 1992 as part of the GSM Phase 1 standard. Its origins lay in test messages sent between engineers setting up the cells and network infrastructure. Growth in usage of SMS was fuelled mainly by children and this led to innovation in mobile phone design and the development of particular language suited to SMS with its constraint on 160 characters per message. By 1999 the SMS market in the European Union alone was estimated to be 1 billion messages per month with growth accelerating as a result of new phones, wireless application protocol and the emerging 2.5 and 3G systems . . . Market estimates suggest that there are around 600 million users of SMS worldwide. Behind SMS is the Multi-media Messaging System (MMS) which offers additional features like audio, image and video messaging; the market estimate is around 130 million, with growth likely to accelerate with widespread adoption of high-capacity 3G technology.

In similar fashion the market for personalized ring tones has also been fuelled by the youth segment. The Yankee group estimated the global ring tone market to be around $2.5 bn in 2004.

Source: Internet Market Research data.

and building close exploratory links to research labs. As with the case of market trend-spotting a key skill here is to improve peripheral vision – not only looking in the places where developments might be expected to occur but also exploring at the edges where something unexpected might take off.[7]

Research has consistently shown that those organizations which adopt an active as opposed to a parochial approach to seeking out links with possible suppliers of technology or information are more successful innovators.[8,9] Possible sources with which links can be made include suppliers, universities, research/technology institutions, other users and producers, trade associations, international bodies (e.g. for standards), etc. The principle behind this is to multiply the range of channels along which technological intelligence can flow.[7] As Box 9.2 shows, the Internet has become a powerful amplifier for making such connections and an increasing number of organizations are developing search strategies (and accompanying 'gatekeeper' skills) to work with it. IBM, for example, uses an approach called 'Webfountain' to help it monitor a wide range of potential triggers. Even the CIA makes use of an internal group called In-Q-Tel to act as a 'venture catalyst' to facilitate trend spotting in key technology areas![10,11]

There is also research to suggest that encouraging the development of an 'invisible college' of contacts between researchers inside and outside the organization is an important source of ideas.[12–14] Recent discussion around the theme of 'communities of practice' has stressed not only the power of formal and informal networking as a source of knowledge but also the importance of bringing unconnected elements together – most new knowledge emerges at such interfaces.[15] In Xerox, for example, 23 000 technical reps from around the world are linked into a network of communities of practice to share knowledge about unusual machine faults and heuristics for finding and fixing them.[10]

Trend-spotting is an approach which can be applied in many areas to pick up on, or bring into sharper focus, trigger signals about innovation. A common experience is the idea of 'tipping points' associated with long-term trends, where an imperceptible movement in opinion or values suddenly flips the rules of the game into something new.[16] For example, concern about the nutritional values in much fast food has been around for some time with growing worries about links to obesity and other health problems. What began as an issue for a relatively small proportion of the population gathered momentum and, with worrying data on obesity levels in the USA and elsewhere emerging, 'tipped' into a mainstream concern around the turn of the twenty-first century. It has had a dramatic effect on the industry and triggered a stream of product and process innovations to realign with popular sentiment. Similar long-term trends and potential for reaching tipping points can be seen in the discussions around energy sources (concerns about running out of oil) and global warming (with rising worries about climate change).

BOX 9.2
CONNECT AND DEVELOP AT PROCTER & GAMBLE

Procter & Gamble spend about $2 bn each year and employ around 7000 people on research to support the business. But these days they use the phrase 'connect and develop' instead of 'research and development' and have set themselves the ambitious goal of sourcing much of their idea input from outside the company. The scale of the challenge is huge; they estimate, for example, that in the 150 core technology areas which they make use of there are more than 1.5 million active researchers outside of P&G. Finding the right needle in a global haystack is a critical strategic challenge.

They achieve this through a variety of links, making particular use of Internet-based sources and employing a number of people whose job it is to act as Internet gatekeepers. These people use sophisticated search and visualization tools to 'mine' information about a wide range of developments in technologies, markets, competitor behaviour, social and political trends, etc. – and to bring it to the notice of others within P&G who may be able to use these signals to trigger innovation.

This search process is complemented by other ways of connecting – for example, an Internet-based business (NineSigma.com) which enables client organizations 'to source innovative ideas, technologies, products and services from outside their organization quickly and inexpensively by connecting them to the very best solution providers from around the world'. They also work with another website – InnoCentive.com – which provides an on-line marketplace where organizations seeking solutions to problems are brought together with scientists and engineers with solutions to offer.

All of this is not to neglect the significant contribution that internal ideas can bring. The company has a wide range of active communities of practice around particular product groups, technologies, market segments, etc. and is able to draw on this knowledge increasingly through the use of intranets. A recent development has been the 'Encore' programme in which retired staff of the company – and potentially those of other companies – can be mobilized to act as knowledge and development resources in an extended innovation network.

The underlying approach is a shift in emphasis, not abandoning internal R&D but complementing it with an extensive external focus. Increasingly they see their task not just as managing 'know-how' but also 'know who'.

Source: Based on reference 17.

Market Forecasting

A wide range of techniques are available for trying to understand the likely dynamics of new markets, running from simple extrapolation of current trends through to complex techniques for handling discontinuous change, such as Delphi panels and scenario writing.[18] (These are described in more detail in Chapter 7.) Such forecasting needs to move beyond sales-related information to include other features which will influence the potential market – for example, demographic, technological, political and environmental issues.[19] For example, the present concern for environmentally friendly 'green' products is likely to increase and will be shaped by a variety of these factors. From this information come valuable clues about the type of performance which the market expects from a particular manufacturer or service provider – and hence the targets for process innovation.

Closely related to this is the idea of identifying markets which do not yet exist but which may emerge as a result of identifiable current trends. For example, the figures on rising levels of obesity indicate a likely growth market in products like healthier foods and in services to support healthier lifestyles. The shift in the age profile of many European countries suggests significant growth in age-related services, not just in care but also in 'lifestyle' offerings like holidays and sports/leisure.

Technological Forecasting

Opportunities arise in part from the continuing advances in knowledge which make new products, services and processes possible. Various techniques exist for exploring technological futures, ranging from simple extrapolation of performance parameters and rates of development to complex, non-linear techniques. Some, like Delphi panels and scenarios, are similar to market forecasting techniques, whilst others are more closely aligned to technological development models.[20,21] An important approach here is the use of S-curves to try and identify the point at which an emerging new technological trajectory takes over from an existing one.[22] But we also need to be careful – as we saw in our earlier discussion of discontinuous innovation, new opportunities may also emerge from making different use of existing technologies configured in new ways.

Integrated Future Search

Although there are well-proven and useful approaches to forecasting technology and market trends, these often make simplifying assumptions about the wider context in which predicted changes might take place. An alternative approach is to take an integrated view of what different futures might look like and then explore innovation triggers within those spaces. Typical of this approach is the work which Shell and other organizations do with scenarios where a number of people work together to build up pictures of alternative parallel futures. These are usually richly woven backgrounds which describe technologies, markets, politics, social values and other elements in the form of a 'storyline' – for example, a recent Shell publication looked at possible scenarios for 2020 in terms of two alternatives: 'business class' and 'prism'. The former offers a vision of 'connected freedom' in which cities and regions become increasingly powerful at the expense of central government, and where there is increasing mobility amongst a 'global elite' whilst the latter describe a different world in which people increasingly look to their roots and re-orientate towards values as the focus around which to organize their lives. Neither are necessarily the 'right' answer to what the world will look like by that time but they do offer a richly described space within which to explore and simulate, and to search for threats and opportunities which might affect the company. In particular they allow an organization to define particular 'domains' – spaces within the bigger scenario where it can think about deploying its particular competencies to advantage – and to carry out a kind of 'targeted hunting' in its search for innovation triggers.[23,24]

Integrated futures exercises of this kind do not have to be organization-specific – indeed, there is much value in getting a diverse picture which brings different perspectives into play and explores the resulting search space from different angles. It is an approach increasingly used at sectoral and national level – for example, in the many 'Foresight' programmes which are in place.[25] In the UK, for example, the Department of Trade and Industry has co-ordinated such activity for over a decade using panels made up of a wide range of interest and expertise. (See http://www.foresight.gov.uk/ for more details on their current programmes.)

Learning from Others

Another group of approaches deals with comparisons against competitor and other organizations. Search techniques of this kind include looking at 'best-practice' demonstration projects and 'reverse engineering'. This theme is discussed more fully in Chapter 4. For example, much of the early growth in Korean manufacturing industries in fields like machine tools came from adopting a strategy of 'copy and develop' – essentially learning (often as a result of taking licences or becoming service agents) by working with established products and understanding how they might be adapted or developed for the local market. Subsequently this learning could be used to develop new generations of products or services.[26,27]

A powerful variation on this theme is the concept of benchmarking.[28] In this process enterprises make structured comparisons with others to try and identify new ways of carrying out particular processes or to explore new product or service concepts. The learning triggered by benchmarking may arise from comparing between similar organizations (same firm, same sector, etc.), or it may come from looking outside the sector but at similar products or processes. For example, Southwest Airlines became the most successful carrier in the USA by dramatically reducing the turnaround times at airports – an innovation which it learned from studying pit stop techniques in the Formula 1 Grand Prix events. Similarly the Karolinska Hospital in Stockholm made significant improvements to its cost and time performance through studying inventory management techniques in advanced factories.[29]

Benchmarking of this kind is increasingly being used to drive change across the public sector, both via 'league tables' linked to performance metrics which aim to encourage fast transfer of good practice between schools or hospitals and also via secondment, visits and other mechanisms designed to facilitate learning from other sectors managing similar process issues such as logistics and distribution.

Involving Stakeholders

Another key approach is to bring in key stakeholders into the process – for example by involving customers in providing information about the kinds of products and services which they require. This can be done through regular surveys, through customer panels and other forms of involvement. In similar fashion there is a growing trend towards shared *process* innovation, where customers and suppliers work together to reduce costs or increase quality or some other performance parameter.[30–32]

Working with users can often offer important insights or new directions. For example, the Danish pharmaceutical company Novo-Nordisk developed a highly successful range of insulin delivery devices to revitalize the market for diabetes treatments by pioneering the 'Novopen' concept. In 1981 the marketing director had read an article in *The Lancet* which described a young English girl with diabetes who each morning filled a disposable syringe with enough insulin for the rest of the day. This enabled her to administer the doses she needed over the course of the day without having to refill the syringe. She had always felt that it was cumbersome and indiscreet to administer a dose of insulin from a vial using a disposable syringe. This led him to ask the development team if it would be possible to produce a device that looked like a fountain pen, was easy to use, which could hold a week's supply of insulin and administer two units of insulin at the touch of a button. The pen had to be simple and discreet, and preferably look like an actual fountain pen. After extensive development work the device was successfully produced and marketed and has become a key product platform for the firm.

An important variation on the theme of involving customers in concept development is the idea of working with *lead users*. In industries like semiconductors and instrumentation, research suggests that the richest understanding of needed new products is held by only a few organizations, who are ahead of the majority of firms in the sector.[30] Equally, finding the most demanding customer in a particular sector is a valuable approach; stretching the concept to meet their needs will ensure that most other potential users come within the envelope.[33]

In similar fashion, successful innovation depends on maintaining a strong user perspective over time; this argues for mechanisms which emphasize continuing interaction rather than a one-off information-gathering exercise. Mechanisms for doing so include involvement of users in the project team and two-way visiting between sites.[34]

Whilst involving users can usually be advantageous it can also have its problems. The 'lead user' approach, for example, is a powerful way of working with an articulate partner to develop innovations which, if they meet lead user needs, will generally be suitable for a much wider range of adopters. The benefits in terms of learning and in being early into the market are powerful – but there is the risk that the lead user requirement may not become the dominant one. Our earlier discussion of disruptive innovation (Chapter 1) highlights the difficulties of working too closely with one group of established users and missing out on signals from the periphery about a different emerging group of users with different needs. The advantages and limitations of user involvement are discussed more extensively in Chapter 7.

BOX 9.3
USER INVOLVEMENT IN INNOVATION – THE COLOPLAST EXAMPLE

One of the key lessons about successful innovation is the need to get close to the customer. At the limit (and as Eric von Hippel and other innovation scholars have noted[31]), the user can become a key part of the innovation process, feeding in ideas and improvements to help define and shape the innovation. The Danish medical devices company, Coloplast, was founded in 1954 on these principles when nurse Elise Sorensen developed the first self-adhering ostomy bag as a way of helping her sister, a stomach cancer patient. She took her idea to a various plastics manufacturers, but none showed interest at first. Eventually one, Aage Louis-Hansen, discussed the concept with his wife, also a nurse, who saw the potential of such a device and persuaded her husband to give the product a chance. Hansen's company, Dansk Plastic Emballage, produced the world's first disposable ostomy bag in 1955. Sales exceeded expectations and in 1957, after having taken out a patent for the bag in several countries, the Coloplast company was established.

continues overleaf

BOX 9.3   *(continued)*

Today the company has subsidiaries in 20 and factories in five countries around the world, with specialist divisions dealing with incontinence care, wound care, skin care, mastectomy care, consumer products (specialist clothing, etc.) as well as the original ostomy care division.

Keeping close to users in a field like this is crucial and Coloplast have developed novel ways of building in such insights by making use of panels of users, specialist nurses and other healthcare professionals located in different countries. This has the advantage of getting an informed perspective from those involved in post-operative care and treatment and who can articulate needs which might for the individual patient be difficult or embarrassing to express. By setting up panels in different countries the varying cultural attitudes and concerns could also be built into product design and development.

An example is the Coloplast Ostomy Forum (COF) board approach. The core objective within COF Boards is to try and create a sense of partnership with key players, either as key customers or key influencers. Selection is based on an assessment of their technical experience and competence but also on the degree to which they will act as opinion leaders and gatekeepers – for example, by influencing colleagues, authorities, hospitals and patients. They are also a key link in the clinical trials process. Over the years Coloplast has become quite skilled in identifying relevant people who would be good COF board members – for example, by tracking people who author clinical articles or who have a wide range of experience across different operation types. Their specific role is particularly to help with two elements in innovation:

- Identifying discussing and prioritizing user needs.
- Evaluating product development projects from idea generation right through to international marketing.

Importantly COF Boards are seen as integrated with the company's product development system and they provide valuable market and technical information into the stage gate decision process. This input is mainly associated with early stages around concept formulation (where the input is helpful in testing and refining perceptions about real user needs and fit with new concepts). There is also significant involvement around project development where involvement is concerned with evaluating and responding to prototypes, suggesting detailed design improvements, design for usability, etc.

Involving Insiders

In looking for signals about possible innovation triggers it is important not to neglect those which originate inside the organization. We have already noted the power of bringing different communities of practice together across organizational boundaries in companies like Xerox and Procter & Gamble. To this we should add the considerable resource which 'ordinary' employees represent in terms of their ideas, particularly for incremental improvement innovations. Tapping into such high involvement innovation potential has been demonstrably helpful to a wide range of organizations; Box 9.4 gives some examples and we return to this theme in more detail in Chapter 11.

BOX 9.4

THE POWER OF 'HIGH INVOLVEMENT' INNOVATION

Innovation can come from many sources but a powerful resource lies in the minds and experiences of existing employees. The following are examples where this potential source has been realized to strategic effect.

In a detailed study of seven leading UK firms in the fast-moving consumer goods (FMCG) sector, Westbrook and Barwise reported a wide range of benefits including:

- Waste reduction of £500k in year 1, for a one-off expense of £100 K.
- A recurrent problem costing over 25 k/year of lost time, rework and scrapped materials eliminated by establishing and correcting root cause.
- 70% reduction in scrap year on year.
- 50% reduction in set-up times, in another case 60 to 90%.
- Uptime increased on previous year by 50% through CI project recommendations.
- £56 k/year overfilling problems eliminated.
- Reduction in raw material and component stocks over 20% in 18 months.
- Reduced labour cost per unit of output from 53 pence to 43 pence.
- Raised service levels (order fill) from 73 to 91%.
- Raised factory quality rating from 87.6 to 89.6%.

- The US financial services group Capital One saw major growth over the period 1999–2002, equivalent to 430% and built a large customer base of around 44 million people. Its growth rate (30% in turnover 2000–2001) makes it one of the most admired and innovative companies in its sector. But, as Wall points

---

continues overleaf

BOX 9.4   (*continued*)

out, 'innovation at Capital One cannot be traced to a single department or set of activities. It's not a unique R&D function, there is no internal think-tank. Innovation is not localized but systemic. It's the lifeblood of this organization and drives its remarkable growth . . . It comes through people who are passionate enough to pursue an idea they believe in, even if doing so means extending well beyond their primary responsibilities.' (Wall, 2002)

- Chevron Texaco is another example of a high-growth company which incorporates – in this case in its formal mission statement – a commitment to high-involvement innovation. It views its 53 000 employees worldwide as 'fertile and largely untapped resources for new business ideas . . . Texaco believed that nearly everyone in the company had ideas about different products the company could offer or ways it could run its business. It felt it had thousands of oil and gas experts inside its walls and wanted them to focus on creating and sharing innovative ideas . . .' (Abraham and Pickett, 2002)

- In implementing high-involvement innovation into a large South African mining company (De Jager, Welgemoed *et al.*, 2004), benefits reported included:
  - Improvements in operating income at one dolomite mine of 23% despite deteriorating market conditions.
  - Increase in truck fleet availability at a large coal mine of 7% (since these are 180 ton trucks the improvement in coal hauled is considerable).
  - Increase in truck utilization of 6% on another iron ore mine.

- Kaplinsky (Kaplinsky, 1994) reports on a series of applications of 'Japanese' manufacturing techniques (including the extensive use of *kaizen* in a variety of developing country factories in Brazil, India, Zimbabwe, Dominican Republic and Mexico. In each case there is clear evidence of the potential benefits which emerge where high-involvement approaches are adopted – although the book stresses the difficulties of creating the conditions under which this can take place.

- Gallagher and colleagues report on a series of detailed case studies of manufacturing and service sector organizations which have made progress towards implementing some form of high-involvement innovation (Gallagher and Austin, 1997). The cases highlight the point that although the sectors involved differ widely – insurance, aerospace, electronics, pharmaceuticals, etc. – the basic challenge of securing high involvement remains broadly similar.

Source: Reference 35. All texts cited are mentioned there.

'Mistakes Management'

Another related source of trigger signals lie in mistakes and apparent failures which may open up completely new directions for innovation. Much depends here on being able to reframe what appears to be a diversion or distraction from a particular planned innovation direction as something which has potential to open a new line of attack. For example, the famous story of 3M's 'Post-it' notes began when a polymer chemist mixed an experimental batch of what should have been a good adhesive but which turned out to have rather weak properties – sticky but not very sticky. This failure in terms of the original project provided the impetus for what has become a billion-dollar product platform for the company. Chesborough calls this process 'managing the false negatives' and draws attention to a number of cases. For example, in the late 1980s, scientists working for the pharmaceutical company Pfizer began testing what was then known as compound UK-92,480 for the treatment of angina. Although promising in the lab and in animal tests, the compound showed little benefit in clinical trials in humans. Despite these initial negative results the team pursued what was an interesting side effect which eventually led to UK-92,480 becoming the blockbuster drug Viagra.[36]

Importantly these examples do not emerge by accident but from an environment in which researchers have the freedom to explore alternative directions and some resources to support this. Serendipity is an important trigger for innovation, but only if the conditions exist to help it emerge. It is also a problem of 'mindset' – as we saw in Chapter 2, organizations have particular ways of framing the world and their activities within it and may dismiss ideas which do not easily fit that frame. For example, Xerox developed many technologies in its laboratories in Palo Alto which did not easily fit their image of themselves as 'the document company'. These included Ethernet (later successfully commercialized by 3Com and others) and PostScript language (taken forward by Adobe Systems). Chesborough reports that 11 of 35 rejected projects from Xerox's labs were later commercialized with the resulting businesses having a market capitalization of twice that of Xerox itself.

Communication and Connection

One last point in this section. Effective signal processing depends not only on picking up signals but also working with the messages and taking them into the organization. An important set of routines is concerned with making sure that the user perspective is communicated to all those different functions and disciplines within the organization and not simply retained as marketing information. Amongst recipes for achieving this are to rotate staff so that they spend some time out working with and listening to customers, and the introduction of the concept that 'everybody is someone's customer'. (Of course, this assumes that the firm is in the right market with the right set of customers to begin with – the risk of getting it wrong strategically will be compounded if the wrong kinds of messages are being communicated.)

An increasing number of tools and structured frameworks are now available for trying to identify, clarify, articulate and communicate 'the voice of the customer' throughout the organization. Based on the principles of quality function deployment (QFD) these tools usually take as their starting point the customer needs as expressed in the customer's own words or images and gradually and systematically decompose them into tasks for the various elements within the organization.[37,38] (QFD and related tools are described in Chapter 7.)

Of particular significance in this context is the role played by various forms of 'gatekeeper' in the organization. This concept – which goes back to the pioneering work of Thomas Allen in his studies within the aerospace industry – relates to a model of communication in which ideas flow via key individuals to those who can make use of them in developing innovation.[13] Gatekeepers are often well positioned in the informal communication networks and have the facility to act as translators and brokers of key information. Studies of the operation of such communication networks stress the informal flows of knowledge and have led to a variety of architectural changes in research laboratories and other knowledge environments. These are essentially variants around the 'village pump' idea where environments are configured to allow plenty of space for informal encounter (such as by the coffee machine or in a relaxation area) where key exchange of information can take place. Significantly the rise of distributed working and the use of virtual teams have spawned an extension to these ideas making use of advanced communications to enable such networking across geographical and organizational boundaries.

9.2 Enabling Strategy-making

Scanning the environment identifies a wide range of potential targets for innovation and effectively answers the question, 'What could we do?' But even the best-resourced organization will need to balance this with some difficult choices about *which* options it will explore – and which it will leave aside. This process should not simply be about responding to what competitors do or customers ask for in the marketplace. Nor should it simply be a case of following the latest technological fashion. Successful innovation strategy requires understanding the key parameters of the competitive game (markets, competitors, external forces, etc.) and also the role which technological knowledge can

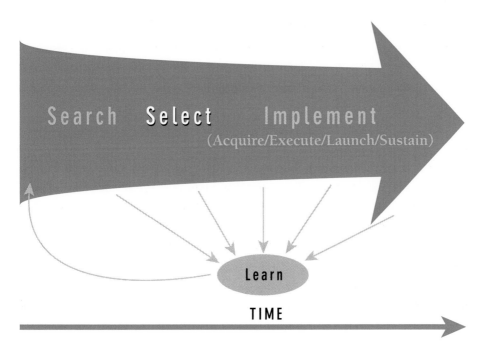

FIGURE 9.3

play as a resource in this game. How can it be accumulated and shared, how can it be deployed in new products/services and processes, how can complementary knowledge be acquired or brought to bear, etc. Such questions are as much about the management of the learning process within the firm as about investments or acquisitions – and building effective routines for supporting this process is critical to success.

Building a strategic framework to guide selection of possible innovation projects is not easy – as we saw in Part II. In a complex and uncertain world it is a nonsense to think that we can make detailed plans ahead of the game and then follow them through in systematic fashion. Life – and certainly organizational life – isn't like that; as John Lennon said, it's what happens when you're busy making other plans!

Equally organizations cannot afford to innovate at random – they need some kind of framework which articulates how they think innovation can help them survive and grow and they need to be able to allocate scarce resources to a portfolio of innovation projects based on this view. It should be flexible enough to help monitor and adapt projects over time as ideas move towards more concrete innovations – and rigid enough to justify continuation or termination as uncertainties and risky guesswork becomes replaced by actual knowledge.

Although developing such a framework is complex we can identify a number of key routines which organizations use to create and deploy such frameworks. These help provide answers to three key questions:

- Strategic analysis – what, realistically, could we do?
- Strategic choice – what are we going to do (and in choosing to commit our resources to that, what will we leave out?)
- Strategic monitoring – over time reviewing to check is this still what we want to do?

Routines to Help Strategic Analysis

Research has repeatedly shown that organizations which simply innovate on impulse are poor performers. For example, a number of studies cite firms which have adopted expensive and complex innovations to upgrade their processes but which have failed to obtain competitive advantage from process innovation.[39,40] By contrast, those which understand the overall business, including their technological competence and their desired development trajectory are more likely to succeed.[41] In similar fashion, studies of product/service innovation regularly point to lack of strategic underpinning as a key problem.[42,43] For this reason many organizations take time – often off-site and away from the day-to-day pressures of their 'normal' operations – to reflect and develop a shared strategic framework for innovation.

The underlying question this framework has to answer is about balancing fit with business strategy – does the innovation we are considering help us reach the strategic goals we have set ourselves (for growth, market share, profit margin, etc.)? – with the underlying competencies – do we know enough about this to pull it off (or if not do we have a clear idea of how we would get hold of and integrate such knowledge)? Much can be gained through taking a systematic approach to answering these questions – a typical approach might be to carry out some form of competitive analysis which looks at the positioning of the organization in terms of its environment and the key forces acting upon competition. Within this picture questions can then be asked about how a proposed innovation might help shift the competitive positioning favourably – by lowering or raising entry barriers, by introducing substitutes to rewrite the rules of the game, etc. A wide range of tools is available to help this process and some are described in Part II of the book.

Many structured methodologies exist to help organizations work through these questions and these are often used to help smaller and less experienced players build management capability. Examples include the SWORD approach to help SMEs find and develop appropriate new product opportunities or the 'Making IT Pay' framework offered by the UK DTI to help firms make strategic process innovation decisions.[44] Increasing emphasis is being placed on the role of intermediaries – innovation consultants and advisors – who can provide a degree of assistance in thinking through innovation strategy – and a number of regional and national government support programmes include this element. Examples include the IRAP programme (developed in Canada but widely used by other countries such as Thailand), the European Union's MINT programme, the TEKES counselling scheme in Finland, the Manufacturing Advisory Service in the UK (modelled in part on the US Manufacturing Extension Service in the USA) and the AMT programme in Ireland.[45–47]

In carrying out such a systematic analysis it is important to build on multiple perspectives. Reviews can take an 'outside in' approach, using tools for competitor and market analysis, or they can adopt an 'inside out' model, looking for ways of deploying competencies. They can build on explorations of the future such as the scenarios described earlier in this chapter, and they can make use of techniques like 'technology road-mapping' to help identify courses of action which will deliver broad strategic objectives.[48] But in the process of carrying out such reviews it is critical to remember that strategy is not an exact science so much as a process of building shared perspectives and developing a framework within which risky decisions can be located.

It is also important not to neglect the need to communicate and share this strategic analysis. Unless people within the organization understand and commit to the analysis it will be hard for them to use it to frame their actions. The issue of strategy *deployment* – communicating and enabling people to use the framework – is essential if the organization is to avoid the risk of having 'know-how' but not 'know why' in its innovation process. Deployment of this kind comes to the fore in the case of focused incremental improvement activities common to implementations of the 'lean' philosophy or of *kaizen*. In principle it is possible to mobilize most of the people in an organization to contribute their ideas and creativity towards continuous improvement, but in practice this often fails. A key issue is the presence – or absence – of some strategic focus within which they can locate their multiple small-scale innovation activities. This requires two key enablers – the creation of a clear and coherent strategy for the business and the deployment of it through a cascade process which builds understanding and ownership of the goals and sub-goals.[35]

This is a characteristic feature of many Japanese *kaizen* systems and may help explain why there is such a strong 'track record' of strategic gains through continuous improvement. In such plants overall business strategy is broken down into focused three-year mid-term plans (MTPs); typically the plan is given a slogan or motto to help identify it. This forms the basis of banners and other illustrations, but its real effect is to provide a backdrop against which efforts over the next three years can be focused. The MTP is specified not just in vague terms but with specific and measurable objectives – often described as pillars. These are, in turn, decomposed into manageable projects which have clear targets and measurable achievement milestones, and it is to these that workplace innovation activities are systematically applied.

Policy deployment of this kind requires suitable tools and techniques and examples include *hoshin* (participative) planning, how–why charts, 'bowling charts' and briefing groups. Chapter 11 picks up this theme in more detail.

Routines to Help Strategic Choice

Even the smallest enterprise is likely to have a number of innovation activities running at any moment. It may concentrate most of its resources on its one major product/ service offering or new process, but alongside this there will be a host of incremental improvements and minor change projects which also consume resources and require monitoring. For giant organizations like P&G or 3M, the range of products is somewhat wider – in 3M's case around 60 000. With pressures on increasing growth through innovation come challenges like 30% of sales to come from products introduced during the past three years – implying a steady and fast-flowing stream of new product/service ideas running through, supported by other streams around process and position innovation. Even project-oriented organizations whose main task might be the construction of a new bridge or office block will have a range of subsidiary innovation projects running at the same time.

As we have seen, the innovation process has a funnel shape with convergence from a wide mouth of possibilities into a much smaller section which represents those projects to which resources will be committed. This poses the question of *which* projects and the subsidiary one of ensuring a balance between risk, reward, novelty, experience and many other elements of uncertainty. The challenge of building a portfolio is as much an issue in non-commercial organizations – for example, should a hospital commit to a new theatre, a new scanner, a new support organization around integrated patient care, or a new sterilization method? No organization can do everything so it must make choices and try to create a broad portfolio which helps with both the 'do what we do better' and the 'do different' agenda.

So what routines can organizations deploy to help them build and manage a portfolio?

Portfolio Management Approaches

There are a variety of approaches which have developed to deal with the question of what is broadly termed 'portfolio management'. These range from simple judgments about risk and reward to complex quantitative tools based on probability theory.[42,49,50] But the underlying purpose is the same – to provide a coherent basis on which to judge which projects should be undertaken, and to ensure a good balance across the port-

TABLE 9.1 Problems arising from poor portfolio management

Without portfolio management there may be . . .	Impacts
No limit to projects taken on	Resources spread too thinly
Reluctance to kill-off or 'de-select' projects	Resource starvation and impacts on time and cost – overruns
Lack of strategic focus in project mix	High failure rates, or success of unimportant projects and opportunity cost against more important projects
Weak or ambiguous selection criteria	Projects find their way into the mix because of politics or emotion or other factors – downstream failure rates high and resource diversion from other projects
Weak decision criteria	Too many 'average' projects selected, little impact downstream in market

folio of risk and potential reward. Failure to make such judgments can lead to a number of problem issues, as Table 9.1 indicates.

In general we can identify three approaches to this problem of building a strategic portfolio – benefit measurement techniques, economic models and portfolio models. Benefit measurement approaches are usually based on relatively simple subjective judgments – for example, checklists which ask whether certain criteria are met or not. More advanced versions attempt some kind of scoring or weighting so that projects can be compared in terms of their overall attractiveness. The main weakness here is that they consider each project in relative isolation.

Economic models attempt to put some financial or other quantitative data into the equation – for example, by calculating a payback time or discounted cash flow arising from the project. Once again these suffer from only treating single projects rather than reviewing a bundle, and they are also heavily dependent on the availability of good financial data – not always the case at the outset of a risky project. The third group – portfolio methods – try to deal with the issue of reviewing across a set of projects and look for balance. A typical example is to construct some form of matrix measuring risk vs. reward – for example, on a 'costs of doing the project' vs. expected returns.

Rather than reviewing projects just on these two criteria it is possible to construct multiple charts to develop an overall picture – for example, comparing the relative familiarity of the market or technology – this would highlight the balance between projects which are in unexplored territory as opposed to those in familiar technical or market areas (and thus with a lower risk). Other possible axes include ease of entry vs. market attractiveness (size or growth rate), the competitive position of the organization

EXAMPLE: FRUIT OF THE LOOM

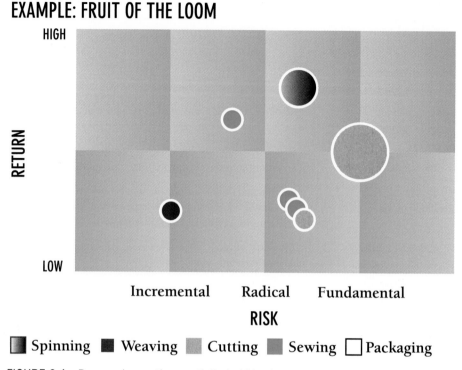

FIGURE 9.4   Process innovation portfolio bubble chart

in the project area vs. the attractiveness of the market, or the expected time to reach the market vs. the attractiveness of the market.

A useful variant on this set of portfolio methods is the 'bubble chart' in which the different projects are plotted but represented by 'bubbles' – circles whose diameter varies with the size of the project (for example, in terms of costs). This approach gives a quick visual overview of the balance of different sized projects against risk and reward criteria. Figure 9.4 shows an example (and see Box 9.5).

However, it is important to recognize that even advanced and powerful screening tools will only work if the corporate will is present to implement the recommended decisions; for example, Cooper and Kleinschmidt found that the majority of firms studied (885) performed poorly at this stage, and often failed to kill off weak concepts.[41]

Building a Business Case

Whilst strategic *selection* involves building a portfolio along the lines outlined above, the process can be influenced by a number of key routines. In particular the development and presentation of a persuasive business case is important and much can be done with tools and techniques to explore and elaborate the core concept. The purpose

---

BOX 9.5

PORTFOLIO MANAGEMENT OF PROCESS INNOVATION IN FRUIT OF THE LOOM

The clothing manufacturer Fruit of the Loom reviewed its worldwide process innovation activities using a portfolio framework to help provide a clearer overview and develop focus. It used simple categories:

- 'incremental' – essentially continuous improvement projects;
- 'radical' – using the same basic technology but with more advanced implementation; and
- 'fundamental' – using different technology – for example, laser cutting instead of mechanical.

Plotting on to a simple, colour-coded bubble chart enabled a quick and easily communicable overview of their strategic innovation portfolio in this aspect of innovation.

Source: Oke, personal communication, 2003.

---

here is to move an outline idea to something with clearer shape and form, on which decisions about resource commitments can be made. This concept can also be tested in the marketplace, explored within design, development and manufacturing, compared with competing offerings, etc. It can be used to provide the vision for the development team; the clearer the vision, the more focused the development activity can be. In the case of process innovation the concept can be tested on the 'internal market' – those users likely to be affected by the change who need to 'buy-in' to it in order for the innovation to succeed. Such investment (what Cooper calls 'buying a look') is usually justified in terms of avoiding problems at a later stage, and in helping refine the specification.[52]

Several techniques are available to support this process – for example, 'product-mapping' and 'focus groups' – and concept testing can be applied to various aspects of the overall product.[53] One area of current concern, for example, is the 'greening' of existing product ranges, and many firms are actively involved in exploring alternative and complementary concepts which stress an environmentally friendly dimension.

Again similar routines can be applied to process innovation. Pilot plant trials are often used in developing and scaling up processes whilst there is increasing use of advanced simulation technology to explore various dimensions with key users. Process-mapping is a widely used technique in ensuring effective design, especially of cross-functional processes where the workflow is not always clearly understood.[54]

Tools for helping here include simulation and prototyping – for example, in introducing new production management software a common practice is to 'walk through' the operation of core processes using computer and organizational simulation. Major developments in recent years have expanded the range of tools available for this exploration in ways which allow much higher levels of experimentation without incurring time or cost penalties. Gann, Salter and Dodgson use the phrase 'think, play, do' to describe an innovation process which, under intensifying pressure to improve efficiency and effectiveness, has adopted a wide range of powerful tools to enable an extended 'play' phase and to postpone final commitment until very late in the process. Examples of such tools include advanced computer modelling which allows for simulation and large-scale experiments, rapid prototyping which offers physical representations of form and substance, and simulation techniques which allow the workings of different options to be explored.[55]

Building Coalitions

Many of the problems in product innovation arise from the multifunctional nature of development and the lack of a shared perspective on the product being developed and/or the marketplace into which it will be introduced. A common problem is that 'X wasn't consulted, otherwise they would have told you that you can't do that . . .' This places a premium on involving all groups at the earliest possible – i.e. the concept definition/product specification – stage. Several structured approaches now exist for managing this, including quality function deployment and functional-mapping.[56] For example, reviewing the possibilities for modular design, for using parts common to other products in a range, for value engineering of key components and for using different assembly techniques offer powerful ways of cutting costs and avoiding delays and problems in development.

The availability of simulation technology, especially computer-aided design, has helped facilitate this kind of early discussion and refinement of the concept. In process innovation the early involvement of key users and the incorporation of their perspectives are strongly associated with improved overall performance and also with acceptability of the process in operation. This methodology has had a strong influence on, for example, the implementation of major integrated computer systems which by definition cut across functional boundaries.[57,58]

Extending this idea of early involvement in concept development, an increasingly important routine is to bring suppliers of components and sub-systems into the discussion. Their specialist expertise can often provide unexpected ways of saving costs and time in the subsequent development and production process. Increasingly, product development is being treated as a co-operative activity in which networks of players, each with a particular knowledge set, are co-ordinated towards a shared objective. Examples include automotive components, aerospace and electronic capital equipment, all of who make growing use of formal supplier involvement programmes.[59,60]

In process innovation this has also proved to be an important routine – both in order to secure acceptance (where early involvement of users is now seen as critical) and also in obtaining improved quality process design.[61] Interaction with outsiders also needs to take account of external regulatory frameworks – for example in product standards, environmental controls, safety legislation. Concept testing (see Box 9.6) can be helped by close involvement with and participation in organizations which have responsibilities in these area.

Routines to Help Strategic Monitoring

Innovation is about uncertainty and its gradual reduction through investment of resources into finding out – more research, development of concepts, testing and feedback, etc. A consequence of this is that decisions should not be seen as irreversible –

BOX 9.6

CONCEPT TESTING OF THE POWERBOOK

An example of concept testing in the marketplace is the development of the Apple Powerbook, widely acknowledged to be one of the most successful and attractive designs in a highly competitive marketplace (estimates suggest there were nearly 400 different machines to choose from). Apple was a comparative latecomer to the market and its first portable computer was a notable flop; the Powerbook was originated in 1990 and launched in late 1991. Its success was partly due to using a product-mapping approach, which focuses less on the product than on how people interact with it. Studies compared the Powerbook with major competitors along 159 dimensions, from opening the box to using it on a tight aeroplane seat. Amongst ideas which emerged from this mapping were the positioning of the trackball in the centre (to suit left- and right-handed users), the design of the hinge to permit different viewing angles, retractable feet to enable a sloping working position on a desk, positioning of the disc drive so that aeroplane seats would not restrict access and strengthened ribs to prevent scuffing and damage. The concept was further tested by a panel of 68 users in six different studies.

Source: *Business Week*, 20/4/1992, pp. 112–113, cited in Gupta.[53]

projects may appear attractive at the outset but as development work takes place there may be unexpected problems, delays or developments. So there is a need to maintain the strategic review throughout the life of a project, monitoring against the original criteria but taking into account the way in which the project and its wider environment changes. Strategic monitoring of this kind is an essential part of the project implementation phase and we will look in detail at routines for maintaining this perspective in that section. Most of these build upon variants of a 'stage gate' project monitoring system which we will discuss shortly.

9.3 Enabling Effective Knowledge Acquisition

This phase involves combining new and existing knowledge (available within and outside the organization) to offer a solution to the problem of innovation. It involves both generation of technological and market knowledge (via research carried out within and outside the organization) and technology transfer (between internal sources or from external sources).

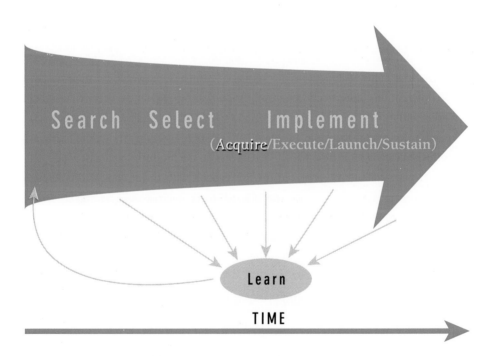

FIGURE 9.5

The challenge in research of this kind is not simply one of putting resources into the system; it is how those resources are used. Effective management of R&D requires a number of organizational routines, including clear strategic direction, effective communication and 'buy-in' to that direction, and integration of effort across different groups. Part II of this book discusses such R&D management challenges in some detail.

But not all firms can afford to invest in R&D; for many smaller firms the challenge is to find ways of using technology generated by others or to complement internally generated core technologies with a wider set drawn from outside (see Chapter 8). Equally, even large organizations with billion-dollar research budgets are increasingly recognizing the need to look outside and build connections in their innovation systems. This places emphasis on the strategy system discussed above – the need to know which to carry out where and the need for a framework to guide policy in this area. Firms can survive even with no in-house capability to generate technology – but to do so they need to have a well-developed network of external sources which can supply it, and the ability to put that externally acquired technology to effective use.

It also requires abilities in finding, selecting and transferring technology in from outside the firm. This is rarely a simple shopping transaction although it is often treated as such; it involves abilities like those listed in Table 9.2.

There are many cautionary tales of technology transfer in which there is a failure of one or more of these abilities leading to ineffective transfer. For example, there is the problem of dependency familiar to many small firms and to those in developing countries with less strong technological traditions. Here what is transferred is the basic product or process technology, but the surrounding package of knowledge about how to use and develop it, and how to internalize it, is missing, so that the firm continues to depend upon the supplier.[62] Negotiation of full technology transfer, including this intangible element, is becoming an important feature of many licensing deals.[63]

TABLE 9.2 Key abilities in technology transfer

Ability	Why?
Building and maintaining a network of technology sources	To ensure a wide range of choice and availability, rather than being forced to take inappropriate solutions
Selecting	To ensure a good fit between internal needs and external offer
Negotiating	To ensure that what is transferred includes the knowledge and experience surrounding the technology and not simply the hardware or licence
Implementing	To ensure the process of transfer is effectively managed
Learning	To ensure that once transferred the development and internalization of the technology takes place

Another common problem with technology transfer is the lack of selection; inexperienced buyers may take on technology which is not suited to their particular needs. For example, a firm needing to improve its manufacturing flexibility may have heard of the concept of a flexible manufacturing system (a combination of computers, machine tools and other equipment). Such systems may offer flexibility under certain conditions – but there are many other alternatives. The ability to make informed choices across a range of technological possibilities and in the face of strong sales pressure is important if the firm is to avoid acquiring what is effectively 'technological jewellery' – a glamorous, high-tech fashion accessory which does not have any value beyond the cosmetic.

The rapid pace of technological change means that firms are increasingly being forced to look at some combination of internal generation and external acquisition. Whether it is product or process technology, the emphasis is increasingly on collaboration and on working with outside sources, whether in universities, research institutes or other firms. As Dodgson points out, these are not always successful and firms are being forced to learn new skills to ensure effective collaboration.[64]

The picture is thus becoming one in which it is not necessary to have all the technological resources in-house but rather one in which the firm knows how, where and when to obtain them from external and complementary sources. This model of the 'extended firm' can be seen in a number of industries where there is close collaboration between suppliers and users of technology – for example, in the automobile or aerospace sectors.[65] The capability to manage such networks can itself be a source of competitive advantage; if the firm can build up a successful network of external resources complementary to its own this may be just as effective as having all the resources in-house. However, a key part of this capability is knowing which elements to outsource and which to retain in-house; the concept of the 'virtual' organization appears to involve considerable risk (see Part II).

A number of tools have been developed to help with strategic decision-making around this issue. Typical of these are those which make some classification of technologies in terms of their open availability and the ease with which they can be protected and deployed to strategic advantage. For example, the consultancy Arthur D. Little uses a matrix which groups technological knowledge into four key groups – base, key, emerging and pacing.[50]

- Base technologies represent those on which product/service innovations are based and which are vital to the business. However, they are also widely known about and deployed by competitors and offer little potential competitive advantage.
- Key technologies represent those which form the core of current products/services or processes and which have a high competitive impact – they are strategically important to the organization and may well be protectable through patent or other form.

- Pacing technologies are those which are at the leading edge of the current competitive game and may be under experimentation by competitors – they have high but as yet unfulfilled competitive potential.
- Emerging technologies are those which are at the technological frontier, still under development and whose impact is promising but not yet clear.

Making this distinction helps identify a strategic for acquisition based on the degree of potential impact plus the importance to the enterprise plus the protectability of the knowledge. For base technologies it may make sense to source outside whereas for key technologies an in-house or carefully selected strategic alliance may make more sense in order to preserve the potential competitive advantage. Emerging technologies may be best served by a watching strategy, perhaps through some pilot project links with universities or technological institutes.

Models of this can be refined – for example, by adding to the matrix information about different markets and their rate of growth or decline. A fast-growing new market may require extensive investment in the pacing technology in order to be able to build on the opportunities being created whereas a mature or declining market may be better served by a strategy which uses base technology to help preserve a position but at low cost.

It is also important to recognize that organizations differ widely in what has been termed their 'absorptive capacity' – that is, their ability to take on and make effective use of new knowledge. Cohen and Levinthal demonstrated that absorptive capacity of this kind is partly linked to the ability to generate new knowledge – in other words, to use research results effectively organizations need to learn the basic skills of actually doing research.[66] Figure 9.6 draws on work by Arnold et al. (1998) which differentiates organizations along these lines. Selection of the right approach to technology transfer depends on where firms are on this model. An important element in innovation support policy in many countries and regions has been the recognition of this problem amongst smaller firms and the development of various kinds of consulting assistance to help firms move up in their capability.[67]

Technological knowledge can come from a variety of sources, as Table 9.3 shows. Each carries with it both strengths and weaknesses; the key management task is to ensure an appropriate match between the sources selected and the context of the firm in terms of its resources and its absorptive capacity. Box 9.7 gives some examples of this process at work.

There are also important implications here for those organizations concerned with supplying technology. Successful transfer of technology will only take place if these factors are taken into account – simply offering technology is not likely to guarantee successful take-up. Instead suppliers are increasingly recognizing the need to offer a

TABLE 9.3 Different mechanisms for acquisition of technology

Mechanism	Strengths	Weaknesses
Mobilizing tacit knowledge	Internal, highly specific knowledge Hard to copy	Hard to mobilize Needs processes to articulate and capture[113]
In-house formal R&D	Strategically directed Under full control Knowledge remains inside the firm Learning by doing	High cost and commitment Risks – no guarantee of success
In-house R&D and network links outside	As above but with less control over knowledge unless there is a clear contract on intellectual property rights	Costs and risks
Reverse engineering	Lower costs Offers insight into competitors' processes and products Knowledge can be inferred, but needs a level of skill to do so	Depends on ability to infer knowledge Knowledge may be protected anyway, e.g. in patent or copyright
Covert acquisition (industrial espionage!) plus internal R&D	Fast access to knowledge and relevance of that knowledge can be managed through internal capability	Illegal Costs of internal R&D
Covert acquisition	Fast access to knowledge	Illegal Risk of not being able to translate external knowledge to internal needs
Technology transfer and absorption	Easier access to knowledge – someone else has developed and packaged it	Costs Risk of not understanding or being able to make full use of technology

Contract R&D	Speed and focus	Costs Lack of control Lack of learning effect – someone else is carrying out the experimentation and learning process May be prohibited from further exploration and learning by terms of licence, etc.
Strategic R&D partnership	Links complementary knowledge sets Enables complex problems to be addressed	Costs Risks in partnership not working Lack of learning since technology development is carried out by other parties
Licensing	Fast access to knowledge	Costs Restricted learning – may also be prohibited by terms of licence
Purchasing	Fast access	Costs Lack of learning
Joint venture	Links complementary knowledge sets Enables complex problems to be addressed	Costs Risks in partnership not working Lack of learning since technology development is carried out by other parties
Acquisition of a company with the knowledge	Fast access to knowledge Control over knowledge	Costs May not be able to absorb knowledge

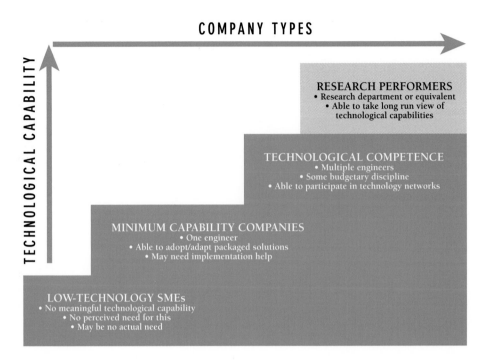

COMPANY TYPES

TECHNOLOGICAL CAPABILITY

RESEARCH PERFORMERS
• Research department or equivalent
• Able to take long run view of
  technological capabilities

TECHNOLOGICAL COMPETENCE
• Multiple engineers
• Some budgetary discipline
• Able to participate in technology networks

MINIMUM CAPABILITY COMPANIES
• One engineer
• Able to adopt/adapt packaged solutions
• May need implementation help

LOW-TECHNOLOGY SMEs
• No meaningful technological capability
• No perceived need for this
• May be no actual need

FIGURE 9.6

package of support which may range from user education (for those with low capability or absorptive capacity) through to various kinds of transfer support to help configure solutions to particular needs.

9.4 Enabling Implementation

The task of making innovation happen – moving from idea through to successful products, services or processes – is essentially one of managing what Wheelwright and Clark call 'the development funnel'. That is, a gradual process of reducing uncertainty through a series of problem-solving stages, moving through the phases of scanning and selecting and into implementation – linking market and technology-related streams along the way (Figure 9.8). The model is equally applicable in thinking about product or process change – the 'market' in process innovation may be one consisting of internal users but they also need to make decisions about whether or not to 'buy in' to the project.

At the outset anything is possible, but increasing commitment of resources during the life of the project makes it increasingly difficult to change direction. Managing

BOX 9.7
MAKING UNEXPECTED CONNECTIONS

Taking this approach can often reveal new ways of dealing with an old and apparently intractable problem. For example, the Cosworth company are a well-known producer of high-performance engines for motor racing and performance car applications. They were seeking a source of aluminium castings which were cheap enough for volume use but of high enough precision and quality for their product; having searched throughout the world they were unable to find any suitable supplier. Either they took the low price route and used some form of die-casting which often lacked the precision and accuracy, or they went along the investment casting route which added significantly to the cost. Eventually they decided to go right back to basics and design their own manufacturing process; they set up a small pilot facility and employed a team of metallurgists and engineers with the brief to come up with an alternative approach that could meet their needs. After three years' work and a very wide and systematic exploration of the problem the team came up with a process which combined conventional casting approaches with new materials (especially a high grade of sand) and other improvements. The breakthrough was, however, the use of an electromagnetic pump which forced molten metal into a shell in such a way as to eliminate the air which normally led to problems of porosity in the final product. This innovation came from well outside the foundry industry, from the nuclear power field where it had been used to circulate the liquid sodium coolant used in the fast breeder reactor programme! The results were impressive; not only did Cosworth meet their own needs, they were also able to offer the service to other users of castings and to license the process to major manufacturers such as Ford and Daimler-Benz.

Consider another example from the motor sport industry: leading race car makers are continually seeking innovation in support of enhanced performance and may take ideas, materials, technology or products from very different sectors. Indeed some have people (called 'technological antennae') whose sole responsibility is to search for new technologies that might be used. For instance, recent developments in the use of titanium components in Formula 1 engines have been significantly advanced by lessons learned about the moulding process from a company producing golf clubs.[68]

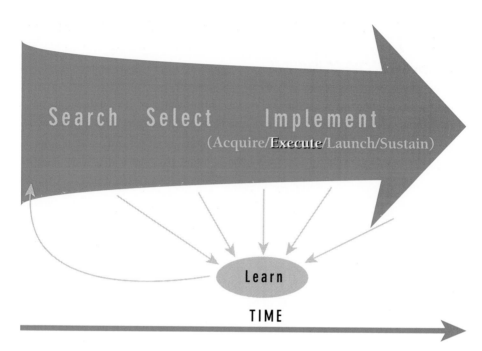

FIGURE 9.7

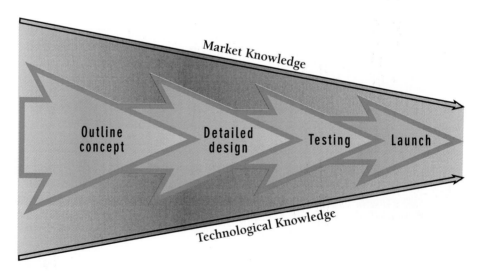

FIGURE 9.8　The development funnel

innovation is a fine balancing act, between the costs of continuing with projects which may not eventually succeed (and which represent opportunity costs in terms of other possibilities) and the danger of closing down too soon and eliminating potentially fruitful options. Taking these decisions can be done on an ad hoc basis but experience suggests some form of structured development system with clear decision points and agreed rules on which to base go/no-go decisions, is a more effective approach.[1]

Successful innovators tend to operate some form of structured, staging process, building on an approach ('stage gates') originally developed by Robert Cooper as a result of his extensive studies of product innovation.[69] This model essentially involves putting in a series of gates at key stages and reviewing the project's progress against clearly defined and accepted criteria. Only if it passes will the gate open – otherwise the project should be killed off or at least returned for further development work before proceeding. (Figure 9.9 shows an example.) Many variations (e.g. 'fuzzy gates') on this approach exist; the important point is to ensure that there is a structure in place which reviews information about both technical and market aspects of the innovation as we move from high uncertainty to high resource commitment but a clearer picture of progress.

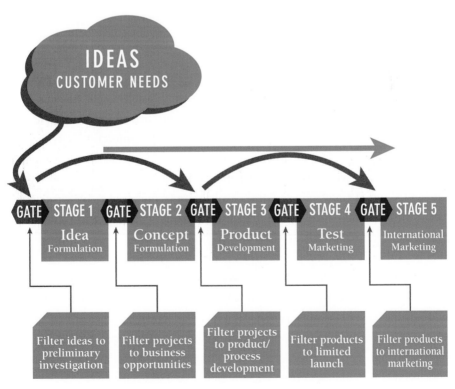

FIGURE 9.9 Accelerating idea to market – the AIM process

Models of this kind have been widely applied in different sectors, both in manufacturing and services.[70-72] We need to recognize the importance here of configuring the practice system to the particular contingencies of the organization – for example, a highly procedural system which works for a global multi-product software and hardware company like Siemens or Lucent will be far too big and complex for many small organizations. And not every project needs the same degree of scrutiny – for some there will be a need to develop parallel 'fast tracks' where monitoring is kept to a light touch to ensure speed and flow in development.

We also need to recognize that the effectiveness of any stage gate system will be limited by the extent to which it is accepted as a fair and helpful framework against which to monitor progress and continue to allocate resources. This places emphasis on some form of shared design of the system – otherwise there is a risk of lack of commitment to decisions made and/or the development of resentment at the progress of some 'pet' projects and the holding back of others.

Emerging 'Good Practice' in Implementation of Innovation Projects

Managing innovation projects is more than simply scheduling resources against time and budget. Although approaches like Gantt charts or critical path analysis can help with the process, the key differentiating feature about innovation projects is their uncertainty. Dealing with unexpected and unpredictable events and gradually bringing projects into being requires high levels of flexibility and creativity – and in particular it involves integrating knowledge sets from across organizational, functional and disciplinary boundaries. Much of the learning about effective implementation routines is based on ways of bringing this shared problem-solving approach to bear in timely and focused fashion on the demands of the project – for example, through using cross-boundary teams, through various forms of parallel or concurrent working, and through the use of simulation and other exploration technologies to anticipate downstream problems and reduce time and resource costs whilst enhancing innovation quality.

Most of these ideas are not in themselves new – for example, Lawrence and Lorsch drew attention to cross-functional team working and co-ordination mechanisms back in the 1960s, whilst Cooper reports on NASA's 'Phased Review Process' as a stage/gate model dating back to the same period.[73] But it can be argued that there is now growing consensus about their integration into a new model of 'good practice' in managing implementation. For example, in a report to the Product Development Management Association Griffin indicated that, across a sample of 380 US firms:[74]

- Progress had been made on reducing new product development (NPD) cycle times. For example, the average time to market for more innovative projects was down to 24 months, a 30% decrease from five years ago.
- The best performers had better commercialization success rates (nearly 80% vs. 53% for the rest) and had higher revenue contribution from new products (nearly 50% vs. 25%).
- Nearly 60% of firms now use some form of stage and gate process but 39% still have no formal process for managing NPD.
- More sophisticated versions of this approach are in use. For example, the best performers were more likely to use a facilitated stage and gate process or a process with flexible gates.
- The best performers were also more likely to drive NPD with a specific strategy.
- Multi-functional teams are in common use with 84% using them for more innovative projects.
- The best companies were more likely to use these team structures on less innovative projects as well.
- Two types of tools were in common use: market research tools to make development more customer driven and engineering tools to automate design, analysis and prototyping.
- The best performers do significantly more qualitative market research and are more likely to use engineering design tools.

Table 9.4 indicates some of the key dimensions of this emergent 'good practice' model. Many of these ideas developed around the creation and development of new physical products, but with the rise in the service economy attention has moved to their application in service innovation. Sectors like financial services or retailing are increasingly concerned with offering variations on their existing range and also totally new service concepts – and with this has come a realization that managing these innovations requires a systematic process.[76] Whilst the applications may differ significantly, the relevance of many of the core principles – and the enabling routines – from physical product innovation is high.[77–79]

Much of what has been learned about effective product development also has relevance for process change. Here success depends not only on good process design – does it do the job better than that which it will replace? – but also on user acceptance – will it actually be used? This depends not only on securing the buy-in of users but also on themes like user involvement (to bring their experience and tacit knowledge to bear), user needs analysis (to ensure the new process takes account of concerns) and user education. These have direct analogues to good practice in product innovation.

TABLE 9.4 Key features of emerging 'good practice' model in product/process development

Theme	Key characteristics
Systematic process for progressing new products	Stage-gate model Close monitoring and evaluation at each stage
Early involvement of all relevant functions	Bringing key perspectives into the process early enough to influence design and prepare for for downstream problems Early detection of problems leads to less rework
Overlapping/parallel working	Concurrent or simultaneous engineering to aid faster development whilst retaining cross-functional involvement
Appropriate project management structures	Choice of structure – e.g. matrix/line/project/heavyweight project management – to suit conditions and task
Cross-functional team working	Involvement of different perspectives, use of team-building approaches to ensure effective team working and develop capabilities in flexible problem-solving
Advanced support tools	Use of tools – such as CAD, rapid prototyping, computer-supported co-operative work aids (e.g. Lotus Notes) – to assist with quality and speed of development
Learning and continuous improvement	Carrying forward lessons learned – via post-project audits, etc. Development of continuous improvement culture

Source: Based on references 43 and 75.

Given their generic nature it will be useful to explore some of the enabling routines for successful implementation in a little more detail.

Early Involvement

Estimates suggest that up to 70% of a product's cost is determined at the design stage, and thus it would seem important to concentrate attention on hammering out any likely problems in manufacturing at this stage. (This represents what some writers are now calling 'learning before doing' and is a powerful source of innovation.[80]) Yet most firms spend less than 5% of their product budget on design and push instead for manufacture as soon as possible. Whilst this may reduce the time to market, it can also have a major effect upon costs. A variety of studies have consistently shown that attention to product simplification and design for manufacture can make substantial savings at later

BOX 9.8
THE TREND TOWARDS MASS CUSTOMIZATION

Whilst Henry Ford was able to build a successful business on the basis of mass production, the world today is very different. Where he had the relative luxury of unsaturated demand, the reverse is true now and markets are increasingly focused on demanding higher levels of customization. The concept of a suit of clothes being made-to-measure is familiar but today we can see the same principles applied to complex assemblies such as motor cars. Mintzberg and Lampel suggest a classification of customization along the following lines:

- Distribution – customers want a say in the delivery schedules and product packaging. For example, Amazon.com can provide individualized delivery and gift-wrapping services for standard consumer goods purchased on-line.

- Assembly – customers want a number of pre-defined options before they purchase the product. These pre-defined options could include colour, size and easily adapted technical specifications which can be configured at the assembly stage. Dell Computers can configure thousands of orders every day and offer individual customers a wide range of options such as multimedia functions, memory and hard drive size, etc. using pre-made components.

- Fabrication – customers want a product based upon a pre-defined design but with unlimited options. Levis Strauss offers a customized tailoring service at a number of retail stores in the USA. Customers are measured at stores in shopping malls and these specifications are forwarded to large-scale production facilities where the jeans are produced. BMW and other luxury car makers are now offering something similar, based on users configuring their options around a basic model platform whilst in the dealership and then having the instructions for manufacturing their choice sent to the major factories. If they wish they can visit to see their car being made or to take delivery.

- Design customization – at the limit customers can participate in the design process, shaping and configuring the ideas before they are turned into products or services. The emergence of technologies and techniques around simulation and rapid prototyping enable much more exploration before commitment but without necessarily imposing a cost penalty.

stages – and the increasing availability of tools such as simulation and rapid prototyping help with this.[81]

The idea of interplay between designers, makers, sellers and users is not new. Indeed, it provides the basis for the product improvement process which operates today in which marketing, in touch with end-users, reports back to the design function information about problems, suggested improvements, etc. Von Hippel and others have drawn attention to this as an important source of innovation. Similarly the experience of making the product on the shop floor eventually finds its way back into the drawing office and leads to suggestions for improvements.[31,82,83]

But this is often a slow and reactive process which, it can be argued, is not very efficient as a mechanism for improving design performance in an era of shortening product life cycles, and increased competition on the basis of non-price factors. By contrast an integrated approach, like that suggested above, is essentially a proactive one in which the design is refined and developed on the basis of 'real-time' interaction so that it is constantly evolving and improving. This can range from consumer products (where firms like Procter & Gamble and Unilever regularly make extensive use of user panels, through to complex products – for example, Eisenhardt and Brown report on a process of constant reconfiguration and development to help firms stay ahead in the turbulent market of semiconductor equipment design and manufacture.[84]

It is a principle which has equal relevance in the service sector. For example, an increasing number of public authorities are putting their service delivery out to management by external firms; some of these contracts last for decades and represent billions of pounds of potential earnings. Outsourcers may have recipes for improving efficiencies and reducing costs in the short term, but their ability to deliver long-term continuing improvements and maintain service quality according to tight service level agreement targets means that they have to engage in a process of shared innovation, a co-evolution of new ideas over the life of the contract.

Pisano has shown that similar patterns operate in process innovation, where such 'learning before doing' is essential in reducing development time and increasing the chances of a successful outcome.[80] Box 9.9 gives an illustration of the value of early and cross-functional involvement.

Concurrent Working

As we have seen the process of taking ideas through to successful innovation reality involves bringing together multiple knowledge sets through a series of phases staged over time. The tendency in organizing this is to run it as a simple sequential process with responsibility moving to different functional groups in the organization as the project progresses. So, for example, in developing a physical product the sequence might be:

BOX 9.9

INTEGRATED DESIGN IN THE MOTOR INDUSTRY

Lamming provides an interesting illustration of this alternative approach in the motor industry. UK companies involved in joint ventures with Japanese partners have sometimes complained about the lack of detail on drawings sent from Japan – assuming that correct product quality can only be achieved through precise (often pedantic) specification at the design and draughting stages prior to 'release' to production. The Japanese method of early (and therefore quick) release from design to production, followed by continuous development of the product by production departments has shown three major areas of benefit, however, which have gradually been realized by the UK partners. First, modification of the product in production ensures continuous improvement in process efficiency (and thus reduction in costs, improvements in quality, etc.). Second, design improvements suggested by production have a natural path for communication, and may be incorporated with a minimum of difficulty. Third, early release of the product enables the designers to concentrate on the replacement model, thus reducing development periods by eliminating 'afterthought'. The apparent 'sloppiness' of some Japanese engineering information (e.g. drawings) is in fact the sign of a more open design authority which is shared for mutual benefit (between departments) rather than jealously guarded for individual 'professional' security. The technique is sometimes referred to by Japanese as using the factory as the laboratory of the designer.[65]

Marketing → R&D → Prototype development → Production → Sales and marketing

Running this kind of process risks missing out on key knowledge input from the different groups and it also fails to take account of the uncertainties and non-linear nature of innovation. It is also by its nature a *slow* process – since things cannot move to the next stage until they have completed a previous one. But in practice there is no reason why many activities cannot be undertaken in parallel and with interaction between different functional groups working concurrently. The metaphor is often used here of moving from a relay race with the baton moving sequentially between players to a rugby team where the ball is passed between players whilst the whole team is moving forward towards the goal.

The concept of 'simultaneous engineering' or 'concurrent engineering' is now widely used as a means to ensure more rapid time to market for new product and service developments. Estimates suggest that being first into the market means a firm obtains a significant (often 50% or better) market share for that product or service, and other benefits include reduced costs (of work hours and inventories) and improved customer relationships because of better, more rapid service. Examples of firms which have made dramatic cuts in their time to market include Honda (which has cut its five-year new car development cycle by 75%), ATT (which also reduced by 50% the time taken to introduce its new cordless telephone), Hewlett-Packard (which reduced the time to develop its printers from 4.5 years to 22 months) and Xerox Corporation which made dramatic reductions in development times, cutting them from six to around three years on a range of office products. 'Cycle compression' is another phrase which is beginning to emerge, especially in the context of long development cycle products like military equipment and aircraft.[85]

The main advantage in such an approach comes from the early identification and resolution of conflicts – it stops the problems being thrown over the fence and demands a co-operative solution. Better information flow also brings in new inputs to design at a stage where they can be used to improve the design – as distinct from the traditional practice of apportioning blame to different areas because they failed to pass on information which could have helped avoid costly design faults.

As with all of these concepts, the important management skill is in knowing where and when they are appropriate. For example, 'cycle compression' approaches may not be a good idea in high-risk projects like construction, whilst (as Chapter 5 discusses) new approaches to project team structure or operation can sometimes be pushed too far and become 'core rigidities' which inhibit effectiveness.

Appropriate Project Structures

Central to effective project management is the need to get a good match between the demands of a development project and the operating structure which enables it. In a 10-year study of 2899 projects, 46% of the projects were successful but project management structure was an important determinant.[86]

We can see the importance of this if we think about the implications of managing cross-boundary teams working in concurrent fashion along the lines outlined above. This kind of approach often runs counter to the 'normal' operating mode of the organization and the project team operates in a significantly different mode. Successful teams stress mutual learning – indeed, some companies prefer to make continuing use of such teams since the investment in group development is so valuable. They need a clear goal towards which they can all work – and the leadership needs full authority to challenge what has traditionally been done. For example, in a report on the development of Nissan's Maxima car the project engineer was able to preserve his design team authority over the heads of board-level management. Such an approach avoids costly last-minute changes to placate senior management – a major difference from US counterparts where senior management often had extensive influence on the process. The benefits of this approach soon became clear: the Maxima was developed within 30 months and also won the industry's top quality award soon after its introduction. Such approaches have since become widespread in the car and other complex product industries where time compression has cut cycle times for new products down to months instead of years.

Traditionally the choice in organizing innovation projects has been between functional teams, cross-functional project teams or some form of matrix between the two. However, in recent studies other models emerge which appear correlated with success under different conditions.[56] Four main types can be identified:

- Functional structure – a traditional hierarchical structure where communication between functional areas is largely handled by function managers and according to standard and codified procedures.
- Lightweight product manager structure – again a traditional hierarchical structure but where a project manager provides an overarching co-ordinating structure to the inter-functional work.
- Heavyweight product manager structure – essentially a matrix structure led by a product (project) manager with extensive influence over the functional personnel involved but also in strategic directions of the contributing areas critical to the project. By its nature this structure carries considerable organizational authority.
- Project execution teams – a full-time project team where functional staff leave their areas to work on the project, under project leader direction.

The choice of which to use depends on the kinds of task being undertaken – for example, heavyweight teams are used when projects involve diverse and uncertain customer needs, whereas projects involving well-understood needs and technology can rely on more traditional approaches. Wheelwright and Clark[56] classify projects into a number of different types, including:

- Derivative projects – involving small changes to existing products or systems.
- Breakthrough projects – those which create new markets or products and require significant resources and a strategic view.
- Platform projects – projects which involve significant incremental improvements but are still linked to same basic platform.
- R&D projects – future-oriented, speculative but exploring where the company might be in five years or more.
- Alliances – cross-company projects, designed to share costs and risks, but also posing problems of co-operation and co-ordination.

Each of these is likely to require different combinations of project structure and team; once again, effective innovation management is strongly linked to understanding the requirements of particular situations and configuring the project accordingly.

Associated with these different structures are different roles for team members and particularly for project managers. For example, the 'heavyweight project manager' has to play several different roles, which include extensive interpreting and communication between functions and players. Similarly, team members have multiple responsibilities.

This implies the need for considerable efforts at team-building and development – for example, to equip the team with the skills to explore problems, to resolve the inevitable conflicts which will emerge during the project, to manage relationships inside and outside the project. Chapter 11 discusses this theme in greater detail.

Team Working

One of the most powerful resources for enabling rapid development is the use of cross-functional teams which contain representatives of all the disciplines involved in the innovation and which has the autonomy to progress the project. Teams of this kind are not formed simply by grouping people together; successful practice involves extensive investments in team-building, providing them with the necessary training to solve problems, to manage conflict, to interact with other parts of the organization and with outside stakeholders, etc.[87,88] Chapter 11 discusses team working in greater detail and Box 9.10 gives an example of the value of developing this resource.

Shared Project Vision

An important issue in running effective innovation projects is to make sure everyone on the team is working towards the same clear goal. Whilst this sounds obvious it is easy to lose the sense of direction or commitment in large and often dispersed teams; conversely being able to provide clear policy deployment can help focus even multiple parallel incremental innovation activities.[89] This need for communicating shared strategic direction and getting 'buy-in' to the overall objectives can be achieved in a number of ways. For example, the original 'skunk works' group within Lockheed Martin had a distinct physical location and a strong sense of team identity focused on their belief in their ability to deliver on what were clear but immensely stretching targets. Similar stories relate to the Apple 'pirates' developing alternative product architectures or Kidder's account of mini-computer innovation projects within Data General.[90,91]

One important way of providing this is to involve them in the process of vision-building, evolving the product concept in the context of a clear understanding of the underlying business drivers and competitive realities. For example, when Digital began

---

BOX 9.10

THE TEAM TAURUS PROJECT

The 'Team Taurus' project at Ford illustrates the potential of this approach. At the time of its development Taurus was a new concept car, which represented a departure both in terms of product and the organization involved in its development. It was a demanding project, requiring more than 4000 newly-designed components and completely new transmission, engine and suspension systems. The core team worked hard to break down the walls which had traditionally separated their activities, and where necessary the team expanded to incorporate other perspectives and ideas – for example, extensively consulting the manufacturing workforce regarding the difficulties in manufacturing the new 'soft shapes' in body components. Whilst computer tools were extensively used, the underlying team integration was critical in resolving the many design changes and challenges. A core value of incorporating the best ideas, rather than the familiar, was developed and this placed stress on open-ended problem-solving rather than using old routines. Team roles were explicitly explored and developed; in particular the team leader became much more of a team captain/facilitator than a member of supervisory management.[92]

Significantly firms need to take care to capture the learning about team organization and management. Whilst the early experiences of developing the Taurus were positive, Walton reports a very different outcome for later generations and one which highlights a lack of effective team-working.[93]

---

to challenge the dominance of Sun Microsystems in the workstation market, part of the vision was the clear need to base operations on the Unix language rather than DEC's proprietary VMS software. This forced a recognition of the need for substantial change in all aspects of the project, from design thinking through to software and hardware design and construction. Ultimately the efforts paid off, because the particular product (the 3100) paved the way for several others in the same series, whilst at the same time creating a new capability in the way in which new products are developed.[92]

Advanced Support Tools

In parallel with development in models for organizing innovation projects there has been a significant increase in the range and power of tools available to support the process. These include:

- Computer-based tools – particularly computer-aided design and manufacturing (CAD/CAM) – permit extensive simulation and shared exploration of concepts, and also accelerate the actual development process by automating key tasks. In addition communicating information in electronic form between design and manufacturing functions significantly reduces the time required in development. Recent developments have extended the concept of computer-supported co-operative work (CSCW) to develop systems which enable communication and interaction between CAD users working at different locations but on the same project.

- Rapid prototyping technologies and approaches are another powerful technological resource in product development. These approaches rely on a variety of techniques, but typically aim to produce a physical model of a concept quickly so that it can be evaluated and explored earlier in the development cycle. For example, the use of polymer resins and computer-controlled shaping equipment can quickly move from a CAD concept to a physical replica of the idea.

- Quality function deployment (QFD) represents a powerful structured methodology for exploring and steering the interaction between different contributors in the product development process. The power of QFD is less in the data representation than in its role as a common structure over which discussion and debate between different functions can take place. It provides a common language and a systematic mechanism for exploring and resolving many of the typical issues. (QFD is discussed in detail in Chapter 7.)

- Design rules treat one of the common problems in product development, which is that design ideas which make sense at one stage pose problems for others further down the chain. For example, positioning a screw in a particular place may look fine on the drawing board or CAD screen, but may pose a major problem in the manufacturing stage, because of difficulties of access, or complexity of operation. This type of difficulty has led to a series of design methodologies governed by various rules to ensure consideration of downstream operations; these are often grouped under the heading of 'design for manufacture'.[94] Amongst considerations are design for ease of assembly, design for speed, design for simplification, design for minimum adjustment, etc. Closely related are rules which aim to reduce cost and simplify component count; these are usually classed under the label 'value engineering' or 'value analysis'.

- Product data management (PDM) systems offer significant improvements to the ways in which specialists share and access information about complex products on which they are working. Studies have found that most engineers are able to spend only 20% of their time – only one day per workweek, on average – actually designing products. On the other hand, they spend nearly twice as much time, approximately 35%, looking for and verifying data regarding design revisions, performance calculations and drawings. Clearly the effective lost productivity here means either that products are late to the market or else they arrive on time but lacking some of the more interesting features which could have been designed in if there were more time.

Tidd and Bodley report on a review of the range of formal tools and techniques available to support the new product development process, and examine the use and usefulness of these by means of a survey of 50 projects in 25 firms. They adopted a four-stage model to examine the process of new product development – idea generation; screening and selection; development; commercialization and review – and identified the effect of project novelty on the frequency of use and perceived usefulness of a range of tools and methods. In terms of usefulness, focus groups, partnering customers and lead users and prototyping were all considered to be more effective for high-novelty projects, and segmentation least useful. Cross-functional development teams were commonplace for all types of project, but were significantly more effective for the high-novelty cases. In addition, many tools rated as useful were not commonly used, and conversely some tools in common use were considered to have low levels of usefulness.[95] Similar findings are reported by the Product Development Management Association in its extensive review of tools for innovation.[96]

9.5 Launch

In parallel with technological aspects of developing an innovation is the process of identifying, exploring and preparing the market for launch of a new product or service (see also Chapter 7). The challenge here is that we are dealing again with uncertainty – even if the product/service is technically excellent there is no guarantee that people will adopt it or continue its use over the long term. Building the 'better mousetrap which no one wants' is one of the prime reasons for failure in new product development.[97] So how can we improve the chances of successful adoption and diffusion? Part of the puzzle can be solved by taking into account the various influences on adoption behaviour, and a wide range of studies on buyer behaviour and influences on the adoption decision are available to help with this.[98] (Chapter 7 goes into more detail.) We can use this information to anticipate likely aspects of the innovation and the ways in which it is launched which will help dispose potential adopters favourably. And we can also deploy strategies like early and active user involvement to help build confidence in emerging innovations.

Enabling routines here include:

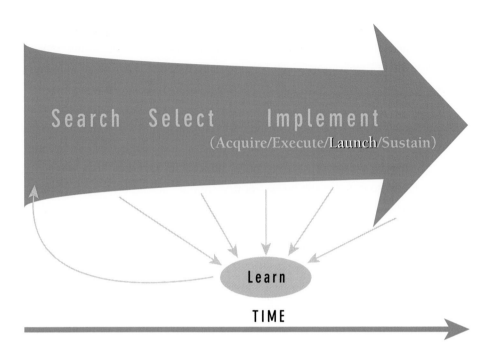

FIGURE 9.10

Customer Testing

This is essentially taking out prototypes of the product to users (or bringing the users in to test them out). It is particularly important in ensuring that the original concept still holds; for example, the Ford Edsel was one of the best-researched product concepts in the company's history. The final product failed disastrously – mainly because the market had moved on in terms of customer tastes in the time it took to develop the production model. Such tests can also be used to explore customer preferences which provide important information about pricing policy, advertising strategy, etc. For more complex products it may be necessary to allow the user an extended period of use to permit learning and familiarization with the product and how to make the best use of it.

Test Marketing

The next step in the process of market development, this involves various kinds of trying out of the marketing strategy. For example, in pre-test marketing potential customers are exposed to the product concept and undertake what is essentially a shopping simulation, using coupons instead of real money. This helps refine marketing plans and predictions; for example, about likely shares. Full test marketing involves a trial sale of the new product to a controlled group of customers; effectively it is a pilot of the full marketing launch. Such testing can reveal actual as opposed to simulated data on acceptance, concerns, adoption rates and speeds, etc. It also offers the opportunity to test different launch strategies – for example, two different regions could be used, each employing a different launch strategy. There is general acceptance of the point that test marketing is worthwhile; although it involves costs in terms of trial production, test administration and data processing it effectively 'buys a look' before the company commits itself fully to launching a new product. It is an approach which works well with products which can be cheaply developed and modified in response to test data.

One approach to test marketing is the concept of alpha, beta and gamma testing, in which different releases of the product under development are issued to controlled groups of users. This strategy allows feedback of positive and negative responses which can be built into the final version; it is simultaneously a good way of enhancing product quality and user need fit whilst also pre-advertising/'warming up' the marketplace. This approach is extensively used in the area of computer software, where user feedback is an important source of product improvement and final development.

Develop a Marketing Strategy

This is essentially similar to other kinds of strategy development, involving a mixture of analysis of the target market and the relevant strengths, weaknesses, threats and opportunities. Typically such an analysis will specify in some detail the particular segments being targeted, the pattern of buyer behaviour which shapes that market, the nature and strength of competition and the overall dynamics of the marketplace. Developing the strategy will also require more detailed projections of market shares and sales, leading to a detailed financial budget. Other aspects include positioning the product against competing offerings and planning the entry, in terms of target pricing, advertising plans, etc. Essentially this involves developing the 'marketing mix' – considering product, promotion, pricing and distribution – which will form the backbone of the marketing effort.

Develop a Marketing Plan

As the name suggests this is a formal documented plan covering objectives, strategies and programmes to underpin the product launch. Once again the key theme of early involvement is relevant here; marketing planning should begin in parallel with product planning once the concept is established. The objectives should be clearly rather than vaguely defined – for example, '. . . to obtain a 20% share of the industrial drill lubricant market in two years' time . . .' – so that it becomes possible to measure progress and see whether the marketing is on course.

Develop a Support Organization

Innovation not only involves extensive interaction and integration with internal functions; it also needs linking with external channels of distribution and promotion. Whether these are owned by the firm or external, the principle of early involvement is still critical. A number of new product launches have failed because levels of service or support have been lacking; this argues for careful selection and involvement of these.[99]

Launching into an Internal Market – Change Management

Within process innovation the question of launch takes on particular significance. Not only must the project be managed along what may be a complex development funnel (as with product innovation) but the market into which it will be launched is an internal one, often involving the same people. This raises the question of change management.

As Voss points out, implementation is often neglected in innovation studies yet it is often the site of the most serious difficulties.[100] Part of the implementation problem arises from a lack of attention at the strategic planning stage to dimensions of the proposed change – for example, ensuring suitable development of both infrastructure and structural elements. Since many process innovations represent major changes in 'the way we do things round here' the question of managing cultural change and overcoming resistance to innovation needs to be addressed, and planning for such organizational development is an important element in manufacturing strategy formulation. To some extent implementation difficulties can be reduced through involving those likely to be affected by the change in some of the strategy formulation and debate.

One way of looking at implementation is to see it involving continuing cycles of mutual adaptation of both technology and organization; this has important implications for reviewing the way in which such change is managed.[101] Trying to introduce changes in the ways in which products are made or services delivered is a co-operative effort which requires inputs of knowledge and expertise from across the organization. Extensive work on successful implementation of IT systems – probably the most widely used class of process innovations over the past 40 years – repeatedly stresses the need to involve users in order to get better designed systems and to get commitment to making them work.[102–104]

Managing organizational change is problematic largely because human beings are programmed to resist or at least be cautious about change. Change is often perceived as threatening, painful, disruptive and sometimes dangerous; resistance has both cognitive and emotional components. Some of this resistance can be dealt with in formal ways – by training, communicating information, etc. – but emotional responses – anxieties about loss of status, power, influence, fear of risk-taking, etc. – cannot be directly. Instead it is necessary to create a climate in which these concerns can be surfaced, issues and conflicts can be addressed and in which individuals can find reassurance.

Organizational development (OD) is a generic term to describe the set of practices and methodologies which have evolved around introducing planned organizational change.[105] OD involves diagnosis and intervention; this is often effected through internal or external change agents who act as catalysts or facilitators of change. There is also a wide variety of tools and techniques for dealing with particular problems – for example, conflict resolution, training in creativity, participation and building commitment. One of the challenges for innovation management in the future is likely to be making more widespread and effective use of OD approaches to manage the continuing development and adaptation of organizations.

Amongst routines associated with effective change management are:

Establish a clear change management strategy at top level (a process which will itself involve considerable challenge and conflict in order to get real agreement and commitment to a common set of goals). Once this has been done, the next stage is to communicate this shared vision to the rest of the organization – essentially this will involve a cascade process down through the organization during which opportunities are set up for others to challenge and take 'ownership' of the same shared vision.

Communication – probably the single most effective key to successful implementation but requires a major effort if it is to succeed. It must be active, not passive, open (rather than allowing information to flow on a 'need-to-know' basis), timely (in advance of change – the informal communication network will disseminate this information anyway and a slow formal system will undermine credibility), and above all, two-way in operation. Unless there are channels through which people can express their responses and ideas and voice their concerns then no amount of top-down communication will succeed in generating commitment. Other features in communication which contribute to success are the use of multiple media – videos, presentations, face-to-face sessions – in addition to traditional memos and notice boards, and the holding of communication meetings on company time – itself a clear expression of commitment to the project.

Early involvement – managers often resist the idea of participation since it appears to add considerably to the time taken to reach a decision or to get something done. But there are two important benefits to allowing participation and allowing it to take place as early as possible in the change process. The first is that without it – even if attempts have been made to consult or to inform – people will not develop a sense of 'ownership' of the project or commitment to it – and may express their lack of involvement later in various forms of resistance. And the second is that involvement and encouragement of participation can make significant improvements in the overall project design.

Create an open climate, in which individual anxieties and concerns can be expressed and the ideas and knowledge held within the organization can be used to positive effect. Once again, this involves generating a sense of 'ownership' of the project and commitment to the shared goals of the whole organization – rather than an 'us and them' climate.

Set clear targets – with major change programmes it is especially important to set clear targets for which people can aim. People need feedback about their performance and the establishment of clear milestones and goals is an important way of providing this. In addition, one of the key features in successful organizational development is to create a climate of continuous improvement in which the achievement of one goal is rewarded but is also accompanied by the setting of the next.

In their work on implementation strategies for advanced manufacturing technology (AMT), Smith and Tranfield identify two discrete phases of target setting – the 'sprint' and the 'performance ratchet'. In the former all the efforts of the organization are focused on achieving a major target – for example, reducing defects by 50% or increasing output by 30% – and everything is directed towards achieving that clear goal. Such a sprint has enormous power in harnessing a widely differing staff to a single goal and brings about the development of a new working culture. But the momentum of such a sprint cannot be maintained for ever and a second mechanism is needed to ensure that the gains made are consolidated.[106]

Invest in training. Traditionally training is often seen as a necessary evil, a cost which must be borne in order that people will be able to push the correct buttons to work a particular new piece of equipment. Successful organizational change depends on viewing training far more as an investment in developing not only specific skills but also in creating an alternative type of organization – one which understands why changes are happening and one which is capable of managing some of the behavioural processes involved in change. This requires a substantial increase in the resources devoted to training, extending them to cover broader kinds of input, much of this devoted to individual development.

BOX 9.11
MANAGING CHANGE AT AMP

AMP is a large manufacturer of components and systems for use in the automotive, medical and aerospace industries. It makes use of a focused innovation initiative called the 'Breakthrough Improvement Event' (BIE) which it runs on regular occasions. It is essentially targeted process re-engineering designed to build high involvement of its workforce in driving out inefficiencies and has been applied across businesses as diverse as manufacturing units and head offices, across its 200 facilities and 40 offices worldwide.

A cross-section of employees from every department is involved with the planning phase, and with the implementation, giving a strong sense of ownership to the changes. The process is structured into phases of data collection, analysis, planning, a final 'Event Week', where the changes are implemented, and a robust review phase at 30-day intervals. Each team is assigned a senior executive to act as coach, facilitator and sponsor.

An example of the process in action is the BIE targeted on the group's UK Head Office where the benefits emerging included an increase in throughput of 43% and of responsiveness by 59% accompanied by a reduction in paperwork of 56%, administrative errors of 34% and overall process cost of 47%. These were against challenging targets set by the company of 50% reduction in all of these areas. Some specific examples help clarify these gains; they achieved:
- Corporate forecast preparation time from five days to three days.
- Customer enquiry-to-quote time from 41 days to 27 days.
- Price approval time from 150 hours to five hours.
- Response time to key customers from 440 hours to one hour.
- Paperwork from approximately 360 000 sheets per annum to 173 000.

Initially, the senior management team set stretch targets and assigned a timescale from project initiation to the Event Week of 17 weeks. Team members from each participating department were identified by team leaders who chose the rest of their team members. Training was a key feature, partly to ensure that all team members 'bought in' to the process; the training pack developed for the BIE showed how processes can be broken down into value-adding and non-value-adding activities, how to quantify each phase and how to approach the elimination of non-value-adding activities. Measurement techniques were also covered. Team members

continues overleaf

BOX 9.11 *(continued)*

---

jointly determined what other training was required; some chose to undertake team-building training while others chose training directly related to the likely changes – for example, PC training.

Teams then carried out extensive data collection and analysis breaking each administrative process down into value- and non-value-adding activities. From this analysis a series of interventions were planned and discussed with key stakeholders to try and ensure commitment to their implementation. During the Event Week changes were implemented by the teams 'around' their colleagues who then moved onto the new systems as they were completed. On the Friday, having made provision for telephone calls to be handled by another site, the entire workforce assembled to see the teams present their results and to celebrate the success. The atmosphere was highly stimulating and provided a significant boost to morale.

A follow-up review process examined each department at 30-, 60- and 90-day intervals after this and most areas achieved or even bettered their targets during the first 30-day period.

The success of the programme came partly because it was built on core principles of organizational development learned by applying the approach elsewhere across the company over several years. Key elements of this are, according to the company:

- commitment;
- planning;
- use of an empowered steering group;
- training;
- teamwork.

Source: Based on 'Best Practice' case study of AMP, CBI/DTI 'Fit for the future', www.dti.gov.uk/bestpractice

---

9.6 Enabling Learning and Re-innovation . . .

The final stage in any innovation process should be one of review of the completed project and an attempt to capture learning from the experience. This is an optional stage and many organizations fail to carry out any kind of review, simply moving on to the next project and running the risk of repeating mistakes made in previous

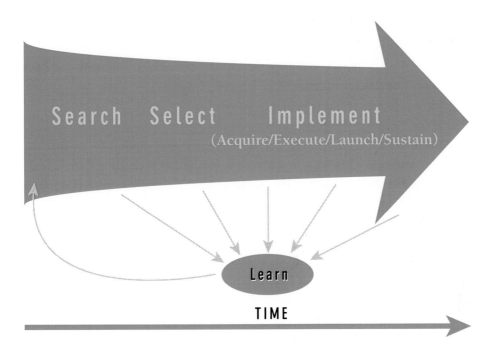

FIGURE 9.11

projects. Others do operate some form of structured review or post-project audit; however, this does not of itself guarantee learning since emphasis may be more on avoiding blame and trying to cover up mistakes.

Effective learning, both in terms of accumulating technological knowledge and knowledge about how to manage the innovation process, requires a commitment to open and informed review. Developing such learning abilities is now seen as a critical role for strategic management.[107–109] We discuss some of the mechanisms which facilitate building a 'learning organization' in Chapter 11, and in Chapter 13 we look at how such an approach can be used to foster continuous development of innovation management capability.

As we saw in Chapter 3, firms which exhibit competitive advantage – the ability to win and to do so continuously – demonstrate 'timely responsiveness and rapid product innovation, coupled with the management capability to effectively co-ordinate and redeploy internal and external competencies'.[110] They possess 'dynamic capability' and developing processes within the organization to enable this is the key. The lack of such capability can explain many failures, even amongst large and well-established organizations. For example, the:

- failure to recognize or capitalize on new ideas which conflict with an established knowledge set – the 'not invented here' problem;[111]

- problem of being too close to existing customers and meeting their needs too well – and not being able to move into new technological fields early enough;[112]
- problem of adopting new technology – following technological fashions – without an underlying strategic rationale;[40]
- problem of lack of codification of tacit knowledge.[113]

The costs of not managing learning – of lacking the dynamic capability – are high. So we need to look hard at the ways in which organizations can learn – and learn to learn in conscious and strategic fashion. This is why routines play such an important role in managing innovation – they represent the firm-specific patterns of behaviours which enable a firm to solve particular problems.[114] In other words, they embody what an organization (and the individuals within it) has learned about how to learn.

We should also recognize the problem of *unlearning*. Not only is learning to learn a matter of acquiring and reinforcing new patterns of behaviour – it is often about forgetting old ones. Letting go in this way is by no means easy, and there is a strong tendency to return to the status quo or equilibrium position – which helps account for the otherwise surprising number of existing players in an industry who find themselves upstaged by new entrants taking advantage of new technologies, emerging markets of new business models. Managing discontinuous innovation requires the capacity to cannibalize and look for ways in which other players will try and bring about 'creative destruction' of the rules of the game. Jack Welch, former CEO of General Electric, is famous for having sent out a memo to his senior managers asking them to tell him how they were planning to destroy their businesses! The intention was not, of course, to execute these plans but rather to use the challenge as a way of focusing on the need to be prepared to let go and rethink – to unlearn.[115]

All of this argues strongly that firms should undertake some form of review of innovation projects in order to help them develop both technological and managerial capability. One way of representing the learning process which can take place in organizations is to use a simple model of a learning cycle. Here learning is seen as requiring:[116]

- Structured and challenging reflection on the process – what happened, what worked well, what went wrong, etc.
- Conceptualizing – capturing and codifying the lessons learned into frameworks and eventually procedures to build on lessons learned.
- Experimentation – the willingness to try and manage things differently next time, to see if the lessons learned are valid.
- Honest capture of experience (even if this has been a costly failure) so we have raw material on which to reflect.

Effective learning from and about innovation management depends on establishing a learning cycle around these themes. To help with the process there is a variety of tools and mechanisms, including post-project reviews, auditing and benchmarking.

Post-project reviews (PPRs) are structured approaches to capturing learning at the end of an innovation project – for example, in a project debrief. On the positive side, they work well when there is a structured framework against which to examine the project, exploring the degree to which objectives were met, the things which went well and those which could be improved, the specific learning points raised and the ways in which they can be captured and codified into procedures which will move the organization forward in terms of managing technology in future.

But such reviews depend on establishing a climate in which people can honestly and objectively explore issues which the project raises. For example, if things have gone badly the natural tendency is to cover up mistakes or try and pass the blame around. Meetings can often degenerate into critical sessions with little being captured or codified for use in future projects.

The other weakness of PPRs is that they are best suited to distinct projects – for example, developing a new product or service or implementing a new process. They are not so useful for the smaller-scale, regular incremental innovation which is often the core of day-to-day improvement activity. Here variations on the standard operating procedures approach can be powerful ways of capturing learning – particularly in translating it from tacit and experiential domains to more codified forms for use by others.[117] They can be simple – for example, in many Japanese plants working on 'total productive maintenance' programmes, operators are encouraged to document the operating sequence for their machinery. This is usually a step-by-step guide, often illustrated with photographs and containing information about 'know why' as well as 'know-how'. This information is usually contained on a single sheet of paper and displayed next to the machine. It is constantly being revised as a result of continuous improvement activities, but it represents the formalization of all the little tricks and ideas which the operators have come up with to make that particular step in the process more effective.

On a larger scale, capturing knowledge into procedures also provides a structured framework within which to operate more effectively. Increasingly organizations are being required by outside agencies and customers to document their processes and how they are managed, controlled and improved – for example, in the quality area under ISO 9000, in the environmental area under ISO 14000 and in an increasing number of customer/supplier initiatives such as Ford's QS9000.

Once again there are strengths and weaknesses in using procedures as a way of capturing learning. On the plus side there is much value in systematically trying to reflect on and capture knowledge derived from experience – it is the essence of the learning cycle. But it only works if there is commitment to learning and a belief in the value of the procedures and their subsequent use. Otherwise the organization simply creates procedures which people know about but do not always observe or use. There is also the risk that, having established procedures, the organization then becomes resistant to changing them – in other words, it blocks out further learning opportunities

Benchmarking is the general name given to a range of techniques which involve comparisons between two examples of the same process so as to provide opportunities for learning.[118–120] Benchmarking can, for example, be used to compare how different companies manage the product development processes; where one is faster than the other there are learning opportunities in trying to understand how they achieve this. Benchmarking works in two ways to facilitate learning. First, it provides a powerful motivator since comparison often highlights gaps which – if they are not closed – might well lead to problems in competitiveness later. And second, it provides a structured way of looking at new concepts and ideas. It can take several forms:

- between similar activities within the same organization;
- between similar activities in different divisions of a large organization;
- between similar activities in different firms within a sector;
- between similar activities in different firms and sectors.

The last group is often the most challenging since it brings completely new perspectives. By looking at, for example, how a supermarket manages its supply chain a manufacturer can gain new insights into logistics. By looking at how an engineering shop can rapidly set up and change over between different products can help a hospital use its expensive operating theatres more effectively.

Benchmarking offers a structured methodology for learning and is widely used by external agencies who see it as a lever with which to motivate particularly smaller enterprises to learn and change.[121]

Related to benchmarking, auditing offers another structured way of reflecting on the process of technological change and how it has been/is being managed. The analogy can be drawn with financial auditing where the health of the company and its various operations can be seen through auditing its books. The principle is simple; using what we know about successful and unsuccessful innovation and the conditions which bring it about, we can construct a checklist of questions to ask of the organization. We can then score its performance against some model of 'best practice' and identify where things could be improved. We can also use real examples and factual data to support the scores we allocate. (An example of a simple innovation audit is given in Chapter 13.)

9.7 Beyond the Steady State – Making it Happen under Discontinuous Conditions

Most of the time the routines we have described work well as a basis for organizing and managing innovation projects which in different ways represent 'doing what we do but better'. This is often the bulk of an organization's innovation activity and paying attention to these factors demonstrably helps create the conditions for success. But – as we saw in Chapter 1 – there are occasions when discontinuous conditions emerge and where existing routines do not always help deal with the new challenge. Indeed, on occasions doing more of the old routines may be positively the wrong thing to do.

As we have seen, organizations that manage steady-state innovation well work closely with customers and suppliers, they make use of sophisticated resource allocation mechanisms to select a strategically relevant portfolio of projects, they use advanced project and risk management approaches in developing new products and processes, and so on. These routines are the product of well-developed adaptive learning processes which give the firm a strong position in managing innovation under steady-state conditions – but they also act as a set of barriers to picking up signals about, and effectively responding to, innovation threats and opportunities associated with discontinuous shifts. Christensen's work on 'the innovator's dilemma' highlights this problem of a virtuous circle which operates in a successful firm and its surrounding value network, and describes in detail the ways in which their markets become disrupted by new entrants.[122]

The problem is further compounded by the networks of relationships the firm has with other firms. Typically, much of the basis of innovation lies at a system level involving networks of suppliers and partners configuring knowledge and other resources to create a new offering. Discontinuous innovation is often problematic because it may involve building and working with a significantly different set of partners than those the firm is accustomed to working with. Whereas 'strong ties' – close and consistent relationships with regular partners in a network – may be important in enabling a steady stream of continuous improvement innovations, evidence suggests that where firms are seeking to do something different they need to exploit much weaker ties across a very different population in order to gain access to new ideas and different sources of knowledge and expertise.[123]

Working 'out of the box' requires a new set of approaches to organizing and managing innovation – for example, how the firm searches for weak signals about potential discontinuities, how it makes strategic choices in the face of high uncertainty, how it resources projects which lie far outside the mainstream of its innovation operations, etc. Established and well-proven routines for 'steady-state' conditions may break down here – for example, an effective 'stage gate' system would find it difficult to deal with high-risk project proposals which lie at the fringes of the firm's envelope of experience. Developing new behaviours more appropriate to these conditions – and then embedding them into routines – requires a different kind of learning – 'generative learning'[124] or 'double loop'.[125]

All of this suggests the need to look for a complementary or parallel set of routines for dealing with discontinuous conditions. It is less easy to prescribe here – not least because, by definition, discontinuity is occasional and so the opportunities for developing routines through trial and error are limited. Nonetheless there are some emerging messages about how to deal with it and we will consider these briefly here.

Search and Scan

The challenge at the start of the innovation process is to search and scan, picking up relevant trigger signals. Most of the time there is an effective filter which channels search activities into spaces where they are likely to be fruitful in helping the organization with its innovation agenda based on 'doing what we do but better'. Under these conditions the search routines described above work well and the space within which it is carried out – the 'selection environment' – is clearly defined. But in the case of discontinuous innovation these signals may lie in unexpected places, often far from the areas covered by the 'normal' radar screen of the organization.

It's a little like the old joke about the drunk who loses his keys somewhere in the distant darkness but who is searching for them under the nearest lamppost – because there is more light there! Familiar search routines are very effective but inevitably will only reveal what the organization expects to find in such familiar territory. The challenge under discontinuous conditions is to develop greater peripheral vision.[126]

The problem is compounded by the fact that sometimes the very routines which help an organization working under steady-state conditions to pick up early on key signals may actively conflict with those designed to help it pick up signals about discontinuities. For example, we have seen that it is good innovation management practice to listen to customers and to respond to their feedback. The problem with this is that customers are often stuck in the same mindset as the firm selling to them: they can see ways of improving on the features of existing products, but they cannot imagine an entirely different solution to their needs. One of the significant themes in Christensen's research on disruptive innovation was that existing 'value networks' – systems of suppliers and customers – effectively conspire with each other so that when a new proposition is put to them involving a different bundle of innovation characteristics they reject it because it doesn't match what they were expecting as a development trajectory.[122]

Developing peripheral vision involves actively searching in unexpected places – for example, by seeking out fringe users and pre-early adopters in the population who have a higher tolerance for failure. These groups are often prepared to explore new areas and will accept that learning is a function of experimentation – an example might be the early beta test users of new software. In technology terms, sources of new signals about emerging technologies may come from technology institutes and suppliers far from the normal focus of attention of the firm. Indeed the good practice precepts around long-term close links with suppliers may be challenged; instead of seeking strategic alliances organizations may need to exploit weaker ties and go for 'strategic dalliances'.[68,127]

The problem is of course not simply about mechanisms for picking up the signals – under these conditions it is also very much about how the organization reacts to them. In some ways organizations behave in a fashion similar to individuals who exhibit what psychologists call 'cognitive dissonance' – that is, they interpret signals coming in as reinforcing what they want to believe. Even if the evidence is strongly in other directions they have the capacity to deceive themselves – in part because they have so much investment (financial, physical and emotional) – in preserving the existing model. Numerous examples exist where such cognitive blinkers stopped otherwise 'smart' organizations from reacting to external challenges as their core competencies became core rigidities.[128] The case of Polaroid's problems with the entry into digital imaging is a good case in point – see Box 5.5 in Part II.[129] In many ways the IBM story is typical – a giant organization which received plenty of early warning about the emerging challenge from decentralized network computing but failed to react fast enough and nearly lost the business as a result.[130] In similar fashion Microsoft were latecomers to the Internet party and it was only the result of a major focused reorientation – a turn of the tanker at high speed steered by Bill Gates himself – that they were able to capitalize on the opportunities posed.

That said, organizations can put in place more open-ended search routines and these include picking up and amplifying weak signals,[131] using multiple and alternative perspectives,[6] using technology antennae (for example via the Internet) to pick up early warning signals, working with fringe users,[132] and deploying future search routines of an open-ended variety (e.g. multiple alternative scenarios[133]). For example, Nokia created an 'insight and foresight' group whose role was to look for changes in the marketplace three to five years out.

There is also scope for using signal generation and processing capacity *within* the organization itself – for example, by exploring more at the periphery of firm – using subsidiaries, joint ventures, distributors as sources of innovation. An example here is Smirnoff Ice – a highly successful European brand for Diageo which originated in Australia as a local product, Stolichnya Lemon Russki, before it was picked up by the corporate marketing department as a product with global potential.[134]

Challenging the way the organization sees things – the corporate mindset – can sometimes be accomplished by bringing in external perspectives. IBM's recovery was due in no small measure to the role played by Lou Gerstner who succeeded at least in part *because* he was a newcomer to the computer industry, and was able to ask the awkward questions that insiders were oblivious to. In similar fashion when Wellington, a Canadian insurance company, embarked on a radical transformation of its business in 1990, the entire management team was brought in from other sectors.[134]

Strategic Selection

Much of the above discussion on corporate mindset creates difficulties downstream in terms of innovation decision-making. If the proposed new project doesn't fit the existing ways of seeing then it has a very poor chance of entering, never mind surviving in the strategic portfolio. The challenge is not simply one of dealing with a risk averse approach which favours doing well understood things as opposed to totally new ones. It is also about the difficulties of predicting where and how novel ideas might move forward. As Christensen points out one of the common problems in the many cases of disruptive innovation which he looked at was that at the outset the new prospects did not offer much apparent market potential. They tended to represent innovation opportunities which were technically risky and which had limited market potential – it was only later, as they grew to multi-million or even -billion dollar businesses that the scale of missed opportunity became apparent. Most well-managed resource allocation systems would tend to favour less radical bets which offered better returns in the short term.[112]

Sometimes strong leadership is critical to carrying the company forward into new territory. When Intel was facing strong competition from Far Eastern producers in the memory chip market Groves and Noyce reported on the need to 'think the unthinkable', i.e. get out of memory production (the business on which Intel had grown up) and to contemplate moving into other product niches. They trace their subsequent success to the point where they found themselves 'entering the void' and creating a new vision for the business.[135]

Routines to help do it include mechanisms for legitimating challenge to the dominant vision. This may come from the top – such as Jack Welch's challenge to 'destroy your business' memo. Perhaps building on their earlier experiences Intel now has a process called 'constructive confrontation', which essentially encourages a degree of dissent. The company has learned to value the critical insights which come from those closest to the action rather than assume senior managers have the 'right' answers every time.[134]

Alternatively many organizations develop a parallel structure or track for ideas which lie outside the mainstream – by setting up some kind of 'dual structure'. These can take many forms, including special project teams, incubators, new venture divisions, corporate venture units and 'skunk works'. Some have more formal status than others, some have more direct power or resource whilst others are dependent on internal sponsors or patrons.

Whatever the particular arrangement the underlying purpose of such dual structures is to protect new and often high-risk ideas from the mainstream organization until they have achieved some measure of commercial viability. But, as Birkinshaw and colleagues point out, such units are hard to manage effectively. Their research suggests that they work best when they have CEO-level support, clear objectives, and their own separate sources of finance and they work least well when parent company managers meddle in the evaluation and selection of ventures, and when they are expected to support multiple (and changing) objectives.[136]

The other issue with such dual structures is the need to bring them back into the mainstream at some point. They can provide helpful vehicles for growing ideas to the point where they can be more fairly evaluated against mainstream criteria and portfolio selection systems, but they need to be seen as temporary rather than permanent mechanisms for doing so. Otherwise there is a risk of separation and at the limit a loss of leverage against the knowledge and other assets of the mainstream organization. (Chapter 12 discusses this theme in more detail.)

Enabling Effective Knowledge Acquisition

As we saw in looking at the search phase, discontinuous innovation requires a more active and extensive search at the periphery and these principles underpin effective knowledge acquisition approaches as well. Building alliances across knowledge boundaries and opening up new space for developing innovation is critical here – but it may require an extensive speculative investment. Unlike the steady state where analytical frameworks such as the Arthur D. Little matrix[50] can help guide strategic knowledge investments, discontinuous innovation involves dealing with much more open-ended and early-stage research. Strategies to deal with it include building extensive network links to key knowledge sources (not just for technology but for markets, for social and political trends, etc.) and participating in risk-sharing strategic research programmes where the investment is primarily buying a stake in a possible future. There has been considerable growth in taking an 'options' approach to R&D management along similar lines to that which operates in financial markets. Another increasingly popular mechanism is the use of small scale start-up firms as pilots and incubators where knowledge can be generated and, if promising, bought into or taken over by larger players – a strategy common in the emerging but high-risk industries of biotechnology and nanotechnology.

Implementation

Again the issue here is that discontinuous innovation projects tend to fall outside the mainstream in terms of resource allocation and project management structures. One way of dealing with progressing such projects is through a separate organization – the dual structure model mentioned earlier. Another is to see the problem as one of managing within a mainstream context but adopting more flexible and entrepreneurial approaches to finding resources, building coalitions of support, etc. Leifer and colleagues report on a series of case studies of such radical innovation and suggest that there are some common features amongst firms with well-developed radical innovation capability including:

- The firm's leadership sets expectations, develops radical innovation culture, establishes facilitative organizational mechanisms and develops goals and reward systems.
- Radical innovation idea hunters seek opportunities.
- Radical hubs of innovation help establish effective evaluation boards that sue appropriate criteria.
- Non-traditional marketing and business development personnel work with radical innovation technical teams to develop the business model.
- Individual managers with authority provide seed funding and internal venture capital.
- The firm adopts a portfolio approach to funding radical innovation projects.
- Radical innovation hubs develop a strategy for identifying, selecting, rewarding and retaining radical innovation champions, experts and team members.
- Relationships between radical innovation activity and internal and external partners are developed at a strategic level.
- Transition team is established to continue application and market development until uncertainty is reduced sufficiently to ensure a successful transition to operating unit.[137]

Learning

Given the relative infrequent occurrence of discontinuous innovation it is difficult to generate sufficient trial and error experience to build robust routines – as is the case with regular steady-state innovation. This puts a premium on capturing learning from such experiences but it also argues for a more flexible approach to carrying through discontinuous innovation projects. The theme of learning and dynamic capability take centre stage here; organizations working with technologies or markets which are not fully formed or understood need to take an approach which is highly experimental, based on a strategy of frequent 'probe and learn' activities. Capturing learning from failure is as important as building on success since early and fast failure can help establish trajectories early. Emphasis shifts to the systemic nature of such innovation – instead of a gap between generators and users of innovation there is much more a joint approach based on co-evolution and gradual emergence.

As we have already noted the learning challenge is also one of letting go as much as acquiring new knowledge and under discontinuous conditions this may lead to a fundamental resetting of the parameters on which the organization has hitherto been based – a process of 're-invention'. This 'generative' or 'double-loop' learning capacity is a key feature of dynamic capability.[125,138]

9.8 Beyond the Boundaries

Throughout the book we have seen the growing importance of viewing innovation as something which needs to be managed at a system level and which is increasingly inter-organizational in nature. The rise of networking, the emergence of small firm clusters, the growing use of 'open innovation' principles and the globalization of knowledge production and application are all indicators of the move to what Rothwell called a 'fifth generation' innovation model. This has a number of implications for the ways in which we deal with the practical organization and management of the process.[9]

The basic model which we have been using throughout the chapter is still relevant but the ways in which the different phases are enabled now needs to build on an increasing network orientation. For example, networking provides a powerful mechanism for extending and covering a richer selection environment and can bring into play a degree of collective efficiency in picking up relevant signals. Strategies like 'connect and develop' are predicated on the potential offered by increasing the range of connections available to an enterprise.

Types of Innovation Networks

Innovation is a social process; it involves people getting together and sharing ideas. Often, this happens informally – it is, after all, the basis for much conference activity and an increasing number of organizations recognize the need to try and allow for informal networking as a means of stimulating creative interchange. But there is an

TABLE 9.5 Typology of innovation networks

Type of innovation network	Primary purpose/innovation target
New product or process development consortium	Sharing knowledge and perspectives to create and market new product or process concept – for example, the Symbian consortium (Sony, Ericsson, Motorola and others) working towards developing a new operating system for mobile phones and PDAs
Sectoral forum	Shared concern to adopt and develop innovative good practice across a sector or product market grouping – for example, in the UK the SMMT Industry Forum or the Logic (Leading Oil and Gas Industry Competitiveness), a gas and oil industry forum
New technology development consortium	Sharing and learning around newly emerging technologies – for example, the pioneering semiconductor research programmes in the USA and Japan
Emerging standards	Exploring and establishing standards around innovative technologies – for example, the Motion Picture Experts Group (MPEG) working on audio and video compression standards
Supply chain learning	Developing and sharing innovative good practice and possibly shared product development across a value chain – for example, the SCRIA initiative in aerospace
Cluster	Regional grouping of companies to gain economic growth through exploiting innovation synergies
Topic network	Mix of firms companies to gain traction on key new technology

increasing trend towards trying to build innovation networks in a purposive fashion – what some researchers call 'engineered' networks.[139] That purpose might be to create a completely new product or process, bringing together radically different combinations of knowledge or it could be a network whose members are adopting and embedding innovative ideas. Players could be linked together by some geographical focus – as in a cluster – or as part of a supply chain trying to develop new ideas along the whole system. What they share is a recognition that they can get traction on some aspect of the innovation problem through networking. Table 9.5 provides an outline typology.

Configuring Innovation Networks

Whatever the purpose in setting it up, actually operating an innovation network is not easy – it needs a new set of management skills. It depends heavily on the type of network and the purposes it is set up to achieve. For example, there is a big difference between the demands for an innovation network working at the frontier where issues of intellectual property management and risk are critical, and one where there is an

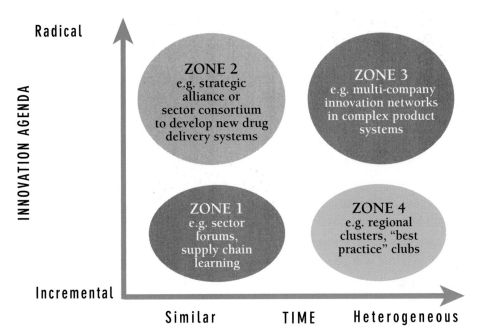

FIGURE 9.12 Types of innovation network

established innovation agenda as might be the case in using supply chains to enhance product and process innovation. We can map some of these different types of innovation network on to a simple diagram (Figure 9.12) which positions them in terms of:

- how radical the innovation target is with respect to current innovative activity;
- the similarity of the participating companies.

Different types of networks have different issues to resolve. For example, in zone 1 we have firms with a broadly similar orientation working on tactical innovation issues. Typically, this might be a cluster or sector forum concerned with adopting and configuring 'good practice' manufacturing. Issues here would involve enabling them to share experiences, disclose information, develop trust and transparency and build a system level sense of shared purpose around innovation.

Zone 2 activities might involve players from a sector working to explore and create new product or process concepts – for example, the emerging biotechnology/pharmaceutical networking around frontier developments and the need to look for interesting connections and synthesis between these adjacent sectors. Here, the concern is exploratory and challenges existing boundaries but will rely on a degree of information-sharing and shared risk-taking, often in the form of formal joint ventures and strategic alliances.

In Zones 3 and 4, the players are highly differentiated and bring different key pieces of knowledge to the party. Their risks in disclosing can be high so ensuring careful IP management and establishing ground rules will be crucial. At the same time, this kind of innovation is likely to involve considerable risk and so putting in place risk and benefit sharing arrangements will also be critical. For example, in a review of 'high value innovation networks' in the UK, researchers from the Advanced Institute of Management Research (AIM)[140] found the following characteristics were important success factors:

- Highly diverse: network partners from a wide range of disciplines and backgrounds who encourage exchanges about ideas across systems.
- Third-party gatekeepers: science partners such as universities but also consultants and trade associations, who provide access to expertise and act as neutral knowledge brokers across the network.
- Financial leverage: access to investors via business angels, venture capitalists firms and corporate venturing which spreads the risk of innovation and provides market intelligence.
- Proactively managed: participants regard the network as a valuable asset and actively manage it to reap the innovation benefits.

Learning to Manage Innovation Networks

We have enough difficulties trying to manage within the boundaries of a typical business. So, the challenge of innovation networks takes us well beyond this. The challenges include:

- how to manage something we don't own or control;
- how to see system-level effects not narrow self-interests;
- how to build trust and shared risk taking without tying the process up in contractual red tape;
- how to avoid 'free riders' and information 'spillovers'.

It's a new game and one in which a new set of management skills becomes important. Innovation networks can be broken down into three stages of a life cycle. Table 9.6 looks at some of the key management questions associated with each stage.

9.9 Summary and Further Reading

Some of the basic approaches to scanning the competitive environment can be found in marketing texts such as Kotler[18] whilst Whiston provides a comprehensive guide to technological forecasting.[21] Some of the tools of forecasting are described well in

TABLE 9.6 Challenges in managing innovation networks

Set-up stage	Operating stage	Sustaining (or closure) stage
Issues here are around providing the momentum for bringing the network together and clearly defining its purpose. It may be crisis triggered – for example, perception of the urgent need to catch up via adoption of innovation. Equally, it may be driven by a shared perception of opportunity – the potential to enter new markets or exploit new technologies. Key roles here will often be played by third parties – network brokers, gatekeepers, policy agents and facilitators	The key issues here are about trying to establish some core operating processes about which there is support and agreement. These need to deal with: • Network boundary management – how the membership of the network is defined and maintained • Decision-making – how (where, when, who) decisions get taken at the network level • Conflict resolution – how conflicts are resolved effectively • Information processing – how information flows among members and is managed • Knowledge management – how knowledge is created, shared and used across the network • Motivation – how members are motivated to join/remain within the network • Risk/benefit-sharing – how the risks and rewards are allocated across members of the network • Co-ordination – how the operations of the network are integrated and co-ordinated	Networks need not last for ever – sometimes they are set up to achieve a highly specific purpose (e.g. development of a new product concept) and once this has been done the network can be disbanded. In other cases there is a case for sustaining the networking activities for as long as members see benefits. This may require periodic review and 're-targeting' to keep the motivation high. For example, CRINE, a successful development programme for the offshore oil and gas industry, was launched in 1992 by key players in the industry such as BP, Shell and major contractors with support from the UK Government with the target of cost reduction. Using a network model, it delivered extensive innovation in product/services and processes. Having met its original cost-reduction targets, the programme moved to a second phase with a focus aimed more at capturing a bigger export share of the global industry through innovation

ref. 20 and ref. 141 and the theme of future scanning is picked up in books by Miles, De Geus and Schwartz.[25,133,142]

Strategic selection and portfolio management are themes explored in depth in ref. 50, ref. 79 and ref. 143 and a detailed selection of cases and seminal papers can be found in ref. 144 and ref. 145. There are a number of prescriptive methodologies for product and process innovation management, including Cooper's work on product innovation,[69] and in ref. 146 whilst ref. 147 provides a similar perspective on process innovation. Terry Hill's description of change in a fictitious organization in *The Strategy Quest* provides a good feel for the strategy implementation process;[148] this echoes the popular work *The Goal* by Goldratt and Cox in describing change in a manufacturing enterprise.[149]

A good overview of the implementation issues can be found in ref. 150 and specific examples of implementation of different technologies – e.g. Internet and IT systems[151–153] – and in different contexts – for example, in social services,[154] in mining,[155] in production environments[156] and in the public sector.[157] Project management texts include a number of standard reference works and handbooks[158,159] whilst the management of specific types of projects – e.g. software development or large-scale complex projects – are covered by refs. 160 to 162. Project teams and the people issues are dealt with in Demarco's book and in a number of other sources.[88,163–168]

Change management is a broad field with many prescriptive texts; Burnes provides a good overview of the key themes whilst Smith and Tranfield and Conner offer some case study-based examples.[106,169,170] A number of books offer perspectives drawn from a long research tradition in organizational development,[105,171,172] two recent practical guides being Senge's *The Dance of Change* and Sharke's *The Change Management Toolkit*.[138,173] Finally, there are some popular books which attempt to extract lessons from companies which have undergone major transformations.[174]

The extended set of routines needed for dealing with discontinuous innovation is discussed in a variety of sources including refs. 7, 11, 136, 175–179.

Inter-organizational innovation routines are picked up in several sources including refs. 15, 138, 180 and 181.

References

1 Rosenau, M. *et al.* (eds) (1996) *The PDMA Handbook of New Product Development*. John Wiley & Sons, Inc., New York.

2 Economist, T. (2000) 'Amazon, the software company', in *The Economist*. 92.

3 Guiltinan, J. and G. Paul (1991) *Marketing Management: Strategies and programs*. 4th edn. McGraw-Hill, New York.

4 Dutta, D. (2004) *Retail @ the speed of fashion*. www.3isite.com/articles/ImagesFashion_Zara_Part_1.pdf.

5 Belussi, F. (1989) 'Benetton – a case study of corporate strategy for innovation in traditional sectors', in Dodgson, M. (ed.), *Technology Strategy and the Firm*. Longman, London.

6 Kurtz, C. and D. Snowden (2003) 'The new dynamics of strategy: sense-making in a complex and complicated world', *IBM Systems Journal*, **42** (3), 462–482.

7 Day, G. and P. Schoemaker (2000) *Wharton on Managing Emerging Technologies*. John Wiley & Sons, Inc., New York.

8 Carter, C. and B. Williams (1957) *Industry and Technical Progress*. Oxford University Press, Oxford.

9 Rothwell, R. (1992) 'Successful industrial innovation: critical success factors for the 1990s'. *R&D Management*, **22** (3), 221–239.

10 Seely Brown, J. (2004) 'Minding and mining the periphery', *Long Range Planning*, **37**, 143–151.

11 Day, G. and P. Schoemaker (2004) 'Driving through the fog: managing at the edge', *Long Range Planning*, **37** (2), 127–142.

12 Urban, G., J. Hauser and N. Dholakia (1987) *Essentials of New Product Management*. Prentice-Hall, Englewood Cliffs, N.J.

13 Allen, T. (1977) *Managing the Flow of Technology*. MIT Press, Cambridge, Mass.

14 Bessant, J. and G. Tsekouras (1999) 'Developing learning networks', *AI and Society*, **15** (2), 22–98.

15 Wenger, E. (1999) *Communities of Practice: Learning, meaning, and identity*. Cambridge University Press, Cambridge.

16 Gladwell, M. (2000) *The Tipping Point: How little things can make a big difference*. Little, Brown, New York.

17 Huston, L. (2004) 'Mining the periphery for new products', *Long Range Planning*, **37**, 191–196.

18 Kotler, P. (2003) *Marketing Management, Analysis, Planning and Control*. 11th edn. Prentice Hall, Englewood Cliffs, N.J.

19 Miles, I. *et al.* (1988) *Information Horizons*. Edward Elgar, London.

20 Millett, S. and E. Honton (1991) *A Manager's Guide to Technology Forecasting and Strategy Analysis Methods*. Battelle Press, Columbus, Ohio.

21 Whiston, T. (1979) *The Uses and Abuses of Forecasting*. Macmillan, London.

22 Foster, R. (1986) *Innovation – The Attacker's Advantage*. Pan Books, London.

23 Shell (2003) *People and Connections*. Shell, London.

24 Shell International (2002) *People and Connections: Global scenarios to 2020*. Shell International Ltd., London.

25 Miles, I. (1997) *Technology Foresight: Implications for social science*. CRIC, University of Manchester, Manchester.

26 Hobday, M. (1995) *Innovation in East Asia – The challenge to Japan*. Edward Elgar, Cheltenham.

27 Kim, L. (1997) *Imitation to Innovation: The dynamics of Korea's technological learning*. Harvard Business School Press, Boston, Mass.

28 Camp, R. (1989) *Benchmarking – the search for industry best practices that lead to superior performance*. Quality Press, Milwaukee, WI.

29 Kaplinsky, R., F. den Hertog and B. Coriat (1995) *Europe's Next Step*. Frank Cass, London.

30 Herstatt, C. and E. von Hippel (1992) 'Developing new product concepts via the lead user method', *Journal of Product Innovation Management*, **9** (3), 213–221.

31 Von Hippel, E. (1988) *The Sources of Innovation*. MIT Press, Cambridge, Mass.

32 Bessant, J., R. Kaplinsky and R. Lamming (2003) 'Putting supply chain learning into practice', *International Journal of Operations and Production Management*, **23** (2), 167–184.

33 Rothwell, R. and P. Gardiner (1983) 'Tough customers, good design', *Design Studies*, **4** (3), 161–169.

34 Treacy, M. and F. Wiersema (1995) *The Discipline of Market Leaders*. Addison-Wesley, Reading, Mass.

35 Bessant, J. (2003) *High Involvement Innovation*. John Wiley & Sons, Ltd, Chichester.

36 Chesborough, H. (2003) 'Managing your false negatives', *Harvard Management Updates*, **8** (8).

37 Shillito, M. (1994) *Advanced QFD: Linking technology to market and company needs*. John Wiley & Sons, Inc., New York.

38 Hauser, J. and D. Clausing (1988) 'The house of quality', *Harvard Business Review*, May–June, 63–73.

39 Ettlie, J. (1988) *Taking Charge of Manufacturing*. Jossey-Bass, San Francisco.

40 Bessant, J. (1991) *Managing Advanced Manufacturing Technology: The challenge of the fifth wave*. NCC-Blackwell, Oxford/Manchester.

41 Cooper, R. and E. Kleinschmidt (1990) *New Products: The key factors in success*. American Marketing Association, Chicago.

42 Griffin, A. *et al.* (1996) *The PDMA Handbook of New Product Development*. John Wiley & Sons, Inc., New York.

43 Ernst, H. (2002) 'Success factors of new product development: a review of the empirical literature', *International Journal of Management Reviews*, **4** (1), 1–40.

44 Carson, J. (1989) *Innovation: A battleplan for the 1990s*. Gower, Aldershot.

45 Bessant, J. (1997) 'Developing technology capability through manufacturing strategy', *International Journal of Technology Management*, **14** (2/3/4), 177–195.

46 DTI (1998) *Making It Fit: Guide to developing strategic manufacturing*. UK Department of Trade and Industry, London.

47 Mills, J. *et al.* (2002) *Creating a Winning Business Formula*. Cambridge University Press, Cambridge.

48 Mills, J. *et al.* (2002) *Competing through Competencies*. Cambridge University Press, Cambridge.

49 Crawford, M. and C. Di Benedetto (1999) *New Products Management*. McGraw-Hill/Irwin, New York.

50 Floyd, C. (1997) *Managing Technology for Corporate Success*. Gower, Aldershot, 228.

51 Cooper, R. (1988) 'The new product process: a decision guide for management', *Journal of Marketing Management*, **3** (3), 238–255.

52 Cooper, R. (1999) 'The invisible success factors in product innovation', *Journal of Product Innovation Management*, **16** (2).

53 Gupta, A. (1994) 'Developing powerful technology and product concepts', in Souder, W. and Sherman, J. (eds), *Managing New Technology Development*. McGraw-Hill, New York.

54 Davenport, T. (1992) *Process Innovation: Re-engineering work through information technology*. Harvard University Press, Boston, Mass., 326.

55 Dodgson, M., D. Gann and A. Salter (2005) *Think, Play, Do: The business of innovation*. Oxford University Press, Oxford.

56 Wheelwright, S. and K. Clark (1992) *Revolutionising Product Development*. Free Press, New York.

57 Mumford, E. (1979) *Designing Human Systems*. Manchester Business School Press, Manchester.

58 Bessant, J. and J. Buckingham (1993) 'Organisational learning for effective use of CAPM', *British Journal of Management*, **4** (4), 219–234.

59 Hines, P. *et al.* (1999) *Value Stream Management: The Development of Lean Supply Chains*. Financial Times Management, London.

60 Rich, N. and P. Hines (1997) 'Supply chain management and time-based competition: the role of the supplier association', *International Journal of Physical Distribution and Logistics Management*, **27** (3/4), 210–225.

61 Legge, K. *et al.* (eds) (1991) *Case Studies in Information Technology, People and Organisations*. Blackwell, Oxford.

62 UNIDO (1995) *Technology Transfer Management*. United Nations Industrial Development Organisation.

63 Saad, M. (1996) 'Development of learning capabilities via the transfer of advanced manufacturing technology: the case of two Algerian firms', *Science, Technology and Development*, **14** (1), 21–35.

64 Dodgson, M. (1993) *Technological Collaboration in Industry*. Routledge, London.

65 Lamming, R. (1993) *Beyond Partnership*. Prentice-Hall, London.

66 Cohen, W. and D. Levinthal (1990) 'Absorptive capacity: a new perspective on learning and innovation'. *Administrative Science Quarterly*, **35** (1), 128–152.

67 Hobday, M., H. Rush and J. Bessant (2005) 'Reaching the innovation frontier in Korea: a new corporate strategy dilemma', *Research Policy*.

68 Mariotti, F. and R. Delbridge (2005) 'A portfolio of ties: managing knowledge transfer and learning within network firms', *Academy of Management Review* (forthcoming).

69 Cooper, R. (2001) *Winning at New Products*. 3rd edn. Kogan Page, London.

70 Bruce, M. and R. Cooper (1997) *Marketing and Design Management*. International Thomson Business Press, London.

71 Bruce, M. and R. Cooper (2000) *Creative Product Design*. John Wiley & Sons, Ltd, Chichester.

72 Bruce, M. and J. Bessant (eds) (2001) *Design in Business*. Pearson Education, London.

73 Lawrence, P. and J. Lorsch (1967) *Organisation and Environment*. Harvard University Press, Cambridge, Mass.

74 Griffin, A. (1998) 'Overview of PDMA survey on best practices', in *Visions*, Product Development Management Association, www.pdma.org.

75 Bessant, J. and D. Francis (1997) 'Implementing the new product development process', *Technovation*, **17** (4), 189–197.

76 Miles, I. (2001) *Services Innovation: A reconfiguration of innovation studies*. University of Manchester Centre for Research on Innovation and Competitiveness, Manchester.

77 Cooper, R. *et al.* (1994) 'What distinguishes the top performing new products in financial services?', *Journal of Product Innovation Management*, **11** (4), 281–300.

78 Edgett, S. and S. Jones (1991) 'New product development in the financial service industry', *Journal of Marketing Management*, **7**, 271–284.

79 Edgett, S., E. Kleinschmidt and R. Cooper (2001) *Portfolio Management for New Products*. Perseus Publishing, New York.

80 Pisano, G. (1996) *The Development Factory: Unlocking the potential of process innovation*. Harvard Business School Press, Boston, Mass.

81 Smith, P. and D. Reinertson (1992) 'Shortening the product development cycle'. *Research Technology Management*, May–June, 44–49.

82 Foxall, G. (1987) 'Strategic implications of user-initiated innovation', in Rothwell, R. and Bessant, J. (eds), *Innovation, Adaptation and Growth*. Elsevier, Amsterdam.

83 De Benedetto, C. (1994) 'Defining markets and users for new technologies', in Souder, W. and Sherman, J. (eds), *Managing New Technology Development*, McGraw-Hill, New York.

84 Eisenhardt, K. and S. Brown (1997) 'The art of continuous change: linking complexity theory and time-paced evolution in relentlessly shifting organizations', *Administrative Science Quarterly*, **42** (1), 1–34.

85 Stalk, G. and T. Hout (1990) *Competing against Time: How time-based competition is reshaping global markets*. Free Press, New York.

86 Souder, W. and J. Sherman (1994) *Managing New Technology Development*. McGraw-Hill, New York.

87 Conway, S. and R. Forrester (1999) *Innovation and Teamworking: Combining perspectives through a focus on team boundaries*. University of Aston Business School, Birmingham.

88 DeMarco, T. and T. Lister (1999) *Peopleware: Productive projects and teams*. Dorset House, New York.

89 Bessant, J. and D. Francis (1999) 'Developing strategic continuous improvement capability', *International Journal of Operations and Production Management*, **19** (11).

90 Rich, B. and L. Janos (1994) *Skunk works*. Warner Books, London.

91 Kidder, T. (1981) *The Soul of a New Machine*. Penguin, Harmondsworth.

92 Jelinek, M. and J. Litterer (1994) 'Organising for technology and innovation', in Souder, W. and Sherman, J. (eds), *Managing New Technology Development*. McGraw-Hill, New York.

93 Walton, M. (1997) *Car: A drama of the American workplace*. Norton, New York.

94 Whitney, J. (1988) 'Manufacturing by design', *Harvard Business Review*, July–August, 83–91.

95 Tidd, J. and K. Bodley (2001) 'The influence of project novelty on the new product development process', *R&D Management*, **32** (2), 127–138.

96 Belliveau, P., A. Griffin, and S. Somermeyer (2002) *The PDMA ToolBook for New Product Development: Expert techniques and effective practices in product development*. John Wiley & Sons, Inc., New York.

97 Cooper, R. (2003) 'Profitable product innovation', in Shavinina, L. (ed.), *International Handbook of Innovation*. Elsevier, New York.

98 Rogers, E. (1984) *Diffusion of Innovation*. Free Press, New York.

99 Calantone, R. and M. Monoya (1994) 'Product launch and follow-on', in Souder, W. and Sherman, J. (eds), *Managing New Technology Development*. McGraw-Hill, New York.

100 Voss, C. (1986) 'Implementation of advanced manufacturing technology', in Voss, C. (ed.), *Managing Advanced Manufacturing Technology*. IFS Publications, Kempston.

101 Leonard-Barton, D. (1988) 'Implementation as mutual adaptation of technology and organization', *Research Policy*, **17**, 251–267.

102 McLoughlin, I. and J. Clark (1994) *Technological Change at Work*. 2nd edn. Open University Press, Milton Keynes.

103 Majchrzak, A. (1988) *The Human Side of Factory Automation*. Jossey-Bass, San Francisco.

104 Clark, J. (1995) *Managing Innovation and Change: People, technology and strategy*. Sage, London.

105 French, W. and C. Bell (1995) *Organisational Development: Behavioural science interventions for organisation improvement*. 4th edn. Prentice-Hall, Englewood Cliffs, N.J.

106 Smith, S. and D. Tranfield (1990) *Managing Change*. IFS Publications, Kempston.

107 Hayes, R., S. Wheelwright and K. Clark (1988) *Dynamic Manufacturing: Creating the learning organisation*. Free Press, New York.

108 Leonard-Barton, D. (1995) *Wellsprings of Knowledge: Building and sustaining the sources of innovation*. Harvard Business School Press, Boston, Mass., 335.

109 Gann, D. and A. Salter (1998) 'Learning and innovation management in project-based, service enhanced firms', *International Journal of Innovation Management*, **2** (4), 431–454.

110 Teece, D., G. Pisano and A. Shuen (1997) 'Dynamic capabilities and strategic management', *Strategic Management Journal*, **18** (7), 509–533.

111 Utterback, J. (1994) *Mastering the Dynamics of Innovation*. Harvard Business School Press, Boston, Mass., 256.

112 Christensen, C. and M. Raynor (2003) *The Innovator's Solution: Creating and sustaining successful growth*. Harvard Business School Press, Boston, Mass.

113 Nonaka, I. (1991) 'The knowledge creating company', *Harvard Business Review*, November–December, 96–104.

114 Nelson, R. and S. Winter (1982) *An Evolutionary Theory of Economic Change*. Harvard University Press, Cambridge, Mass.

115 Welch, J. (2001) *Jack! What I've learned from leading a great company and great people*. Headline, New York.

116 Kolb, D. and R. Fry (1975) 'Towards a theory of applied experiential learning', in Cooper, C. (ed.), *Theories of Group Processes*, John Wiley & Sons, Ltd, Chichester.

117 Nonaka, I., S. Keigo and M. Ahmed (2003) 'Continuous innovation: the power of tacit knowledge', in Shavinina, L. (ed.), *International Handbook of Innovation*. Elsevier, New York.

118 Barnes, J. and M. Morris (1999) *Improving Operational Competitiveness through Firm-level Clustering: A case study of the KwaZulu-Natal Benchmarking Club*. School of Development Studies, University of Natal, Durban, South Africa.

119 Oliver, N. (1996) *Benchmarking Product Development*. University of Cambridge, Cambridge.

120 Rush, H. *et al.* (1993) *Benchmarking R&D institutes*. Centre for Research in Innovation Management, University of Brighton.

121 Zairi, M. (1996) *Effective Benchmarking: Learning from the best*. Chapman & Hall, London.

122 Christenson, C. (1997) *The Innovator's Dilemma*. Harvard Business School Press, Cambridge, Mass.

123 Philips, W. *et al.* (2004) *Promiscuous Relationships: Discontinuous innovation and the role of supply networks*, IPSERA Conference, Naples, Italy.

124 Senge, P. (1990) 'The leader's new work: building learning organisations', *Sloan Management Review*, **32** (1), 7–23.

125 Argyris, C. and D. Schon (1970) *Organizational Learning*. Addison Wesley, Reading, Mass.

126 Winter, S. (2004) 'Specialised perception, selection and strategic surprise: learning from the moths and the bees', *Long Range Planning*, **37**, 163–169.

127 Phillips, W. *et al.* (2004) *Promiscuous Relationships: Discontinuous innovation and the role of supply networks*, IPSERA Annual Conference, Naples, Italy.

128 Leonard, D. (1992) 'Core capabilities and core rigidities: a paradox in new product development', *Strategic Management Journal*, **13**, 111–125.

129 Tripsas, M. and G. Gavetti (2000) 'Capabilities, cognition and inertia: evidence from digital imaging', *Strategic Management Journal*, **21**, 1147–1161.

130 Garr, D. (2000) *IBM Redux: Lou Gerstner and the business turnaround of the decade*. Harper-Collins, New York.

131 Day, G. and P. Schoemaker (2004) 'Driving through the fog: managing at the edge', *Long Range Planning*, **37**, 127–142.

132 Moore, G. (1999) *Crossing the Chasm: Marketing and selling high-tech products to mainstream customers*. Harper Business, New York.

133 de Geus, A. (1996) *The Living Company*. Harvard Business School Press, Boston, Mass.

134 Bessant, J., J. Birkinshaw and R. Delbridge (2004) 'One step beyond – building a climate in which discontinuous innovation will flourish', in *People Management*, **10** (3), 28–32.

135 Groves, A. (1999) *Only the Paranoid Survive*. Bantam Books, New York.

136 Buckland, W., A. Hatcher and J. Birkinshaw (2003) *Inventuring: Why big companies must think small*. McGraw Hill Business, London.

137 Leifer, R. *et al.* (2000) *Radical Innovation*. Harvard Business School Press, Boston, Mass.

138 Senge, P. (1999) *The Dance of Change: Mastering the twelve challenges to change in a learning organisation*. Doubleday, New York.

139 Conway, S. and F. Steward (1998) 'Mapping innovation networks', *International Journal of Innovation Management*, **2** (2), 165–196.

140 AIM (2004) *I-works: How high value innovation networks can boost UK productivity*. ESRC/EPSRC Advanced Institute of Management Research, London.

141 Vanston, J. (1998) *Technology Forecasting: An aid to effective technology management*. Technology Futures Inc., Austin, Texas.

142 Schwartz, P. (1996) *The Art of the Long View: Planning for the future in an uncertain world*. Doubleday, New York.

143 Roussel, P., K. Saad and T. Erickson (1991) *Third Generation R&D: Matching R&D projects with corporate strategy*. Harvard Business School Press, Cambridge, Mass.

144 Burgelman, R., C. Christensen and S. Wheelwright (eds) (2004) *Strategic Management of Technology and Innovation*. 4th edn. McGraw Hill Irwin, Boston.

145 Schilling, M. (2005) *Strategic Management of Technological Innovation*. McGraw Hill, New York.

146 Robert, M. (1995) *Product Innovation Strategy – Pure and simple*. McGraw Hill, New York.

147 Brown, S. *et al.* (2005) *Strategic Operations Management*. 2nd edn. Butterworth Heinemann, Oxford.

148 Hill, T. (1995) *The Strategy Quest*. Pitman, London.

149 Goldratt, E. and J. Cox (2004) *The Goal*. North River Press, New York.

150 Rhodes, E. and D. Wield (1994) *Implementing New Technologies*. Basil Blackwell, Oxford.

151 Clements, R. (1999) *IS Manager's Guide to Implementing and Managing Internet Technology*. Prentice Hall, Englewood Cliffs, NJ.

152 Goetsch, D. and S. Davis (1995) *Implementing Total Quality*. Prentice Hall, Englewood Cliffs, NJ.

153 Grover, V. and S. Jeong (1995) 'The implementation of business process re-engineering', *Journal of management Information Systems*, **12** (1), 109–144.

154 Schoech, D. (1999) *Human Services Technology: Understanding, designing and implementing computer and internet application in social services*. Haworth, New York.

155 De Jager, B. *et al.* (2004) 'Enabling continuous improvement – an implementation case study', *International Journal of Manufacturing Technology Management*, **15** (4), 315–324.

156 Voss, C., D. Twigg and G. Winch (1991) *CAD Implementation and Organisational change*. University of Warwick, Coventry.

157 Albury, D. (2004) *Innovation in the Public Sector*. Strategy Unit, Cabinet Office, London.

158 Lock, D. (1997) *Project Management*. Prentice-Hall, New Jersey.

159 Dinsmore, P. (ed.) (1993) *The AMA Handbook of Project Management*. AMACOM, New York.

160 Rea, K. and B. Lientz (1998) *Breakthrough Technology Project Management*. Academic Press, New York.

161 Rush, H., T. Brady and M. Hobday (1997) *Learning between Projects in Complex Systems*. Centre for the Study of Complex Systems.

162 Gann, D. and A. Salter (2000) 'Innovation in project-based, service-enhanced firms: the construction of complex products and systems', *Research Policy*, **29**, 955–972.

163 Kharbanda, O. and M. Stallworthy (1990) *Project Teams*. NCC-Blackwell, Manchester.

164 Smith, P. and E. Blanck (2002) 'From experience: leading dispersed teams', *Journal of Product Innovation Management*, **19**, 294–304.

165 Duarte, D. and N. Tennant Snyder (1999) *Mastering Virtual Teams*. Jossey-Bass, San Francisco.

166 Katzenbach, J. and D. Smith (1992) *The Wisdom of Teams*. Harvard Business School Press, Boston, Mass.

167 Jarvenpaaa, S. and D. Leidner (1999) 'Communication and trust in global virtual teams', *Organization Science*, **10** (6), 791–815.

168 Sapsed, J. *et al.* (2002) 'Teamworking and knowledge management: a review of converging themes', *International Journal of Management Reviews*, **4** (1).

169 Burnes, B. (1992) *Managing Change*. Pitman, London.

170 Conner, D. and D. Conner (1992) *Managing at the Speed of Change*. Villard Books, New York.

171 Bennis, W. (2000). *Managing the Dream: Reflections on leadership and change*. Perseus Books, New York.

172 Hesselbein, F., M. Goldsmith and R. Beckhard (eds) (1997) *Organization of the Future*. Jossey-Bass/The Drucker Foundation, San Francisco.

173 Sharke, G. (1999) *The Change Management Toolkit*. Winhope Press, New York.

174 Tichy, N. and S. Sherman (1999) *Control Your Destiny or Someone Else Will: Lessons in mastering change*. Harper Business, New York.

175 Francis, D., J. Bessant and M. Hobday (2003) 'Managing radical organisational transformation', *Management Decision*, **41** (1), 18–31.

176 Hamel, G. (2000) *Leading the Revolution*. Harvard Business School Press, Boston, Mass.

177 Foster, R. and S. Kaplan (2002) *Creative Destruction*. Harvard University Press, Cambridge.

178 Dvir, R., F. Lettice and P. Thomond (2004) *Are You Ready to Disrupt It?* Innovation Ecology, Tel Aviv.

179 Trott, P. (2002) 'When market research may hinder the development of discontinuous new products', in Sundbo, J. and Fugelsang, L. (eds), *Innovation as Strategic Reflexivity*. Routledge, London.

180 Dyer, J. and K. Nobeoka (2000) 'Creating and managing a high-performance knowledge-sharing network: the Toyota case', *Strategic Management Journal*, **21** (3), 345–367.

181 Loudon, A. (2001) *Webs of Innovation*. FT. Com., London.

Learning Through Corporate Ventures

In Chapter 9 we examined the processes necessary to develop new products and processes within the existing corporate environment, based on core competencies identified in Chapter 6. In this chapter we explore how firms develop technologies, products and processes outside their existing core competencies. Specifically, we will discuss the role of corporate ventures in the development of new technologies, products and businesses. The key questions are:

- What is the purpose of the corporate venture?
- What is the best way to structure, manage and grow a corporate venture?
- What is the profile of a typical corporate entrepreneur?
- How do technologies and markets affect the opportunities for entrepreneurship?

10.1 What is a Corporate Venture?

An organization that seeks to apply its competencies to a new market or business, or needs to acquire new competencies to respond to potentially disruptive innovation has three options:[1]

1. Attempt to change the competencies and culture within the existing organizational structure and processes.
2. Acquire an organization that has the necessary competencies.
3. Develop a separate organization within itself, with different structures, processes and cultures.

The first option, the objective of change management, is time-consuming and the outcomes uncertain. Some approaches to change management were discussed in Chapter 9. The second option requires a suitable acquisition to exist, is expensive and the outcomes variable. In most cases the organization acquired takes time to become integrated, and in many cases key personnel are lost in the process.[2] The third option is usually called corporate venturing or internal corporate venturing to distinguish it from venturing which takes the form of investments in smaller companies. If managed effectively, a corporate venture has the resources of a large organization and the entrepreneurial benefits of a small one. The Internet Bubble of the late 1990s produced an ill-timed bandwagon for corporate venturing in large established companies in the information and communications technology sector as they attempted to capture some of the rapid growth of the dot.com start-up firms: in 1996 Nortel Networks created the Business Ventures Programme (see Box 10.1), in 1997 Lucent established the Lucent New Ventures Group, in 2000 Ericsson formed Ericsson Business Innovation and British Telecom formed Brightstar.

BOX 10.1
CORPORATE VENTURING AT NORTEL NETWORKS

Nortel Networks is a leader in a high-growth, high-technology sector, and around a quarter of all its staff are in R&D, but it recognizes that it is extremely difficult to initiate new businesses outside the existing divisions. Therefore in December 1996 it created the Business Ventures Programme (BVP) to help to overcome some of the structural shortcomings of the existing organization, and identify and nurture new business ventures outside the established lines of business: 'The basic deal we're offering employees is an extremely exciting one. What we're saying is "Come up with a good business proposal and we'll fund and support it. If we believe your business proposal is viable, we'll provide you with the wherewithal to realize your dreams."' The BVP provides:

- guidance in developing a business proposal;
- assistance in obtaining approval from the board;
- an incubation environment for start-ups;
- transition support for longer-term development.

The BVP selects the most promising venture proposals which are then presented jointly by the BVP and employee(s) to the advisory board. The advisory board applies business and financial criteria in its decision whether to accept, reject or seek further development, and if accepted the most appropriate executive sponsor, structure and level of funding. The BVP then helps to incubate the new venture, including staff and resources, objectives and critical milestones. If successful, the BVP then assists the venture to migrate into an existing business division, if appropriate, or creates a new line or business or spin-off company:

> The programme is designed to be flexible. Among the factors determining whether or not to become a separate company are the availability of key resources within Nortel, and the suitability of Nortel's existing distribution channels . . . Nortel is not in this programme to retain 100% control of all ventures. The key motivators are to grow equity by maximizing return on investment, to pursue business opportunities that would otherwise be missed, and to increase employee satisfaction.

In 1997 the BVP attracted 112 business proposals, and given the staff and financial resources available aimed to fund up to five new ventures. The main problems experienced have been the reaction of managers in established lines of business to proposals outside their own line of:

> At the executive council level, which represents all lines of business, there is a lot of support . . . where it breaks down in terms of support is more in the political infrastructure, the middle to low management executive level where they feel threatened by it . . . the first stage of our marketing plan is just titled 'overcoming internal barriers'. That is the single biggest thing we've had to break through.

Initially, there was also a problem capturing the experience of ventures that failed to be commercialized:

> Failures were typically swept under the rock, nobody really talked about them . . . that is changing now and the focus is on celebrating our failures as well as our successes, knowing that we have learned a lot more from failure than we do from success. Start-up venture experience is in high demand. Generally, it's the projects that fail, not the people.

A corporate venture differs from conventional R&D and product development activities in its objectives and organization (see Box 10.2). The former seeks to exploit existing technological and market competencies, whereas the primary function of a new venture is to learn new competencies. In practice the distinction may be less clear. For example, technical staff at 3M are expected to use 15% of their time exploring ideas outside their existing projects, and at the corporate level this translates into a target of at least 25% of sales from products less than five years old. In Chapter 6 we reviewed the arguments for different structures for, and location of, R&D activities. The most effective organization and management of a new venture will depend on two dimensions: the strategic importance of the venture for corporate development, and its proximity to the core technologies and business.[3]

BOX 10.2
MATSUSHITA RESEARCH INSTITUTE TOKYO (MRIT)

Matsushita Research Institute Tokyo (MRIT) blurs the distinction between central research laboratory, venture group and joint venture organization. Matsushita is one of the largest manufacturers of consumer electronics in the world, and ranked in the global top 10 investors in R&D: $3.2 bn in 1994. Trade names include Panasonic and Technics.

Matsushita has a more conventional central research facility at its headquarters in Osaka, which employs 230 researchers working on basic research, and a dedicated corporate product development division that employs 3000 researchers. However, MRIT has a different role. MRIT is organized as a separate limited company, wholly owned by Matsushita, which is an unusual structure for a Japanese research organization. It is located in Tokyo and employs almost 250 scientists and engineers. About three-quarters of its funding is from its parent company, and the remainder from research contracts and government projects. It was established to support diversification through the creation and incubation of new technologies and markets. To this end, MRIT has developed a number of new technologies and identified new applications for existing technologies, partly through its own efforts, partly via government programmes, and partly through collaboration with academic and business partners in Japan, Europe and the USA.

MRIT has worked on advanced materials and surface science with Imperial College and Cambridge University in the UK, and on optical computing and neural networks with MIT and the University of California in the USA. MRIT is also a leading participant in the Japanese government's national research programme on micro-machines. None of these programmes has immediate relevance to existing businesses within Matsushita, but MRIT seeks to develop fundamental expertise in such emerging technologies in order to identify and grow future new business opportunities.

Typically, top management has risen through the ranks of the organization, and therefore will be familiar with the evaluation of proposals related to the existing lines of business. However, by definition, new venture proposals are likely to require assessment of new technologies and/or markets. The following checklist can be used to assess the strategic importance of a new venture:

- Would the venture maintain our capacity to compete in new areas?
- Would it help create new defensible niches?

- Would it help identify where not to go?
- To what extent could it put the firm at risk?
- How and when could the firm exit from the venture?

Assessment of the second dimension, the proximity to existing skills and capabilities, is more difficult. On the one hand, a new venture may be driven by newly developed skills and capabilities, but on the other a new venture may drive the development of new skills and capabilities. The former is consistent with an 'incremental' strategy in which diversification is a consequence of evolution, the latter with a 'rational' strategy which begins with the identification of new market opportunity. The relative merits and implications of these contrasting approaches were discussed in detail in Chapter 3.

Whatever the primary motive for establishing a new venture, the proposal should identify potential opportunities for positive synergies across existing technologies, products or markets. A checklist for assessing the proximity of the venture proposal to existing skills and capabilities would include:

- What are the key capabilities required for the venture?
- Where, how and when is the firm going to acquire the capabilities, and at what cost?
- How will these new capabilities affect current capabilities?
- Where else could they be exploited?
- Who else might be able to do this, perhaps better?

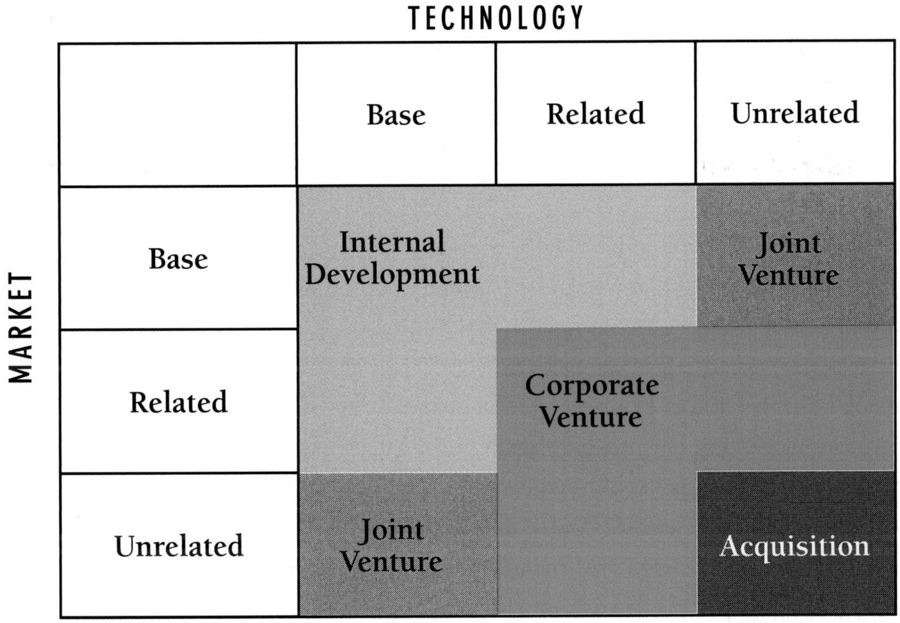

FIGURE 10.1   Role of corporate venturing
Source: Adapted from Burgelman (1984, p. 544).[3]

Assessment of a new venture along these two dimensions will help determine the organization and management of the venture. In particular, the strategic importance will determine the degree of administrative control required, and the proximity to existing skills and capabilities will determine the degree of operational integration that is desirable. In general, the greater the strategic importance, the stronger the administrative linkages between the corporation and venture. Similarly, the closer the skills and capabilities are to the core activities, the greater the degree of operational integration necessary for reasons of efficiency. Putting the two dimensions together creates a number of different options for the organization and management of a new venture (Figure 10.1). In this chapter we explore the design and management of internal corporate ventures. Chapter 8 examined the role and management of joint ventures and alliances.

10.2 Reasons for Corporate Venturing

The management structures and processes necessary for routine operations are very different from those required to manage innovation. The pressures of corporate long-range strategic planning on the one hand, and the short-term financial control on the other, combine to produce a corporate environment that favours carefully planned and stable growth based on incremental developments of products and processes:

- Budgeting systems favour short-term returns on incremental improvements.
- Production favours efficiency rather than innovation.
- Sales and marketing are organized and rewarded on the basis of existing products and services.

Such an environment is unlikely to be conducive to radical innovation. An internal corporate venture attempts to exploit the resources of the large corporation, but provide an environment more conducive to radical innovation. A corporate venture is likely to be necessary when a firm attempts to enter a new market or to develop a new technology. The key factors that distinguish a potential new venture from the core business are risk, uncertainty, newness and significance. A corporate venture is a separate organization or system designed to be consistent with the needs of new, high-risk, but potentially high-growth businesses. The term 'intrapreneurship' is sometimes used to describe such entrepreneurial activity within a corporate setting. However, it is not sufficient to promote entrepreneurial behaviour within a large organization. Entrepreneurial behaviour is not an end in itself, but must be directed and translated into desired business outcomes. Entrepreneurial behaviour is not associated with superior organizational performance, unless it is combined with an appropriate strategy in a heterogeneous or uncertain environment.[4] This suggests the need for clear strategic objectives for corporate venturing and appropriate organizational structures and processes to achieve those objectives.

There are a wide range of motives for establishing corporate ventures:[5]

- Grow the business.
- Exploit underutilized resources.
- Introduce pressure on internal suppliers.
- Divest non-core activities.
- Satisfy managers' ambitions.
- Spread the risk and cost of product development.
- Combat cyclical demands of mainstream activities.
- Diversify the business.
- Develop new technological or market competencies.

We will discuss each of these motives in turn, and provide examples. The first three are primarily operational, the remainder primarily strategic.

To Grow the Business

The desire to achieve and maintain expected rates of growth is probably the most common reason for corporate venturing, particularly when the core businesses are maturing. Depending upon the time frame of the analysis, between only 5% and 13% of firms are able to maintain a rate of growth above the rate of growth in GNP (gross national product).[6] However, for pressure to achieve this for publically listed firms is significant, as financial markets and investors expect the maintenance or improvement of rates of growth. The need to grow underlies many of the other motives for corporate venturing.

As a specific issue to be considered here, there is the drive to achieve growth in a corporate whose primary markets are maturing. For example, the UK water companies have a fully mature customer base, the regions are insular in their operation and the only opportunity for growth is beyond this mainstream and local activity, a situation which is driving their diversification into related and unrelated areas. Two water companies were recently reported to have shown interest in Tarmac's divestment of Econowaste, which could fetch up to £100 m. In seeking to grow the business, or make better commercial use of existing expertise, one has to challenge the more normal small scale of response to such initiatives.

Often the push is to analyse new or adapted products, processes or techniques, and relatively little emphasis is put on the expansion of the business considered as a whole.

Severn Trent International, in contrast, has been set up with the task of taking Severn Trent Water's core business capabilities in water and waste water management to new international markets. This direct transfer of skills must surely imply less uncertainty than venturing into both new technologies and new markets.

To Exploit Underutilized Resources in New Ways

This includes both technological and human resources. Typically, a company has two choices where existing resources are underutilized – either to divest and outsource the process or to generate additional contribution from external clients. However, if the company wants to retain direct and in-house control of the technology or personnel it can form an internal venture team to offer the service to external clients. For example, at Cadbury Schweppes the Information Technology Division was extracted from Group Management Services forming a separate business unit to supply the internal needs of the group – Cadbury Limited and Coca-Cola and Schweppes Beverages Limited. With the mechanisms in place for operation as a trading entity ITnet Limited began to seek and develop external clients, who now provide a significant proportion of its revenues.

To Introduce Competitive Pressure on to Internal Suppliers

This is a common motive, given the current fashion for outsourcing and market testing internal services. When a business activity is separated to introduce competitive pressure a choice has to be made – whether the business is to be subjected to the reality of commercial competition, or just to learn from it. If the corporate clients are able to go so far as to withdraw a contract, which is not conducive to learning, the business should be sold to allow it to compete for other work. For example, General Motors exposed its dominant supplier Delco to such competitive pressure by requiring it to earn a certain proportion of its sales from external sales.

To Divest Non-core Activities

Much has been written of the benefits of strategic focus, 'getting back to basics', and creating the 'lean' organization–rationalization which prompts the divestment of those activities which can be outsourced. However, this process can threaten the skill diversity required for an ever-changing competitive environment. New ventures can provide a mechanism to release peripheral business activities, but to retain some management control and financial interest.

To Satisfy Managers' Ambitions

As a business activity passes through its life cycle it will require different management styles to bring out the maximum gain. This may mean that the management team responsible for a business area will need to change, whether between conception to growth, growth to maturity or maturity to decline phases. A paradoxical situation often arises because of the changing requirements of a business area: top managers in place who are ambitious and want to see growth, and managing businesses which are reaching the limits of that growth. To retain the commitment of such managers the corporation will have to create new opportunities for change or expansion. These managers are not only potential facilitators for venture opportunities, but potential creators of venture opportunities. For example, Intel has long had a venture capital programme that invests in related external new ventures, but in 1998 it established the New Business Initiative to bootstrap new businesses developed by its staff: 'They saw that we were putting a lot of investment into external companies and said that we should be investing in our own ideas . . . our employees kept telling us they wanted to be more entrepreneurial.' The initiative invests only in ventures unrelated to the core microprocessor business, and in 1999 attracted more than 400 proposals, 24 of which are being funded.

To Spread the Risk and Cost of Product Development

Two situations are possible in this case: (1) where the technology or expertise needs to be developed further before it can be applied to the mainstream business or sold to current external markets, or (2) where the volume sales on a product awaiting development must sell to a target greater than the existing customer groups to be financially justified. In both cases the challenge is to understand how to venture outside current served markets. Too often, when the existing customer base is not ready for a product, the research unit will just continue their development and refinement process. If intermediary markets were exploited these could contribute to the financial costs of development, and to the maturing of the final product.

To Combat Cyclical Demands of Mainstream Activities

In response to the problem of cyclical demand Boeing set up two groups, Boeing Technology Services (BTS) and Boeing Associated Products (BAP), specifically with the function of keeping engineering and laboratory resources more fully employed when its own requirements waned between major development programmes. The remit for BTS was 'to sell off excess engineering laboratory capacity without a detrimental impact on schedules or commitments to major Boeing product-line activities'; it has stuck carefully to this charter, and been careful to turn off such activity when the mainstream business requires the expertise. BAP was created to commercially exploit Boeing inventions that are usable beyond their application to products manufactured by Boeing. About 600 invention disclosures are submitted by employees each year, and these are reviewed in terms of their marketability and patentability. Licensing agreements are used to exploit these inventions, 259 agreements are currently active. Beyond the financial benefits to the company and to the employees of this programme it is seen to foster the innovation spirit within the organization.

To Learn About the Process of Venturing

Venturing is a high-risk activity because of the level of uncertainty attached, and one cannot expect to understand the management process as one does for the mainstream business. If a learning exercise is to be undertaken, and a particular activity is to be chosen for this process, it is critical that goals and objectives are set, including a review schedule. This is important not just for the maximum benefit to be extracted but for the individuals who will pioneer that venture. For example, NEES Energy, a subsidiary of New England Electric Systems Inc., was set up to bring financial benefits, but was also expected to provide a laboratory to help the parent company learn about starting new ventures.[7]

Many companies develop hobby-size business activities to provide this 'learning by doing', but seldom is a time limit set on this learning stage, and as a consequence, no decision is formally made for the venture activities to be considered 'proper businesses'. The implications of this practice are to drain the enterprising managers of their enthusiasm and erode the value of potential opportunities.

To Diversify the Business

Whilst the discussion so far has implied that business development would be on a relatively small scale, this need not be the case. Corporate ventures are often formed in an effort to create new businesses in a corporate context, and therefore represent an attempt to grow via diversification. Therefore a decline in the popularity of internal ventures is associated with an emphasis on greater corporate focus and greater efficiency. For example, the identification and re-engineering of *existing* business processes became fashionable in the mid-1990s, but as firms have begun to exhaust the benefits of this approach they are now exploring options for creating *new* businesses. Such diversification may be vertical, that is downstream or upstream of the current process in order to capture a greater proportion of the value added; or horizontal, that is by exploiting existing competencies across additional product markets.

For example, the fossil fuel energy sector has been facing for a long time the threat of extinction – coal, gas, oil reserves being exhausted. The cloud that has hung on the horizon is the collapse of its core business, and the temptation is therefore to diversify into other business areas that will ensure the continued existence of the firm. BP, Elf, Shell, Standard Oil and Total all set up corporate venturing initiatives when oil companies were first experiencing declining margins and an unfavourable world economic climate, with diversification as their main objective. These initiatives were primarily exercised through investment in external opportunities, although some have pursued venture opportunities based on internal technologies and expertise.

To Develop New Competencies

Growth and diversification are generally based on the exploitation of existing competencies in new products markets, but a corporate venture can also be used as an opportunity for learning new competencies. Organizational learning has four components:[8]

* knowledge acquisition;
* information interpretation;
* information distribution;
* organizational memory.

An organization can acquire knowledge by experimentation, which is a central feature of formal R&D and market research activities. However, different functions and divisions within a firm will develop particular frames of reference and filters based on their experience and responsibilities, and these will affect how they interpret information. Greater organizational learning occurs when more varied interpretations are made, and a corporate venture can better perform this function as it is not confined to the needs of existing technologies or markets.

Similarly, a corporate venture can act as a broker or clearing house for the distribution of information within the firm. In practice, large organizations often do not know what they know. Many firms now have databases and groupware to help store, retrieve and share information, but such systems are often confined to 'hard' data. As a result, functional groups or business units with potentially synergistic information may not be aware of where such information could be applied. Organizational learning occurs when more of an organization's components obtain new knowledge and recognize it as of potential use.

Organizational memory is the process by which knowledge is stored for future use. Such information is stored either in the memories of members of an organization, or in the operating procedures and routines of the organization. The former suffers from all of the shortcomings of human memory, with the additional organizational problem

TABLE 10.1 Objectives of corporate venturing in the UK

Objective	Mean rank*
1. Long-term growth	4.58
2. Diversification	3.50
3. Promote entrepreneurial behaviour	2.68
4. Exploit in-house R&D	2.23
5. Short-term financial returns	2.08
6. Reduce/spread cost of R&D	1.81
7. Survival	1.76

($n = 90$). * Scale: $1 =$ minimum, $5 =$ maximum importance.
Source: From Withers (1997) *Window on Technology: Corporate venturing in practice.* Withers, London.

TABLE 10.2 Comparison of motives for internal corporate venturing in the USA and Japan

	US firms ($n = 43$)	Japanese firms ($n = 149$)
To meet strategic goals	76	73
Maturity of the base business	70	57
To provide challenges to managers*	46	15
To survive	35	28
To develop future managers*	30	17
To provide employment*	3	24

* Denotes statistically significant difference. Reprinted by permission of Harvard Business School Press. Source: From *Corporate Venturing: Creating new businesses within the firm* by Block, Z. and I. MacMillan. NIA, Boston, 1993. Copyright ©1993 by the President and Fellows of Harvard College: all rights reserved.

of personnel loss or turnover. Organizational procedures and routines may provide a more robust memory, but difficulty in anticipating future needs means that much non-routine information is never stored in this way. Over time these routines create and are reinforced by artefacts such as organizational structures, procedures and policies which suit existing technologies and markets, but make it difficult to store and retrieve non-routine information. A corporate venture can act as a repository for such knowledge.

Organizational learning is more difficult in conventional product development activities because of the cost and time pressures. For example, there will be a trade-off between reducing the time and cost of development of a specific product, and documenting what has been learnt for future development projects.

In practice, the primary motives for establishing a corporate venture are strategic: to meet strategic goals and long-term growth in the face of maturity in existing markets (Table 10.1). However, personnel issues are also important. Sectorial and national differences exist. In the USA, new ventures are also used to stimulate and develop entre-preneurial management, and in Japan they help provide employment opportunities for managers and staff relocated from the core businesses (Table 10.2). Nonetheless, the primary objectives are strategic and long term, and therefore warrant significant management effort and investment.

10.3 Managing Corporate Ventures

A corporate venture is rarely the result of a spontaneous act or serendipity. Corporate venturing is a process that has to be managed. The management challenge is to create an environment that encourages and supports entrepreneurship, and to identify and support potential entrepreneurs (see Box 10.1). In essence, the venturing process is simple, and consists of identifying an opportunity for a new venture, evaluating that opportunity and subsequently providing adequate resources to support the new venture. There are six distinct stages, divided between definition and development.[9]

Definition Stages

1. Establish an environment that encourages the generation of new ideas and the identification of new opportunities, and establish a process for managing entre-preneurial activity.
2. Select and evaluate opportunities for new ventures, and select managers to imple-ment the venturing programme.
3. Develop a business plan for the new venture, decide the best location and organ-ization of the venture and begin operations.

Development Stages

4. Monitor the development of the venture and venturing process.
5. Champion the new venture as it grows and becomes institutionalized within the corporation.
6. Learn from experience in order to improve the overall venturing process.

Identification of Corporate Ventures

Creating an environment which is conducive to entrepreneurial activity is the most important, but most difficult stage. Superficial approaches to creating an entrepreneurial culture can be counterproductive. Instead, venturing should be the responsibility of the entire corporation, and top management should demonstrate long-term commitment to venturing by making available sufficient resources and implementing the appropriate processes.

The conceptualization stage consists of the generation of new ideas and identification of opportunities that might form the basis of a new business venture. The interface between R&D and marketing is critical during the conceptualization stage, but the scope of new venture conceptualization is much broader than the conventional activities of the R&D or marketing functions, which understandably are constrained by the needs of existing businesses. At this stage three basic options exist:

1. Rely on R&D personnel to identify new business opportunities based on their technological developments, that is, essentially a 'technology-push' approach.
2. Rely on marketing managers to identify opportunities, and direct the R&D staff into the appropriate development work, essentially a 'market-pull' approach.
3. Encourage marketing and R&D personnel to work together to identify opportunities.

The technology-push approach has been described as being 'first generation R&D', the 'market-pull' strategy as 'second generation' and the close coupling 'third generation', the implication being that firms should progress to close coupling.[10] The issue of strategic positioning was discussed in detail in Chapter 6. In theory, the third option is most desirable as it should encourage the coupling of technological possibilities and market opportunities at the concept stage, before substantial resources are committed to evaluation and development. However, in practice technology push appears to be the dominant strategy. This is because at the conceptualization stage highly specialized technical knowledge is required about what is feasible and what is not, and therefore what the characteristics of the final product are likely to be. Nevertheless, R&D personnel may become locked into a specific technical solution or address the needs of atypical users. Therefore management must ensure that R&D personnel are sufficiently flexible to modify or drop their proposals should technical issues or market requirements dictate.

Drucker identifies a number of sources of ideas and opportunities, and argues that the search process should be systematic rather than relying on serendipity.[11] He suggests seven common sources of opportunities which should be monitored on a routine basis:
- demographic changes;
- new knowledge;
- incongruities (i.e. gaps between expectations and reality);

- changes in industry or market structure;
- unexpected successes or failures;
- process needs;
- changes in perception.

Other sources of ideas include trade shows and exhibitions and trade journals. In the specific case of new business ventures there are four primary sources of ideas:

- the 'bright idea';
- customers requesting a new product or service;
- internal analysis of a company's competencies and business processes;
- scanning of external opportunities in related technologies, markets or services.

Contrary to popular perceptions, the 'bright idea' is the least common and most risky source of new business ventures, because the other sources are more directly stimulated by a market need, technological expertise or both together. These can be the initiative of either someone at operational or managerial level; the former may have difficulties finding an effective champion, whereas the latter may be too powerful, having the influence to force through an idea before it is exhaustively tested. A balance needs to be achieved between screening and championing the proposal. In contrast, a business venture based on a customer request has the highest chance of success as a potential market is to some extent predetermined. However, such ventures are typically based on an adaptation or extension of an existing product or service, and therefore less likely to spawn radical new businesses. These tend to be bottom-up initiatives, and the most difficult problem is to decide how the potential new business relates to the existing business or division. By far the two most promising corporate ventures are the result of systematic scanning of the internal and external environments, a process we advocated in Chapter 2.

Scanning the internal environment   The systematic scanning of the organization involves the search and evaluation of shelved concepts, novel combinations of existing technologies and new applications for existing competencies. This type of activity is sometimes referred to as 'knowledge management', and is based on the notion that large organizations typically do not know what they know, and fail to exploit existing knowledge. For example, Boisot developed the concept of C-space (culture space) to analyse the flow of knowledge within and between organizations.[12] It consists of two dimensions: *codification*, the extent to which information can be easily expressed, and *diffusion*, the extent to which information is shared by a given population. Using this framework he proposes a social learning cycle which involves four stages: scanning, problem-solving, diffusion and absorption (Figure 10.2).

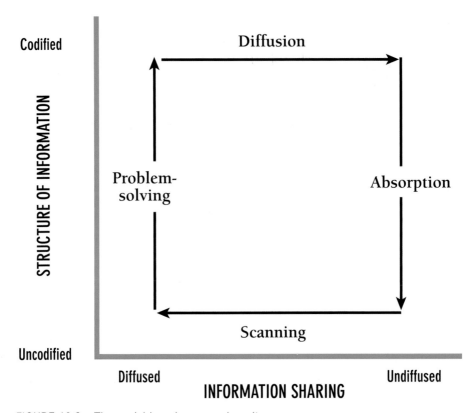

FIGURE 10.2 The social learning curve in culture space

Source: Griffiths, D. and M. Boisot (2000) 'Are there any competencies out there? Identifying and using technical competencies', in Tidd, J. (ed.), *From Knowledge Management to Strategic Competence.* Imperial College Press, London, pp. 199–228.

Knowledge management has now become a staple of the consultancy industry. It is possible to identify two distinct approaches to knowledge management. The first is based on investments in IT, usually based on intranet technology. The second approach is more people and process-based, and attempts to encourage staff to identify, store, share and use information throughout the organization. Research suggests that as in previous cases of process innovation, the benefits of the technology are not fully realized unless the organizational aspects are first dealt with. A wide range of knowledge management strategies are possible (Table 10.3). No single approach will be appropriate in all circumstances, and the optimum strategy will depend on the existing organizational structure, processes and culture, as well as the availability of resources and urgency of action. In an R&D context, Coombs identifies 'knowledge management practices' which include mapping knowledge relationships, managing intellectual property, information management and human resource management.[13] Another approach advocates the use of formally appointed 'knowledge brokers' to sys-

TABLE 10.3 Knowledge management strategies

Strategy	Characteristics	Requirements	Risks
Ripple	Bottom-up, continuous improvement, e.g. quality management	Process tools, sustained motivation	Isolation from technical excellence
Integration	Integration of functional knowledge within processes, e.g. product development	Improved interfaces, early involvement, overlapping phases	Conformity, co-ordination burden
Embedding	Coupling of systems, products and services, e.g. enterprise resource planning (ERP)	Common information systems and technology, motivation and rewards	Loss of autonomy, system complexity
Bridge	New knowledge by novel combination of existing competencies, e.g. architectural innovations	Common language and objectives	High control needs, technical feasibility, market failure
Transfer	Exploiting existing knowledge in a new context, e.g. related diversification	New market knowledge	Inappropriate technology, customer support and service

Source: Adapted from Friso den Hertog, J. and E. Huizenga (2000) *The Knowledge Enterprise.* Imperial College Press, London.

tematically scavenge the organization for old or unused ideas, to pass these around the organization and imagine their application in different contexts.[14] For example, Hewlett-Packard created a SpaM group to help identify and share good practice among its 150 business divisions. Before the new group was formed, divisions were unlikely to share information because they often competed for resources and were measured against each other. Similarly, Skandia, a Swedish insurance company active in overseas markets, attempts to identify, encourage and measure its intellectual capital, and has appointed a 'knowledge manager' who is responsible for this. The company has developed a set of indicators that it uses both to manage knowledge internally, and for external financial reporting. However, improved communication is not an end in itself. It is desirable to distinguish between more general experiments to improve the work environment or to make it more democratic, from those practices proven to improve performance. For instance, simply creating a participative climate at work will not generate new business ventures, whereas improved internal communications combined with management support and a critical assessment of new business proposals will help create new businesses.[15]

Scanning the external environment   Scanning the external environment consists of searching, filtering and evaluating potential opportunities from outside the organization, including related and emerging technologies, new market and services, which can be exploited by applying or combining with existing competencies. A study of corporate entrepreneurship in 169 companies concluded that 'opportunity recognition, which is a precursor to entrepreneurial behaviour, is often associated with a flash of genius, but in reality is probably more often the end result of a laborious process of environmental scanning'.[16] External scanning can be conducted at various levels. It can be an operational initiative with market- or technology-focused managers becoming more conscious of new developments within their own environments, or a top-driven initiative where venture managers or professional capital firms are used to monitor and invest in potential opportunities. For example, the TRIZ system developed by Genrich Altshuller identifies standard solutions to common technical problems distilled from an analysis of 1.5 million patents, and applies these in different contexts. Many leading companies use the system, including 3M, Rolls-Royce and Motorola.[17] Venture capital firms can help firms to monitor the external environment without distraction, and to take equity stakes in potential partners fairly anonymously. This practice is common in the pharmaceutical industry, where firms use a range of strategies to tap into the knowledge of biotechnology firms, including direct investment, licensing deals and indirect investment through professionally managed venture funds. Direct investments are favoured for technologies of high strategic importance, licensing for process and product developments, and indirect investments for windows on emerging technologies.[18]

Development of Corporate Ventures

Having identified the potential for a new venture, a product champion must convince higher management that the business opportunity is both technically feasible and commercially attractive and therefore justifies development and investment. Potential corporate entrepreneurs face significant political barriers:

- They must establish their legitimacy within the firm by convincing others of the importance and viability of the venture.
- They are likely to be short of resources, but will have to compete internally against established and powerful departments and managers.
- As advocates of change and innovation, they are likely to face at best organizational indifference, and at worst hostile attacks.

To overcome these barriers a potential venture manager must have political and social skills, in addition to a viable business plan. In addition, the product champion must be able to work effectively in a non-programmed and unpredictable environment. This contrasts with much of the R&D conducted in the operating divisions which is likely to be much more sequential and systematic. Therefore a product champion requires dedication, flexibility and luck to manage the transition from product concept to corporate venture, in addition to sound technical and market knowledge. The product champion is likely to require a complementary organizational champion, who is able to relate the potential venture to the strategy and structure of the corporation. A number of key roles must be filled when a new venture is established:[19]

- The technical innovator, who was responsible for the main technological development.
- The business innovator or venture manager, who is responsible for the overall progress of the venture.
- The product champion, who promotes the venture through the early critical stages.
- The executive champion or organizational champion, who acts as a protector and buffer between the corporation and venture.
- A high-level executive responsible for evaluating, monitoring and authorizing resources for the venture, but not the operation of specific ventures.

A new venture requires two types of skill: the technical knowledge necessary to develop the product, process or knowledge base; and the management expertise necessary to communicate and sell to the markets and parent organization (Table 10.4). The dilemma that has to be resolved in each case is whether to allow and develop technical experts to play a role in selling the product or managing the business, or to place managers above their heads to take the baton on.

To take project managers to venture manager status is often dangerous. Whilst these individuals understand the product fully, they may have difficulties in maximizing the cost/price differential, perhaps not always realizing the commercial value of the product

TABLE 10.4 Systematic differences between R&D and marketing personnel

	R&D personnel	Marketing personnel
Work environment		
Structure	Well defined	Ill defined
Methods	Scientific and codified	Ad hoc and intuitive
Data	Systematic and objective	Unsystematic and subjective
Pressures	Internal: How long will it take?	External: How long do we have?
Professional orientation		
Assumptions	Serendipity	Planning
Goals	New ideas: Can it be improved?	Big ideas: Does it work?
Performance criteria	Technical quality	Commercial value
Education and experience	Deep and focused	Broad

and being less experienced in the negotiation process. It can be equally difficult to identify a manager who can communicate the product characteristics to customers with real needs, relay those needs to the product development team, and communicate and justify venture management needs to the corporate centre.

Assessing the Venture

The most appropriate filter to apply to a potential venture will depend on the motive for venturing. Roberts illustrates the point:

> The best time to detect if a CEO has a strategy or not is to observe the management team at work when trying to evaluate opportunities, especially those somewhat remote from the current business. On these occasions, we noticed that when faced with unfamiliar opportunities, management would put them through a hierarchy of different filters. The ultimate filter was always a fit between the products, customers and markets that the opportunity brought and one key element, or driving force, of the business. This is a clear signal that management had a sound filter for its decision.[20]

Without the type of strategic filters discussed in Chapter 6, managers are forced to rely on narrow financial methods of evaluation, such as potential sales growth, margins or net present value. The seduction to new opportunities is highest when a company is cash rich. For example, as illustrated in Chapter 4, Daimler Benz in the mid-1980s began an acquisition trail of many unrelated businesses, which have subsequently been disposed of. This distraction to top management may also be a contributing factor for Mercedes Benz losing its hold in its home market to BMW. Similarly, Sony and Matsushita have made questionable decisions to diversify into US film production, respectively Columbia and MCA.

In assessing any venture it is essential to specify the purpose and the criteria for success in the new market, business or technology. Ultimately the style of assessment adopted will depend on the size of the potential venture, the abilities of the people who currently understand the product and whether new partners or managers are expected to be introduced following assessment. See Box 10.3 for a description of how Lucent Technologies approached this. A plan needs to be written by the managers involved in the venture, in part to test whether they understand the business as well as the technology. It is essential for in-house managers to be fully involved in the market research. The use of market research consultants should be limited to providing a first pass of potential markets. No one can know the product better, especially if it is new, and has niche applications, than the people who have worked on its development, and whose future careers may depend on it.

The purpose and nature of a business plan for a new venture differ from that for established businesses. The main purpose of the venture plan is to establish if and how

BOX 10.3
LUCENT'S NEW VENTURE GROUP

Lucent Technologies was created in 1996 from the break-up of the famous Bell Labs of AT&T. Lucent established the New Venture Group (NVG) in 1997 to explore how better to exploit its research talent by exploiting technologies which did not fit any of Lucent's current businesses, its mission was to '. . . leverage Lucent technology to create new ventures that bring innovations to market more quickly . . . to create a more entrepreneurial environment that nurtures and rewards speed, teamwork, and prudent risk-taking'. At the same time it took measures to protect the mainstream research and innovation processes within Lucent from the potential disruption NVG might cause. To achieve this balance, at the heart of the process are periodic meetings between NVG managers and Lucent researchers, where ideas are 'nominated' for assessment. These nominated ideas are first presented to the existing business groups within Lucent, and this creates pressure on the existing business groups to make decisions on promising technologies, as the vice president of the NVG notes: 'I think the biggest practical benefit of the (NVG) group was increasing the clockspeed of the system.'

If the nominated idea is not supported or resourced by any of the businesses, the NVG can develop a business plan for the venture. The business plan would include an exit strategy for the venture, ranging from an acquisition by Lucent, external trade sale, IPO (initial public offering), or license. The initial evaluation stage typically takes two to three months and costs US$50 000 to $100 000. Subsequent stages of internal funding reached $1 m. per venture, and in later stages in many cases external venture capital firms are involved to conduct 'due diligence' assessments, contribute funds and management expertise. By 2001, 26 venture companies had been created by the NVG, and included 30 external venture capitalists who invested more than $160m. in these ventures. Interestingly, Lucent re-acquired at market prices three of the new ventures NVG had created, all based on technologies that existing Lucent businesses had earlier turned down. This demonstrates one of the benefits of corporate venturing – capturing *false negatives* – projects which were initially judged too weak to support, and that are rejected by the conventional development processes. However, following the fall in telecom and other technology equity prices, in 2002 Lucent sold its 80% interest in the remaining ventures to an external investor group for under $100 m.

Source: Chesbrough, H. (2003) *Open Innovation*, Harvard Business School Press.

TABLE 10.5   Components of a typical business plan for a new venture

1. Description of the proposed business, including its objectives and characteristics.
2. Strategic relationship between the new business and the parent firm.
3. The target markets, including size, trends, reasons for purchase and specific target customers.
4. Assessment of the present and anticipated competition.
5. Human, physical and financial resources required.
6. Financial projections, including assumptions and sensitivity analysis.
7. Well-defined milestones and go/no-go conditions.
8. Principal risks and how they will be managed.
9. Definition of failure, and conditions under which the venture should be terminated.
10. Description of the venture's management and compensation required.

to conduct the new business, and to attract key personnel and resources. The purpose of a plan for an existing business is to monitor and control performance. The technical and commercial aspects of a new venture plan will have much greater uncertainty than that for existing businesses. There are 10 essential elements of a new venture plan (Table 10.5).

Every new venture is an industrial experiment, and therefore the experiment must be designed to allow the assumptions and risks to be evaluated. The plan will be based on assumptions concerning the technology, product, market, economy, competition and the business environment, and given the inherent uncertainty of a new venture it is essential that testing of these assumptions is included in the go/no-go criteria. Similarly, contingency plans should be included to minimize any technological, market, management or financial risk.

The main criteria for assessing the business plan for a corporate venture are strategic fit and potential to enhance competitive position. But beyond such basic requirements, there appear to be significant differences between the criteria applied by American and Japanese firms (Table 10.6). American firms typically expect a high return on investment from corporate ventures, and therefore favour technologies that have high potential for premium pricing or rapid growth. Japanese firms appear to favour ventures that use related technologies to create new markets. This suggests that American firms view such ventures as opportunities to generate cash for the core business, whereas the Japanese see them as an opportunity to create new businesses. There is relatively little research on corporate venturing in Europe, but the experience of corporate ventures in the UK suggests a pattern closer to that of the American experience.

Following development of an initial plan for the project, milestones for progress assessment should be developed and agreed upon by those responsible for monitoring and managing the venture. Milestones will ideally be a mix of financial and strategic.

TABLE 10.6 Criteria for selecting corporate ventures

	USA (n = 39)	Japan (n = 126)
Strategic fit	4.1	3.9
Competitive advantage	4.0	3.8
Potential return on investment*	3.9	3.6
Existence of market*	3.9	4.4
Potential sales	3.9	3.9
Risk/reward ratio	3.8	3.6
Presence of product champion	3.6	4.0
Synergy	3.5	3.7
Opportunity to create new market*	3.1	3.8
Closeness to present technology*	2.9	3.5
Patentability*	2.3	2.9

1 = unimportant, 5 = critical. * Denotes statistically significant difference.
Source: Reprinted by permission of Harvard Business School Press. From *Corporate Venturing: Creating new businesses within the firm*, by Block, Z. and I. MacMillan. Boston, Mass., 1993. Copyright ©1993 by the President and Fellows of Harvard College; all rights reserved.

Strategic milestones can focus managers more on the long-term business direction rather than the (often) short-term financial criteria. Examples of milestones for the initial stages of a venture include: achievement of product development within time stated; product having been demonstrated on at least three sites; a sale to an important customer has been finalized.

10.4 Structures for Corporate Ventures

The choice of location and structure for a new venture will depend on a number of factors, the most fundamental being how close the activities are to the core business. How close a venture's focal activity is to the parent firm's technology, products and markets will determine the learning challenges the venture will face and the most appropriate linkages with the parent. In practice, there is likely to be some trade-off between the desire to optimize learning and the desire to optimize the use of existing resources. The venture will need to acquire resources, know-how and information from the corporate parent, get sufficient attention and commitment, but at the same time be protected politically and allowed optimal access to the target market. Consideration of these sometimes conflicting requirements will determine the best location and structure for the venture.

The classic study by Burgelman and Sayles of six internal ventures within a large

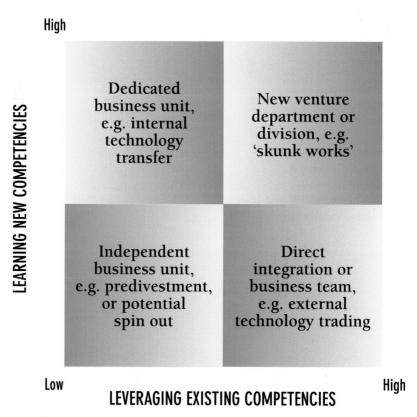

FIGURE 10.3 The structure of a corporate venture depends on the balance between the desire to learn and leverage competencies

Source: Adapted from Tidd, J. and S. Taurins (1999) 'Learn or leverage? Strategic diversification and organizational learning through corporate ventures', *Creativity and Innovation Management*, **8** (2), 122–129.

American corporation demonstrated the managerial and administrative difficulties of establishing and managing internal ventures.[21] The study confirmed that no single organizational solution is optimal, and that different structures and processes are required in different circumstances. The choice of structure will depend on the level and urgency of the venturing activity, the nature and number of ventures to be established, and the corporate culture and experience. More fundamentally, it will depend on the balance between the desire to learn new competencies and the need to leverage existing competencies (Figure 10.3). For example, in e-business established firms are faced with the decision whether to develop separate businesses to exploit the opportunities, or to fully integrate e-business with the existing business. Neither strategy nor structure appears to be inherently superior, and depends on a consideration of the relatedness of the assets, operations, management and brand.[22] Design options for corporate ventures include:

- direct integration with existing business;
- integrated business teams;
- a dedicated staff function to support efforts company-wide;
- a separate corporate venturing unit, department or division;
- divestment and spin-off.

Each structure will demand different methods of monitoring and management – that is, procedures, reporting mechanisms and accountability. These choices are illustrated by our own study of corporate venturing in the UK, as described in the following sections.

Direct Integration

Direct integration as an additional business activity is the preferred choice where radical changes in product or process design are likely to impact immediately on the mainstream operations and if the people involved in that activity are inextricably involved in day-to-day operations. For example, many engineering-based companies have introduced consultancy to their business portfolio, and in other technical organizations with large laboratory facilities these too have been sold out for analysis of samples, testing of materials, etc. In such cases it is not possible to outsource such activities because the same personnel and equipment are required for the core business.

The Natural History Museum (NHM) provides a good example. In the late 1980s funding was withdrawn from the museum, and it had to continue to reduce its costs and to find new sources of revenue. A principal aim was to generate sufficient revenue streams to avoid a massive reduction in research capacity, and avoid further redundancies. The NHM's resource of some 300 scientists has been developed in a commercial capacity from ground level up. The transformation from research to consultancy work required a complete cultural change. Heads of department ('keepers') were made responsible for identifying and developing the consultant activity, in addition to their normal operational management duties. Given the pervasive nature of the scientific expertise scattered throughout the organization and across departments, an integrated approach to venturing was the only realistic option. Science consultancy is now earning in excess of £1 m. per annum, and the museum has been able to buy new capital equipment to support its work.

Integrated Business Teams

Integrated business teams are most appropriate where the expertise will have been nurtured within the mainstream operations, and may support or require support from those operations for development. Strategically, the product is sufficiently related to the mainstream business's key technologies or expertise that the centre wishes to retain some control. This control may either be to protect the knowledge that is intrinsic in the activity or to ensure a flow-back of future development knowledge. A business team of secondees is established to co-ordinate sourcing of both internal and external clients, and is usually treated as a separate accounting entity in order to ease any subsequent transition to a special business unit.

An example of this type of venture is provided by the development of an expert system within Welsh Water. The original system for planning water distribution and supply was devised by Cardiff University, but the system was co-developed and proto-typed by a new business team working with the mainstream divisions. The new business team was created and supervised by the Enterprise Division, although it worked closely with the other divisions at operational level. The development team consisted of secondees released from other divisions for a duration of either six or 12 months. When the product was fully developed, it was passed to the Enterprise Division for sale externally.

New Ventures Department

A new ventures department is a group separate from normal line management that facilitates external trading. It is most suitable when projects are likely to emerge from the operational business on a fairly frequent basis and when the proposed activities may be beyond current markets or the type of product package sold is different. This is the most natural way for the trading of existing expertise to be developed when it lies fragmented through the organization, and each source is likely to attract a different type of customer. The group has responsibility for marketing, contracting and negotiation, but technical negotiation and supply of services take place at operational level.

Imperial College of Science, Technology and Medicine is a good example. Two separate venture organizations exist. One, IMPEL, is responsible for patenting and licensing technologies and products that emerge, and the other, ICON, for contracting out expertise or processes. ICON combs the organization for potential leading-edge process-oriented technologies, which are supported by available and willing expertise, and attempts to match these with market opportunities. ICON also offers solutions to problems identified by potential clients. For example, ICON manages external access to the ion implantation facility at Imperial, which has been applied to completely new market applications.

Techniques here are built very much on internal competencies that could not be separated from the mainstream activities and could not really be spun out or licensed to another company. In fact, development cycles are so short that it would not be realistic to spin off an individually developed technique. There are important additional benefits to the college through this activity: companies that initially commission a consultancy project may subsequently fund a related research project; also, researchers are brought far closer to the needs of technology users.

New Venture Division

A new venture division provides a safe haven where a number of projects emerge throughout the organization, and allows separate administrative supervision. Strategically, top management can retain a certain level of control until greater clarity on each venture's strategic importance is understood, but the efficiency of the mainstream business needs to be maintained without distraction, so some autonomy is required. Operational links are loose enough to allow information and know-how to be exchanged with the corporate environment. The origins of such a division vary:

- An effort to bring existing technologies and expertise throughout the company together for adaptation to new or existing markets.
- To combine research from different fields or locations, to accelerate the development of new products.
- To purchase or acquire expertise currently outside of the business for application to internal operations, or to assist new developments.
- To examine new market areas as potential targets for existing or adapted products within the current portfolio.

Where a critical mass of projects exist, a separate new venture division allows greater focus on the external environment, and the distance from the core corporation facilitates a global and cross-divisional view to be taken. Unfortunately, the division can often become a kind of dustbin for every new opportunity, and therefore it is critical to define the limits of its operation and its mission, in particular the criteria for termination or continued support of specific projects.

For example, British Gas established a division to exploit internal and external technologies. The division took technology that existed within the company and sought to exploit it commercially, in terms of a licence, new application or market. By means of a venture capital fund, the division also identified external technologies that might enhance the company's mainstream technical knowledge.

Similarly, Rolls-Royce Business Ventures Limited was set up to exploit Rolls-Royce technology in new product areas. Its mission was to lay the foundations for new businesses which had a good fit with the technology and skills available to the parent company, but which were outside the mandate of the parent company's existing business groups. The company, which was given its own site, has now been wound up. Nevertheless, it established two spin-off companies during its short life: Stresswave Technology Limited and Reflex Manufacturing Systems Limited.

Special Business Units

Special dedicated new business units are wholly owned by the corporation. High strategic relevance requires strong administrative control. Businesses like this tend to come about because the activity is felt to have enough potential to stand alone as a profit centre, and can thus be assessed and operated as a separate business entity. The requirement is that key people can be identified and extracted from their mainstream operational role.

For the business to succeed under the total ownership and control of a large corporate it must be capable of producing significant revenue streams in the medium term. On average, the critical mass appears to be around 12% of total corporate turnover, in a £200 m. business perhaps £20 m., but in some cases the threshold for a separate unit is much higher. A potential new business must not only be judged on its relative size or profitability, but more importantly, by its ability to sustain its own development costs. For example, a profitable subsidiary may never achieve the status of a separate new business if it cannot support its own product development.

However, physically separating a business activity does not ensure autonomy. The greatest impediment to such a unit competing effectively in the market is a cosy corporate mentality. If managers of a new business are under the impression that the corporate parent will always assist, provide business and second its expertise and services at non-market rates, that business may never be able to survive commercial pressures. Conversely, if the parent plans to retain total ownership, the parent cannot realistically treat that unit independently. For example, a company which had been set up as a special business unit was undermined when the parent placed an order from an alternative supplier: the venture lost a large proportion of annual revenues, but more importantly that business lost all credibility in the eyes of its other customers. In 2000 Unilever established a ventures unit to invest in Internet businesses, established by staff or external start-ups in related businesses, and for the first time has allowed staff to take a financial interest in the new ventures.

Independent Business Units

Differing degrees of ownership will determine the administrative control over independent business units, ranging from subsidiary to minority interest. Control would only be exercised through a board presence if that were held. There are two reasons for establishing an independent business as opposed to divisionalizing an activity: to focus on the core business by removing the managerial and technical burden of activities unrelated to the mainstream business; or to facilitate learning from external sources in the case of enabling technologies or activities. This structure has benefits for both parent and venture:

- Defrayed risk for parent, greater freedom for venture.
- Less supervisory requirement for parent, less interference for venture.
- Reduced management distraction for parent, and greater focus for venture.
- Continued share of financial returns for parent, greater commitment from managers of the venture.
- Potential for flow-back or process improvements or product developments for parent, and learning for the venture.

Whilst the mainstream corporation may not be short of cash for investment, bringing in external funding provides additional advantages, such as sharing of risk with other parties and insulating the venture from changes in corporate investment policy. When release of ownership is allowed, venture capital firms can provide valuable help. For example, CharterRail, the spin-off from GICN, was provided with new sources of capital, and assisted in developing its management team and board by the venture capital investors.

By releasing some ownership, executives of the new business can hold a share of the business, indeed this could be a requirement of their involvement. Securing a capital commitment from these key people ensures their personal commitment to its commercial success. In spin-offs that have emerged from Thorn EMI, GICN and ICI, an equity interest has been maintained in each, but a board presence in only two. Part ownership does not have to involve the centre managerially, but when control is reduced the centre must perceive the holding more as an investment than a subsidiary.

The assignment of technical personnel is one of the most difficult problems when establishing an independent business unit. If the individuals necessary to co-ordinate future product development are unwilling to leave the relative security and comfort of a large corporate facility, which is understandable, the new business may be stopped in its tracks. It is critical to identify the most desirable individuals for such an operation, assessed in terms of their technical ability and personal characteristics. It is also important to assess the effect of these individuals leaving the mainstream development operations, as the capability of the parent's operations could be easily damaged.

Nurtured Divestment

Nurtured divestment is appropriate where an activity is not critical to the mainstream business. The product or service has most likely evolved from the mainstream, and whilst supporting these operations it is not essential for strategic control. The design option provides a way for the corporate to release responsibility for a particular business area. External markets may be built up prior to separation, giving time to identify which employees should be retained by the corporate and providing a period of acclimatization for the venture. The parent may or may not retain some ownership.

Siemens has been very successful at developing nurtured divestments. It made a 25% stake in Tele-Processing Systems GmbH, which was started in 1986 by three Siemens employees who had developed products for the remote operation of computers, network auditing and in-house private branch exchanges (PBXs). Similarly, Siemens took a 15% stake in Micro Quartz GmbH which was set up in 1988 in order to take a technology using fused silicon tubes for the chromatography of gas to the manufacturing level, and 38% of ECT GmbH, a spin-off based on electron beam manufacturing equipment developed in-house.

Instead of a formal equity stake, a parent may support a spin-off by contracted or seconded technologies or expertise. Manpower may be seconded and technologies or know-how may be sold through some legal agreement, whether licence or sale with a confidentiality tie. For example, British Gas developed a real-time expert system by forming a club of 35 companies. Once the prototype was ready, a new company was formed with the backing of a small number of the original club companies to market the product. British Gas provided the chairman and technical director of the new company. People who had been seconded from British Gas subsequently joined the new venture.

Complete Spin-off

No ownership is retained by the parent corporation in the case of a complete spin-off. This is essentially a divest option, where the corporation wants to pass over total responsibility for activity, commercially and administratively. This may be due to strategic unrelatedness or strategic redundancy, as a consequence of changing corporate strategic focus. A complete spin-off allows the parent to realize the hidden value of the venture, and allows senior management of the parent to focus on their main business.

For example, in 1991 Quaker Oats Co. spun off Fisher Price Inc. Quaker management saw this as allowing the company to concentrate its efforts and resources on its core grocery business and to give shareholders more choice as to the industry segments in which they invest. The toy operations were separated from the divisional structure to become a separate entity. Once rationalized as a separate business, shares were issued to Quaker Oats shareholders on a one-for-five basis, and existing shareholders were effectively used as buyers, with no need for formal divestment. In this way Fisher Price stock was separately floated and market capitalizations were enhanced beyond the previous sum of the whole.

In addition to having the most appropriate structure for corporate venturing, Tushman and O'Reilly identify three other organizational aspects that have to be managed to achieve what they call the 'ambidextrous' organization – the co-existence of young, entrepreneurial, risky ventures with the more established, proven operations:[23]

- *Articulating a clear, emotionally engaging and consistent vision.* This helps to provide a strategic anchor for the diverse demands of the mainstream and venture businesses.
- *Building a senior team with diverse competencies.* The composition and demography of the senior team is critical. Homogeneity typically results in greater consensus, faster decision-making, and easier execution, but lowers levels of creativity and innovation, whereas heterogeneity can cause conflicts, but promotes more diverse perspectives. To achieve a balance, they suggest homogeneity by tenure/length of service, but diversity in backgrounds and perspectives. Alternatively, senior teams can be relatively homogenous, but have more diverse middle management teams reporting to them.
- *Developing healthy team processes.* The need for creativity needs to be balanced with the need for execution, and team members must be able to resolve conflicts and to collaborate.

However, there is disagreement in the literature regarding the influences of the degree of integration of corporate ventures and the effects on their subsequent success. A study of almost 100 corporate ventures in Canada provided strong support for the need for high levels of integration between the corporate parent and the ventures. It found that the success of a venture was associated with a strong relationship with the corporate parent – specifically use of the parent firm's systems and resources – and conversely that the autonomy of ventures was associated with lower performance of the venture.[24] This appears to contradict the more general body of research which suggests that the managerial independence of ventures is associated with success. For example, a study of spin-offs from Xerox found that those ventures with high levels of funding and senior management from the parent were less successful than those funded more by professional venture capitalists and outside management.[25] One reason for this disagreement might be the period of assessment and measures of success: the Canadian study used the achievement of milestones as the measure of success, and the average

age of the new ventures was less than five years; the Xerox study used two measures of success, average rates of growth and financial market value of the ventures, and assessed these over 20 years. In any case, this reflects the real difficulty of getting the right balance between autonomy and integration, as one study found:

> Internal entrepreneurs are faced with two choices: either go underground or spin-off a new venture, with or without the blessing of the parent company . . . it is therefore advisable to spin-off a company in agreement with the parent that contributes technology, personnel and possibly cash, in exchange for minority equity participation. The parent can hold one or more seats on the board of directors, provide advice, networking, and marketing support, share its R&D and pilot production facilities etc, *but must refrain from interfering with management* . . . continued cooperation with the parent also carries a price . . . with a seat on the board the parent is able to monitor and influence the evolution of the technology, and more importantly of the market. [26] (emphasis added)

This is critical as the Xerox study found that the eventual successful business models developed by the spin-offs evolved substantially from the initial plans at formation, were very different to the business models of the parent company, and involved significant experimentation to explore the technologies and markets.

10.5 Learning Through Internal Ventures

The success of corporate venturing varies enormously between firms, but on average around half of all new ventures survive to become operating divisions, which suggests that venturing may be a less risky strategy for diversification than acquisition or merger. Typically, a venture will achieve profitability within two to three years, and almost half are profitable within six years. However, the profitability of the overall corporate venturing process may be lower due to the effect of a few large failures. Four factors appear to characterize firms that are consistently successful at corporate venturing:

1. Distinguish between bad decisions and bad luck when assessing failed ventures.
2. Measure a venture's progress against agreed milestones, and if necessary redirect.
3. Terminate a venture when necessary, rather than make further investments.
4. View venturing as a learning process, and learn from failures as well as successes.

There are two main causes of failure of internal ventures: strategic reversal and the emergence trap.[22] Strategic reversal occurs because of a conflict between the timescales of the new venture and the parent organization. An internal venture may be set up for a number of reasons: to support a strategy of diversification, because of a risk-taking top management, an excess of corporate cash or a decline in the firm's main line of business. Whatever the reasons, the internal or external environment is unlikely to remain stable for the life of the new venture. A change of climate can result in the premature termination of a venture. Even normal business cycles may affect the fortunes of a new venture. For example, there appears to be a strong correlation between changes in corporate profits and the number of new ventures set up.[27]

The other, more subtle cause of venture failure is the emergence trap. As a venture expands it may lead to internal territorial infringements, and success leads to jealousy and may result in attempts to undermine the venture. Differences between the culture and style of managers in the parent firm and new venture are likely to amplify these problems (Table 10.7). In particular, new venture divisions are highly visible and represent a concentration of expenditure, and are therefore more vulnerable to changes in corporate performance or management sentiment.

In practice there is a trade-off between rapid growth and learning. A new venture will not have an indefinite period in which to prove itself, and in most cases corporate management will set high targets for growth and financial return in order to offset the risk and uncertainty inherent in a new venture. If successful, the venture will quickly achieve a track record and therefore attract further support from corporate management, resulting in a virtuous spiral of growth and investment. Conversely, if the venture fails to deliver early growth in sales or returns, it may be starved of further support, thus increasing the likelihood of subsequent failure, a vicious spiral of low investment and decline. There are a number of ways to help avoid these problems:[28]

- Make corporate and divisional managers aware of the long-term benefits of venture operations.
- Clearly specify the functions, procedures, boundaries and rewards of venture management.
- Establish a limited number of ventures with independent budgets.
- Establish and maintain multiple sources of sponsorship for ventures.

TABLE 10.7 Potential sources of conflict between corporate and new venture management

Corporate management	New venture management
Modest uncertainty	Major technical and market uncertainties
Emphasis on detailed planning	Emphasis on opportunistic risk-taking
Negotiation and compromise	Autonomous behaviour
Corporate interests and rules	Individualistic and ad hoc
Homogeneous culture and experience	Heterogeneous backgrounds

Therefore it is critical to define the purpose of a new venture, in order to apply the most appropriate financial and organizational structures. Firms may organize and manage new ventures in order to maximize exploitation of existing know-how, or to optimize learning, but not both. Therefore it is critical to define clearly scope and focal activity of a new venture, so that the appropriate linkages to other functions can be established (see Box 10.4). Where the primary motive is learning, systems and structures to support the new business must be established, rather than simply capitalizing on rapid sales growth. The precise structure and linkages with the parent firm will depend on the relatedness of product and process technologies and product markets (Table 10.8).

The failure of the parent company to define and articulate the role of the venture is the proximate cause of most difficulties experienced with corporate ventures. Such conflicts can be minimized by ensuring that the primary motive for the venture is made explicit, and communicated to both corporate and venture management. In this way

---

BOX 10.4
OXFORD UNIVERSITY PRESS'S ENTRY INTO MULTIMEDIA[29]

The Oxford University Press (OUP) is the largest university press in the world with a total in 1995 of just under 3000 employees and over 3000 titles published. It is not a company in the normal sense and has no shareholders, being legally a department of the University of Oxford, owned by the 'Chancellor, Masters, and Scholars of the University of Oxford'.

An electronic publishing department was created in 1986, which then consisted of an R&D manager and a secretary. A separate department was established because of the specialist skills required for the development and marketing of electronic products. Initially, the department had been subsidized from the Press's other activities. It started as a service centre for the publishing departments, but soon became an independent publishing unit, which has been a source of some tension with established departments. The performance of the existing book departments is assessed on turnover, so there is little financial reward for handing over material to electronic publishing, which is only required to pay a royalty to the originating department.

The aim was to exploit the skills of existing book staff and assets of the Press to differentiate the electronic products, but at the same time to develop new skills in electronic publishing. The background of the present staff is varied and includes book publishing, magazine publishing, software publishing, programming and television programming. The markets and distribution channels are also different from those for book products, so the electronic publishing department recruited its own dedicated marketing and sales staff for direct sales.

Electronic publishing differs from book publishing in a number of ways. Overall, there are many more activities and a much greater degree of involvement is required during software development. Another difference from books is that the changing technologies and markets lead to a need to bring out products much more quickly than is the norm with books. For example, planning has to be 10 years in advance for some books, such as the *Oxford English Dictionary*, whereas the aim is to publish most electronic titles within two years. Multimedia products also need after-sales support, unlike books. The complexity of the development process, the fast-changing technology and markets and the lack of experience with similar products make forecasting sales extremely difficult and the current financial reporting systems inappropriate. Finally, electronic publishing also costs a lot more money. Microsoft's Encarta was reputed to have cost $5m. and OUP was reputed to be spending more than £1m. on its *Children's Encyclopaedia*. Although the Press has published electronic titles for a number of years, these have been mainly text based and did not include video and animation. Early titles included the *Oxford English Dictionary*, the Bodleian pre-1920 catalogue, a series of medical databases, scientific and econometric software. Most have been derived from existing book products although a few have been developed independently. OUP's first true multimedia project was the *Oxford Children's Encyclopaedia*, launched in 1997. One of the main aims of the project was to establish a presence in the multimedia market. OUP recognizes that it cannot compete with companies like Microsoft in the number of software products being developed and the investment in each, but wants a presence in the market. It was a four-year project with about 150 different authors, about 60 subject specialists and six major consultants. One of the main learning outcomes was the development process required, including video and animation production.

The long-term aim is gradually to transfer responsibility to the book people, but this was seen as being too early given the stage of development of the market and the technology. The first stage was learning about the market and developing the necessary skills. The second stage is the dissemination of that learning to the other divisions. The third and final stage of complete transfer of responsibility to individual divisions was not envisaged to occur until well into the twenty-first century.

TABLE 10.8 New venture relatedness, focus and linkages with parent

Venture type	Relatedness of:			Focal activity of venture	Linkages with parent firm
	Product technology	Process technology	Product market		
Product development	Low	Low	High	Development and production	Marketing
Technological innovation	Low	High	High	R&D	Research, marketing and production
Market diversification	High	High	Low	Branding and marketing	Development and production
Technology commercialization	High	Low	Low	Marketing and production	Development
Blue-sky	Low	Low	Low	Development, production and marketing	Finance

TABLE 10.9 Motives, structure and management of corporate ventures

Primary motive	Preferred structure	Key management task
Satisfy managers' ambition	Integrated business team	Motivation and reward
Spread cost and risk of development	Integrated business team	Resource allocation
Exploit economies of scope	Micro-venture department	Reintegration of venture
Learn about venturing	New venture division	Develop new skills
Diversify the business	Special business unit	Develop new assets
Divest non-core activities	Independent business unit	Management of intellectual property rights

Source: Adapted from Tidd, J. and S. Taurins (1999) 'Learn or leverage? Strategic diversification and organisational learning through corporate ventures', *Creativity and Innovation Management*, **8** (2), 122–129.

the most appropriate structure and management processes can be developed. Table 10.9 suggests the most appropriate links between the motives, structure and management of internal corporate ventures.

It is very difficult in practice to assess the success of corporate venturing. Simple financial assessments are usually based on some comparison of the investments made by the corporate parent and the subsequent revenue streams or market valuation of the ventures. Both of the latter are highly sensitive to the timing of the assessment. For example, at the height of the Internet Bubble, financial market valuations suggested corporate venture returns of 70% or more, whereas a few years later these paper returns no longer existed. For example, a study of 35 spin-offs from Xerox over a period of 22 years reveals that the aggregate market value of these spin-offs exceeded those of the parent by a factor of two by 2001, and by a factor of five at the peak of the previous stock market bubble.[30] Assessment of the strategic benefits of corporate venturing is not much easier, but provided the time frames are sufficiently long these can be identified. An historical analysis of the development and commercialization of superconductor technologies at General Electric between 1960 and 1990 reveals how the technology began in internal research and development, but reached a point at which there was deemed to be insufficient market potential to justify any further internal investment. Two GE operating businesses were offered the technology, but declined to fund further development. Rather than abandon the technology altogether, in 1971 GE established a 40% owned venture called Intermagnetics General Corp. (IGC) to develop the technology further. GE became a major customer of IGC as demand for the technology grew in its Medical Systems business due to the growth of MRI (Magnetic Resonance Imaging). However, by 1983 the need for the technology has become so central to GE business, that GE had to redevelop its own core competencies in the field.[31]

10.6 Summary and Further Reading

In this chapter we have explored the rationale, characteristics and management of internal corporate venturing. A firm may establish a corporate venture for a number of reasons, primarily to maintain or improve its competitiveness by exploiting existing processes or exploring more attractive product markets. However, a new venture also represents an opportunity to grow new businesses based on new technologies, products or markets. In such cases more strategic measures of success are needed.

Like any new business, a corporate venture requires a clear business plan and strong champion or intrapreneur who must identify the opportunity for a new venture, raise the finance and manage the development and growth of the business. Like any entrepreneur, the intrapreneur must be highly motivated and will demand a high level of autonomy. However, unlike his counterpart, the corporate intrapreneur requires a high degree of political and social skill. This is because the corporate entrepreneur has the advantage of the financial, technical and marketing resources of the parent firm, but must deal with internal politics and bureaucracy.

There are many books and journal articles on the more general subject of entrepreneurship, but relatively little has been produced on the more specific subject of intrapreneurship or corporate venturing. On the subject of internal corporate venturing Burgelman and Sayles's *Inside Corporate Innovation* (Macmillan, London, 1986) remains the best combination of theory and case studies, but the more recent book by Block and MacMillan, *Corporate Venturing: Creating new businesses within the firm* (Harvard Business School Press, 1995), provides a better review of research on internal corporate ventures. More recent books which include some interesting examples of venturing in the information and telecommunications sectors are *Webs of Innovation* by Alexander Loudon (FT.com, 2001), which despite its title has several chapters related to venturing, and Henry Chesbrough's *Open Innovation* (Harvard Business School Press, 2003), which includes case studies of the usual suspects such as IBM, Xerox, Intel and Lucent. The book *Inventuring* by W. Buckland, A. Hatche and J. Bikinshaw (McGraw-Hill, 2003) is also a good review of corporate venture initiatives, including those at GE, Intel and Lucent, which suggest a range of successful venture models and common reasons for failure.

References

1 Christensen, C. (2000) *The Innovator's Dilemma*, 2nd edn. HarperCollins, New York.
2 Ernst, H. and J. Vitt (2000) 'The influence of corporate acquisitions on the behaviour of key inventors', *R&D Management*, **30** (2), 105–119.
3 Burgelman, R. (1984) 'Managing the internal corporate venturing process', *Sloan Management Review*, **25** (2), Winter, 33–48.
4 Dess, G., G. Lumpkin and J. Covin (1997) 'Entrepreneurial strategy making and firm performance', *Strategic Management Journal*, **18** (9), 677–695.
5 Tidd, J. and S. Taurins (1999) 'Learn or leverage? Strategic diversification and organisational learning through corporate ventures', *Creativity and Innovation Management*, **8** (2), 122–129.
6 Christensen, C. and M. Raynor (2003) *The Innovator's Solution: Creating and sustaining successful growth*. Harvard Business School Press, Boston, Mass.
7 Kanter, R. (1985) 'Supporting innovation and venture development in established companies', *Journal of Business Venturing*, **1**, 47–60.
8 Huber, G. (1996) 'Organizational learning: the contributing processes and the literatures', in Cohen, M. and Sproull, L. (eds), *Organizational Learning*. Sage, London, 124–162.
9 Block, Z. and I. MacMillan (1993) *Corporate Venturing: Creating new businesses within the firm*. Harvard Business School Press, Boston, Mass.
10 Roussel, P., K. Saad and T. Erickson (1991) *Third Generation R&D: Managing the link to corporate strategy*. Harvard Business School Press, Boston, Mass.
11 Drucker, P. (1985) *Innovation and Entrepreneurship*. Harper & Row, New York.
12 Griffiths, D. and M. Biosot (2000) 'Are there any competencies out there?', in Tidd, J. (ed.), *From Knowledge Management to Strategic Competence*. Imperial College Press, London, 199–228.
13 Coombs, R. and R. Hull (1998) 'Knowledge management practices and path-dependency in innovation', *Research Policy*, **27**, 237–253.
14 Hargadon, A. and R. Sutton (2000) 'Building an innovation factory', *Harvard Business Review*, May–June, 157–166.
15 Kivimaki, M. *et al.* (2000) 'Communication as a determinant of organizational innovation', *R&D Management*, **30** (1), 33–42.

16 Barringer, B. and A. Bluedorn (1999) 'The relationship between corporate entrepreneurship and strategic management', *Strategic Management Journal*, **20**, 421–444.

17 Altshuller, G. (1996) *And Suddenly the Inventor Appeared: TRIZ, the theory of inventive problem solving*. Blackwell, Oxford.

18 Tidd, J. and S. Barnes (1999) 'Spin-in or spin-out? Corporate venturing in life sciences', *International Journal of Entrepreneurship and Innovation*, **1** (2), 109–116.

19 Maidique, M. (1980) 'Entrepreneurs, champions and technological innovation', *Sloan Management Review*, **21** (2), Winter, 59–76.

20 Roberts, M. (1992) 'The do's and don'ts of strategic alliances', *Journal of Business Strategy*, March/April, 50–53.

21 Burgelman, K. and L. Sayles (1986) *Inside Corporate Innovation*. Macmillan, London.

22 Gulati, R. and J. Garino (2000) 'Get the right mix of bricks and clicks', *Harvard Business Review*, May–June, 107–166.

23 Tushman, M. and C. O'Reilly (2002) *Winning through Innovation: A practical guide to leading organizational change and renewal*. Harvard Business School Press, Boston, Mass.

24 Thornhill, S. and Amit, R. (2000) 'A dynamic perspective of external fit in corporate venturing', *Journal of Business Venturing*, **16**, 25–50.

25 Chesbrough, H. (2002) 'The governance and performance of Xerox's technology spin-off companies', *Research Policy*, **32**, 403–421.

26 Abetti, P. (2002) 'From science to technology to products and profits: superconductivity at General Electric and Intermagnetics General (1960–1990)', *Journal of Business Venturing*, **17**, 83–98.

27 Martin, M. (1994) *Managing Innovation and Entrepreneurship in Technology*. John Wiley & Sons, Inc., New York.

28 Fast, N. (1979) 'A visit to a new venture graveyard', *Research Management*, **22** (2), 18–22.

29 Chen, S. (1997) 'Core capabilities and core rigidities in the multi-media industry', unpublished Ph.D. thesis, The Management School, Imperial College, University of London.

30 Chesbrough, H. (2002) 'The governance and performance of Xerox's technology spin-off companies', *Research Policy*, **32**, 403–421.

31 Abetti, P. (2002) 'From science to technology to products and profits: superconductivity at General Electric and Intermagnetics General (1960–1990)', *Journal of Business Venturing*, **17**, 83–98.

Part V

CREATING THE INNOVATIVE
ORGANIZATION

Innovations – as we have seen – do not emerge in a vacuum, and one important influence on success and failure is the organizational context in which they are created and implemented. Organizations differ widely, but we have learned over time about some of the factors which make for a more or less supporting context. These include the structure of the organization, the roles played by key individuals, the training and development of staff, the way in which work is organized (teamwork, projects, etc.), the extent to which people are involved in innovation, and how the organization itself goes about learning and sharing knowledge. Chapter 11 reviews the various themes which contribute to creating an innovative organization.

Chapter 12 looks at the special case of creating a new organization, specifically small innovative enterprises. Much is expected of such innovative firms as a source of employment and growth, and as sources of additional revenue for universities and other so-called incubator organizations. However, experience suggests that the failure rate amongst this group is very high, and we identify the key management issues that must be addressed at each stage of the development of such enterprises.

Building the Innovative Organization

'Innovation has nothing to do with how many R&D dollars you have. . . . it's not about money. It's about the people you have, how you're led, and how much you get it.'

(Steve Jobs, interview with *Fortune Magazine*, 1998[1])

'People are our greatest asset.' This phrase – or variations on it – has become one of the clichés of management presentations, mission statements and annual reports throughout the world. Along with concepts like 'empowerment' and 'team working', it expresses a view of people being at the creative heart of the enterprise. But very often the reader of such words – and particularly those 'people' about whom they are written – may have a more cynical view, seeing organizations still operating as if people were part of the problem rather than the key to its solution.

In the field of innovation this theme is of central importance. It is clear from a wealth of psychological research that every human being comes with the capability to find and solve complex problems, and where such creative behaviour can be harnessed amongst a group of people with differing skills and perspectives extraordinary things can be achieved. We can easily think of examples. At the individual level, innovation has always been about exceptional characters who combine energy, enthusiasm and creative insight to invent and carry forward new concepts. James Dyson with his alternative approaches to domestic appliance design; Spence Silver, the 3M chemist who discovered the non-sticky adhesive behind 'Post-it' notes; and Shawn Fanning, the young programmer who wrote the Napster software and almost single-handedly shook the foundations of the music industry, are good illustrations of this.

Innovation is increasingly about teamwork and the creative combination of different disciplines and perspectives. Whether it is in designing a new car in half the time usually taken, bringing a new computer concept to market, establishing new ways of delivering old services like banking, insurance or travel services, or in putting men and women routinely into space, success comes from people working together in high-performance teams.

This effect, when multiplied across the organization, can yield surprising results. In his work on US companies, Pfeffer notes the strong correlation between proactive people management practices and the performance of firms in a variety of sectors.[2] A comprehensive review for the UK Chartered Institute of Personnel and Development suggested that '. . . more than 30 studies carried out in the UK and US since the early 1990s leave no room to doubt that there is a correlation between people management and business performance, that the relationship is positive, and that it is cumulative: the more and the more effective the practices, the better the result'.[3] Similar studies confirm the pattern in German firms.[4] In a knowledge economy where creativity is at a premium, people really are the most important assets which a firm possesses. The management challenge is how to go about building the kind of organizations in which such innovative behaviour can flourish.

This chapter deals with the creation and maintenance of an innovative organizational context, one whose structure and underlying culture – pattern of values and beliefs – support innovation. It is easy to find prescriptions for innovative organizations which highlight the need to eliminate stifling bureaucracy, unhelpful structures, brick walls blocking communication and other factors stopping good ideas getting through. But we must be careful not to fall into the chaos trap – not all innovation works in organic, loose, informal environments or 'skunk works' – and these types of organization can sometimes act against the interests of successful innovation. We need to determine appropriate organization – that is, the most suitable organization given the operating contingencies. Too little order and structure may be as bad as too much.

Equally, 'innovative organization' implies more than a structure; it is an integrated set of components which work together to create and reinforce the kind of environment which enables innovation to flourish. Studies of innovative organizations have been extensive, although many can be criticized for taking a narrow view, or for placing too much emphasis on a single prescription like 'team working' or 'loose structures'. Nevertheless it is possible to draw out from these a set of components which appear linked with success; these are outlined in Table 11.1, and explored in the subsequent discussion.

11.1 Shared Vision, Leadership and the Will to Innovate

Innovation is essentially about learning and change and is often disruptive, risky and costly. So, as Box 11.1 shows, it is not surprising that individuals and organizations develop many different cognitive, behavioural and structural ways of reinforcing the status quo. Innovation requires energy to overcome this inertia, and the determination to change the order of things. We see this in the case of individual inventors who champion their ideas against the odds, in entrepreneurs who build businesses through risk-

TABLE 11.1 Components of the innovative organization

Component	Key features	Example references
Shared vision, leadership and the will to innovate	Clearly articulated and shared sense of purpose Stretching strategic intent 'Top management commitment'	5–9
Appropriate structure	Organization design which enables creativity, learning and interaction. Not always a loose 'skunk works' model; key issue is finding appropriate balance between 'organic and mechanistic' options for particular contingencies	10–17
Key individuals	Promoters, champions, gatekeepers and other roles which energize or facilitate innovation	18–23
Effective team working	Appropriate use of teams (at local, cross-functional and inter-organizational level) to solve problems. Requires investment in team selection and building	24–29
Continuing and stretching individual development	Long-term commitment to education and training to ensure high levels of competence and the skills to learn effectively	30–32
Extensive communication	Within and between the organization and outside. Internally in three directions – upwards, downwards and laterally	18, 33–35
High involvement in innovation	Participation in organization-wide continuous improvement activity	36–39
External focus	Internal and external customer orientation. Extensive networking	19, 40–43
Creative climate	Positive approach to creative ideas, supported by relevant motivation systems	44–47
Learning organization	High levels of involvement within and outside the firm in proactive experimentation, finding and solving problems, communication and sharing of experiences and knowledge capture and dissemination	48–52

taking behaviour and in organizations which manage to challenge the accepted rules of the game.

The converse is also true – the 'not invented here' problem, in which an organization fails to see the potential in a new idea, or decides that it does not fit with their current pattern of business. In other cases the need for change is perceived, but the strength or saliency of the threat is underestimated. We saw in Chapter 6 the difficulties which IBM experienced in the early 1990s in responding to the emerging 'client-server' and network shift in computing away from mainframes, and this is a good example of a firm which believed it had seen and assessed the threat but which nearly

> ## BOX 11.1
> ### MISSING THE BOAT . . .
>
> On 10 March 1875 Alexander Graham Bell called to his assistant, 'Mr Watson, come here, I want you' – the surprising thing about the exchange being that it was the world's first telephone conversation. Excited by their discovery, they demonstrated their idea to senior executives at Western Union. Their written reply, a few days later, suggested that 'after careful consideration of your invention, which is a very interesting novelty, we have come to the conclusion that it has no commercial possibilities . . . we see no future for an electrical toy . . .' Within four years of the invention there were 50 000 telephones in the USA and within 20 years there were 5 million. In the same time the company which Bell formed, American Telephone and Telegraph (ATT) over the next 20 years grew to become the largest corporation in the USA, with stock worth $1000/share. The original patent (number 174455) became the single most valuable patent in history.[53]

drove the company out of business. Similarly, General Motors found it difficult to appreciate and interpret the information about Japanese competition, preferring to believe that their access in US markets was due to unfair trade policies rather than recognizing the fundamental need for process innovation which the 'lean manufacturing' approach pioneered in Japan was bringing to the car industry.[54] Christensen, in his studies of disk drives, and Tripsas and Gravetti, in their analysis of the problems Polaroid faced in making the transition to digital imaging, provide powerful evidence to show the difficulties established firms have in interpreting the signals associated with a new and potentially disruptive technology.[55,56]

This is also where the concept of 'core rigidities', introduced in Chapter 5, becomes important.[50] We have become used to seeing core competencies as a source of strength within the organization, but the downside is that the mindset which is being highly competent in doing certain things can also block the organization from changing its mind. Thus ideas which challenge the status quo face an uphill struggle to gain acceptance; innovation requires considerable energy and enthusiasm to overcome barriers of this kind. One of the concerns in successful innovative organizations is finding ways to ensure that individuals with good ideas are able to progress them without having to leave the organization to do so.[23] Chapter 10 discusses this theme (of 'intrapreneurship') in more detail.

Changing mindset and refocusing organizational energies requires the articulation of a new vision, and there are many cases where this kind of leadership is credited with starting or turning round organizations. Examples include Jack Welch of GE, Bill Gates (Microsoft), Steve Jobs (Pixar/Apple), Andy Groves (Intel) and Richard Branson (Virgin).[57,58] Whilst we must be careful of vacuous expressions of 'mission' and 'vision', it is also clear that in cases like these there has been a clear sense of, and commitment to, shared organizational purpose arising from such leadership.

'Top management commitment' is a common prescription associated with successful innovation; the challenge is to translate the concept into reality by finding mechanisms which demonstrate and reinforce the sense of management involvement, commitment, enthusiasm and support. In particular, there needs to be long-term commitment to major projects, as opposed to seeking short-term returns. Since much of innovation is about uncertainty, it follows that returns may not emerge quickly and that there will be a need for 'patient money'. This may not always be easy to provide, especially when demands for shorter-term gains by shareholders have to be reconciled with long-term technology development plans. One way of dealing with this problem is to focus not only on returns on investment but on other considerations like future market penetration and growth or the strategic benefits which might accrue to having a more flexible or responsive production system (Chapter 6 discusses this theme in more detail). Boxes 11.2 and 11.3 provide examples of such leadership.

Part of this pattern is also top management acceptance of risk. Innovation is inherently uncertain and will inevitably involve failures as well as successes. Successful technology management thus requires that the organization be prepared to take risks and to accept failure as an opportunity for learning and development. This is not to say that unnecessary risks should be taken – rather, as Robert Cooper suggests, the inherent uncertainty in innovation should be reduced where possible through the use of information collection and research.[59]

One last point – we should not confuse leadership and commitment with always being the active change agent. In many cases innovation happens in spite of the senior management within an organization, and success emerges as a result of guerrilla tactics rather than a frontal assault on the problem. Much has been made of the dramatic turnaround in IBM's fortunes under the leadership of Lou Gerstner who took the ailing giant firm from a crisis position to one of leadership in the IT services field and an acknowledged pioneer of e-business. But closer analysis reveals that the entry into e-business was the result of a bottom-up team initiative led by a programmer called Dave Grossman. It was his frustration with the lack of response from his line managers that eventually led to the establishment of a broad coalition of people within the company who were able to bring the idea into practice and establish IBM as a major e-business leader. The message for senior management is as much about leading through creating space and support within the organization as it is about direct involvement.

BOX 11.2
INNOVATION LEADERSHIP AND CLIMATE

Organizations have traditionally conceived of leadership as an heroic attribute, appointing a few 'real' leaders to high-level senior positions in order to get them through difficult times. However, many observers and researchers are becoming cynical about this approach and are beginning to think about the need to recognize and utilize a wider range of leadership practices. Leadership needs to be conceived of as something that happens across functions and levels. New concepts and frameworks are needed in order to embrace this more inclusive approach to leadership.

For example, there is a great deal of writing about the fundamental difference between leadership and management. This literature abounds and has generally promoted the argument that leaders have vision and think creatively ('doing different'), while managers are merely drones and just focus on doing things better. This distinction has led to a general devaluation of management. Emerging work on styles of creativity and management suggests that it is useful to keep preference distinct from capacity. Creativity is present both when doing things differently and doing things better. This means that leadership and management may be two constructs on a continuum, rather than two opposing characteristics.

Our particular emphasis is on resolving the unnecessary and unproductive distinction that is made between leadership and management. When it comes to innovation and transformation, organizations need both sets of skills. We develop a model of innovation leadership that builds on past work, but adds some recent perspectives from the fields of change and innovation management, and personality and social psychology. This multidimensional view of leadership raises the issue of context as an important factor, beyond concern for task and people. This approach suggests the need for a third factor in assessing leadership behaviour, in addition to the traditional concerns for task and people. Therefore we integrate three dimensions of leadership: concern for task, concern for people, and concern for change.

One of the most important roles that leaders play within organizational settings is to create the climate for innovation. We identify the critical dimensions of the climate for innovation, and suggest how leaders might nurture these. By using a Situational Outlook Questionnaire (SOQ) as a diagnostic, we identify nine dimensions to help decide what kind of interventions might be helpful in establishing the appropriate context for innovation.

From S. Isaksen and J. Tidd (2006) *Meeting the Innovation Challenge: Leadership for Transformation and Growth*. John Wiley & Sons, Ltd.

BOX 11.3

THE VISION THING – HOW LEADERSHIP CONTRIBUTES TO TRANSFORMATIONAL CHANGE

Moving from a diverse and clumsy conglomerate with origins in the wood and paper industry to the market leader position in mobile telephones is not easy. Yet the story of Nokia is one of managed transformation from a nineteenth-century timber firm to the fifth largest company in Europe, with 44 000 people employed in 14 countries, over a third of whom work on R&D or product design. Much of this transition – which, like many transformations, contained an element of luck – is attributed to the energy and vision of the CEO, Jorma Ollila, who took up this role in 1992 from the mobile phone division.

The transition was not easy – a series of problems, including logistics and availability of chips meant that the phone division made serious losses in 1995 and the stock value was cut in half. In order to meet this challenge Ollila effectively 'bet the company' disposing of almost all of its non-telecoms businesses (which ranged from television sets to toilet paper!) so that by 1995 90% of Nokia was concerned with telecommunications.

A similar pattern can be seen with the case of Siemens. Again with roots in the nineteenth century, Siemens grew to be one of the great names in electrical engineering and a major force in the German economy. But recent years have seen concerns about the company, criticizing it for a lack of focus and for being slow and unresponsive. Faced with this developing picture the company appointed a new board member in 1998 – Edward Krubasik – who came from outside the firm. Restructuring under his leadership has led to the divestment of nearly £10bn of old businesses and to the repositioning of Siemens as a major IT and software player. In 1999 profits surged and the sales price tripled and 60% of the business is concerned with software. Perhaps most significant as an indicator of this new vision is the fact that Siemens employed 27 000 software engineers in 2000 – more even than Microsoft![60]

11.2 Appropriate Organization Structure

No matter how well developed the systems are for defining and developing innovative products and processes they are unlikely to succeed unless the surrounding organizational context is favourable. Achieving this is not easy; it involves creating the organizational structures and processes which enable technological change to thrive. For example, rigid hierarchical organizations in which there is little integration between functions and where communication tends to be top-down and one-way in character are unlikely to be very supportive of the smooth information flows and cross-functional co-operation recognized as being important success factors.

Much of the literature recognizes that organizational structures are influenced by the nature of tasks to be performed within the organization. In essence the less programmed and more uncertain the tasks, the greater the need for flexibility around the structuring of relationships.[12] For example, activities like production, order processing, purchasing, etc. are characterized by decision-making which is subject to little variation. (Indeed in some cases these decisions can be automated through employing particular decision rules embodied in computer systems, etc.) But others require judgment and insight and vary considerably from day to day – and these include those decisions associated with innovation. Activities of this kind are unlikely to lend themselves to routine, structured and formalized relationships, but instead require flexibility and extensive interaction. Several writers have noted this difference between what have been termed 'programmed' and 'non-programmed' decisions and argued that the greater the level of non-programmed decision-making, the more the organization needs a loose and flexible structure.[16]

Considerable work was done on this problem in the late 1950s by researchers Tom Burns and George Stalker who outlined the characteristics of what they termed 'organic' and 'mechanistic' organizations.[14] The former are essentially environments suited to conditions of rapid change whilst the latter are more suited to stable conditions; although these represent poles on an ideal spectrum they do provide useful design guidelines about organizations for effective innovation. Other studies include those of Rosabeth Moss-Kanter,[61] and Hesselbein et al.[17]

We should be careful, however, in assuming that innovation is simply confined to R&D laboratories (where versions of this form of organization have often been present). Increasingly, innovation is becoming a corporate-wide task, involving production, marketing, administration, purchasing and many other functions; this provides strong pressure for widespread organizational change towards more organic models.

The relevance of Burns and Stalker's model can be seen in an increasing number of cases where organizations have restructured to become less mechanistic. For example, General Electric in the USA underwent a painful but ultimately successful transformation, moving away from a rigid and mechanistic structure to a looser and decentralized form.[58] ABB, the Swiss–Swedish engineering group, developed a particular approach to their global business based on operating as a federation of small businesses, each of which retained much of the organic character of small firms.[8] Other examples of radical changes in structure include the Brazilian white goods firm Semco and the Danish hearing aid company Oticon.[62,63] But again we need to be careful – what works under one set of circumstances may diminish in value under others. Whilst models such as that deployed by ABB helped at the time, later developments meant that these proved less appropriate and were insufficient to deal with new challenges emerging elsewhere in the business.

Related to this work has been another strand which looks at the relationship between different environments and organizational form. Once again the evidence suggests that the higher the uncertainty and complexity in the environment, the greater the need for flexible structures and processes to deal with it.[64,65] This partly explains why some fast-growing sectors – for example, electronics or biotechnology, are often associated with more organic organizational forms, whereas mature industries often involve more mechanistic arrangements.

One important study in this connection was that originally carried out by Lawrence and Lorsch looking at product innovation. Their work showed that innovation success in mature industries like food packaging and growing sectors like plastics depended on having structures which were sufficiently differentiated (in terms of internal specialist groups) to meet the needs of a diverse marketplace. But success also depended on having the ability to link these specialist groups together effectively so as to respond quickly to market signals; they reviewed several variants on co-ordination mechanisms, some of which were more or less effective than others. Better co-ordination was associated with more flexible structures capable of rapid response.[66]

We can see clear application of this principle in the current efforts to reduce 'time to market' in a range of businesses.[67] Rapid product innovation and improved customer responsiveness are being achieved through extensive organizational change programmes involving parallel working, early involvement of different functional specialists, closer market links and user involvement, and through the development of team working and other organizational aids to co-ordination.

Another strand of work which has had a strong influence on the way we think about organizational design was that originated by Joan Woodward associated with the nature of the industrial processes being carried out.[11] Her studies suggested that structures varied between industries with a relatively high degree of discretion (such as small batch manufacturing) through to those involving mass production where more hierarchical and heavily structured forms prevailed. Significantly, the process industries, although also capital intensive, allowed a higher level of discretion.

Other variables and combinations which have been studied for their influence on structure include size, age and company strategy.[68,69] The extensive debate on organization structure began to resolve itself into a 'contingency' model in the 1970s. In essence this view argues that there is no single 'best' structure, but that successful organizations tend to be those which develop the most suitable 'fit' between structure and operating contingencies. For example, it makes sense to structure an operation like McDonald's in a mechanistic and highly controlled form, in order to be able to replicate this model across the world and be able to deliver similar standards of product and service. But trying to develop a new computer operating system or genetically engineer a new drug would not be possible in such a structure.

Similarly, structures which enable a large international firm to carry out R&D simultaneously in several countries and to integrate their efforts through a series of procedures would be largely irrelevant and excessively bureaucratic for a small, high-tech start-up firm.

The Canadian writer Henry Mintzberg drew much of the work on structure together and proposed a series of archetypes which provide templates for the basic structural configurations into which firms are likely to fall.[15] These categories – and their implications for innovation management – are summarized in Table 11.2. Box 11.4 gives an example of the importance of organizational structure and the need to find appropriate models.

The increasing importance of innovation and the consequent experience of high levels of change across the organization have begun to pose a challenge for organizational structures normally configured for stability. Thus traditional machine bureaucracies – typified by the car assembly factory – are becoming more hybrid in nature, tending towards what might be termed a 'machine adhocracy' with creativity and flexibility (within limits) being actively encouraged. The case of 'lean production' with its emphasis on team working, participation in problem-solving, flexible cells and flattening of hierarchies is a good example, where there is significant loosening of the original model to enhance innovativeness.[70]

The key challenge here for managing innovation remains one of *fit* – of getting the most appropriate structural form for the particular circumstances. Another view of structure is that it is an artefact of what people believe and how they behave; if there is good fit, structure will enable and reinforce innovative behaviour. If it is contradictory to these beliefs – for example, restricting communication, stressing hierarchy – then it is likely to act as a brake on creativity and innovation.

11.3 Key Individuals

Another important element is the presence of key enabling figures. The uncertainty and complexity involved in innovation mean that many promising inventions die before they make it to the outside world. One way round this problem is if there is a key individual (or sometimes a group of people) who is prepared to champion its cause and to

TABLE 11.2 Mintzberg's structural archetypes

Organization archetype	Key features	Innovation implications
Simple structure	Centralized organic type – centrally controlled but can respond quickly to changes in the environment. Usually small and often directly controlled by one person. Designed and controlled in the mind of the individual with whom decision-making authority rests. Strengths are speed of response and clarity of purpose. Weaknesses are the vulnerability to individual misjudgment or prejudice and resource limits on growth	Small start-ups in high technology – 'garage businesses' are often simple structures. Strengths are in energy, enthusiasm and entrepreneurial flair – simple structure innovating firms are often highly creative. Weaknesses are in long-term stability and growth, and over-dependence on key people who may not always be moving in the right business direction
Machine bureaucracy	Centralized mechanistic organization, controlled centrally by systems. A structure designed like a complex machine with people seen as cogs in the machine. Design stresses the function of the whole and specialization of the parts to the point where they are easily and quickly interchangeable. Their success comes from developing effective systems which simplify tasks and routinize behaviour. Strengths of such systems are the ability to handle complex integrated processes like vehicle assembly. Weaknesses are the potential for alienation of individuals and the build-up of rigidities in inflexible systems	Machine bureaucracies depend on specialists for innovation, and this is channelled into the overall design of the system. Examples include fast food (McDonald's), mass production (Ford) and large-scale retailing (Tesco), in each of which there is considerable innovation, but concentrated on specialists and impacting at the system level. Strengths of machine bureaucracies are their stability and their focus of technical skills on designing the systems for complex tasks. Weaknesses are their rigidities and inflexibility in the face of rapid change, and the limits on innovation arising from non-specialists
Divisionalized form	Decentralized organic form designed to adapt to local environmental challenges. Typically associated with larger organizations, this model involves specialization into semi-independent units. Examples would be strategic business units or operating divisions. Strengths of such a form are the ability to attack particular niches (regional, market,	Innovation here often follows a 'core and periphery' model in which R&D of interest to the whole organization, or of a generic nature is carried out in central facilities whilst more applied and specific work is carried out within the divisions. Strengths of this model include the ability to concentrate on

continues overleaf

TABLE 11.2 *(continued)*

Organization archetype	Key features	Innovation implications
	product, etc.) whilst drawing on central support. Weaknesses are the internal frictions between divisions and the centre	developing competency in specific niches and to mobilize and share knowledge gained across the rest of the organization. Weaknesses include the 'centrifugal pull' away from central R&D towards applied local efforts and the friction and competition between divisions which inhibits sharing of knowledge
Professional bureaucracy	Decentralized mechanistic form, with power located with individuals but co-ordination via standards. This kind of organization is characterized by relatively high levels of professional skills, and is typified by specialist teams in consultancies, hospitals or legal firms. Control is largely achieved through consensus on standards ('professionalism') and individuals possess a high degree of autonomy. Strengths of such an organization include high levels of professional skill and the ability to bring teams together	This kind of structure typifies design and innovation consulting activity within and outside organizations. The formal R&D, IT or engineering groups would be good examples of this, where technical and specialist excellence is valued. Strengths of this model are in technical ability and professional standards. Weaknesses include difficulty of managing individuals with high autonomy and knowledge power
Adhocracy	Project type of organization designed to deal with instability and complexity. Adhocracies are not always long-lived, but offer a high degree of flexibility. Team-based, with high levels of individual skill but also ability to work together. Internal rules and structure are minimal and subordinate to getting the job done. Strengths of the model are its ability to cope with high levels of uncertainty and its creativity. Weaknesses include the inability to work together effectively due to unresolved conflicts, and a lack of control due to lack of formal structures or standards	This is the form most commonly associated with innovative project teams – for example, in new product development or major process change. The NASA project organization was one of the most effective adhocracies in the programme to land a man on the moon; significantly the organization changed its structure almost once a year during the 10-year programme, to ensure it was able to respond to the changing and uncertain nature of the project. Strengths of adhocracies are the high levels of creativity and

TABLE 11.2   (*continued*)

Organization archetype	Key features	Innovation implications
		flexibility – the 'skunk works' model advocated in the literature. Weaknesses include lack of control and over-commitment to the project at the expense of the wider organization
Mission-oriented	Emergent model associated with shared common values. This kind of organization is held together by members sharing a common and often altruistic purpose – for example, in voluntary and charity organizations. Strengths are high commitment and the ability of individuals to take initiatives without reference to others because of shared views about the overall goal. Weaknesses include lack of control and formal sanctions	Mission-driven innovation can be highly successful, but requires energy and a clearly articulated sense of purpose. Aspects of total quality management and other value-driven organizational principles are associated with such organizations, with a quest for continuous improvement driven from within rather than in response to external stimulus. Strengths lie in the clear sense of common purpose and the empowerment of individuals to take initiatives in that direction. Weaknesses lie in over-dependence on key visionaries to provide clear purpose, and lack of 'buy-in' to the corporate mission

provide some energy and enthusiasm to help it through the organizational system. Such key figures or project champions have been associated with many famous innovations – for example, the development of Pilkington's float glass process or Edwin Land and the Polaroid photographic system.[9,22,71–72] Box 11.5 gives an example.

There are, in fact, several roles which key figures can play which have a bearing on the outcome of a project. First, there is the source of critical technical knowledge – often the inventor or team leader responsible for an invention. They will have the breadth of understanding of the technology behind the innovation and the ability to solve the many development problems likely to emerge in the long haul from laboratory or drawing board to full scale. The contribution here is not only of technical

BOX 11.4
THE EMERGENCE OF MASS PRODUCTION

Perhaps the most significant area in which there is a change of perspective is in the role of human resources. Early models of organization were strongly influenced by the work of Frederick Taylor and his principles of 'scientific management'. These ideas – used extensively in the development of mass production industries like automobile manufacture – essentially saw the organization problem as one which required the use of analytical methods to arrive at the 'best' way of carrying out the organization's tasks. This led to an essentially mechanistic model in which people were often seen as cogs in a bigger machine, with clearly defined limits to what they should and shouldn't do. The image presented by Charlie Chaplin in *Modern Times* was only slightly exaggerated; in the car industry the average task cycle for most workers was less than two minutes.

The advantages of this system for the mass production of a small range of goods were clear; productivity increases often ran into three figures with the adoption of this approach. For example, Ford's first assembly line, installed in 1913 for flywheel assembly, saw the assembly time fall from 20 man-minutes to five, and by 1914 three lines were being used in the chassis department to reduce assembly time from around 12 hours to less than two. But its limitations lay in the ability of the system to change, and in the capacity for innovation. By effectively restricting innovation to a few specialists, an important source of creative problem-solving, in terms of product and process development, was effectively cut off.

The experience of Ford and others highlights the point that there is no single 'best' kind of organization; the key is to ensure congruence between underlying values and beliefs and the organization which enables innovative routines to flourish. For example, whilst the 'skunk works' model may be appropriate to US product development organizations, it may be inappropriate in Japan where a more disciplined and structured form is needed. Equally some successful innovative organizations are based on team working whereas others are built around key individuals – in both cases reflecting underlying beliefs about how innovation works in those particular organizations. Similarly successful innovation can take place within strongly bureaucratic organizations just as well as in those in which there is a much looser structure – providing that there is underlying congruence between these structures and the innovative behavioural routines.

BOX 11.5
BAGS OF IDEAS – THE CASE OF JAMES DYSON

In October 2000 the air inside Court 58 of the Royal Courts of Justice in London rang with terms like 'bagless dust collection', 'cyclone technology', 'triple vortex' and 'dual cyclone' as one of the most bitter of patent battles in recent years was brought to a conclusion. On one side was Hoover, a multinational firm with the eponymous vacuum suction sweeper at the heart of a consumer appliance empire. On the other a lone inventor – James Dyson – who had pioneered a new approach to the humble task of house cleaning and then seen his efforts threatened by an apparent imitation by Hoover. Eventually the court ruled in Dyson's favour.

This represented the culmination of a long and difficult journey which Dyson travelled in bringing his ideas to a wary marketplace. It began in 1979 when Dyson was using, ironically, a Hoover Junior vacuum cleaner to dust the house. He was struck by the inefficiency of a system which effectively reduced its capability to suck the more it was used since the bag became clogged with dust. He tried various improvements such as a finer mesh filter bag but the results were not promising. The breakthrough came with the idea of using industrial cyclone technology applied in a new way – to the problem of domestic cleaners.

Dyson was already an inventor with some track record and one of his products was a wheelbarrow which used a ball instead of a front wheel. In order to spray the black dust paint in a powder coating plant they had installed a cyclone – a well-established engineering solution to the problem of dust extraction. Essentially a mini-tornado is created within a shell and the air in the vortex moves so fast that particles of dust are forced to the edge where they can be collected whilst clean air moves to the centre. Dyson began to ask why the principle could not be applied in vacuum cleaners – and soon found out why. His early experiments – with the Hoover – were not entirely successful but eventually he applied for a patent in 1980 for a vacuum cleaning appliance using cyclone technology.

It took another four years and 5127 prototypes and even then he could not patent the application of a single cyclone since that would only represent an improvement on an existing and proven technology. He had to develop a dual cyclone system which used the first to separate out large items of domestic refuse – cigarette ends, dog hairs, cornflakes, etc. – and the second to pick up the finer dust particles. But having proved the technology he found a distinct cold shoulder on the part of the existing vacuum cleaner industry represented by firms

continues overleaf

BOX 11.5   (*continued*)

like Hoover, Philips and Electrolux. In typical examples of the 'not invented here' effect they remained committed to the idea of vacuum cleaners using bags and were unhappy with bagless technology. (This is not entirely surprising since suppliers such as Electrolux make a significant income on selling the replacement bags for their vacuum cleaners.)

Eventually Dyson began the hard work of raising the funds to start his own business – and it gradually paid off. Launched in 1993 – 14 years after the initial idea – Dyson now runs a design-driven business worth around £530m. and has a number of product variants in its vacuum cleaner range; other products under development aim to re-examine domestic appliances like washing machines and dishwashers to try and bring similar new ideas into play. The basic dual cyclone cleaner was one of the products identified by the UK Design Council as one of its 'millennium products'.

Perhaps the greatest accolade though is the fact that the vacuum cleaner giants like Hoover eventually saw the potential and began developing their own versions. Although Hoover lost the case they are planning to appeal, arguing that their version used a different technology developed for the oil and gas industry by the UK research consultancy BHR. Whoever wins, Dyson has once again shown the role of the individual champion in innovation – and that success depends on more than just a good idea. Edison's famous comment that it is '1% inspiration and 99% perspiration' seems an apt motto here!

knowledge; it also involves inspiration when particular technological problems appear insoluble, and it involves motivation and commitment.

Influential though such technical champions might be, they may not be able to help an innovation progress unaided through the organization. Not all problems are technical in nature; other issues such as procuring resources or convincing sceptical or hostile critics elsewhere in the organization may need to be dealt with. Here our second key role emerges – that of organizational sponsor.

Typically this person has power and influence and is able to pull the various strings of the organization (often from a seat on the board); in this way many of the obstacles to an innovation's progress can be removed or the path at least smoothed. Such sponsors do not necessarily need to have a detailed technical knowledge of the innovation (although this is clearly an asset), but they do need to believe in its potential.

Recent exploration of the product development process has highlighted the important role which is played by the team members and in particular the project team leader. There are close parallels to the champion model; influential roles range from what Clark and Fujimoto call 'heavyweight' project managers who are deeply involved and have the organizational power to make sure things come together, through to the 'lightweight' project manager whose involvement is more distant. Research on Japanese product development highlights the importance of the *shusha* or team leader; in some companies (such as Honda) the *shusha* is empowered to override even the decisions and views of the chief executive![73] The important message here is to match the choice of project manager type to the requirements of the situation – and not to use the 'sledgehammer' of a heavyweight manager for a simple task.

Key roles are not just on the technical and project management side; studies of innovation (going right back to Project SAPPHO and its replications) also highlighted the importance of the 'business innovator', someone who could represent and bring to bear the broader market or user perspective.[19]

Although innovation history is full of examples where such key individuals – acting alone or in tandem – have had a marked influence on success, we should not forget that there is a downside as well. Negative champions – project assassins – can also be identified, whose influence on the outcome of an innovation project is also significant but in the direction of killing it off. For example, there may be internal political reasons why some parts of an organization do not wish for a particular innovation to progress – and through placing someone on the project team or through lobbying at board level or in other ways a number of obstacles can be placed in its way. Equally, our technical champion may not always be prepared to let go of their pet idea, even if the rest of the organization has decided that it is not a sensible direction in which to progress. Their ability to mobilize support and enthusiasm and to surmount obstacles within the organization can sometimes lead to wrong directions being pursued, or the continued chasing up what many in the organization see as a blind alley.

One other type of key individual is worth mentioning, that of the 'technological gatekeeper'. Innovation is about information and, as we saw above, success is strongly associated with good information flow and communication. Research has shown that such networking is often enabled by key individuals within the organization's informal structure who act as 'gatekeepers' – collecting information from various sources and passing it on to the relevant people who will be best able or most interested to use it. Thomas Allen, working at MIT, made a detailed study of the behaviour of engineers during the large-scale technological developments surrounding the Apollo rocket programme. His studies highlighted the importance of informal communications in successful innovation, and drew particular attention to gatekeepers – who were not always in formal information management positions but who were well connected in the informal social structure of the organization – as key players in the process.[18]

This role is becoming of increasing importance in the field of knowledge management where there is growing recognition that enabling effective sharing and communication of valuable knowledge resources is not simply something which can be accomplished by advanced IT and clever software – there is a strong interpersonal element.[74,75] Such approaches become particularly important in distributed or virtual teams where 'managing knowledge spaces' and the flows across them are of significance.[76]

11.4 Stretching Training and Development

A core characteristic associated with high-performance organizations is the extent to which they commit to training and development. Studies at national, sector and individual company level repeatedly stress the relationship between investment of this kind and innovation capability.[3,10,31,77] The argument here is that the ability of an organization to make the best use of new equipment or to produce products and services with novelty in design, quality or performance depends to a large extent on the knowledge and skills of those involved in producing such innovations.

Equipping people with the skills they need to understand and operate new equipment, procedures or concepts is an important step, but training and development can take a broader role. It has considerable potential, for example, as a motivator – people value the experience of acquiring new skills and abilities, and also feel valued as part of the organization. For example, a recent survey of continuous improvement in the UK (essentially concerned with increasing levels of participation in innovation) found that the opportunity for personal development was ranked higher than financial motivators as a reward mechanism.[78]

Training and development are also essential complements to enabling people to take on more responsibility and demonstrate more initiative – so-called 'empowerment' exercises. Harnessing creativity and encouraging experiment depend on people having the necessary skills and confidence to deploy them – and this places considerable emphasis on long-term training and development strategy.[4,79]

Training is also valuable as part of a wider change programme. When major innovations are introduced people often resist the change for a variety of reasons, not all of which are rational or clearly articulated.[80,81] A major component of this is often the sense that the innovation will require abilities or skills which the individual does not possess, or pose challenges which are not fully understood. Training – not only in the narrow sense of 'know-how' but also a component of education around the strategic rationale for the change (the 'know why') – can provide a powerful lubricant for oiling the wheels of such innovation programmes.[82] For example, in studies of the introduction of computer-aided enterprise management systems (ERP), the presence, or absence, of such extensive training (beyond the minimum usually offered by suppliers of such systems) was a major determinant of successful implementation.[83,84]

One other aspect of training and development concerns its use to develop the habit of learning. A core element in any 'learning organization' will be the continual discovery and sharing of new knowledge – in other words, a continuing and shared learning process.[48] But to put this in place requires that employees understand how to learn; an increasing number of organizations have recognized that this is not an automatic process and have begun implementing training programmes designed less to equip employees with skills than to engender the habit of learning. So, for example, some firms offer access to courses in foreign languages, hobby skills and other activities unrelated to work but with the twin aims of motivating staff and getting them back into the habit of learning.

11.5 High Involvement in Innovation

Whereas innovation is often seen as the province of specialists in R&D, marketing, design or IT, the underlying creative skills and problem-solving abilities are possessed by everyone. If mechanisms can be found to focus such abilities on a regular basis across the entire company, the resulting innovative potential is enormous. Although each individual may only be able to develop limited, incremental innovations, the sum of these efforts can have far-reaching impacts.[39]

A good illustration of this is the 'quality miracle' which was worked by Japanese manufacturing industry in the post-war years, and which owed much to what they term *kaizen* – continuous improvement. Firms like Toyota and Matsushita receive millions of suggestions for improvements every year from their employees – and the vast majority of these are implemented.[85,86] Western firms have done much to close this gap in recent years. For example, a study of firms in the UK which have acquired the 'Investors in People' (IiP) award (an externally assessed review of employee involvement practices) shows a correlation between this and higher business performance. Such businesses have a higher rate of return on capital (RRC), higher turnover/sales per employee and have higher profits per employee.

Individual case studies confirm this pattern in a number of countries. As one UK manager[3] put it, 'Our operating costs are reducing year on year due to improved efficiencies. We have seen a 35% reduction in costs within two and a half years by improving quality. There are an average of 21 ideas per employee today compared to nil in 1990. Our people have accomplished this.' Box 11.6 gives another example.

TABLE 11.3 Performance of IiP companies against others[2]

	Average company	Investors company	Gain
RRC	9.21%	16.27%	77%
Turnover/sales per employee	£64912	£86625	33%
Profit per employee	£1815	£3198	76%

BOX 11.6
HIGH INVOLVEMENT IN INNOVATION

At first sight XYZ systems does not appear to be anyone's idea of a 'world class' manufacturing outfit. Set in a small town in the Midlands with a predominately agricultural industry, XYZ employs around 30 people producing gauges and other measuring devices for the forecourts of filling stations. Their products are used to monitor and measure levels and other parameters in the big fuel tanks underneath the stations, and on the tankers which deliver to them. Despite their small size (although they are part of a larger but decentralized group) XYZ have managed to command around 80% of the European market. Their processes are competitive against even large manufacturers; their delivery and service level the envy of the industry. They have a fistful of awards for their quality and yet manage to do this across a wide range of products some dating back 30 years, which still need service and repair. They use technologies from complex electronics and remote sensing right down to basics – they still make a wooden measuring stick, for example.

Their success can be gauged from profitability figures but also from the many awards, which they receive and continue to receive as one of the best factories in the UK.

Yet if you go through the doors of XYZ you would have to look hard for the physical evidence of how they achieved this enviable position. This is not a highly automated business – it would not be appropriate. Nor is it laid out in modern facilities; instead they have clearly made much of their existing environment and organized it and themselves to best effect.

Where does the difference lie? Fundamentally in the approach taken with the workforce. This is an organization where training matters – investment is well above the average and everyone receives a significant training input, not only in their own particular skills area but also across a wide range of tasks and skills. One consequence of this is that the workforce are very flexible; having been trained to carry out most of the operations, they can quickly move to where they are most needed. The payment system encourages such co-operation and team working, with its simple structure and emphasis on payment for skill, quality and team working. The strategic targets are clear and simple, and are discussed with everyone before being broken down into a series of small manageable improvement projects in a process of policy deployment. All around the works there are copies of the 'bowling chart' which sets out simply – like a tenpin bowling score sheet – the tasks to be worked on as improvement projects and how they could contribute to the overall strategic aims of the business. And if they achieve or exceed those strategic targets – then everyone gains thorough a profit-sharing and employee ownership scheme.

Being a small firm there is little in the way of hierarchy but the sense of team working is heightened by active leadership and encouragement to discuss and explore issues together – and it doesn't hurt that the operations director practices a form of MBWA – management by walking about!

Perhaps the real secret lies in the way in which people feel enabled to find and solve problems, often experimenting with different solutions and frequently failing – but at least learning and sharing that information for others to build on. Walking round the factory it is clear that this place isn't standing still – whilst major investment in new machines is not an everyday thing, little improvement projects – *kaizens* as they call them – are everywhere. More significant is the fact that the operations director is often surprised by what he finds people doing – it is clear that he has not got a detailed idea of which projects people are working on and what they are doing. But if you ask him if this worries him the answer is clear – and challenging. 'No, it doesn't bother me that I don't know in detail what's going on. They all know the strategy, they all have a clear idea of what we have to do (via the "bowling charts". They've all been trained, they know how to run improvement projects and they work as a team. And I trust them . . .'

Although high-involvement schemes of this kind received considerable publicity in the late twentieth century, associated with total quality management and lean production, they are not a new concept. For example, Denny's Shipyard in Dumbarton, Scotland had a system which asked workers (and rewarded them for) 'any change by which work is rendered either superior in quality or more economical in cost' – back in 1871. John Patterson, founder of the National Cash Register Company in the USA, started a suggestion and reward scheme aimed at harnessing what he called 'the hundred headed brain' around 1894.

Since much of such employee involvement in innovation focuses on incremental change it is tempting to see its effects as marginal. Studies show, however, that when taken over an extended period it is a significant factor in the strategic development of the organization.[87]

Underpinning such continuous incremental innovation is the organizational culture to support and encourage over the long term. This simple point has been recognized in a number of different fields, all of which converge around the view that higher levels of participation in innovation represents a competitive advantage. For example:

- In the field of quality management it became clear that major advantages could accrue from better and more consistent quality in products and services. Crosby's work on quality costs suggested the scale of the potential savings (typically 20–40% of total sales revenue), and the experience of many Japanese manufacturers during the post-war period provide convincing arguments in favour of this approach.[88–91]

- The concept of 'lean thinking' has diffused widely during the past 20 years and is now applied in manufacturing and services as diverse as chemicals production, hospital management and supermarket retailing.[92] It originally emerged from detailed studies of assembly plants in the car industry which highlighted significant differences between the best and the average plants along a range of dimensions, including productivity, quality and time. Efforts to identify the source of these significant advantages revealed that the major differences lay not in higher levels of capital investment or more modern equipment, but in the ways in which production was organized and managed.[54] The authors of the study concluded:

  > . . . our findings were eye-opening. The Japanese plants require one-half the effort of the American luxury-car plants, half the effort of the best European plant, a quarter of the effort of the average European plant, and one-sixth the effort of the worst European luxury car producer. At the same time, the Japanese plant greatly exceeds the quality level of all plants except one in Europe – and this European plant required four times the effort of the Japanese plant to assemble a comparable product . . .

  Central to this alternative model was an emphasis on team working and participation in innovation.

- The principles underlying 'lean thinking' had originated in experiences with what were loosely called 'Japanese manufacturing techniques'.[93] This bundle of approaches (which included umbrella ideas like 'just-in-time' and specific techniques like poke yoke) were credited with having helped Japanese manufacturers gain significant competitive edge in sectors as diverse as electronics, motor vehicles and steel making.[94,95] Underpinning these techniques was a philosophy which stressed high levels of employee involvement in the innovation process, particularly through sustained incremental problem-solving – *kaizen*.[36,96]

The transferability of such ideas between locations and into different application areas has also been extensively researched. It is clear from these studies that the principles of 'lean' manufacturing can be extended into supply and distribution chains, into product development and R&D and into service activities and operations.[97–99] Nor is there any particular barrier in terms of national culture; high-involvement approaches to innovation have been successfully transplanted to a number of different locations (see Box 11.7).[63,100,101]

So there is a considerable weight of experience now available to support the view that enhanced performance can and does result from increasing involvement in innovation through 'high-involvement innovation' (HII). But there is also a secondary effect which should not be underestimated; the more people are involved in change, the more receptive they become to change itself. Since the turbulent nature of most organizational environments is such that increasing levels of change are becoming the norm, involvement of employees in HII programmes may provide a powerful aid to effective management of change.

BOX 11.7
DIFFUSION OF HIGH-INVOLVEMENT INNOVATION

How far has this approach diffused? Why do organizations choose to develop it? What benefits do they receive? And what barriers prevent them moving further along the road towards high involvement?

Questions like these provided the motivation for a large survey carried out in a number of European countries and replicated in Australia during the late 1990s.[4] It was one of the fruits of a co-operative research network which was established to share experiences and diffuse good practice in the area of high involvement innovation. The survey involved over 1000 organizations in a total of seven

continues overleaf

BOX 11.7   (*continued*)

countries and provides a useful map of the take-up and experience with high-involvement innovation. (The survey only covered manufacturing although follow-up work is looking at services as well). Some of the key findings were:

- Overall around 80% of organizations were aware of the concept and its relevance, but its actual implementation, particularly in more developed forms, involved around half of the firms.

- The average number of years which firms had been working with high involvement innovation on a systematic basis was 3.8, supporting the view that this is not a 'quick fix' but something to be undertaken as a major strategic commitment. Indeed, those firms which were classified as 'CI innovators' – operating well-developed high-involvement systems – had been working on this development for an average of nearly seven years.

- High involvement is still something of a misnomer for many firms, with the bulk of efforts concentrated on shop-floor activities as opposed to other parts of the organization. There is a clear link between the level of maturity and development of high involvement here – the 'CI innovators' group was much more likely to have spread the practices across the organization as a whole.

- Motives for making the journey down this road vary widely but cluster particularly around the themes of quality improvement, cost reduction and productivity improvement.

- In terms of the outcome of high-involvement innovation there is clear evidence of significant activity, with an average per capita rate of suggestions of 43/year of which around half were actually implemented. This is a difficult figure since it reflects differences in measurement and definition but it does support the view that there is significant potential in workforces across a wide geographical range – it is not simply a Japanese phenomenon. Firms in the sample also reported indirect benefits arising from this including improved morale and motivation, and a more positive attitude towards change.

- What these suggestions can do to improve performance is, of course, the critical question and the evidence from the survey suggests that key strategic targets were being impacted upon. On average improvements of around 15% were reported in process areas like quality, delivery, manufacturing lead time, and overall productivity, and there was also an average of 8% improvement in the area of product cost. Of significance is the correlation between performance improvements reported and the maturity of the firm in terms of high-involvement behaviour. The 'CI innovators' – those which had made most progress towards establishing high involvement as 'the way we do things around here' were also the group with the largest reported gains – averaging between 19 and 21% in the above process areas.

Performance areas (% change)	UK	SE	N	NL	FI	DK	Australia	Average across sample (n = 754 responses)
Productivity improvement	19	15	20	14	15	12	16	15
Quality improvement	17	14	17	9	15	15	19	16
Delivery performance improvement	22	12	18	16	18	13	15	16
Lead time reduction	25	16	24	19	14	5	12	15
Product cost reduction	9	9	15	10	8	5	7	8

- Almost all high-involvement innovation activities take place on an 'in-line' basis – that is, as part of the normal working pattern rather than as a voluntary 'off-line' activity. Most of this activity takes place in some form of group work although around a third of the activity is on an individual basis.
- To support this there is widespread use of tools and techniques particularly those linked to problem-finding and -solving which around 80% of the sample reported using. Beyond this there is extensive use of tools for quality management, process-mapping and idea generation, although more specialized techniques like statistical process control or quality function deployment are less widespread. Perhaps more significant is the fact that even with the case of general problem-finding and -solving tools only around a third of staff had been formally trained in their use.

Growing recognition of the potential has moved the management question away from whether or not to try out employee involvement to one of 'how to make it happen?' The difficulty is less about getting started than about keeping it going long enough to make a real difference. Many organizations have experience of starting the process – getting an initial surge of ideas and enthusiasm during a 'honeymoon' period – and then seeing it gradually ebb away until there is little or no HII activity. We shouldn't really be surprised at this – clearly if we want to change the ways in which people think and behave on a long-term-basis then it's going to need a strategic development programme to make it happen. A quick 'sheep dip' of training plus a bit of enthusiastic arm-waving from the managing director isn't likely to do much in the way of fundamentally changing 'the way we do things around here' – the underlying culture – of the organization.

A Roadmap for the Journey

Research on implementing HII suggests that there are a number of stages in this journey, progressing in terms of the development of systems and capability to involve people but also progressing in terms of the bottom line benefits which can be expected.[39] Each of these takes time to move through, and there is no guarantee that organizations will progress to the next level. Moving on means having to find ways of overcoming the particular obstacles associated with different stages (see Figure 11.1).

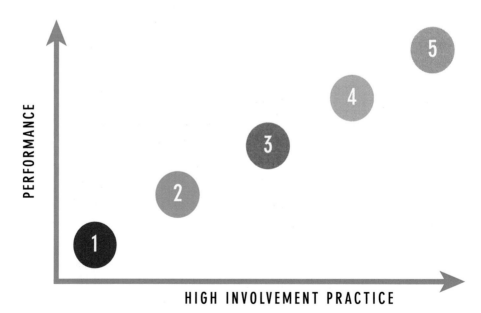

FIGURE 11.1   The five-stage high-involvement innovation model

The first stage – level 1 – is what we might call 'unconscious HII'. There is little, if any, HII activity going on, and when it does happen it is essentially random in nature and occasional in frequency. People do help to solve problems from time to time – for example, they will pull together to iron out problems with a new system or working procedure, or getting the bugs out of a new product. But there is no formal attempt to mobilize or build on this activity, and many organizations may actively restrict the opportunities for it to take place. The normal state is one in which HII is not looked for, not recognized, not supported – and often, not even noticed. Not surprisingly, there is little impact associated with this kind of change.

Level 2, on the other hand, represents an organization's first serious attempts to mobilize HII. It involves setting up a formal process for finding and solving problems in a structured and systematic way – and training and encouraging people to use it. Supporting this will be some form of reward/recognition arrangement to motivate and encourage continued participation. Ideas will be managed through some form of system for processing and progressing as many as possible and handling those which cannot be implemented. Underpinning the whole set-up will be an infrastructure of appropriate mechanisms (teams, task forces or whatever), facilitators and some form of steering group to enable HII to take place and to monitor and adjust its operation over time. None of this can happen without top management support and commitment of resources to back that up.

Level 2 is all about establishing the habit of HII within at least part of the organization. It certainly contributes improvements but these may lack focus and are often concentrated at a local level, having minimal impact on more strategic concerns of the organization. The danger in such HII is that, once having established the habit of HII, it may lack any clear target and begin to fall away. In order to maintain progress there is a need to move to the next level of HII – concerned with strategic focus and systematic improvement.

Level 3 involves coupling the HII habit to the strategic goals of the organization such that all the various local level improvement activities of teams and individuals can be aligned. In order to do this two key behaviours need to be added to the basic suite – those of strategy deployment and of monitoring and measuring. Strategy (or policy) deployment involves communicating the overall strategy of the organization and breaking it down into manageable objectives towards which HII activities in different areas can be targeted. Linked to this is the need to learn to monitor and measure the performance of a process and use this to drive the continuous improvement cycle.

Level 3 activity represents the point at which HII makes a significant impact on the bottom line – for example, in reducing throughput times, scrap rates, excess inventory, etc. It is particularly effective in conjunction with efforts to achieve external measurable standards (such as ISO 9000) where the disciplines of monitoring and measurement provide drivers for eliminating variation and tracking down root cause problems. The majority of 'success stories' in HII can be found at this level – but it is not the end of the journey.

One of the limits of level 3 HII is that the direction of activity is still largely set by management and within prescribed limits. Activities may take place at different levels, from individuals through small groups to cross-functional teams, but they are still largely responsive and steered externally. The move to level 4 introduces a new element – that of 'empowerment' of individuals and groups to experiment and innovate on their own initiative.

Clearly this is not a step to be taken lightly, and there are many situations where it would be inappropriate – for example, where established procedures are safety critical. But the principle of 'internally directed' HII as opposed to externally steered activity is important, since it allows for the open-ended learning behaviour which we normally associate with professional research scientists and engineers. It requires a high degree of understanding of, and commitment to, the overall strategic objectives, together with training to a high level to enable effective experimentation. It is at this point that the kinds of 'fast-learning' organizations described in some 'state-of-the-art' innovative company case studies can be found – places where everyone is a researcher and where knowledge is widely shared and used.

Level 5 is a notional end-point for the journey – a condition where everyone is fully involved in experimenting and improving things, in sharing knowledge and in creating an active learning organization. Table 11.4 illustrates the key elements in each stage. In the end the task is one of building a shared set of values which bind people in the organization together and enable them to participate in its development. As one manager put it in a UK study, '. . . we never use the word empowerment! You can't empower people – you can only create the climate and structure in which they will take responsibility . . .'[33] Box 11.8 gives another example.

11.6 Effective Team Working

'It takes five years to develop a new car in this country. Heck, we won World War 2 in four years . . .' Ross Perot's critical comment on the state of the US car industry in the late 1980s captured some of the frustration with existing ways of designing and building cars. In the years that followed significant strides were made in reducing the development cycle, with Ford and Chrysler succeeding in dramatically reducing time and improving quality. Much of the advantage was gained through extensive team-working; as Lew Varaldi, project manager of Ford's Team Taurus project put it, '. . . it's amazing the dedication and commitment you get from people . . . we will never go back to the old ways because we know so much about what they can bring to the party . . .'[103]

TABLE 11.4   Stages in the evolution of HII capability

Stage of development	Typical characteristics
1. 'Natural'/background HII	Problem-solving random No formal efforts or structure Occasional bursts punctuated by inactivity and non-participation Dominant mode of problem-solving is by specialists Short-term benefits No strategic impact
2. Structured HII	Formal attempts to create and sustain HII Use of a formal problem-solving process Use of participation Training in basic HII tools Structured idea management system Recognition system Often parallel system to operations
3. Goal-oriented HII	All of the above, plus formal deployment of strategic goals Monitoring and measurement of HII against these goals In-line system
4. Proactive/empowered HII	All of the above, plus responsibility for mechanisms, timing, etc., devolved to problem-solving unit Internally directed rather than externally directed HII High levels of experimentation
5. Full HII capability – the learning organization	HII as the dominant way of life Automatic capture and sharing of learning Everyone actively involved in innovation process Incremental and radical innovation

These days cars – and an increasing range of complex products and projects – are designed and built with lead times measured in months not years, and with increasing 'stretch' in terms of features and functionality. Complex service packages are designed and delivered in highly customized fashion and configured and reconfigured to suit the changing needs of a wide range of users. Public sector services like utilities, transport, healthcare and policing are all adapting to deal with radical new demands and emerging challenges. All of this puts a premium on the kind of team working suggested above which builds on the principle that innovation is primarily about combining different perspectives in solving problems.

Experiments indicate that groups have more to offer than individuals in terms of both fluency of idea generation and in flexibility of solutions developed. Focusing this potential on innovation tasks is the prime driver for the trend towards high levels of team working – in project teams, in cross-functional and inter-organizational problem-

---

BOX 11.8
CREATING HIGH INVOLVEMENT INNOVATION CONDITIONS

Dutton Engineering does not, at first sight, seem a likely candidate for world class. A small firm with 28 employees, specializing in steel cases for electronic equipment, it ought to be amongst the ranks of hand-to-mouth metal-bashers of the kind which you can find all round the world. Yet Dutton has been doubling its turnover, sales per employee have doubled in an eight-year period, rejects are down from 10% to 0.7%, and over 99% of deliveries are made within 24 hours – compared to only 60% being achieved within one week a few years ago. This transformation has not come overnight – the process started in 1989 – but it has clearly been successful and Dutton are now held up as an example to others of how typical small engineering firms can change.

At the heart of the transformation which Ken Lewis, the original founder and architect of the change, has set in train is a commitment to improvements through people. The workforce are organized into four teams who manage themselves, setting work schedules, dealing with their own customers, costing their own orders and even setting their pay! The company has moved from traditional weekly pay to a system of 'annualized hours' where they contract to work for 1770 hours in year – and tailor this flexibly to the needs of the business with its peaks and troughs of activity. There is a high level of contribution to problem-solving, encouraged by a simple reward system which pays £5–15 for bright ideas, and by a bonus scheme whereby 20% of profits are shared.[102]

---

solving groups and in cells and work groups where the focus is on incremental, adaptive innovation.

Considerable work has been done on the characteristics of high-performance project teams for innovative tasks, and the main findings are that such teams rarely happen by accident.[104,105] They result from a combination of selection and investment in team-building, allied to clear guidance on their roles and tasks, and a concentration on managing group process as well as task aspects. A variety of studies have been carried out aimed at identifying key drivers and barriers to effective performance, and they share the conclusion that effective team-building is a critical determinant of project success.[29] For example, research within the Ashridge Management College team working programme developed a model for 'superteams' which includes components of building and managing the internal team and also its interfaces with the rest of the organization.[26]

Holti and Neumann provide a useful summary of the key factors involved in developing team working.[106] Although there is considerable current emphasis on team working, we should remember that teams are not always the answer. In particular, there are dangers in putting nominal teams together where unresolved conflicts, personality clashes, lack of effective group processes and other factors can diminish their effectiveness. Tranfield *et al.* look at the issue of team working in a number of different contexts and highlight the importance of selecting and building the appropriate team for the task and the context.[107]

Teams are increasingly being seen as a mechanism for bridging boundaries within the organization – and indeed, in dealing with inter-organizational issues. Cross-functional teams can bring together the different knowledge sets needed for tasks like product development or process improvement – but they also represent a forum where often deep-rooted differences in perspectives can be resolved.[108] Lawrence and Lorsch in their pioneering study of differentiation and integration within organizations found that interdepartmental clashes were a major source of friction and contributed much to delays and difficulties in operations. Successful organizations were those which invested in multiple methods for integrating across groups – and the cross-functional team was one of the most valuable resources.[66] But, as we indicated above, building such teams is a major strategic task – they will not happen by accident, and they will require additional efforts to ensure that the implicit conflicts of values and beliefs are resolved effectively.

Teams also provide a powerful enabling mechanism for achieving the kind of decentralized and agile operating structure which many organizations aspire to. As a substitute for hierarchical control, self-managed teams working within a defined area of autonomy can be very effective. For example, Honeywell's defence avionics factory reports a dramatic improvement in on-time delivery – from below 40% in the 1980s to 99% in 1996 – to the implementation of self-managing teams.[109] In the Netherlands one of the most successful bus companies is Vancom Zuid-Limburg which has improved both price and non-price performance and has high customer satisfaction ratings. Again they attribute this to the use of self-managing teams and to the reduction in overhead costs which results. In their system one manager supervises over 40 drivers where the average for the sector is a ratio of 1:8. Drivers are also encouraged to participate in problem-finding and solving in areas like maintenance, customer service and planning.[110]

Key elements in effective high performance team working include:

- clearly defined tasks and objectives;
- effective team leadership;
- good balance of team roles and match to individual behavioural style;
- effective conflict resolution mechanisms within the group;
- continuing liaison with external organization.

Teams typically go through four stages of development, popularly known as 'forming, storming, norming and performing'.[111] That is, they are put together and then go through a phase of resolving internal differences and conflicts around leadership, objectives, etc. Emerging from this process is a commitment to shared values and norms governing the way the team will work, and it is only after this stage that teams can move on to effective performance of their task.

Central to team performance is the make-up of the team itself, with good matching between the role requirements of the group and the behavioural preferences of the individuals involved. Belbin's work has been influential here in providing an approach to team role matching. He classifies people into a number of preferred role types – for example, 'the plant' (someone who is a source of new ideas), 'the resource investigator', 'the shaper' and the 'completer/finisher'. Research has shown that the most effective teams are those with diversity in background, ability and behavioural style. In one noted experiment highly talented but similar people in 'Apollo' teams consistently performed less well than mixed, average groups.[25]

With increased emphasis on cross-boundary and dispersed team activity, a series of new challenges are emerging. In the extreme case a product development team might begin work in London, pass on to their US counterparts later in the day who in turn pass on to their far Eastern colleagues – effectively allowing a 24-hour non-stop development activity. This makes for higher productivity potential – but only if the issues around managing dispersed and virtual teams can be resolved. Similarly the concept of sharing knowledge across boundaries depends on enabling structures and mechanisms.[75,76,112]

11.7 Creative Climate

'Microsoft's only factory asset is the human imagination.'
(Bill Gates)

Many great inventions came about as the result of lucky accidental discoveries – for example, Velcro fasteners, the adhesive behind 'Post-it' notes or the principle of float glass manufacturing. But as Louis Pasteur observed, 'chance favours the prepared mind' and we can usefully deploy our understanding of the creative process to help set up the conditions within which such 'accidents' can take place.

Two important features of creativity are relevant in doing this. The first is to recognize that creativity is an attribute which everyone possesses – but their preferred style of expressing it varies widely.[113] Some people are comfortable with ideas which challenge the whole way in which the universe works, whilst others prefer smaller increments of change – ideas about how to improve the jobs they do or their working environment in small incremental steps. (This explains in part why so many 'creative' people – artists, composers, mad scientists – are also seen as 'difficult' or living outside the conventions of acceptable behaviour.) This has major implications for how we manage creativity within the organization; innovation, as we have seen, involves bringing something new into widespread use, not just inventing it. Whilst the initial flash may require a significant creative leap, much of the rest of the process will involve hundreds of small problem-finding and -solving exercises – each of which needs creative input. And though the former may need the skills or inspiration of a particular individual the latter require the input of many different people over a sustained period of time. Developing the light bulb or the Post-it note or any successful innovation is actually the story of the combined creative endeavour of many individuals.

Organizational structures are the visible artefacts of what can be termed an innovative culture – one in which innovation can thrive. Culture is a complex concept, but it basically equates to the pattern of shared values, beliefs and agreed norms which shape behaviour – in other words, it is 'the way we do things round here' in any organization. Schein suggests that culture can be understood in terms of three linked levels, with the deepest and most inaccessible being what each individual believes about the world – the 'taken for granted' assumptions. These shape individual actions and the collective and socially negotiated version of these behaviours defines the dominant set of norms and values for the group. Finally, behaviour in line with these norms creates a set of artefacts – structures, processes, symbols, etc. – which reinforce the pattern.[114]

Given this model it is clear that management cannot directly change culture – but it can intervene at the level of artefacts – by changing structures or processes – and by providing models and reinforcing preferred styles of behaviour. Such 'culture change' actions are now widely tried in the context of change programmes towards total quality management and other models of organization which require more participative culture.

A number of writers have looked at the conditions under which creativity thrives or is suppressed.[44,47] Kanter[6] provides a list of environmental factors which contribute to stifling innovation; these include:

- dominance of restrictive vertical relationships;
- poor lateral communications;
- limited tools and resources;
- top-down dictates;
- formal, restricted vehicles for change;
- reinforcing a culture of inferiority (i.e. innovation always has to come from outside to be any good);

- unfocused innovative activity;
- unsupporting accounting practices.

The effect of these is to create and reinforce the behavioural norms which inhibit creativity and lead to a culture lacking in innovation. It follows from this that developing an innovative climate is not a simple matter since it consists of a complex web of behaviours and artefacts. And changing this culture is not likely to happen quickly or as a result of single initiatives (such as restructuring or mass training in a new technique).

Instead, building a creative climate involves systematic development of organizational structures, communication policies and procedures, reward and recognition systems, training policy, accounting and measurement systems and deployment of strategy. Mechanisms for doing so in various different kinds of organizations and in different national cultures are described by a number of authors including Cook, Rickards and Ekvall.[45,46,115]

Of particular relevance in this area is the design of effective reward systems. Many organizations have reward systems which reflect the performance of repeated tasks, rather than encourage the development of new ideas. Progress is associated with 'doing things by the book' rather than challenging and changing things. By contrast, innovative organizations look for ways to reward creative behaviour, and to encourage its emergence. Examples of reward systems include the establishment of a 'dual ladder' which enables technologically innovative staff to progress within the organization without needing to move across to management posts.[116]

One aspect of this worth highlighting concerns the emerging idea of 'intrapreneurship' – internal entrepreneurship.[23] In an organization with a supportive and innovative culture, individuals with bright ideas can progress them with support and encouragement from the system. For example, in 3M the culture encourages individuals to follow up interesting ideas and allows them up to 15% of their time for such activity. If things look promising, there are internal venture funds to enable a more thorough exploration – and if the individual thinks they can build a business out of the idea, 3M will back it and give them the responsibility to run it. In this way the company has grown through organic, 'intrapreneurial' means. Chapter 10 explores this theme in more detail.

11.8 External Focus

A recurring theme in this book has been the extent to which innovation has become an open process involving richer networks across and between organizations. This high-

lights a long-established characteristic of successful innovating organizations – an orientation which is essentially open to new stimuli from outside.[40] Whether the signals
are of threats or opportunities, these organizations have approaches which pick them
up and communicate them through the organization. Developing a sense of external
orientation – for example, towards key customers or sources of major technological
developments – and ensuring that this pervades organizational thinking at all levels are
of considerable importance in building an innovative organization.

One of the consistent themes in the literature on innovation success and failure concerns the need to understand user needs. Developing this sense of customer requirements is essential in dealing with the external marketplace as we saw in Chapter 7,
but it is also a principle which can be usefully extended within the organization. By
developing a widespread awareness of customers – both internal and external – quality
and innovation can be significantly improved. This approach – which forms one of the
cornerstones of 'total quality management' thinking – contrasts sharply with the
traditional model in which problems were passed on between sequential elements in
the innovation process, and where there was no provision for feedback or mutual
adjustment.[117]

Of course, not all industries have the same degree of customer involvement – and
in many the dominant focus is more on technology. This does not mean that customer
focus is an irrelevant concept; the issue here is one of building relationships which
enable clear and regular communication, providing inputs for problem-solving and
shared innovation.[118]

But the idea of extending involvement goes far beyond customers and end-users. As
we have seen repeatedly open innovation requires building such relationships with an
extended cast of characters, including suppliers, collaborators, competitors, regulators
and multiple other players.[119] We will look in more detail at the implications for this
'beyond the boundaries' aspect of the innovative organization at the end of this chapter.

11.9 Extensive Communication

Closely linked to the external focus point in the previous section is the requirement for
extensive communication. By this we mean communication which is multidirectional
(up, down and laterally) and which makes use of multiple channels and media. Many
problems occur in the innovation process through failures in communication, particularly between different functional elements in the process. Developing mechanisms for
resolving conflicts and improving clarity and frequency of communication across such
interfaces are critical to innovation success, particularly since so much problem-solving

depends on combining different knowledge sets which may be widely distributed across the organization.

Whilst organizations may benefit from 'gatekeeper' figures who can channel and focus communication, there is also a need for more structured approaches.

Mechanisms for enhancing communication include:

- job rotation and secondment;
- cross-functional teams and projects;
- policy-deployment and review sessions;
- team briefings;
- multiple media – video, notice boards, e-mail, intranets, etc.

This latter theme is becoming highly relevant as organizations recognize the need to think actively about knowledge management. Innovation is increasingly seen as requiring the creating, combination, sharing and deployment of knowledge – and this places strong emphasis on the channels and mechanisms which are used for communication. New technologies such as 'groupware' and intranets are being widely used to facilitate this, but attention is also being paid to the social organization of knowledge networks.[49,75,120]

11.10 The Learning Organization

Much recent discussion has focused on the concept of 'learning organizations', seeing knowledge as the basis for competition in the twenty-first century. Mobilizing and managing knowledge becomes a primary task and many of the recipes offered for achieving this depend upon mobilizing a much higher level of participation in innovative problem-solving and on building such routines into the fabric of organizational life.[32,48,121]

One way of looking at innovation is as a learning cycle, involving a process of experiment, experience, reflection and consolidation. Managing the process is primarily a function of the creation of conditions under which learning opportunities emerge and are exploited. A key determinant or relative success or failure is the ability to manage this learning cycle in explicit form – for example, in the development of new products or the implementation of new process technology.[122,123]

Crucially, what is learned and developed with each cycle of innovation is not only technological knowledge to add to the firm-specific knowledge base (in formal and tacit form), but also knowledge about how to manage the process itself. To take an analogy, human beings not only acquire new content of knowledge as they grow, but they also learn to learn; some develop more effective learning strategies than others, whilst for others it is a case of 'some people never learn'. Over time, successful innovators review and build upon particular courses of action and internalize particular routines for managing the innovation challenge – for example, ways of getting close to users, ways of managing projects, ways of harnessing and sharing information, ways of exciting and supporting intrapreneurship, etc.

There is, of course, no guarantee that the cycle will be completed; evidence suggests many organizations repeat mistakes and fail to learn from earlier problems in innovation. This has prompted much discussion about the process of post-innovation auditing and consolidation – and the need to move beyond blame accounting or simple financial or cost–benefit auditing to more comprehensive reviews. There is also the question of 'unlearning' – of eliminating those routines which do not contribute to success. Part of learning new tricks is the ability to forget old ones – and the reason old dogs are difficult to teach is that the old patterns are very deeply embedded.[13,55,56]

Organizations do not learn, it is the people within them who do; routines are thus directed at creating the stage on which they act and the scripts they work to. We are interested in the routines which the organization develops to enable the learning process, and in particular in the ways in which individual and shared learning can be mobilized. For example, Garvin[48] suggests the following mechanisms as important:

- training and development of staff;
- development of a formal learning process based on a problem-solving cycle;
- monitoring and measurement;
- documentation;
- experiment;
- display;
- challenge existing practices;
- use of different perspectives;
- reflection – learning from the past.

These represent powerful tools for sustaining a single-loop learning approach to steady-state innovation. But inevitably the context within which such activities take place is constantly changing and there is a need for a second-order learning approach which retunes the organization's learning routines to deal with new challenges. In particular we will look in the next sections at the implications for building and sustaining an innovative organization in extended – beyond boundaries – mode and under discontinuous – beyond the steady state – conditions.

11.11 Beyond the Boundaries

All of the above discussion presumes that the organization in question is a single entity, a group of people organized in a particular fashion towards some form of collective purpose. But increasingly we are seeing the individual enterprise becoming linked with others in some form of collective – a supply chain, an industrial cluster, a co-operative learning club or a product development consortium. Studies exploring this aspect of inter-firm behaviour include learning in shared product development projects,[124,125] in complex product system configuration,[126] in technology fusion,[127] in strategic alliances,[128-130] in regional small firm clusters,[131-133] in sector consortia,[134] in 'topic networks',[135] and in industry associations.[136,137]

Consider some examples:

- Studies of 'collective efficiency' have explored the phenomenon of clustering in a number of different contexts.[138,139] From this work it is clear that the model is not just confined to parts of Italy, Spain and Germany, but diffused around the world – and under certain conditions, extremely effective. For example, one town (Sialkot) in Pakistan plays a dominant role in the world market for specialist surgical instruments made of stainless steel. From a core group of 300 small firms, supported by 1500 even smaller suppliers, 90% of production (1996) was exported and took a 20% share of the world market, second only to Germany. In another case the Sinos valley in Brazil contains around 500 small-firm manufacturers of specialist, high-quality leather shoes. Between 1970 and 1990 their share of the world market rose from 0.3 to 12.5% and they now export some 70% of total production. In each case the gains are seen as resulting from close interdependence in a co-operative network.
- Similarly, there has been much discussion about the merits of technological collaboration, especially in the context of complex product systems development.[140-142] Innovation networks of this kind offer significant advantages in terms of assembling different knowledge sets and reducing the time and costs of development – but are again often difficult to implement.[125,127]
- Much has been written on the importance of developing co-operative rather than adversarial supply chain relationships.[143] But it is becoming increasingly clear that the kind of 'collective efficiency' described above can operate in this context and contribute not only to improved process efficiency (higher quality, faster speed of response, etc.) but also to shared product development. The case of Toyota is a good illustration of this – the firm has continued to stay ahead despite increasing catch-up efforts on the part of Western firms and the consolidation of the industry. Much of this competitive edge can be attributed to its ability to create and maintain a high-performance knowledge-sharing network.[144]

- Networking represents a powerful solution to the resource problem – no longer is it necessary to have all the resources for innovation (particularly those involving specialized knowledge) under one roof provided you know where to obtain them and how to link up with them. The emergence of powerful information and communications technologies have further facilitated the move towards 'open innovation' and 'virtual organizations' are increasingly a feature of the business landscape.[145]

- Networking is increasingly offered as a way forward for industrial development. National and regional governments are trying to emulate the cluster effect which has proved so successful amongst small firms in Italy and Spain. Large firms are trying to develop their supply chains towards greater efficiency through supply chain learning and development programmes. Complex product systems are increasingly the subject of shared development projects amongst consortia of technology providers. But experience and research suggest that without careful management of these – and the availability of a shared commitment to deal with them – the chances are that such networks will fail to perform effectively.[43]

- Studies of learning behaviour in supply chains suggest considerable potential – one of the most notable examples being the case of the *kyoryokukai* (supplier associations) of Japanese manufacturers in the second half of the twentieth century.[144,146,147] Hines reports on other examples of supplier associations (including those in the UK), which have contributed to sustainable growth and development in a number of sectors particularly engineering and automotive.[143] Imai, in describing product development in Japanese manufacturers observes: '[Japanese firms exhibit] an almost fanatical devotion towards learning – both within organizational membership and with outside members of the inter-organizational network.'[36] Later, Lamming[98] (p. 206) identifies such learning as a key feature of lean supply, linking it with innovation in supply relationships. Marsh and Shaw describe collaborative learning experiences in the wine industry including elements of SCL, whilst the AFFA study reports on other experiences in the agricultural and food sector in Australia.[148,149] Case studies of SCL in the Dutch and UK food industries, the construction sector and aerospace provide further examples of different modes of SCL organisation.[150–152] Humphrey *et al.* describe their emergence in a developing country context (India).[153]

But obtaining the benefits of networking is not an automatic process – it requires considerable efforts in the area of co-ordination. Effective networks have what systems theorists call 'emergent properties' – that is, the whole is greater than the sum of the parts. But the risk is high that simply throwing together a group of enterprises will lead to sub-optimal performance with the whole being considerably less than the sum of the parts due to friction, poor communications, persistent conflicts over resources or objectives, etc.

Research on inter-organizational networking suggests that a number of core processes need managing in a network, effectively treating it as if it were a particular form of organization.[154] For example, a network with no clear routes for resolving conflicts is likely to be less effective than one which has a clear and accepted set of norms – a 'network culture' – which can handle the inevitable conflicts which emerge.

Building and operating networks can be facilitated by a variety of enabling inputs – for example, the use of advanced information and communications technologies may have a marked impact on the effectiveness with which information processing takes place. In particular research highlights a number of enabling elements which help build and sustain effective networks, including:

- *Key individuals* – creating and sustaining networks depends on putting energy into their formation and operation. Studies of successful networks identify the role of key figures as champions and sponsors, providing leadership and direction, particularly in the tasks of bringing people together and giving a system-level sense of purpose.[155] Increasingly the role of 'network broker' is being played by individuals and agencies concerned with helping create networks on a regional or sectoral basis.

- *Facilitation* – another important element is providing support to the process of networking but not necessarily acting as members of the network. Several studies indicate that such a neutral and catalytic role can help, particularly in the set-up stages and in dealing with core operating processes like conflict resolution.

- *Key organizational roles* – mirroring these individual roles are those which are played by key organizations – for example, a regional development agency organizing a cluster or a business association bringing together a sectoral network. Gereffi and others talk about the concept of network governance and identify the important roles played by key institutions such as major customers in buyer-driven supply chains.[156,157] Equally their absence can often limit the effectiveness of a network – for example, in research on supply chain learning the absence of a key governor limited the extent to which inter-organizational innovation could take place[158] (see Box 11.9).

11.12 Beyond the Steady State

As we saw in Chapter 1, increasingly organizations have to deal with the challenge not only of managing innovation in the 'steady state' doing what they do, but better, but also under discontinuous 'do different' conditions. Research suggests that those organizations which are able to thrive and exploit innovative opportunities under these conditions are agile, fast-moving and tolerant of high levels of risk and failure. For this

---

BOX 11.9
LEARNING NETWORKS

One approach to capturing learning in structured fashion is to work with other organizations or individuals, sharing perspectives and challenging complacent views and ideas. Shared learning works on all stages of the learning cycle and provides an extra source of traction in moving around it successfully. For example, it brings together different backgrounds and consequently different concepts and theories which may offer fresh insights. It offers the chance for shared or multiple experiments – 'Let's try something different this time . . .' – and it provides a supportive but challenging focus for reflection. It is harder to avoid confronting difficult but important issues if there is a group of peers pushing and challenging for answers.

One of the main attributes of learning networks is that they help maintain the learning cycle by inputs at all its key stages – challenging reflection, introduction of new concepts, shared experimentation and exchange of different experiences. Do they work? Increasing evidence suggests that this form of inter-organizational innovation can bring significant benefits, especially in terms of incremental, 'do better' innovation.[159]

Learning networks of this kind can take many forms. Sometimes they are informal groupings of managers who get together to share experiences and discuss possible solutions on a regular basis. Increasingly such networks are taking on a more structured form, aimed at helping capture and share learning around particular themes or amongst particular groups. Examples of these could be as below:

Type	Learning target	Examples
Professional	Increased professional knowledge and skill = better practice	Professional institution
Sector-based – association interests of firms with common interests in the development of a sector	Improved competence in some aspect of competitive performance – e.g. technical knowledge	Trade association Sector-based research organization
Topic-based	Improved awareness/knowledge of a particular field – e.g. a new technology or technique in which many firms have an interest	'Best practice' clubs

continues overleaf

BOX 11.9   (continued)

Type	Learning target	Examples
Region-based	Improved knowledge around themes of regional interest – for example, SMEs learning together about how to export, diffuse technology, etc.	'Clusters' and local learning co-operatives
Supplier- or value-stream-based	Learning to achieve standards of 'best practice' in quality, delivery, cost reduction, etc.	Particular firms supplying to a major customer or members of a shared value stream
Government-promoted networks	National or regional initiatives to provide upgrades in capacity – knowledge about technology, exporting, marketing, etc.	Regional development agencies, extension services, etc.
Task support networks	Similar to professional networks, aimed at sharing and developing knowledge about how to do a particular – especially novel – task	Practitioner networks

reason it is often new entrant firms which do well and existing incumbents tend to suffer when discontinuities emerge. This does not mean that new entrants always win but rather that we need two different organizational models to deal with the two different sets of conditions. Table 11.5 outlines these.

This poses two related challenges for innovation management. For new entrant enterprises trying to take advantage of new opportunities opened up by discontinuity the issues are particularly about using their agility to probe, learn and reconfigure in search of the dominant design which will eventually emerge. But doing so is resource-intensive and hard to sustain; for this reason many new entrant firms fail because they back the wrong horse or run out of resources. If they do manage to ride the emerging wave until a dominant design emerges they then face the challenge of building a business – essentially moving into the 'mature phase' by building routines and structures to support incremental development.

At the other extreme established organizations run the risk of being too slow to respond or too set in their organizational ways to manage the transition effectively. Their challenge is to find ways of retaining both sets of characteristics within an organization

TABLE 11.5 Two different innovation management archetypes

	Type 1 – Steady state-archetype	Type 2 – Discontinuous innovation-archetype
Interpretive schema – how the organization sees and makes sense of the world	There is an established set of 'rules of the game' by which other competitors also play. Particular pathways in terms of search and selection environments and technological trajectories exist and define the 'innovation space' available to all players in the game	No clear 'rules of the game' – these emerge over time but cannot be predicted in advance. Need high tolerance for ambiguity – seeing multiple parallel possible trajectories. 'Innovation space' defined by open and fuzzy selection environment. Probe and learn experiments needed to build information about emerging patterns and allow dominant design to emerge.
	Strategic direction is highly path-dependent	Highly path-independent
Strategic decision-making	Makes use of decision-making processes which allocate resources on the basis of risk management linked to the above 'rules of the game'. (Does the proposal fit the business strategic directions? Does it build on existing competence base?) Controlled risks are taken within the bounds of the 'innovation space'. Political coalitions are significant influences maintaining the current trajectory	High levels of risk-taking since no clear trajectories – emphasis on fast and lightweight decisions rather than heavy commitment in initial stages. Multiple parallel bets, fast failure and learning as dominant themes. High tolerance of failure but risk is managed by limited commitment. Influence flows to those prepared to 'stick their neck out' – entrepreneurial behaviour
Operating routines	Operates with a set of routines and structures/procedures which embed them which are linked to these 'risk rules' – for example, stage gate monitoring and review for project management. Search behaviour is along defined trajectories and uses tools and techniques for R&D, market research, etc. which assume a known space to be explored – search and selection environment. Network building to support innovation – e.g. user involvement, supplier partnership, etc. – is on basis of developing close and strong ties	Operating routines are open ended, based around managing emergence. Project implementation is about 'fuzzy front end', light touch strategic review and parallel experimentation. Probe and learn, fast failure and learn rather than managed risk. Search behaviour is about peripheral vision, picking up early warning through weak signals of emerging trends. Linkages are with heterogeneous population and emphasis less on established relationships than on weak ties

– ambidextrous capability – and there are many different approaches to dealing with the challenge. At one end of the spectrum are models which spin or split off the 'type 2' activities in some form of separate venture, whilst at the other there are models which attempt to build ambidextrous capability within the organization – for example through the intrapreneurship model of 3M and others (see Box 11.10). We discuss this theme in more detail in Chapter 12.

If we revisit our model of the innovative organization we can see some specific themes relevant to building such capability (see Table 11.6).

BOX 11.10
BUILDING AN INNOVATIVE ORGANIZATION – THE CASE OF 3M

3M is a well-known organization employing around 70 000 people in around 200 countries across the world. Its $15bn of annual sales come from a diverse product range involving around 50 000 items serving multiple markets but building on core technical strengths, some of which like coatings can be traced back to the company's foundation. The company has been around for just over 100 years and during that period has established a clear reputation as a major innovator. Significantly they paint a consistent picture in interviews and in publications – innovation success is a consequence of creating the culture in which it can take place – it becomes 'the way we do things around here' in a very real sense. This philosophy is borne out in many anecdotes and case histories – the key to their success has been to create the conditions in which innovation can arise from any one of a number of directions, including lucky accidents, and there is a deliberate attempt to avoid putting too much structure in place since this would constrain innovation.

Elements in this complex web include:

- Recognition and reward – throughout the company there are various schemes which acknowledge innovative activity – for example, their Innovator's Award which recognizes effort rather than achievement.
- Reinforcement of core values – innovation is respected – for example, there is a 'hall of fame' whose members are elected on the basis of their innovative achievements.
- Sustaining 'circulation' – movement and combination of people from different perspectives to allow for creative combinations – a key issue in such a large and dispersed organization.

- Allocating 'slack' and permission to play – allowing employees to spend a proportion of their time in curiosity-driven activities which may lead nowhere but which have sometimes given them breakthrough products.
- Patience – acceptance of the need for 'stumbling in motion' as innovative ideas evolve and take shape. Breakthroughs like Post-its and 'Scotchgard' were not overnight successes but took two to three years to 'cook' before they emerged as viable prospects to put into the formal system.
- Acceptance of mistakes and encouragement of risk-taking – a famous quote from a former CEO – is often cited in this connection: 'Mistakes will be made, but if a person is essentially right, the mistakes he or she makes are not as serious, in the long run, as the mistakes management will make if it's dictatorial and undertakes to tell those under its authority exactly how they must do their job . . . Management that is destructively critical when mistakes are made kills initiative, and it is essential that we have many people with initiative if we are to continue to grow.'
- Encouraging 'bootlegging' – giving employees a sense of empowerment and turning a blind eye to creative ways which staff come up with to get around the system – acts as a counter to rigid bureaucratic procedures.
- Policy of hiring innovators – recruitment approach is looking for people with innovator tendencies and characteristics.
- Recognition of the power of association – deliberate attempts not to separate out different functions but to bring them together in teams and other groupings.
- Encouraging broad perspectives – for example, in developing their overhead projector business it was close links with users made by getting technical development staff to make sales calls that made the product so user-friendly and therefore successful.
- Strong culture dating back to 1951 of encouraging informal meetings and workshops in a series of groups, committees, etc., under the structural heading of the Technology Forum – established 'to encourage free and active interchange of information and cross-fertilization of ideas'. This is a voluntary activity although the company commit support resources – but it enables a company-wide 'college' with fluid interchange of perspectives and ideas.
- Recruiting volunteers – particularly in trying to open up new fields; involvement of customers and other outsiders as part of a development team is encouraged since it mixes perspectives.

TABLE 11.6 Components of the innovative organization under discontinuous conditions

Component	Key features	Example references
Shared vision, leadership and the will to innovate	Top level support for difficult decisions or radical new directions Different perspectives – often coming from outside the organization or the sector. Willingness to let go of the past	7, 160–162
Appropriate structure	Type 1 and type 2 models and finding balance – the ambidextrous challenge Corporate venturing models, skunk works and other modes	163, 164
Key individuals	Key roles of gatekeepers to extend peripheral vision and champions to promote risk-taking. New roles to facilitate internal venturing – for example, Shell's 'Gamechangers' Emphasis on intrapreneurship	23, 165
Effective team working	Increasing emphasis on bringing together different perspectives and fast forming temporary teams. Increasing boundary-crossing within and between organizations – virtual and dispersed team working	75, 76, 112, 166
Continuing and stretching individual development	Training to think and work 'out of the box' Development of alternative perspectives via formal training, secondment, etc.	46, 167
Extensive communication	Need to develop channels for unorthodox ideas to flow Need capacity to deal with 'off-message' signals	168, 169
High involvement in innovation	Internal programmes which seek out and capture new ideas from across the organization and harness entrepreneurial energy to take them forward	37, 39, 164
External focus	Extensive networking required to extend peripheral vision Move beyond existing and effective value networks to open up new options Open innovation	119, 170, 171
Creative climate	Fostering open environment receptive to new and often challenging ideas Development of intrapreneurship rather than forcing people to leave to exploit new opportunities which they see and believe in	44, 47
Learning organization	Increasing emphasis on 'probe and learn' and high failure/fast learning Extending learning across boundaries and into networks	51, 158, 172–174

11.13 Summary and Further Reading

The field of organizational behaviour is widely discussed and there are some good basic texts, including references 15, 175, and 176. Specific issues surrounding the development of innovative organizations are well treated by references 17, 177–179 whilst the core text by Burns and Stalker is still valuable reading.[14] Peters is typical of the 'chaos' school of thought which opposes bureaucratic forms[103] but other studies suggest that this needs tempering with some elements of structure. Discussion of extended and virtual organizational forms can be found in references 180 and 181 and the challenge of becoming 'ambidextrous' is raised in reference 182.

Many books and articles look at specific aspects – for example, the development of creative climates,[44–46,167,183–185] team working[27,75,76,106,108,186] or continuous improvement.[37–39] Increasingly, the theme of organizational learning is being seen as central, and a number of good descriptive and prescriptive texts are available including references 87, 187–190. Case studies of innovative organizations focus on many of the issues highlighted in this chapter; good examples include references 6, 145, 160, 191–193. In particular, a number of books discuss the relationship between management and shop floor and the potential for more participative forms.[179,193–197]

The 'beyond boundaries' issue of networking is covered by several writers including references 42, 43, 127, 133, 142, 198–202.

Finally the theme of 'beyond the steady state' and its implications for organizing for innovation is picked up in several pieces including references 7, 160, 182, 203–205. We address the relationships between leadership, innovation and organizational renewal more fully in our book *Meeting the Innovation Challenge: Leadership for Transformation and Growth*, by Scott Isaksen and Joe Tidd (John Wiley & Sons, Ltd, Chichester, 2006).

References

1 Kirkpatrick, D. (1998) 'The second coming of Apple', *Fortune*, **138**, 90.
2 Pfeffer, J. (1998) *The Human Equation: Building profits by putting people first*. Harvard Business School Press, Boston, Mass.
3 Caulkin, S. (2001) *Performance through People*. Chartered Institute of Personnel and Development, London.
4 Huselid, M. (1995) 'The impact of human resource management practices on turnover, productivity and corporate financial performance', *Academy of Management Journal*, **38**, 647–656.
5 Kay, J. (1993) *Foundations of Corporate Success: How business strategies add value*. Oxford University Press, Oxford, 416.
6 Kanter, R. (ed) (1997) *Innovation: Breakthrough thinking at 3M, DuPont, GE, Pfizer and Rubbermaid*. Harper Business, New York.

7 Hamel, G. (2000) *Leading the Revolution*. Harvard Business School Press, Boston, Mass.

8 Champy, J. and N. Nohria (eds) (1996) *Fast Forward*. Harvard Business School Press, Cambridge, Mass.

9 Hesselbein, F., M. Goldsmith and I. Somerville (eds) (2002) *Leading for Innovation*. Jossey Bass, San Francisco.

10 Pfeffer, J. (1994) *Competitive Advantage through People*. Harvard Business School Press, Boston, Mass.

11 Woodward, J. (1965) *Industrial Organisation: Theory and practice*. Oxford University Press, Oxford.

12 Thompson, J. (1967) *Organisations in Action*. McGraw-Hill, New York.

13 Peters, T. (1997) *The Circle of Innovation*. Coronet, London.

14 Burns, T. and G. Stalker (1961) *The Management of Innovation*. Tavistock, London.

15 Mintzberg, H. (1979) *The Structuring of Organisations*. Prentice-Hall, Englewood Cliffs, NJ.

16 Perrow, C. (1967) 'A framework for the comparative analysis of organisations', *American Sociological Review*, **32**, 194–208.

17 Hesselbein, F., M. Goldsmith and R. Beckhard (eds) (1997) *Organization of the Future*. Jossey-Bass/The Drucker Foundation, San Francisco.

18 Allen, T. (1977) *Managing the Flow of Technology*. MIT Press, Cambridge, Mass.

19 Rothwell, R. (1992) 'Successful industrial innovation: critical success factors for the 1990s'. *R&D Management*, **22** (3), 221–239.

20 Rubenstein, A. (1994) 'Ideation and entrepreneurship', in Souder, W. and Sherman, J. (eds), *Managing New Technology Development*. McGraw-Hill, New York.

21 Bess, J. (1995) *Creative R&D Leadership: Insights from Japan*. Quorum Books, Westport, CT.

22 Dyson, J. (1997) *Against the Odds*. Orion, London.

23 Pinchot, G. (1999) *Intrapreneuring in Action – Why you don't have to leave a corporation to become an entrepreneur*. Berrett-Koehler Publishers, New York.

24 Kharbanda, O. and M. Stallworthy (1990) *Project Teams*. NCC-Blackwell, Manchester.

25 Belbin, M. (2004) *Management Teams – Why they succeed or fail*. Butterworth-Heinemann, London.

26 Bixby, K. (1987) *Superteams*. Fontana, London.

27 Katzenbach, J. and D. Smith (1992) *The Wisdom of Teams*. Harvard Business School Press, Boston, Mass.

28 Francis, D. and D. Young (1988) *Top Team Building*. Gower, Aldershot.

29 Thamhain, H. and D. Wilemon (1987) 'Building high performing engineering project teams', *IEEE Transactions on Engineering Management*, **EM-34** (3), 130–137.

30 Pedler, M., T. Boydell and J. Burgoyne (1991) *The Learning Company: A strategy for sustainable development*. McGraw-Hill, Maidenhead.

31 Jarvis, V. and S. Prais (1995) *The Quality of Manufactured Products in Britain and Germany*. National Institute of Economic and Social Research.

32 Leonard-Barton, D. (1992) 'The organisation as learning laboratory', *Sloan Management Review*, **34** (1), 23–38.

33 DTI (1997) *Competitiveness through Partnerships with People*. Department of Trade and Industry, London.

34 Spence, W. (1994) *Innovation: The communication of change in ideas, practices and products*. Chapman & Hall, London.

35 De Meyer, A. (1985) 'The flow of technological innovation in an R&D department', *Research Policy*, **4**, 315–328.

36 Imai, K. (1987) *Kaizen*. Random House, New York.

37 Boer, H. *et al.* (1999) *CI Changes: From suggestion box to the learning organisation.* Ashgate, Aldershot.

38 Schroeder, M. and A. Robinson (1993) 'Training, continuous improvement and human relations: the US TWI programs and Japanese management style', *California Management Review,* **35** (2).

39 Bessant, J. (2003) *High Involvement Innovation.* John Wiley & Sons, Ltd, Chichester.

40 Carter, C. and B. Williams (1957) *Industry and Technical Progress.* Oxford University Press, Oxford.

41 Brown, S. *et al.* (2005) *Strategic Operations Management.* 2nd edn. Butterworth Heinemann, Oxford.

42 Wenger, E. (1999) *Communities of Practice: Learning, meaning, and identity.* Cambridge University Press, Cambridge.

43 Best, M. (2001) *The New Competitive Advantage.* Oxford University Press, Oxford.

44 Leonard, D. and W. Swap (1999) *When Sparks Fly: Igniting creativity in groups.* Harvard Business School Press, Boston, Mass.

45 Cook, P. (1999) *Best Practice Creativity.* Gower, Aldershot.

46 Rickards, T. (1997) *Creativity and Problem Solving at Work.* Gower, Aldershot.

47 Amabile, T. (1998) 'How to kill creativity', *Harvard Business Review,* September/October, 77–87.

48 Garvin, D. (1993) 'Building a learning organisation', *Harvard Business Review,* July/August, 78–91.

49 Nonaka, I., S. Keigo and M. Ahmed (2003) 'Continuous innovation: the power of tacit knowledge', in Shavinina, L. (ed.), *International Handbook of Innovation.* Elsevier, New York.

50 Leonard-Barton, D. (1995) *Wellsprings of Knowledge: Building and sustaining the sources of innovation.* Harvard Business School Press, Boston, Mass. 335.

51 Senge, P. (1999) *The Dance of Change: Mastering the twelve challenges to change in a learning organisation.* Doubleday, New York.

52 Argyris, C. and D. Schon (1970) *Organizational Learning.* Addison Wesley, Reading, Mass.

53 Bryson, B. (1994) *Made in America.* Minerva, London.

54 Womack, J., D. Jones and D. Roos (1991) *The Machine that Changed the World.* Rawson Associates, New York.

55 Christenson, C. (1997) *The Innovator's Dilemma.* Harvard Business School Press, Cambridge, Mass.

56 Tripsas, M. and G. Gavetti (2000) 'Capabilities, cognition and inertia: evidence from digital imaging', *Strategic Management Journal,* **21,** 1147–1161.

57 'Patience of Jobs pays off', (1995) *Financial Times,* London, 7.

58 Moody, F. (1995) *I Sing the Body Electronic.* Hodder & Stoughton, London.

59 Cooper, R. (2001) *Winning at New Products.* 3rd edn. Kogan Page, London.

60 Francis, D., J. Bessant and M. Hobday (2003) 'Managing radical organisational transformation', *Management Decision,* **41** (1), 18–31.

61 Kanter, R. (1984) *The Change Masters.* Unwin, London.

62 Semler, R. (1993) *Maverick.* Century Books, London.

63 Kaplinsky, R., F. den Hertog and B. Coriat (1995) *Europe's Next Step.* Frank Cass, London.

64 Miles, R. and C. Snow (1978) *Organisational Strategy, Structure and Process.* McGraw-Hill, New York.

65 Lawrence, P. and P. Dyer (1983) *Renewing American Industry.* Free Press, New York.

66 Lawrence, P. and J. Lorsch (1967) *Organisation and Environment.* Harvard University Press, Cambridge, Mass.

67 Stalk, G. and T. Hout (1990) *Competing against Time: How time-based competition is reshaping global markets*. Free Press, New York.

68 Child, J. (1980) *Organisations*. Harper & Row, London.

69 Greiner, L. (1972) 'Evolution and revolution as organisations grow', *Harvard Business Review*, July/August, 37–46.

70 Adler, P. (1992) 'The learning bureaucracy: NUMMI', in Staw, B. and Cummings, L. (eds), *Research in Organisational Behaviour*. JAI Press, Greenwich, CT.

71 Nayak, P. and J. Ketteringham (1986) *Breakthroughs: How leadership and drive create commercial innovations that sweep the world*. Mercury, London.

72 Kidder, T. (1981) *The Soul of a New Machine*. Penguin, Harmondsworth.

73 Clark, K. and T. Fujimoto (1992) *Product Development Performance*. Harvard Business School Press, Boston, Mass.

74 Blackler, F. (1995) 'Knowledge, knowledge work and organisations', *Organization Studies*, **16** (6), 1021–1046.

75 Sapsed, J. *et al.* (2002) 'Teamworking and knowledge management: a review of converging themes', *International Journal of Management Reviews*, **4** (1).

76 Duarte, D. and N. Tennant Snyder (1999) *Mastering Virtual Teams*. Jossey-Bass, San Francisco.

77 Prais, S. (1995) *Productivity, Education and Training*. Cambridge University Press, Cambridge.

78 Readman, J. and J. Bessant (2004) 'What have UK firms learnt about implementing continuous improvement? Results of the UK CI survey 2003' in *5th CINET Conference*, Sydney, Australia.

79 CIPD (2001) *Raising UK productivity: why people management matters*. Chartered Institute of Personnel and Development, London.

80 Burnes, B. (1992) *Managing Change*. Pitman, London.

81 Smith, S. and D. Tranfield (1990) *Managing Change*. IFS Publications, Kempston.

82 Walton, R. (1986) *Human Resource Practices for Implementing AMT: Report of the Committee on Effective Implementation of AMT*. National Research Council, Washington, DC.

83 Bessant, J. and J. Buckingham (1993) 'Organisational learning for effective use of CAPM', *British Journal of Management*, **4** (4), 219–234.

84 Swan, J., S. Newell and P. Clark (1993) 'The importance of user design in the appropriation of new information technologies: the example of PICS', *International Journal of Operations and Production Management*, **13** (2), 4–22.

85 Kaplinsky, R. (1994) *Easternization: The spread of Japanese management techniques to developing countries*. Frank Cass, London.

86 Schroeder, D. and A. Robinson (1991) 'America's most successful export to Japan – continuous improvement programmes', *Sloan Management Review*, **32** (3), 67–81.

87 Figuereido, P. (2001) *Technological Learning and Competitive Performance*. Edward Elgar, Cheltenham.

88 Deming, W. (1986) *Out of the Crisis*. MIT Press, Cambridge, Mass.

89 Crosby, P. (1977) *Quality is Free*. McGraw-Hill, New York.

90 Dertouzos M, R. Lester and L. Thurow (1989) *Made in America: Regaining the productive edge*. MIT Press, Cambridge, Mass.

91 Garvin, D. (1988) *Managing Quality*. Free Press, New York.

92 Womack, J. and D. Jones (1997) *Lean Thinking*. Simon & Schuster, New York.

93 Schonberger, R. (1982) *Japanese Manufacturing Techniques: Nine hidden lessons in simplicity*. Free Press, New York.

94 Shingo, S. (1983) *A Revolution in Manufacturing: The SMED system*. Productivity Press, Cambridge, Mass.

95 Suzaki, K. (1988) *The New Manufacturing Challenge*. Free Press, New York.

96 Lillrank, P. and N. Kano (1990) *Continuous Improvement; Quality control circles in Japanese industry*. University of Michigan Press, Ann Arbor.

97 Caffyn, S. (1998) 'Continuous improvement in the new product development process', in *Centre for Research in Innovation Management*, University of Brighton, Brighton.

98 Lamming, R. (1993) *Beyond Partnership*. Prentice-Hall, London.

99 Owen, M. and J. Morgan (2000) *Statistical Process Control in the Office*. Greenfield Publishing, Kenilworth.

100 Ishikure, K. (1988) 'Achieving Japanese productivity and quality levels at a US plant', *Long Range Planning*, **21** (5), 10–17.

101 Wickens, P. (1987) *The Road to Nissan: Flexibility, quality, teamwork*. Macmillan, London.

102 Lewis, K. and S. Lytton (2000) *How to Transform Your Company*. Management Books 2000, London.

103 Peters, T. (1988) *Thriving on Chaos*. Free Press, New York.

104 Forrester, R. and A. Drexler (1999) 'A model for team-based organization performance', *Academy of Management Executive*, **13** (3), 36–49.

105 Conway, S. and R. Forrester (1999) *Innovation and Teamworking: Combining perspectives through a focus on team boundaries*. University of Aston Business School, Birmingham.

106 Holti, R., J. Neumann and H. Standing (1995) *Change Everything at Once: The Tavistock Institute's guide to developing teamwork in manufacturing*. Management Books 2000, London.

107 Tranfield, D. *et al.* (1998) 'Teamworked organisational engineering: getting the most out of teamworking', *Management Decision*, **36** (6), 378–384.

108 Jassawalla, A. and H. Sashittal (1999) 'Building collaborative cross-functional new product teams', *Academy of Management Executive*, **13** (3), 50–53.

109 DTI (1996) *UK Software Purchasing Survey*. Department of Trade and Industry, London.

110 Van Beusekom, M. (1996) *Participation Pays! Cases of successful companies with employee participation*. Netherlands Participation Institute, The Hague.

111 Tuckman, B. and N. Jensen (1977) 'Stages of small group development revisited', *Group and Organizational Studies*, **2**, 419–427.

112 Smith, P. and E. Blanck (2002) 'From experience: leading dispersed teams', *Journal of Product Innovation Management*, **19**, 294–304.

113 Kirton, M. (1989) *Adaptors and Innovators*. Routledge, London.

114 Schein, E. (1984) 'Coming to a new awareness of organisational culture', *Sloan Management Review*, Winter, 3–16.

115 Ekvall, G. (1990) 'The organizational culture of idea management', in Henry, J. and Walker, D. (eds), *Managing Innovation*. Sage, London, 73–80.

116 Badawy, M. (1997) *Developing Managerial Skills in Engineers and Scientists*. John Wiley & Sons, Inc., New York.

117 Oakland, J. (1989) *Total Quality Management*. Pitman, London.

118 Schonberger, R. (1990) *Building a Chain of Customers*. Free Press, New York.

119 Chesborough, H. (2003) *Open Innovation: The new imperative for creating and profiting from technology*. Harvard Business School Press, Boston, Mass.

120 Swan, J. (2003) 'Knowledge, networking and innovation: developing an understanding of process', in Shavinina, L. (ed)., *International Handbook of Innovation*. Elsevier, New York.

121 Senge, P. (1990) *The Fifth Discipline*. Doubleday, New York.

122 Bowen, K. *et al.* (1994) *The Perpetual Enterprise Machine: Seven keys to corporate renewal through successful product and process development*. Oxford University Press, New York.

123 Maidique, M. and B. Zirger (1985) 'The new product learning cycle', *Research Policy*, **14** (6), 299–309.

124 Bozdogan, K. (1998) 'Architectural innovation in product development through early supplier integration', *R&D Management*, **28** (3), 163–173.

125 Oliver, N. and M. Blakeborough (1998) 'Innovation networks: the view from the inside', in Grieve Smith, J. and Michie, J. (eds), *Innovation, Co-operation and Growth*. Oxford University Press, Oxford.

126 Miller, R. (1995) 'Innovation in complex systems industries: the case of flight simulation', *Industrial and Corporate Change*, **4** (2).

127 Tidd, J. (1997) 'Complexity, networks and learning: integrative themes for research on innovation management', *International Journal of Innovation Management*, **1** (1), 1–22.

128 Simonin, B. (1999) 'Ambiguity and the process of knowledge transfer in strategic alliances', *Strategic Management Journal*, **20**, 595–623.

129 Szulanski, G. (1996) 'Exploring internal stickiness: impediments to the transfer of best practice within the firm', *Strategic Management Journal*, **17**, Winter, 5–9.

130 Hamel, G., Y. Doz and C. Prahalad (1989) 'Collaborate with your competitors – and win', *Harvard Business Review*, **67** (2), 133–139.

131 Schmitz, H. (1998) 'Collective efficiency and increasing returns', *Cambridge Journal of Economics*, **23** (4), 465–483.

132 Nadvi, K. and H. Schmitz (1994) *Industrial Clusters in Less Developed Countries: Review of experiences and research agenda*. Institute of Development Studies, Brighton.

133 Keeble, D. and F. Williamson (eds) (2000) *High Technology Clusters, Networking and Collective Learning in Europe*. Aldershot, Ashgate.

134 DTI/CBI (2000) *Industry in Partnership*. Department of Trade and Industry/Confederation of British Industry, London.

135 Bessant, J. (1995) 'Networking as a mechanism for technology transfer: the case of continuous improvement', in Kaplinsky, R., den Hertog, F. and Coriat, B. (eds), *Europe's Next Step*, Frank Cass, London.

136 Semlinger, K. (1995) 'Public support for firm networking in Baden-Wurttemburg', in Kaplinsky, R. den Hertog, F. and Coriat, B. (eds), *Europe's Next Step*. Frank Cass, London.

137 Keeble, D. *et al*. (1999) 'Institutional thickness in the Cambridge region', *Regional Studies*, **33** (4), 319–332.

138 Piore, M. and C. Sabel (1982) *The Second Industrial Divide*. Basic Books, New York.

139 Nadvi, K. (1997) *The Cutting Edge: Collective efficiency and international competitiveness in Pakistan*. Institute of Development Studies, University of Sussex.

140 Dodgson, M. (1993) *Technological Collaboration in Industry*. Routledge, London.

141 Hobday, M. (1996) *Complex Systems vs Mass Production Industries: A new innovation research agenda*. Complex Product Systems Research Centre, Brighton.

142 Marceau, J. (1994) 'Clusters, chains and complexes: three approaches to innovation with a public policy perspective', in Rothwell, R. and Dodgson, M. (eds), *The Handbook of Industrial Innovation*. Edward Elgar, Aldershot.

143 Hines, P. (1999) *Value Stream Management: The development of lean supply chains*. Financial Times Management, London.

144 Dyer, J. and K. Nobeoka (2000) 'Creating and managing a high-performance knowledge-sharing network: the Toyota case', *Strategic Management Journal*, **21** (3), 345–367.

145 Dell, M. (1999) *Direct from Dell*. HarperCollins, New York.

146 Hines, P. (1994) *Creating World Class Suppliers: Unlocking mutual competitive advantage*. Pitman, London.

147 Cusumano, M. (1985) *The Japanese Automobile Industry: Technology and management at Nissan and Toyota*. Harvard University Press, Cambridge, Mass.

148 AFFA (1998) *Chains of Success*. Department of Agriculture, Fisheries and Forestry – Australia (AFFA), Canberra.

149 Marsh, I. and B. Shaw (2000) 'Australia's wine industry: collaboration and learning as causes of competitive success', in *Working Paper*. Australian Graduate School of Management, Melbourne.

150 AFFA (2000) *Supply Chain Learning: Chain reversal and shared learning for global competitiveness*. Department of Agriculture, Fisheries and Forestry – Australia (AFFA), Canberra.

151 Fearne, A. and D. Hughes (1999) 'Success factors in the fresh produce supply chain: insights from the UK', *Supply Management*, **4** (3).

152 Dent, R. (2001) *Collective Knowledge Development, Organisational Learning and Learning Networks: An integrated framework*. Economic and Social Research Council, Swindon.

153 Humphrey, J., R. Kaplinsky and P. Saraph (1998) *Corporate Restructuring: Crompton Greaves and the challenge of globalisation*. Sage Publications, New Delhi.

154 Bessant, J. and G. Tsekouras (2001) 'Developing learning networks', *AI and Society*, **15** (2), 82–98.

155 Barnes, J. and M. Morris (1999) *Improving Operational Competitiveness through Firm-level Clustering: A case study of the KwaZulu-Natal Benchmarking Club*. School of Development Studies, University of Natal, Durban, South Africa.

156 Kaplinsky, R., M. Morris and J. Readman (2003) 'The globalisation of product markets and immiserising growth: lessons from the South African furniture industry', *World Development*, **30** (7), 1159–1178.

157 Gereffi, G. (1994) 'The organisation of buyer-driven global commodity chains: how U.S. retailers shape overseas production networks', in Gereffi, G. and Korzeniewicz, P. (eds), *Commodity Chains and Global Capitalism*. Praeger, London.

158 Bessant, J., R. Kaplinsky and R. Lamming (2003) 'Putting supply chain learning into practice', *International Journal of Operations and Production Management*, **23** (2), 167–184.

159 Bessant, J., M. Morris and R. Kaplinsky (2003) 'Developing capability through learning networks', *International Journal of Technology Management and Sustainable Development*, **2** (1).

160 Foster, R. and S. Kaplan (2002) *Creative Destruction*. Harvard University Press, Cambridge.

161 Welch, J. (2001) *Jack! What I've learned from leading a great company and great people*. Headline, New York.

162 Groves, A. (1999) *Only the Paranoid Survive*. Bantam Books, New York.

163 Buckland, W., A. Hatcher and J. Birkinshaw (2003) *Inventuring: Why big companies must think small*. McGraw Hill Business, London.

164 Bessant, J., J. Birkinshaw and R. Delbridge (2004) 'One step beyond – building a climate in which discontinuous innovation will flourish', *People Management*, 28–32.

165 Day, G. and P. Schoemaker (2004) 'Driving through the fog: managing at the edge', *Long Range Planning*, **37**, 127–142.

166 Rich, B. and L. Janos (1994) *Skunk Works*. Warner Books, London.

167 Kingdon, M. (2002) *Sticky Wisdom – How to start a creative revolution at work*. Capstone, London.

168 Winter, S. (2004) 'Specialised perception, selection and strategic surprise: learning from the moths and the bees', *Long Range Planning*, **37**, 163–169.

169 Prahalad, C. (2004) 'The blinders of dominant logic', *Long Range Planning*, **37** (2), 171–179.

170 Huston, L. (2004) 'Mining the periphery for new products', *Long Range Planning*, **37**, 191–196.

171 Seely Brown, J. (2004) 'Minding and mining the periphery', *Long Range Planning*, **37**, 143–151.

172 Allen, P. (2001) 'A complex systems approach to learning, adaptive networks', *International Journal of Innovation Management*, **5**, 149–180.

173 Ayas, K. (1997) *Design for Learning for Innovation*. Eburon, Delft.

174 Mariotti, F. and R. Delbridge (2005) 'A portfolio of ties: managing knowledge transfer and learning within network firms', *Academy of Management Review* (forthcoming).

175 Morgan, G. (1986) *Images of Organisation*. Sage, London.

176 Buchanan, D. and A. Huczynski (1997) *Organizational Behaviour: An introductory text*. Prentice-Hall, London.

177 Leifer, R. *et al.* (2000) *Radical Innovation*. Harvard Business School Press, Boston, Mass.

178 Kanter, R. (1996) *World Class*. Simon & Schuster, New York.

179 Clark, P. and N. Staunton (1989) *Innovation and Technology in Organisations*. Routledge, London.

180 Chesborough, H. and D. Teece (1996) 'When is virtual virtuous?', *Harvard Business Review*, January.

181 Hamel, G. (1999) 'Bringing Silicon Valley inside', *Harvard Business Review*, September.

182 Tushman, M. and C. O'Reilly (1996) *Winning through Innovation*. Harvard Business School Press, Boston, Mass.

183 Claxton, G. (2001) 'The innovative mind', in Henry, J. (ed.), *Creative Management*. Sage, London.

184 De Bono, E. (1993) *Serious Creativity*. HarperCollins, London.

185 Sternberg, R. (ed) (1999) *Handbook of Creativity*. Cambridge University Press, Cambridge.

186 DeMarco, T. and T. Lister (1999) *Peopleware: Productive projects and teams*. Dorset House, New York.

187 De Geus, A. (1996) *The Living Company*. Harvard Business School Press, Boston, Mass.

188 Pisano, G. (1994) 'Knowledge, integration and the locus of learning: an empirical analysis of process development', *Strategic Management Journal*, **15**, 85.

189 Prokesch, S. (1997) 'Unleashing the power of learning', *Harvard Business Review*, September/October, 147–168.

190 Saad, M. (2000) *Development through Technology Transfer*. Intellect Publishers, Bristol.

191 Gundling, E. (2000) *The 3M Way to Innovation: Balancing people and profit*. Kodansha International, New York.

192 Baden-Fuller, C. and M. Pitt (1996) *Strategic Innovation*. Routledge, London.

193 Weisberg, R. (2003) 'Case studies of innovation: ordinary thinking, extraordinary outcomes', in Shavinina, L. (ed.), *International Handbook of Innovation*. Elsevier, New York.

194 Clark, J. (ed) (1993) *Human Resource Management and Technical Change*. Sage, London.

195 Preece, D. (1995) *Organisations and Technical Change*. Routledge Series in the Management of Technology, Bessant, J. and D. Preece (eds), Routledge/International Thompson, London.

196 McCloughlin, I. (1998) 'Creative technological change', in Preece, D. and Bessant, J. (eds), *The Management of Technology and Innovation*. Routledge, London.

197 Dawson, P. (1996) 'Technology and quality: change in the workplace', in Bessant, J. and Preece, D. (eds), *Management of Technology and Innovation*. International Thomson Business Press, London.

198 Swann, P., M. Prevezer and D. Stout (eds) (1998) *The Dynamics of Industrial Clustering*. Oxford University Press, Oxford.

199 Morosini, P. (2004) 'Industrial clusters, knowledge integration and performance', *World Development*, **32** (2).

200 Conway, S. and F. Steward (1998) 'Mapping innovation networks', *International Journal of Innovation Management*, **2** (2), 165–196.

201 Grandori, A. and G. Soda (1995) 'Inter-firm networks: antecedents, mechanisms and forms', *Organization Studies*, **16** (2), 183–214.

202 Loudon, A. (2001) *Webs of Innovation*. FT.Com, London.

203 Day, G. and P. Schoemaker (2000) *Wharton on Managing Emerging Technologies*. John Wiley & Sons Inc., New York.

204 Jones, T. (2002) *Innovating at the Edge*. Butterworth Heinemann, London.

205 Von Stamm, B. (2003) *The Innovation Wave*. John Wiley & Sons, Ltd, Chichester.

Creating Innovative New Firms

Much of what we have discussed so far has been directed to the issue of managing innovation in large, complex organizations. Most of the management structures and processes we have identified are equally applicable to smaller organizations, but other factors are unique to smaller firms, such as funding and growth. Contrary to popular belief, the majority of small firms are not particularly innovative, so here we examine the nature of innovative small firms and the issues particular to their creation, management and growth.

12.1 Sources of New Technology-based Firms

In Chapter 5 we identified four categories of small firms, including specialist suppliers, supplier dominated, superstars and NTBFs. There is a wealth of research on small and medium-sized enterprises (SMEs), and in particular their role in innovation and economic growth. However, we are concerned here with a subset of SMEs which are based on new technologies. Such firms differ from other SMEs because typically they are established by highly qualified personnel, require large amounts of capital and are characterized by greater technical and market risk.

Much of what we know about new technology-based firms (NTBFs) is based on the experience of firms in the USA, in particular, the growth of biotechnology, semiconductor and software firms. Many of these originated from a parent or 'incubator' organization, typically either an academic institution or large well-established firm. Examples of university incubators include Stanford which spawned much of Silicon Valley, the Massachusetts Institute of Technology (MIT) which spawned Route 128 in Boston, and Imperial and Cambridge in the UK. MIT in particular has become the archetype academic incubator, and in addition to the creation of Route 128, its alumni have established some 200 NTBFs in northern California, and account for more than a fifth of employment in Silicon Valley.[1] The so-called MIT model has been adopted worldwide, so far with limited success. For example, in 1999 Cambridge University in the UK formed a UK Government-sponsored joint venture with MIT to help develop spin-offs in the UK. However, to put such initiatives into perspective, Hermann Hauser, a venture capitalist, notes 'Stanford alumni have produced companies worth a trillion dollars. MIT half a trillion dollars. If Cambridge is getting to $20 bn we will be lucky.' One reason is the differences in scale. Mike Lynch, founder of the software company Autonomy, observes, 'Silicon Valley is 60 miles long and in the last few months there will have been 70 to 80 money raisings in the $50 million to $200 million range. In Cambridge we might think of one, perhaps.'

Examples of large incubator firms include the Xerox PARC and Bell Laboratories in the USA which spawned Fairchild Semiconductor which in turn led to numerous spin-offs including Intel, Advanced Memory Systems, Teledyne and Advanced Micro-Devices. Similarly, Engineering Research Associates (ERA) led to more than 40 new firms, including Cray, Control Data Systems, Sperry and Univac (see Box 12.1). In many cases, incubator firms provide the technical entrepreneurs, and the associated academic institutions provide the additional qualified manpower.

NTBF spin-offs tend to cluster around their respective incubator organizations, forming regional networks of expertise. The firms tend to remain close to their parents for a number of technical and personal reasons. Most NTBFs retain contacts with their parent organizations to gain financial and technical support, and are often reluctant to disrupt their social and family lives whilst establishing a new venture. Perhaps surprisingly, the mortality rate of NTBFs is lower than that of most types of new firm, around 20 to 30% in 10 years compared to more than 80% for other types of new business.[2] One explanation for the higher survival rate of NTBFs is that the barriers to entry are higher than for many other businesses, in terms of expertise and capital. Therefore those NTBFs which are able to overcome such barriers are more likely to survive. The concentration of start-ups in a region can create positive feedback, through demonstration effects and by increasing the demand for, and experience of, supporting institutions, such as venture capitalists, legal services and contract research and production, thereby improving the environment and probability of success of subsequent start-ups. Failures are an inherent part of such a system, and provided a steady stream of new venture proposals exists and venture capitalists maintain diverse investment portfolios and are ruthless with failed ventures, the system continues to learn from both good and bad investments.

However, the unique circumstances of the US environment in the 1970s and 1980s question the generalizability of the lessons of Silicon Valley and Route 128. Specifically, the role of the defence industry investment, liberal tax regimes and sources of venture capital were unique. In addition, it is important to distinguish the evolutionary growth of such regional clusters of NTBFs, from more recent attempts to establish science parks based around universities. For example, success of science parks in Europe and Asia in the 1990s, and other attempts to emulate the early US experiences, has been limited.[3] A survey comparing high-technology firms located on and off university science parks

---

BOX 12.1

SPIN-OFF COMPANIES FROM XEROX'S PARC LABS

Xerox established its Palo Alto Research Center (PARC) in California in 1970. PARC was responsible for a large number of technological innovation in the semiconductor lasers, laser printing, Ethernet networking technology and web indexing and searching technologies, but it is generally acknowledged that many of its most significant innovations were the result of individuals who left the company and firms which spun-off from PARC, rather than developed via Xerox itself. For example, many of the user-interface developments at Apple originated at Xerox, as did the basis of Microsoft's Word package. By 1998 Xerox PARC had spun-out 24 firms, including 10 which went public such as 3Com, Adobe, Documentum, and SynOptics. By 2001 the value of the spin-off companies was more than twice that of Xerox itself.

A debate continues to the reasons for this, most attributing the failure to retain the technologies in-house to corporate ignorance and internal politics. However, most of the technologies did not simply 'leak out', but instead were granted permission by Xerox, which often provided non-exclusive licenses and an equity stake in the spin-off firms. This suggests that Xerox's research and business managers saw little potential for exploiting these technologies in its own businesses. One of the reasons for the failure to commercialize these technologies in-house was that Xerox had been highly successful with its integrated product-focused strategy, which made it more difficult to recognize and exploit potential new *businesses*.

Source: Chesbrough H. (2003) *Open Innovation: The new imperative for creating and profiting from technology.* Harvard Business School Press, Boston, Mass.

---

concluded that there were no statistically significant differences between their technological inputs, such as expenditure on R&D, and outputs, such as new products and patents.[4]

12.2 University Incubators

The creation and sharing of intellectual property is a core role of a university, but managing it for commercial gain is a different challenge. Most universities with significant commercial research contracts understand how to license, and the roles of all parties – the academics, the university and the commercial organization – are relatively clear. In particular, the academic will normally continue with the research whilst possibly having a consultancy arrangement with the commercial company. However, forming an independent company is a different matter. Here both the university and the scientist must agree that spin-out is the most viable option for technology commercialization and must negotiate a spin-out deal. This may include questions of, for example, equity split, royalties, academic and university investment in the new venture, academic secondment, identification and transfer of intellectual property and use of university resources in the start-up phase. In short, it is complicated. As Chris Evans, founder of Chiroscience and Merlin Ventures notes: 'Academics and universities . . . have no management, no muscle, no vision, no business plan and that is 90% of the task of exploiting science and taking it to the market place. There is a tendency for universities to think, "we invented the thing so we are already 50% there". The fact is they are 50% to nowhere' (*Times Higher*, 27 March 1998). A characteristically provocative statement, but it does highlight the gulf between research and successful commercialization. Many universities have accepted and followed the fashion for the commercial exploitation of technology, but typically put too much emphasis on the importance of the technology and ownership of the intellectual property, and 'fail to recognize the importance and sophistication of the business knowledge and expertise of management and other parties who contribute to the non-technical aspects of technology shaping and development . . . the linear model gives no insight into the interplay of technology push and market pull'.[5]

Since the mid-1980s the role of universities in the commercialization of technology has increased significantly. For example, the number of patents granted to US universities doubled between 1984 and 1989, and doubled again between 1989 and 1997. In 1979 the number of patents granted to US universities was only 264, compared to 2436 in 1997. There are a number of explanations for this significant increase in patent activity. Changes in government funding and intellectual property law played a role, but detailed analysis indicates that the most significant reason was technological opportunity. For example, changes in funding and law in the 1980s clearly encouraged many more universities to establish licensing and technology transfer departments, but the impact of these has been relatively small. For example, there is strong evidence that the scientific and commercial quality of patents has fallen since the mid-1980s as a result of these policy changes, and that the distribution of activity has a very long tail. Measured in terms of the number of patents held or exploited, or by income from patent and software licenses, commercialization of technology is highly concentrated in a small number of elite universities which were highly active prior to changes to funding policy and law: the top 20 US universities account for 70% of the patent activity.[6] Moreover, at each of these elite universities a very small number of key patents account for most of the licensing income, the five most successful patents typically account for 70 to 90% of total income.[7] This suggests that a (rare) combination of research excellence and critical mass is required to succeed in the commercialization of technology. Nonetheless, technological opportunity has reduced some of the barriers to commercialization. Specifically, the growing importance of developments in the biosciences and software present new opportunities for universities to benefit from the commercialization of technology.

University spin-outs are an alternative to exploitation of technology through licensing, and involve the creation of an entirely new venture based upon intellectual property developed within the university. Estimates vary, but between 3 and 12% of all technologies commercialized by universities are via new ventures. As with licensing, the propensity and success of these ventures varies significantly. For example, MIT and Stanford University each create around 25 new start-ups each year, whereas Columbia and Duke Universities rarely generate any start-up companies. Studies in the USA suggest that the financial returns to universities are much higher from spin-out companies than from the more common licensing approach. One study estimated that the average income from a university licence was $63 832, whereas the average return from a university spin-out was more than 10 times this – $692 121. When the extreme cases were excluded from the sample, the return from spin-outs was still $139 722, more than twice that for a licence.[8] Apart from these financial arguments, there are other reasons why forming a spin-out company may be preferable to licensing technology to an established company:

- No existing company is ready or able to take on the project on a licensing basis.
- The invention consists of a portfolio of products or is an 'enabling technology' capable of application in a number of fields.
- The inventors have a strong preference for forming a company and are prepared to invest their time, effort and money in a start-up.

As such they involve the 'academic entrepreneur' more fully in the detail of creating and managing a market entry strategy than is the case for other forms of commercialization. They also require major career decisions for the participants. Consequently, they highlight most clearly the dilemmas faced as the scientist tries to manage the interface between academe and industry. The extent to which an individual is motivated to attempt the launch of a venture depends upon three related factors – antecedent influences, the incubator organization and environmental factors:

- *Antecedent influences*, often called the 'characteristics' of the entrepreneur, include genetic factors, family influences, educational choices; and previous career experiences all contribute to the entrepreneur's decision to start a venture.

- *Individual incubator experiences* immediately prior to start-up include the nature of the physical location, the type of skills and knowledge acquired, contact with possible fellow founders, the type of new venture or small business experience gained.
- *Environmental factors* include economic conditions, availability of venture capital, entrepreneurial role models, availability of support services.

There are relatively few data on the characteristics of the academic entrepreneur, partly due to the low numbers involved, but also because the traditional context within which they have operated, particularly as they apply to IPR and equity sharing, has meant that many have been unwilling to be researched. It is also probable that this is compounded by inadequate university data capture systems. Nevertheless, it is clear that in the USA, scientists and engineers working in universities have long become disposed towards the commercialization of research. A study of American universities in 1990 observed: 'Over the last eight years we have seen increasing legitimising of university–industry research interactions.'[9] A study of 237 scientists working in three large national laboratories in the USA found clear differences between the levels of education in inventors in national laboratories and those in a study of technical entrepreneurs from MIT.[10] The study found significant differences between entrepreneurs and non-entrepreneurs in terms of situational variables such as the level of involvement in business activities outside the laboratory or the receipt of royalties from past inventions. A study of scientists in four research institutes in the UK identified a relationship between attitudes to industry, number of industry links and commercial activity.[11] This begs the question: what is the direction of causation? Do entrepreneurial researchers seek more links outside the organization, or do more links encourage entrepreneurial behaviour?

Entrepreneurs, academic or otherwise, require a supportive environment. Surveys indicate that two-thirds of university scientists and engineers now support the need to commercialize their research, and half the need for start-up assistance.[12] There are two levels of analysis of the university environment, the formal institutional rules, policies and structures, and the 'local norms' within the individual department. There are a number of institutional variables which might influence academic entrepreneurship:

1. Formal policy and support for entrepreneurial activity from management.
2. Perceived seriousness of constraints to entrepreneurship, e.g. IPR issues.
3. Incidence of successful commercialization, which demonstrates feasibility and provides role models.

Formal policies to encourage and support entrepreneurship can have both intended and unintended consequences. For example, a university policy of taking an equity stake in new start-ups in return for paying initial patenting and licensing expenses seems to result in a higher number of start-ups, whereas granting generous royalties to academic entrepreneurs appears to encourage licensing activity, but tends to suppress significantly the number of start-up companies.[13] Similarly, encouraging commercially oriented, or industry-funded research, appears to have no effect on the number of start-ups, whereas a university's intellectual eminence has a very strong positive effect.[13] A reason for the former effect is that typically such research restricts the ownership of formal intellectual property, and narrows the choice of route to market. There are two reasons for the former effect: more prestigious universities typically attract better researchers and higher funding; and other commercial investors use the prestige or reputation of the institution as a signal or indicator of quality. In addition, some very common university policies appear to have little or no positive effect on the number of subsequent success of start-ups, including university incubators and local venture capital funding.[13] Moreover, badly targeted and poorly monitored financial support may encourage 'entrepreneurial academics', rather than academic entrepreneurs – scientists in the public sector who are not really committed to creating start-ups, but rather are seeking alternative support for their own research agendas.[14] This can result in start-ups with little or no growth prospects, remaining in incubators for many years.

A survey of 778 life scientists working in 40 US universities concluded that developing formal policies may send a signal, but the effect on individual behaviour depends very much on whether these policies are reinforced by behavioural expectations.[15] They found that individual characteristics and local norms appear to be equally effective predictors of entrepreneurial activity, but only provided 'weak and unsystematic predictions of the forms of entrepreneurship'. Where successful, this can create a virtuous circle, the demonstration effect of a successful spin-out encouraging others to try. This leads to clusters of spin-outs in space and time, resulting in entrepreneurial departments or universities, rather than isolated entrepreneurial academics. Local norms or culture at the departmental level will influence the effectiveness of formal policies by providing a strong mediating effect between the institutional context and individual perceptions. Local norms evolve through self-selection during recruitment, resulting in staff with similar personal values and behaviour, and reinforced by peer pressure or behavioural socialization resulting in a convergence of personal values and behaviour. However, there is a real potential conflict between the pursuit of knowledge and its commercial exploitation, and a real danger of lowering research standards exists. Therefore it is essential to have explicit guidelines for the conduct of business in a university environment:[16]

1. Specific guidelines on the use of university facilities, staff and students and intellectual property rights.

2. Specific guidelines for, and periodic reviews of, the dual employment of scientist-entrepreneurs, including permanent part-time positions.

3. Mechanisms to resolve issues of financial ownership and the allocation of research contracts between the university and the venture.

A recent study of nine university spin-off companies in the UK identified a number of common stages of development, each demanding different capabilities, resources and support:[17]

- *Research phase* – all of the academic entrepreneurs were at the forefront of their respective fields, were focused on their research, respected by their academic communities and had high levels of publication. This contributes to the generation of know-how and the likelihood of generating more formal intellectual property.

- *Opportunity framing phase* – the development of an understanding of how best to create commercial value from the science. In most cases the opportunities are defined imprecisely, targeted ambiguously and prove impracticable. In particular, there is a need to define the complementary resources necessary for commercialization, including human, financial, physical and technological resources. Therefore the framing process is usually iterative and slow, taking many months or even years.

- *Pre-organization phase* – decisions made at this early stage often have a significant impact upon the entire future success of the venture, since they direct the path of development and constrain future options. At this stage access to networks of expertise and prior entrepreneurial experience are critical.

- *Re-orientation phase* – once the venture has gained sufficient resource and credibility to start-up, the venture must 'repackage' its technology and acquire new information and resources to create something of value to some target customer group.

- *Sustainable returns phase* – with an emphasis on business capabilities, winning orders, selling products or services, and making a return. This demands professional management, greater financial resources and a broader range of capabilities.

At each of these stages there are different significant challenges to overcome in order to make a successful transition to the next stage, what the researchers call 'critical junctures':

- *Opportunity recognition* – at the interface of the research and opportunity framing phases. This requires the ability to connect a specific technology or know-how to a commercial application, and is based on a rather rare combination of skill, experience, aptitude, insight and circumstances. A key issue here is the ability to synthesize scientific knowledge and market insights, which increases with the entrepreneur's social capital – linkages, partnerships and other network interactions.

- *Entrepreneurial commitment* – acts and sustained persistence that bind the venture champion to the emerging business venture. This often demands difficult personal decisions to be made – for example, whether or not to remain an academic – as well as evidence of direct financial investments to the venture.

- *Venture credibility* – is critical for the entrepreneur to gain the resources necessary to acquire the finance and other resources for the business to function. Credibility is a function of the venture team, key customers and other social capital and relationships. This requires close relationships with sponsors, financial and other, to build and maintain awareness and credibility. Lack of business experience, and failure to recognize their own limitations are a key problem here. One solution is to hire the services of a 'surrogate entrepreneur'. As one experienced entrepreneur notes, 'The not so smart or really insecure academics want their hands over everything. These prima donnas make a complete mess of things, get nowhere with their companies and end up disappointed professionally and financially.'

In the UK, the Lambert Review of Business–University Collaboration reported in December 2003. It reviewed the commercialization of intellectual property by universities in the UK, and also made international comparisons of policy and performance. The UK has a similar pattern of concentration of activity as the USA: in 2002 80% of UK universities made no patent applications, whereas 5% filed 20 or more patents; similarly, 60% of universities issued no new licenses, but 5% issued more than 30. However, in the UK there has been a bias towards spin-outs rather than licensing, which the Lambert Report criticizes. It argues that spin-outs are often too complex and unsustainable, and of low quality – a third in the UK are fully funded by the parent university and attract no external private funding. In 2002 universities in the UK created over 150 new spin-out firms, compared to almost 500 by universities in the USA; the respective figures for new licenses that year were 648 and 4058. As a proportion of R&D expenditure, this suggests that British universities place greater emphasis on spin-outs than their North American counterparts, and less on licensing. Lambert argues that universities in the UK may place too high a price on their intellectual property, and that contracts often lack clarity of ownership. Both of these problems discourage businesses from licensing intellectual property from universities, and may encourage universities to commercialize their technologies through wholly owned spin-outs.

12.3 Profile of a Technical Entrepreneur

This section draws on the research by Roberts,[1] who studied 156 NTBFs which were spin-offs from MIT in the USA (herein referred to as 'the US study'), and Oakey,[18] who examined 131 NTBFs in the UK, and more general research on NTBFs (herein referred to as 'the UK study').

FIGURE 12.1   Factors affecting the decision to establish an NTBF

The creation of an NTBF is the interaction of individual skills and disposition and the technological and market characteristics. The US study emphasizes the role of personal characteristics, such as family background, goal orientation, personality and motivation, whereas the UK study stresses the role of technological and market factors. The decision to start an NTBF typically begins with a desire to gain independence and to escape the bureaucracy of a large organization, whether it be in the public or private sector. Thus the background, psychological profile and work and technical experience of a technical entrepreneur all contribute to the decision to form an NTBF (Figure 12.1).

Much of the American research on NTBFs, and more general studies of entrepreneurs, tends to emphasize the background and characteristics of a typical entrepreneur. Factors found to affect the likelihood of establishing a venture include:

- family background;
- religion;
- formal education and early work experience;
- psychological profile.

A number of studies confirm that both family background and religion affect an individual's propensity to establish a new venture. A significant majority of technical entrepreneurs have a self-employed or professional parent. Studies indicate that between 50 and 80% have at least one self-employed parent. For example, the US study found that four times as many technical entrepreneurs have a parent who is a professional, compared with other groups of scientists and engineers. The most common explanation for this observed bias is that the parent acts as a role model and may provide support for self-employment.

The effect of religious background is more controversial, but it is clear that certain religions are over-represented in the population of technical entrepreneurs. For example, in the USA and Europe Jews are more likely to establish an NTBF, and Chinese are more likely to in Asia. Whether this observed bias is the result of specific cultural or religious norms, or the result of minority status, is the subject of much controversy but little research. The US study suggests that dominant cultural values are more important than minority status, but even this work indicates that the effect of family background is more significant than religion. In any case, and perhaps more importantly, there appears to be no significant relationship between family and religious background and the subsequent probability of success of an NTBF.

Education and training are major factors that distinguish the founders of NTBFs from other entrepreneurs. The median level of education of technical entrepreneurs in the US study was a master's degree and, with the important exception of biotechnology-based NTBFs, a doctorate was superfluous. Significantly, the levels of education of technical entrepreneurs do not differentiate them from other scientists and engineers. However, potential technical entrepreneurs tend to have higher levels of productivity than their technical work colleagues, measured in terms of papers published or patents granted: 6.35 versus 2.2 papers on average; and 1.6 patents versus 0.05. This suggests that potential entrepreneurs may be more driven than their corporate counterparts.

In addition to a master's-level education, on average, a technical entrepreneur will have around 13 years of work experience before establishing an NTBF. In the case of Route 128, the entrepreneurs' work experience is typically with a single incubator organization, whereas technical entrepreneurs in Silicon Valley tend to have gained their experience from a larger number of firms before establishing their own NTBF. This suggests that there is no ideal pattern of previous work experience. However, experience of development work appears to be more important than work in basic research. As a result of the formal education and experience required, a typical technical entrepreneur will be aged between 30 and 40 years when establishing their first NTBF. This is relatively late in life compared to other types of venture, and is due to a combination of ability and opportunity. On the one hand, it typically takes between 10 and 15 years for a potential entrepreneur to attain the necessary technical and business experience. On the other hand, many people begin to have greater financial and family responsibilities at this time. Thus there appears to be a window of opportunity to start an NTBF some time in the mid-thirties. Moreover, different fields of technology have different entry and growth potential. Therefore the choice of a potential entrepreneur will be constrained by the dynamics of the technology and markets. The capital requirements, product lead times and potential for growth are likely to vary significantly between sectors.

Much of the research on the psychology of entrepreneurs is based on the experience of small firms in the USA, so the generalizability of the findings must be questioned. However, in the specific case of technical entrepreneurs there appears to be some consensus regarding the necessary personal characteristics. The two critical requirements appear to be an internal locus of control and a high need for achievement. The former characteristic is common in scientists and engineers, but the need for high levels of achievement is less common. Entrepreneurs are typically motivated by a high need for achievement (so-called 'n-Ach'), rather than a general desire to succeed. This behaviour is associated with moderate risk-taking, but not gambling or irrational risk-taking. A person with a high n-Ach:

* Likes situations where it is possible to take personal responsibility for finding solutions to problems;
* Has a tendency to set challenging but realistic personal goals and to take calculated risks;
* Needs concrete feedback on personal performance.

However, the US study of almost 130 technical entrepreneurs and almost 300 scientists and engineers found that not all entrepreneurs have high n-Ach, only some do. Technical entrepreneurs had only moderate n-Ach, but low need for affiliation (n-Aff). This suggests that the need for independence, rather than success, is the most significant motivator for technical entrepreneurs. Technical entrepreneurs also tend to have an internal locus of control. In other words, technical entrepreneurs believe that they have personal control over outcomes, whereas someone with an external locus of control believes that outcomes are the result of chance, powerful institutions or others. More sophisticated psychometric techniques such as the Myers–Briggs type indicators (MBTI) confirm the differences between technical entrepreneurs and other scientists and engineers.

Numerous surveys indicate that around three-quarters of technical entrepreneurs claim to have been frustrated in their previous job. This frustration appears to result from the interaction of the psychological predisposition of the potential entrepreneur and poor selection, training and development by the parent organization. Specific events may also trigger the desire or need to establish an NTBF, such as a major reorganization or downsizing of the parent organization.

12.4 The Business Plan

The primary reason for developing a formal business plan for a new venture is to attract external funding. However, it serves an important secondary function. A business plan can provide a formal agreement between founders regarding the basis and future development of the venture. A business plan can help reduce self-delusion on the part of the founders, and avoid subsequent arguments concerning responsibilities and rewards. It can help to translate abstract or ambiguous goals into more explicit operational needs, and support subsequent decision-making and identify trade-offs. Of the factors *controllable* by entrepreneurs, business planning has the most significant positive effect on new venture performance. However, there are of course many *uncontrollable* factors, such as market opportunity, which have an even more significant influence on performance.[19] As we noted in Chapter 3, Pasteur's advice still applies, '. . . chance favours only the prepared mind.'

No standard business plan exists, but in many cases venture capitalists will provide a pro forma for the business plan. Typically a business plan should be relatively concise, say no more than 10 sides, begin with an executive summary, and include sections on the product, markets, technology, development, production, marketing, human resources, financial estimates with contingency plans, and the timetable and funding requirements. Most business plans submitted to venture capitalists are strong on the technical considerations, often placing too much emphasis on the technology relative to other issues. As Roberts notes, 'Entrepreneurs propose that they can do *it* better than anyone else, but may forget to demonstrate that anyone wants *it*.'[1] He identifies a number of common problems with business plans submitted to venture capitalists: marketing plan, management team, technology plan and financial plan.

There were found to be serious inadequacies in all four of these areas, but the worst were in marketing and finance. Less than half of the plans examined provided a detailed marketing strategy, and just half included any sales plan. Three-quarters of the plans failed to identify or analyse any potential competitors. As a result most business plans contain only basic financial forecasts, and just 10% conducted any sensitivity analysis on the forecasts. The lack of attention to marketing and competitor analysis is particularly problematic as research indicates that both factors are associated with subsequent success.

For example, the UK study found that in the early stages NTBFs rely too much on a few major customers for sales, and are therefore vulnerable. In the extreme case, half of NTBFs rely on a single customer for more than half of their first-year sales. This over-dependence on a small number of customers has three major drawbacks:

1. Vulnerability to changes in the strategy and health of the dominant customer.
2. A loss of negotiating power, which may reduce profit margins.
3. Little incentive to develop marketing and sales functions, which may limit future growth.

Funding

NTBFs are different from other new ventures in that there is often no marketable product available before or shortly after formation. Therefore, initial funding of the venture cannot normally be based on cash flow derived from early sales. The precise cash flow profile will be determined by a number of factors, including development time and cost, and the volume and profit margin of sales. Different development and sales strategies exist, but to some extent these factors are determined by the nature of the technology and markets (Figure 12.2a–c).

For example, biotechnology ventures typically require more start-up capital than electronics or software-based ventures, and have longer product development lead times. Therefore, from the perspective of a potential entrepreneur, the ideal strategy would be to conduct as much development work as possible within the incubator organization before starting the new venture. However, there are practical problems with this strategy, in particular ownership of the intellectual property on which the venture is to be based.

(a) Research-based venture e.g. biotechnology

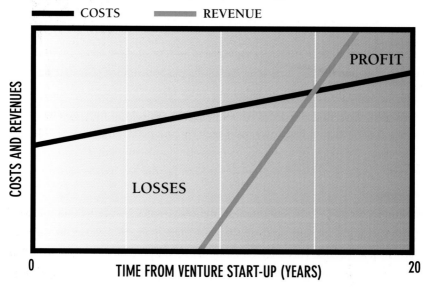

(b) Development-based venture e.g. electronics

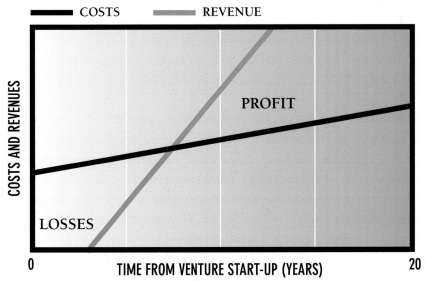

(c) Production-based venture e.g. software

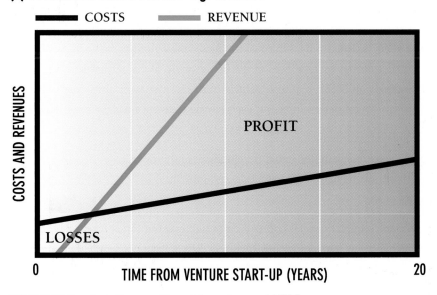

FIGURE 12.2 Cash flow profiles of three types of NTBF

Given their strong desire for independence, most entrepreneurs seek to avoid external funding for their ventures. However, in practice this is not always possible, particularly in the latter growth stages. The initial funding required to form an NTBF includes the purchase of accommodation, equipment and other start-up costs, plus the day-to-day running costs such as salaries, heating, light and so on. Research in the USA and UK suggests that most NTBFs begin life as part-time ventures, and are funded by personal savings, loans from friends and relatives, and bank loans, in that order. Around half also receive some funding from government sources, but in contrast receive next to nothing from venture capitalists. Venture capital is typically only made available at later stages to fund growth on the basis of a proven development and sales record.

Research in the USA suggests that the initial capital needed to start an NTBF is relatively modest, typically less than $50 000 and in almost half of the cases less than $10 000 (1990 US dollars). However, the extent of the need for external funding will depend on the nature of the technology and the strategy of the NTBF. For example, an electronics or software-based venture will also demand high initial funding if a strategy of aggressive growth is to be achieved. The UK study shows that both the amount and source of initial funding for the formation of an NTBF vary considerably. For example, as software-based ventures typically require less start-up capital than either electronics or biotechnology ventures, it is more common for such firms to rely solely on personal funding. Biotechnology firms tend to have the highest R&D costs, and consequently most require some external funding. In contrast, software firms typically require little R&D investment, and are less likely to seek external funds. The UK study found that almost three-quarters of the software firms were funded by profits after three years, whereas only a third of the biotechnology firms had achieved this.

The initial funding to establish an NTBF is rarely a major problem. However, Drucker suggests an NTBF requires financial restructuring every three years.[20] Other studies identify stages of development, each having different financial requirements:

1. Initial financing for launch.
2. Second-round financing for initial development and growth.
3. Third-round financing for consolidation and growth.
4. Maturity or exit.

In general, professional financial bodies are not interested in initial funding because of the high risk and low sums of money involved. It is simply not worth their time and effort to evaluate and monitor such ventures. However, as the sums involved are relatively small – typically of the order of tens of thousands of pounds – personal savings, remortgages and loans from friends and relatives are often sufficient. In contrast, third-round finance for consolidation is relatively easy to obtain, because by that time the venture has a proven track record on which to base the business plan, and the venture capitalist can see an exit route.

Venture capitalists are keen to provide funding for a venture with a proven track record and strong business plan, but in return will often require some equity or management involvement. Moreover, most venture capitalists are looking for a means to make capital gains after about five years. However, almost by definition technical entrepreneurs seek independence and control, and there is evidence that some will sacrifice growth to maintain control of their ventures. For the same reason, few entrepreneurs are prepared to 'go public' to fund further growth. Thus many entrepreneurs will choose to sell the business and found another NTBF. In fact, the typical technical entrepreneur establishes an average of three NTBFs. Therefore the biggest funding problem for an NTBF is likely to be for the second-round financing to fund development and growth. This can be a time-consuming and frustrating process to convince venture capitalists to provide finance. The formal proposal is critical at this stage. Professional investors will assess the attractiveness of the venture in terms of the strengths and personalities of the founders, the formal business plan and the commercial and technical merits of the product, probably in that order.

Corporate Venture Funding

A survey of corporate funding of NTBFs in the UK found that around 15% of large companies had made investments in external new ventures, mainly in their own sector.[21] As with internal corporate venturing (see Chapter 10), the funding of *external* ventures by large corporations is cyclical, reflecting the business environment. For example, surveys suggest that in 1998 the number of major corporations funding external ventures was around 110, but by 2000 this had grown to 350.[22] The typical investment (in 1997) was in excess of £500 000, and the investing companies preferred ventures requiring additional capital for expansion, rather than funds for start-up or early development. The most common problems encountered were agreement of the rate of return and details of corporate representation in the venture. The average period of investment was five to seven years, and corporate investors typically demanded a rate of return of 20 to 30%, which compares favourably with professional venture capitalists required returns of around 75%. Regarding professional venture capitalists, Figure 12.3 highlights two important issues. First, that the availability of venture capital varies worldwide, as we discussed in Chapter 4, and such disparities tend to be self-reinforcing as potential new ventures relocate to seek funding. The second point to note is the strong bias for finance for expansion, rather than start-ups, which is most significant in the UK. This creates a potential venture funding gap, between the initial, usually self-financed stage, and the first involvement of professional venture capital. In the UK this gap is in the region of £200 000 to £750 000.[23]

Corporate investment in new ventures is increasingly popular in high-technology sectors, where large firms do not have access to all technologies in-house, and where emerging technologies remain unproven.[24] Investments in small biotechnology

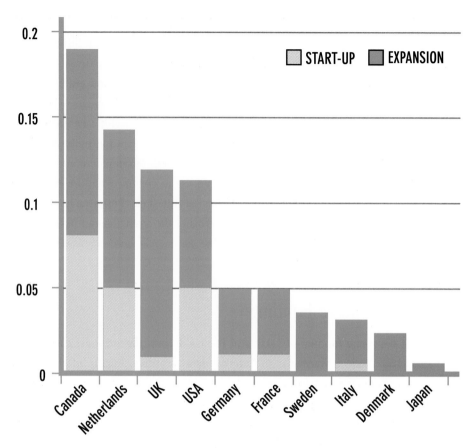

FIGURE 12.3 Venture capital as a percentage of GDP, 1997.
Sources: CVCA (Canada), NVCA (USA), MITI (Japan), OECD (other nations).

companies by pharmaceutical companies can be direct, or indirect investment through specialist venture funds (see Box 12.2). Direct investment is preferred where there is a high probability of technological success which is likely to impact the product pipeline in the near term. Indirect investments are concerned more with gaining windows on a range of early-stage technologies with the *potential* to impact the future direction of the product pipeline.[25] There has been a marked increase in the number of pharmaceutical companies investing through specialist venture funds, recent examples being Novartis (Novartis Ventures) and Bayer (Bayer Innovation). At the same time, pharmaceutical companies and their venture funds appear to be investing increasingly in independent seed capital funds focused on early-stage biotechnology, such as UK Medical Ventures (UK), New Medical Technologies (Switzerland) and Medical Technology Partners (USA). The precise objectives of such funds vary, but all share a common emphasis on strategic issues rather than purely financial. A principal investment criterion is 'no fit, no deal', the decision to invest being largely strategic, to 'scout for "out there" science'. The alternative mode of indirect venturing is participation in

> **BOX 12.2**
> **JOHNSON AND JOHNSON DEVELOPMENT CORPORATION**
>
> Johnson and Johnson Development Corporation (JJDC) is an independent venture capital firm within the Johnson and Johnson group of companies, and aims to identify and fund new technologies and businesses in the pharmaceutical and healthcare sector. JJDC was established in the USA 25 years ago and has since invested in more than 300 start-up businesses worldwide. In 1997 it created a dedicated European division, Johnson and Johnson Development Capital. Both companies exploit the scientific and market know-how of Johnson and Johnson and typically invest alongside professional venture capital firms in ventures in the start-up and early growth stages.

independent seed capital funds targeted at early-stage investments. This included direct investment in seed capital funds by the corporate parent, or indirect investment in seed funds via corporate venture funds. As may be expected, objectives include providing windows on early-stage technologies; for example, one company invested in a health informatics fund in order to place representatives on the board of the seed fund to gain competence in the field. Another reason for investing is to access 'deal flow' – that is the opportunity to participate *directly* in subsequent rounds of funding beyond the seed capital stage. This was aimed at investing more directly in later stage technologies should they appear to be strategically important for the corporate parent. Clearly then, the goals of pharmaceutical firms' investments in new ventures are fundamentally different from those of professional venture capital firms. A similar strategy applies in other sectors, such as information and communications technology (see Box 12.3). The goals of corporate venture funds are largely strategic, focusing on technology and potential new products, whereas the goals of venture capitalists are (rightly) purely financial.

Venture Capital

Whilst there is general agreement about the main components of a good business plan, there are some significant differences in the relative weights attributed to each component. General venture capital firms typically only accept 5% of the technology ventures they are offered, and the specialist technology venture funds are even more selective, accepting around 3%. The main reasons for rejecting technology proposals compared to more general funding proposals are the lack of intellectual property, the skills of the management team, and size of the potential market. A survey of venture capitalists in

---

BOX 12.3

REUTERS' CORPORATE VENTURE FUNDS

Reuters established its first fund for external ventures, Greenhouse 1, in 1995. It has since added a further two venture funds, which aim to invest in related businesses such as financial services, media, and network infrastructure. By 2001 it had invested US$432m. in 83 companies, and these investments contributed almost 10% to its profits. However, financial return was not the primary objective of the funds. For example, it invested $1m. in Yahoo in 1995, and consequently Yahoo acquired part of its content from Reuters. This increased the visibility of Reuters in the growing Internet markets, particularly in the USA where it was not well known, and resulted in other portals following Yahoo's lead with content from Reuters. By 2001 Reuters' content was available on 900 Web services, and had an estimated 40 million users per month.

Source: Loudon, A. (2001) *Webs of Innovation: The networked economy demands new ways to innovate*. FT.com, Pearson Education, Harlow.

North America, Europe and Asia found major similarities in the criteria used, but also identified several interesting differences in the weights attached to some criteria (Table 12.1). The criteria are similar to those discussed earlier, grouped into five categories:

1. the entrepreneur's personality;
2. the entrepreneur's experience;
3. characteristics of the product;
4. characteristics of the market;
5. fnancial factors.

Overall, the study confirmed the importance of a bundle of personal, market and financial factors, which were consistently ranked as being most significant: a proven ability to lead others and sustain effort; familiarity with the market; and the potential for a high return within 10 years (see Box 12.4). The personality and experience of the entrepreneurs were consistently ranked as being more important than either product or market characteristics, or even financial considerations. However, there were a number of significant differences between the preferences of venture capitalists from different regions. Those from the USA placed greater emphasis on a high financial return and

TABLE 12.1 Criteria used by venture capitalists to assess proposals

Criteria	European (n = 195)	American (n = 100)	Asian (n = 53)
Entrepreneur able to evaluate and react to risk	3.6	3.3	3.5
Entrepreneur capable of sustained effort	3.6	3.6	3.7
Entrepreneur familiar with the market	3.5	3.6	3.6
Entrepreneur demonstrated leadership ability*	3.2	3.4	3.0
Entrepreneur has relevant track record*	3.0	3.2	2.9
Product prototype exists and functions*	3.0	2.4	2.9
Product demonstrated market acceptance*	2.9	2.5	2.8
Product proprietary or can be protected*	2.7	3.1	2.6
Product is 'high technology'*	1.5	2.3	1.4
Target market has high growth rate*	3.0	3.3	3.2
Venture will stimulate an existing market	2.4	2.4	2.5
Little threat of competition within three years	2.2	2.4	2.4
Venture will create a new market*	1.8	1.8	2.2
Financial return >10 times within 10 years*	2.9	3.4	2.9
Investment is easily made liquid* (e.g. made public or acquired)	2.7	3.2	2.7
Financial return >10 times within 5 years*	2.1	2.3	2.1

1 = irrelevant, 2 = desirable, 3 = important, 4 = essential. * Denotes significant at the 0.05 level.
Source: Adapted from Knight, R. (1992) 'Criteria used by venture capitalists', in Khalil, T. and Bayraktar, B. (eds), *Management of Technology III: The key to global competitiveness*. Industrial Engineering & Management Press, Georgia, 574–583.

liquidity than their counterparts in Europe or Asia, but less emphasis on the existence of a prototype or proven market acceptance. Perhaps surprisingly, all venture capitalists are adverse to technological and market risks. Being described as a 'high-technology' venture was rated very low in importance by the US venture capitalists, and the European and Asian venture capitalists rated this characteristic as having a negative influence on funding. Similarly, having the potential to create an entirely new market was considered a drawback. In short, venture capitalists are not particularly adventurous, as the examples of Chiroscience and Autonomy demonstrate (Boxes 12.5 and 12.6).

BOX 12.4
ANDREW RICKMAN AND BOOKHAM TECHNOLOGY

Andrew Rickman founded Bookham Technology in 1988, aged 28. Rickman has a degree in mechanical engineering from Imperial College London, a Ph.D. in integrated optics from Surrey University, an MBA and has worked as a venture capitalist. Unlike many technology entrepreneurs, he did not begin with the development of a novel technology and then seek a means to exploit it. Instead, he first identified a potential market need for optical switching technology for the then fledgling optical fibre networks, and then developed an appropriate technological solution. The market for optical components is growing fast as the use of Internet and other data-intensive traffic grows. Rickman aimed to develop an integrated optical circuit on a single chip to replace a number of discrete components such as lasers, lenses and mirrors. He chose to use silicon rather than more exotic materials to reduce development costs and exploit traditional chip production techniques. The main technological developments were made at Surrey University and the Rutherford Appleton Laboratory, where he had worked, and 27 patents were granted and a further 140 applied for. Once the technology had been proven, the company raised US$110m. over several rounds of funding from venture capitalist 3i, and leading electronics firms Intel and Cisco. The most difficult task was scale-up and production: 'Taking the technology out of the lab and into production is unbelievably tough in this area. It is infinitely more difficult than dreaming up the technology.' Bookham Technology floated in London and on the Nasdaq in New York in April 2000 with a market capitalization of more than £5bn, making Andrew Rickman, with 25% of the equity, a paper billionaire. Bookham is based in Oxford, and employs 400 staff. The company acquired the optical component businesses of Nortel and Marconi in 2002, and in 2003 the US optical companies Ignis Optics and New Focus, and the latter included chip production facilities in China. This puts Bookham in the top three in the global optoelectronics sector.

A study of venture capitalists in the UK compared attitudes to funding technology ventures over a 10-year period, and found that investment in technology-based firms as a percentage of total venture capital had increased from around 11% in 1990 to 25% by 2000 (by value).[26] Of the total venture capital investment in UK NTBFs of £1.6bn in the year 2000, 30% was for early-stage funding (by value, or 47% by number of firms), 47% for expansion (by value, or 47% by number of firms), and the rest for management buy-outs (MBO). This increase was due to a combination of the growth of

BOX 12.5
MIKE LYNCH AND AUTONOMY

Mike Lynch founded the software company Autonomy in 1994, a spin-off from his first start-up Neurodynamics. Lynch, a grammar-school graduate, studied information science at Cambridge where he carried out Ph.D. research on probability theory. He rejected a conventional research career as he had found his summer job at GEC Marconi a 'boring, tedious place'. In 1991, aged 25, he approached the banks to raise money for his first venture, Neurodynamics, but 'met a nice chap who laughed a lot and admitted that he was only used to lending money to people to open newsagents'. He subsequently raised the initial £2000 from a friend of a friend. Neurodynamics developed pattern recognition software which it sold to specialist niche users such as the UK police force for matching fingerprints and identifying disparities in witness statements, and banks to identify signatures on cheques.

Autonomy was spun off in 1994 to exploit applications of the technology in Internet, intranet and media sectors, and received the financial backing of venture capitalists Apax, Durlacher and Enic. Autonomy was floated on the Easdaq in July 1998, on the Nasdaq in 1999, and in February 2000 was worth US$5bn, making Lynch the first British software billionaire. Autonomy creates software which manages unstructured information, which accounts for 80% of all data. The software applies Bayesian probabilistic techniques to identify patterns of data or text, and compared to crude keyword searches can better take into account context and relationships. The software is patented in the USA, but not in Europe as patent law does not allow patent protection of software. The business generates revenues through selling software for cataloguing and searching information direct to clients such as the BBC, Barclays, BT, Eli Lilly, General Motors, Merrill Lynch, News Corporation, Nationwide, Procter & Gamble and Reuters. In addition, it has more than 50 licence agreements with leading software companies to use its technology, including Oracle, Sun and Sybase. A typical licence will include a lump sum of US$100000 plus a royalty on sales of 10–30%. By means of such licence deals Autonomy aims to become an integral part of a range of software and the standard for intelligent recognition and searching. In the financial year ending March 2000 the company reported its first profit of US$440000 on a turnover of $11.7m. The company employs 120 staff, split between Cambridge in the UK and Silicon Valley, and spends 17% of its revenues on R&D. In 2004 sales were around $60m., with an average licence costing $360000, and high gross margins of 95%. New customers include AOL, BT, CitiBank, Deutsche Bank, Ford, the 2004 Greek Olympics, and the defence agencies in the USA, Spain, Sweden and Singapore. Repeat customers accounted for 30% of sales.

BOX 12.6
CHRIS EVANS AND CHIROSCIENCE

Chiroscience plc is one of the nearly 20 biotechnology firms founded by the micro-biologist/entrepreneur Chris Evans. Evans, Ph.D. and since OBE, formed his first new venture, Enzymatix Ltd, in 1987, aged 30. His business plan was rejected by venture capitalists, so he was forced to sell his house for £40000 to raise the initial finance. Subsequent finance of £1m. was provided by the commodities group Berisford International, but following financial problems in the property market, the company was divided into Celsis plc, which makes contamination testing equipment, and Chiroscience, which exploits chiral technology, the basis of which is that most molecules have mirror images that have different properties, essentially a right-hand sense and a left-hand sense. Isolating the more effective mirror image in an existing drug formulation can improve its efficacy, or reduce unwanted side effects.

Chiroscience was formed in 1990, other directors being recruited from large established pharmaceutical firms such as Glaxo, SmithKline Beecham and Zeneca. The company was floated on the London Stock Exchange in 1994. This was only possible because in 1992 the Stock Exchange relaxed its requirements for market entry, and no longer required three consecutive years' profits before listing. The biotechnology company applies chiral technology to the purification of existing drugs and design of new drugs. Chiroscience has three potential applications of chiral technology: first, and most immediately, the improvement of existing drugs by isolating the most effective sense of molecules; second, the development of alternative processes for the production of existing drugs as they come off-patent and finally, the design of new drugs by means of single isomer technology.

Chiroscience was the first British biotechnology firm to be granted approval for sale of a new product, Dexketoprofen, in 1995. This is a non-steroidal anti-inflammatory drug, based on a right-handed version of the older drug ketoprofen. The drug is marketed by the Italian firm Menarini. Chiroscience has been involved in a number of collaborative development and marketing deals. In 1995 it formed an alliance with the Swedish pharmaceutical group Pharmacia, to develop and market its local anaesthetic, Levobupivacaine. It also forged a more general strategic alliance with Medeva, the pharmaceutical group which performs no primary research, but specializes in taking products to market.

Biotechnology stocks are more volatile than most other investments, and it is difficult to use conventional techniques to assess their current value or future potential. Expenditure on R&D in the initial years typically results in significant losses, and sales may be negligible for up to 10 years. Therefore there are no price–earnings ratios or future revenues to discount. For example, in its first two years after flotation Chiroscience reported cumulative losses of £3.7m., due largely to research spending of £12.4m. Nevertheless, Chiroscience has outperformed the financial markets, and most other biotechnology stock. The company was floated in 1994 at 150p, and quickly fell to below 100p. However, by December 1995 shares had reached 364p. As a result, Chris Evans's personal fortune was estimated to have reached £50m. by 1995.

In January 1999 Chiroscience merged with Celltech to form Celltech Chiroscience, which subsequently acquired Medeva to become the Celltech Group. The new company has some 400 research staff, an R&D budget of £51m. and adds much-needed sales and marketing competencies with a sales force of 550. Celltech Group is three times the size of Chiroscience, and reached a market capitalization of £3bn in 2000. It is one of the few British biotechnology companies to gain regulatory approval for its products in the USA, and the first to achieve profitability. Sir Chris Evans (he was knighted in 2001) now runs the biotechnology venture capital firm Merlin Biosciences.

specialist technology venture capitalists, and greater interest by the more general venture capital firms. As venture capital firms have gained experience of this type of funding, and the opportunities for flotation have increased due to the new secondary financial markets in Europe such as the AIM, TechMARK and Neur Markt, their returns on investment have increased significantly. In the 1980s returns to UK early-stage technology investments were under 10%, compared to venture capital norms of twice that, but by 2000 the returns of technology ventures increased to almost 25%, which is higher than all other types of venture investment. However, this recent growth in venture capital funding of NTBFs needs to be put into perspective. Although the UK has the most advanced venture capital community in Europe, venture capital still only accounts for between 1 and 3% of the external finance raised by small firms.

An important issue is the influence of venture capitalists on the success of NTBFs. They can play two distinct roles. The first, to identify or select those NTBFs that have the best potential for success – that is, 'picking winners' or 'scouting'. The second role is to help develop the chosen ventures, by providing management expertise and access to resources other than financial – that is, a 'coaching' role. Distinguishing between the effects of these two roles is critical for both the management of and policy for NTBFs. For managers, it will influence the choice of venture capital firm, and for policy, the balance between funding and other forms of support. A study of almost 700 biotechnology firms over 10 years provides some insights to these different roles.[27] It found that when selecting start-ups to invest in, the most significant criteria used by venture capitalists were a broad, experienced top management team, a large number of recent patents, and downstream industry alliances (but not upstream research alliances, which had a negative effect on selection). The strongest effect on the decision to fund was the first criterion, and the human capital in general. However, subsequent analysis of venture performance indicates that this factor has limited effect on performance, and that the few significant effects are split equally between improving and impeding the performance of a venture. The effects of technology and alliances on subsequent performance are much more significant and positive. In short, in the *selection* stage, venture capitalists place too much emphasis on human capital, specifically the top management team. In the development or coaching stages, venture capitalists do contribute to the success of the chosen ventures, and tend to introduce external professional management much earlier than in NTBFs not funded by venture capital. Taken together, this suggests that the coaching role of venture capitalists is probably as important, if not more so, than the funding role, although policy interventions to promote NTBFs often focus on the latter.

12.5 Growth and Performance of Innovative Small Firms

We discussed innovation strategy and the positioning of small firms in Chapters 3 and 4. Here we examine what is known about the subsequent growth and performance of innovative small firms.

There has been a great deal of economic and management research on small firms, but much of this has been concerned with the contribution all types of small firms make to economic, employment or regional development. Relatively little is known about innovation in small firms, or the more narrow issue of the performance of NTBFs. As demonstrated by the preceding discussion, almost all research on innovative small firms has been confined to a small number of high-technology sectors, principally microelectronics and more recently biotechnology. A notable exception is the survey of 2000 SMEs conducted by the Small Business Research Centre in the UK. The survey found that 60% of the sample claimed to have introduced a major new product or service innovation in the previous five years.[28] Whilst this finding demonstrates that the management of innovation is relevant to the majority of small firms, it does not tell us much about the significance of such innovations, in terms of research and investment, or subsequent market or financial performance.

Research over the past decade or so suggests that the innovative activities of SMEs exhibit broadly similar characteristics across sectors.[29] They:

- are more likely to involve product innovation than process innovation;
- are focused on products for niche markets, rather than mass markets;
- will be more common amongst producers of final products, rather than producers of components;
- will frequently involve some form of external linkage;
- tend to be associated with growth in output and employment, but not necessarily profit.

The limitations of a focus on product innovation for niche or intermediate markets were discussed earlier, in particular problems associated with product planning and marketing, and relationships with lead customers and linkages with external sources of innovation. These topics were discussed in Part III, but a number of issues are specific to smaller firms. Where an SME has a close relationship with a small number of customers, it may have little incentive or scope for further innovation, and therefore will pay relatively little attention to formal product development or marketing. Therefore SMEs in such dependent relationships are likely to have limited potential for future growth, and may remain permanent infants or subsequently be acquired by competitors or customers.[30–32] Moreover, an analysis of the growth in the number of NTBFs suggests that the trend has as much to do with negative factors, such as the downsizing of larger firms, as it does with more positive factors such as start-ups.[33]

Innovative SMEs are likely to have diverse and extensive linkages with a variety of external sources of innovation, and in general there is a positive association between the level of external scientific, technical and professional inputs and the performance of an SME.[34] The sources of innovation and precise types of relationship vary by sector, but links with contract research organizations, suppliers, customers and universities are consistently rated as being highly significant, and constitute the 'social capital' of the firm. However, such relationships are not without cost, and the management and exploitation of these linkages can be difficult for an SME, and overwhelm the limited technical and managerial resources of SMEs.[35] As a result, in some cases the cost of collaboration may outweigh the benefits[36] and in the specific case of collaboration between SMEs and universities there is an inherent mismatch between the short-term, near-market focus of most SMEs and the long-term, basic research interests of universities.[37]

In terms of innovation, the performance of SMEs is easily exaggerated. Early studies based on innovation counts consistently indicated that when adjusted for size, smaller firms created more new products than the larger counterparts. However, methodological shortcomings appear to undermine this clear message. When the divisions and subsidiaries of larger organizations are removed from such samples,[38] and the innovations weighted according to their technological merit and commercial value, the relationship between firm size and innovation is reversed: larger firms create proportionally more significant innovations than SMEs.[39] The amount of expenditure by SMEs on design and engineering has a positive effect on the share of exports in sales,[40] but formal R&D by SMEs appears to be only weakly associated with profitability,[41] and is not correlated with growth.[42] Similarly, the high growth rates associated with NTBFs are not explained by R&D effort,[43] and investment in technology does not appear to discriminate between the success and failure of NTBFs.[32] Instead, other factors have been found to have a more significant effect on profitability and growth, in particular the contributions of technically qualified owner managers and their scientific and engineering staff, and attention to product planning and marketing.[44] A large study of start-ups in Germany found that the founder's level of management experience was a significant predictor of the growth of a venture. However, innovation, broadly defined, was found to be statistically three times more important to growth than founder attributes or any other of the factors measured.[45] Another study, of Korean technology start-ups, also found that innovativeness, defined as a propensity to engage in new idea generation, experimentation and R&D, was associated with performance. So was proactiveness, defined as the firm's approach to market opportunities through active market research and the introduction of new products and services.[46] The same study found also found that what it referred to as sponsorship-based linkages had a positive effect on performance. This included links with venture capital firms, which reinforces the developmental role these can play, as discussed earlier.

The size and location of NTBFs also has an effect on performance. Geographic closeness increases the likelihood of informal linkages and encourages the mobility of skilled labour across firms. However, the probability of a start-up benefiting from such local knowledge exchanges appears to decrease as the venture grows.[47] This growing inability to exploit informal linkages is a function of organizational size, not the age of the venture, and suggests that as NTBFs grow and become more complex, they begin to suffer many of the barriers to innovation discussed in Chapter 2, and therefore the explicit processes and tools to help overcome these, discussed in Chapter 9, become more relevant. This interpretation is reinforced by other, cross-sectional research. Larger

SMEs are associated with a greater spatial reach of innovation-related linkages, and with the introduction of more novel product or process innovations for international markets. In contrast, smaller SMEs are more embedded in local networks, and are more likely to be engaged in incremental innovations for the domestic market.[48] It is always difficult to untangle cause and effect relationships from such associations, but it is plausible that as the more innovative start-ups begin to outgrow the resources of their local networks, they actively replace and extend their networks, which both creates the opportunity and demand for higher levels of innovation. Conversely, the less innovative start-ups fail to move beyond their local networks, and therefore are less likely to have either the opportunity or need for more radical innovation.

In short, most of what we have discussed in this book about managing innovation is relevant to innovative small firms, but more research needs to be done on how to manage the particular problems they face. As we have argued throughout the book, we believe that there is a generic core innovation process which represents management 'best practice', but that this has to be modified in different organizational, technological and market contingencies. In the specific case of small innovative firms, best practice would include scanning the external environment and developing appropriate relationships with sources of innovation and lead users. However, different contingencies will demand different innovation strategies. For example, a study of 116 software start-ups identified five factors that affected success: level of R&D expenditure, how radical new products were, the intensity of product upgrades, use of external technology and management of intellectual property.[49] In contrast, a study of 94 biotechnology start-ups found that three factors were associated with success: location within a significant concentration of similar firms, quality of scientific staff (measured by citations) and the commercial experience of the founder.[50] The number of alliances had no significant effect on success, and the number of scientific staff in the top management team had a negative association, suggesting that the scientists are best kept in the laboratory. Other studies of biotechnology start-ups confirm this pattern, and suggest that maintaining close links with universities reduces the level of R&D expenditure needed, increases the number of patents produced, and moderately increases the number of new products under development. However, as with more general alliances, the *number* of university links has no effect on the success or performance of biotechnology start-ups, but the *quality* of such relationships does.[51]

Such sector-specific studies confirm that the environment in which small firms operate significantly influences both the opportunity for innovation, in a technological and market sense, and the most appropriate strategy and processes for innovation. For example, a NTBF may have a choice of whether to use its intellectual assets by translating its technology into product and services for the market, or alternatively it may exploit these assets through a larger, more established firm, through licensing, sale of IPR or by collaboration. We discussed the more general issues involved in this decision in Chapter 8, but more specifically the NTBF needs to consider two environmental factors:[52]

- *Excludability* – to what extent the NTBF can prevent or limit competition from incumbents who develop similar technology?
- *Complementary assets* – to what extent do the complementary assets – production, distribution, reputation, support, etc. – contribute to the value proposition of the technology?

Combining these two dimensions creates four strategy options:

- *Attacker's advantage* – where incumbent's complementary assets contribute little or no value, and the start-up cannot preclude development by the incumbent (e.g. where formal intellectual property is irrelevant, or enforcement poor), NTBFs will have an opportunity to disrupt established positions, but technology leadership is likely to be temporary as other NTBFs and incumbents respond, resulting in fragmented niche markets in the longer term. This pattern is common in computer components businesses.
- *Ideas factory* – in contrast, where incumbents control the necessary complementary assets, but the NTBF can preclude effective development of the technology by incumbents, co-operation is essential. The NTBF is likely to focus on technological leadership and research, with strong partnerships downstream for commercialization. This pattern tends to reinforce the dominance of incumbents, with the NTBFs failing to develop or control the necessary complementary assets. This pattern is common in biotechnology.
- *Reputation-based* – where incumbents control the complementary assets, but the NTBF cannot prevent competing technology development by the incumbents, NTBFs face a serious problem of disclosure and other contracting hazards from incumbents. In such cases NTBF will need to seek established partners with caution, and attempt to identify partners with a reputation for fairness in such transactions. Cisco and Intel have both developed such a reputation, and are frequently approached by NTBFs seeking to exploit their technology. This pattern is common in capital-intensive sectors such as aerospace and automobiles. However, these sectors have a lower 'equilibrium', as established firms have a reputation for expropriation, therefore discouraging start-ups.
- *Greenfield* – where incumbents assets are unimportant, and the NTBF can preclude effective imitation, there is the potential for the NTBF to dominate an emerging business. Competition or co-operation with incumbents are both viable strategies, depending upon how controllable the technology is – for example, through establishing standards or platforms, and where value is created in the value chain.

12.6 Summary and Further Reading

In this chapter we have explored the rationale, characteristics and management of new technology-based ventures. Typically an entrepreneur establishes a new technology-based venture primarily to achieve independence, and venture capitalists will aim to make a five- to tenfold gain on their investment within five to seven years. There are important similarities in the organization and management of corporate ventures examined in Chapter 10. Both require a clear business plan and strong product champion or entrepreneur who must identify the opportunity for a new venture, raise the finance and manage the development and growth of the business. Thus the individuals involved in internal and external new ventures are likely to have similar backgrounds, levels of education and personalities; they tend to be highly motivated and demand a high level of autonomy.

The main differences between the two types of entrepreneur are the need for affiliation and interaction, and the degree of political and social skills needed. Corporate entrepreneurs seek higher levels of affiliation and need greater social skills than their independent counterparts. This is because the corporate entrepreneur has the advantage of the financial, technical and marketing resources of the parent firm, but must deal with internal politics and bureaucracy. The external entrepreneur must raise finance and develop functional expertise, but has the advantage of independence and managerial and technical autonomy. Both types of new venture represent an opportunity to grow new businesses based on new technology.

There are many books and journal articles on the more general subject of entrepreneurship, but relatively little has been produced on the more specific subject of new technology-based entrepreneurism. E. B. Roberts's *Entrepreneurs in High Technology: Lessons from MIT and beyond* (Oxford University Press, 1991) is an excellent study of the MIT experience, although perhaps places too much emphasis on the characteristics of individual entrepreneurs. For a broader analysis of technology ventures in the USA see Martin Kenny (ed.), *Understanding Silicon Valley: Anatomy of an entrepreneurial region* (Stanford University Press, 2000). Ray Oakey's *High-technology New Firms* (Paul Chapman, 1995) is a similar study of NTBFs in the UK, but places greater emphasis on how different technologies constrain the opportunities for establishing NTBFs, and affect their management and success. For a review of recent research on the broader issue of innovative small firms see 'Small firms, R&D, technology and innovation: a literature review' by Kurt Hoffman *et al.*, published in *Technovation,* **18** (1), 39–55, 1998. A special issue of the *Strategic Management Journal* (volume 22, July 2001) examined entrepreneurial strategies, and includes a number of papers on technology-based firms, and a special issue of the journal *Research Policy* (volume 32, 2003) features papers on technology spin-offs and start-ups.

References

1 Roberts, E. (1991) *Entrepreneurs in High Technology: Lessons from MIT and beyond*. Oxford University Press, Oxford.

2 Martin, M. (1994) *Managing Innovation and Entrepreneurship in Technology*. John Wiley & Sons, Inc., New York.

3 Massey, D., D. Wield and P. Quintas (1991) *High-Tech Fantasies: Science parks in society, science and space*. Routledge, London.

4 Westhead, P. (1997) 'R&D 'inputs' and 'outputs' of technology-based firms located on and off science parks', *R&D Management*, 27 (1), 45–61.

5 Bower, J. (2003) 'Business model fashion and the academic spin out firm', *R&D Management*, **33** (2), 97–106.

6 Henderson, R., A. Jaffe and M. Trajtenberg (1998) 'Universities as a source of commercial technology: a detailed analysis of university patenting 1965–1988', *Review of Economics and Statistics*, 119–127.

7 Mowery, D. *et al.* (2001) 'The growth of patenting and licensing by U.S. universities: an assessment of the effects of the Bayh–Dole Act of 1980', *Research Policy*, **30**.

8 Bray, M. and J. Lee (2000) 'University revenues from technology transfer: Licensing fees versus equity positions', *Journal of Business Venturing*, 15, 385–392.

9 Peters, L. and H. Etzkowitz (1990) 'University–industry connections and academic values', *Technology in Society*, **12**, 427–440.

10 Kassicieh, S., R. Radosevich and J. Umbarger (1996) 'A comparative study of entrepreneurship incidence among inventors in national laboratories', *Entrepreneurship Theory and Practice*, Spring, 33–49.

11 Butler, S. and S. Birley (1999) 'Scientists and their attitudes to industry links', *International Journal of Innovation Management*, **2** (1), 79–106.

12 Lee, Y. (1996) 'Technology transfer and the research university: a search for the boundaries of university–industry collaboration', *Research Policy*, **25**, 843–863.

13 Di Gregorio, D. and S. Shane (2003) 'Why do some universities generate more start-ups than others?', *Research Policy*, **32**, 209–227.

14 Meyer, M. (2004) 'Academic entrepreneurs or entrepreneurial academics? Research-based ventures and public support mechanisms', *R&D Management*, **33** (2), 107–115.

15 Seashore, L. *et al.* (1989) 'Entrepreneurs in academe: an exploration of behaviors among life scientists', *Administrative Science Quarterly*, **34**, 110–131.

16 Samson, K. and M. Gurdon (1993) 'University scientists as entrepreneurs: a special case of technology transfer and high-tech venturing', *Technovation*, **13** (2), 63–71.

17 Vohora, A., M. Wright and A. Lockett (2004) 'Critical junctures in the development of university high-tech spinout companies', *Research Policy*, **33**, 147–175.

18 Oakey, R. (1995) *High-technology New Firms*. Paul Chapman, London.

19 Delmar, F. and S. Shane (2003) 'Does business planning facilitate the development of new ventures?', *Strategic Management Journal*, **24**, 1165–1185.

20 Drucker, P. (1985) *Innovation and Entrepreneurship*. Harper & Row, New York.

21 Withers (1997) *Window on Technology: Corporate venturing in practice*. Withers, London.

22 Loudon, A. (2001) *Webs of Innovation: The networked economy demands new ways to innovate*. FT.com, Pearson Education, Harlow.

23 Harding, R. (2000) 'Venture capital and regional development: towards a venture capital system', *Venture Capital*, **2** (4), 287–311.

24 Binding, K., C. McCubbin and L. Doyle (1998) *Technology Transfer in the UK Life Sciences*. Arthur Andersen, London.

25 Tidd, J. and S. Barnes (1999) 'Spin-in or spin-out? Corporate venturing in life sciences', *International Journal of Entrepreneurship and Innovation*, **1** (2), 109–116.

26 Lockett, A., G. Murray and M. Wright (2002) 'Do UK venture capitalists still have a bias against investment in new technology firms?', *Research Policy*, **31**, 1009–1030.

27 Baum, J. and B. Silverman (2004) 'Picking winners or building them? Alliance, intellectual and human capital as selection criteria in venture financing and performance of biotechnology startups', *Journal of Business Venturing*, **19**, 411–436.

28 Small Business Research Centre (1992) *The State of British Enterprise: Growth, innovation and competitiveness in small and medium sized firms*. SBRC, Cambridge.

29 Hoffman, K. *et al.* (1998) 'Small firms, R&D, technology and innovation in the UK: a literature review', *Technovation*, **18** (1), 39–55.

30 Calori, R. (1990) 'Effective strategies in emerging industries', in Loveridge, R. and Pitt, M. (eds), *The Strategic Management of Technological Innovation*. John Wiley & Sons, Ltd, Chichester, 21–38.

31 Walsh, V., J. Niosi and P. Mustar (1995) 'Small firms formation in biotechnology: a comparison of France, Britain and Canada', *Technovation*, **15** (5), 303–328.

32 Westhead, P., D. Storey and M. Cowling (1995) 'An exploratory analysis of the factors associated with survival of independent high technology firms in Great Britain', in Chittenden, F., Robertson, M. and Marshall, I. (eds), *Small Firms: Partnership for growth in small firms*. Paul Chapman, London, 63–99.

33 Tether, B. and D. Storey (1998) 'Smaller firms and Europe's high technology sectors: a framework for analysis and some statistical evidence', *Research Policy*, **26**, 947–971.

34 MacPherson, A. (1997) 'The contribution of external service inputs to the product development efforts of small manufacturing firms', *R&D Management*, **27** (2), 127–143.

35 Rothwell, R. and M. Dodgson (1993) 'SMEs: their role in industrial and economic change', *International Journal of Technology Management*, Special Issue, 8–22.

36 Moote, B. (1993) *Financial Constraints to the Growth and Development of Small High Technology Firms*. Small Business Research Centre, University of Cambridge; Oakey, R. (1993) 'Predatory networking: the role of small firms in the development of the British biotechnology industry', *International Small Business Journal*, **11** (3), 3–22.

37. Storey, D. (1992) 'United Kingdom: case study', in OECD, *Small and Medium Sized Enterprises, Technology and Competitiveness*. OECD, Paris; Tang, N. *et al.* (1995) 'Technological alliances between HEIs and SMEs: examining the current evidence', in Bennett, D. and Steward, F. (eds), *Proceedings of the European Conference on the Management of Technology: Technological innovation and global challenges*. Aston University, Birmingham.

38 Tether, B. (1998) 'Small and large firms: sources of unequal innovations?', *Research Policy*, **27**, 725–745.

39 Tether, B., J. Smith and A. Thwaites (1997) 'Smaller enterprises and innovations in the UK: the SPRU Innovations Database revisited', *Research Policy*, **26**, 19–32.

40 Strerlacchini, A. (1999) 'Do innovative activities matter to small firms in non-R&D-intensive industries?', *Research Policy*, **28**, 819–832.

41 Hall, G. (1991) 'Factors associated with relative performance amongst small firms in the British instrumentation sector', Working Paper No. 213, Manchester Business School.

42 Oakey, R., R. Rothwell and S. Cooper (1988) *The Management of Innovation in High Technology Small Firms*. Pinter, London.

43 Keeble, D. (1993) *Regional Influences and Policy in New Technology-based Firms: Creation and growth*. Small Business Research Centre, University of Cambridge.

44 Dickson, K., A. Coles and H. Smith (1995) 'Scientific curiosity as business; an analysis of the scientific entrepreneur', paper presented at the 18th National Small Firms Policy and Research

Conference, Manchester; Lee, J. (1993) 'Small firms' innovation in two technological settings', *Research Policy*, **24**, 391–401.

45 Bruderl, J. and P. Preisendorfer (2000) 'Fast-growing businesses', *International Journal of Sociology*, **30**, 45–70.

46 Lee, C., K. Lee and J. Pennings (2001) 'Internal capabilities, external networks, and performance: a study of technology-based ventures', *Strategic Management Journal*, **22**, 615–640.

47 Almeida, P., G. Dokko and L. Rosenkopf (2003) 'Startup size and the mechanisms of external learning: Increasing opportunity and decreasing ability?', *Research Policy*, **32**, 301–315.

48 Freel, M. (2003) 'Sectoral patterns of small firm innovation, networking and proximity', *Research Policy*, **32**, 751–770.

49 Zahra, S. and W. Bogner (2000) 'Technology strategy and software new ventures performance', *Journal of Business Venturing*, **15** (2), 135–173.

50 Deeds, D., D. DeCarolis and J. Coombs (2000) 'Dynamic capabilities and new product development in high technology ventures: an empirical analysis of new biotechnology firms', *Journal of Business Venturing*, **15** (3), 211–229.

51 George, G., S. Zahra and D. Robley Wood (2002) 'The effects of business-university alliances on innovative output and financial performance: a study of publicly traded biotechnology companies', *Journal of Business Venturing*, **17**, 577–609.

52 Gans, J. and S. Stern (2003) 'The product and the market for "ideas": commercialization strategies for technology entrepreneurs', *Research Policy*, **32**, 333–350.

Part VI

ASSESSING AND IMPROVING INNOVATION MANAGEMENT PERFORMANCE

In this final section we briefly review the main themes explored in the book, and draw them together into a simple framework for reviewing and assessing innovation management in any specific organization. We adopt the concepts of auditing and benchmarks as aids to learning and improvement, and conclude with a short checklist of questions as a self-assessment exercise.

An Integrative Approach to Innovation Management

13.1 Key Themes

Throughout the book we have tried to focus on a number of key themes surrounding innovation management. We can summarize these as follows:

- Learning and adaptation are essential in an inherently uncertain future – thus innovation is an imperative.
- Innovation is about interaction of technology, market and organization.
- Innovation can be linked to a generic process which all enterprises have to find their way through.
- Different firms use different routines with greater or lesser degrees of success. There are general recipes from which general suggestions for effective routines can be derived – but these must be customized to particular organizations and related to particular technologies and products.
- Routines are learned patterns of behaviour which become embodied in structures and procedures over time. As such they are hard to copy and highly firm-specific.
- Innovation management is the search for effective routines – in other words, it is about managing the learning process towards more effective routines to deal with the challenges of the innovation process.

We have also argued that innovation management is not a matter of doing one or two things well, but about good all-round performance. There are no, single, simple magic bullets but a set of learned behaviours. In particular we have identified four clusters of behaviour which we feel represent particularly important routines (see Figure 13.1).

- Successful innovation is strategy-based.
- Successful innovation depends on effective internal and external linkages.
- Successful innovation requires enabling mechanisms for making change happen.
- Successful innovation only happens within a supporting organizational context.

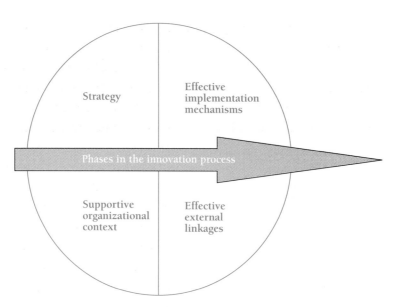

FIGURE 13.1

In the *strategy* domain there are no simple recipes for success but a capacity to learn from experience and analysis is essential. Research and experience point to three essential ingredients in innovation strategy:

1. The *position* of the firm, in terms of its products, processes, technologies and the national innovation system in which it is embedded. Although a firm's technology strategy may be influenced by a particular national system of innovation it is not determined by it.
2. The technological *paths* open to the firm given its accumulated competencies. Firms follow technological trajectories, each of which has distinct sources and directions of technological change and which define key tasks for strategy. We identify five generic trajectories in the book.
3. The organizational *processes* followed by the firm in order to integrate strategic learning across functional and divisional boundaries.

Within the area of *linkages*, developing close and rich interaction with markets, with suppliers of technology and other organizational players, is of critical importance. Linkages offer opportunities for learning – from tough customers and lead users, from competitors, from strategic alliances and from alternative perspectives. The theme of 'open innovation' is increasingly becoming recognized as relevant to an era in which networking and inter-organizational behaviour is the dominant mode of operation.

In order to succeed organizations also need *effective implementation mechanisms* to move innovations from idea or opportunity through to reality. This process involves systematic problem-solving and works best within a clear decision-making framework which should help the organization to stop as well as to progress development if things are going wrong. It also requires skills in project management, control under uncertainty and parallel development of both the market and the technology streams. And it needs to pay attention to managing the change process itself, including anticipating and addressing the concerns of those who might be affected by the change.

Finally, innovation depends on having *a supporting organizational context* in which creative ideas can emerge and be effectively deployed. Building and maintaining such organizational conditions are a critical part of innovation management, and involve working with structures, work organization arrangements, training and development, reward and recognition systems and communication arrangements. Above all, the requirement is to create the conditions within which a learning organization can begin to operate, with shared problem identification and solving and with the ability to capture and accumulate learning about technology and about management of the innovation process.

Throughout the book we have tried to consider the implications of managing innovation as a generic process but also to look at the ways in which approaches need to take into account two key challenges in the twenty-first century – those of managing 'beyond the steady state' and 'beyond boundaries'. The same basic recipe still applies but there is a need to configure established approaches and to learn to develop new approaches to deal with these challenges.

13.2 Learning to Manage Innovation

As we suggested, developing innovation management involves a learning process concerned with building and integrating key behaviours into effective routines. Such a learning process can be assisted by inputs to the learning cycle through:

- Experience-sharing, learning from and through the experience of others of both success and failure.
- Introducing new concepts, new ideas about tools and techniques.
- Experimenting, trying different approaches to the basic problem of innovation management.
- Structured reflection, examining and reviewing how innovation is currently managed.

Benchmarking (in its various forms) can help this learning process in several ways – in particular it offers a powerful motivator for change since unfavourable comparisons are hard to ignore. But it can also offer valuable clues about how to manage key processes in different fashions. Such learning can come not only from direct comparisons between similar firms but also between firms in different sectors carrying out broadly similar processes. For example, as mentioned in Chapter 9, Southwest Airlines has achieved an enviable record for its turnaround speed at airport terminals. It drew inspiration from watching how industry carried out rapid changeover of complex machinery between tasks – and, in turn, those industries learned from watching activities like pit-stop procedures in the Grand Prix motor racing world.[1] In similar fashion Kaplinsky reports on dramatic productivity and quality improvements in the healthcare sector, drawing on lessons originating in inventory management systems in manufacturing and retailing.[2]

But perhaps its most valuable contribution is as a framework for structured review and reflection on how the organization currently performs. Such auditing does not necessarily have to be done in comparison with another organization but can usefully be done against ideal-type or normative models of good practice. This found particular expression during the 'quality revolution' of the 1990s where benchmarking frameworks such as the Malcolm Baldrige Award in the USA, the Deming Prize in Japan and the European Quality Award all used sophisticated benchmarking frameworks.[3] The approach has been extended to a number of other domains – for example, software development processes which have been benchmarked against a capability/maturity model developed by the Software Engineering Institute at Carnegie Mellon University.[4] In the UK a framework for benchmarking and auditing manufacturing performance was developed and offered as a national service, with special emphasis on assisting smaller firms improve their performance.[5,6]

This auditing approach has considerable potential relevance for the practice of innovation management and a number of frameworks have been developed. Back in the 1980s the UK National Economic Development Office developed an 'innovation management tool kit' which has been updated and adapted for use as part of a European programme aimed at developing better innovation management amongst SMEs. Another framework, originally developed at London Business School, was promoted by the UK Department of Trade and Industry[7] and led on to the development of a series of frameworks including the 'living innovation' model which was jointly promoted with the Design Council.[8] Francis offers an overview of a number of these.[9] Other frameworks have been developed which cover particular aspects of innovation management, such as continuous improvement and product development.[10,11] With the increasing use of the Internet have come a number of sites which offer interactive frameworks for assessing innovation management performance as a first step towards organization development.*

In each case the purpose of such auditing is not to score points or win prizes but to enable the operation of an effective learning cycle through adding the dimension of structured reflection. It is the process of regular review and discussion which is important rather than detailed information or exactness of scores. The point is not simply to collect data but to use these measures to drive improvement of the innovation process and the ways in which it is managed. As the quality guru, W. Edwards Deming, pointed out, 'If you don't measure it you can't improve it!' (see Box 13.1).

In reviewing innovative performance we can look at a number of possible measures and indicators:

- Measures of specific *outputs* of various kinds – for example, patents and scientific papers as indicators of knowledge produced, or number of new products introduced (and percentage of sales and/or profits derived from them) as indicators of product innovation success.[12]
- Output measures of operational or process elements, such as customer satisfaction surveys to measure and track improvements in quality or flexibility.[13,14]
- Output measures which can be compared across sectors or enterprises – for example, cost of product, market share, quality performance, etc.

BOX 13.1
MEASURING INNOVATION

The problem with audit frameworks and benchmarks of this kind is that they often provide an indication of how a system and its components are performing but they fail to take into account the final piece of the puzzle – why are they successful? For example, in the total quality field in the USA much interest was shown in the self-assessment framework surrounding the Malcolm Baldrige Award, and many firms used this benchmarking and assessment framework to improve their quality performance. However, one of the winners, Florida Power and Light, whilst undoubtedly doing many of the right things, was none the less forced into receivership; this prompted the addition of a performance category to the assessment framework.

---

* See, for example, www.innovationdoctor.htm, www.thinksmart.htm, www.jpb.com/services/audit.php, www.innovation-triz.com/innovation/, www.cambridgestrategy.com/page_c5_summary.htm, and www.innovationwave.com/

- Output measures of strategic success, where the overall business performance is improved in some way and where at least some of the benefit can be attributed directly or indirectly to innovation – for example, growth in revenue or market share, improved profitability, higher value added.[15]

We could also consider a number of more specific measures of the internal workings of the innovation process or particular elements within it. For example:

- Number of new ideas (product/service/process) generated at start of innovation system.
- Failure rates – in the development process, in the marketplace.
- Number or percentage of overruns on development time and cost budgets.
- Customer satisfaction measures – was it what the customer wanted?
- Time to market (average, compared with industry norms).
- Development human-hours per completed innovation.
- Process innovation average lead time for introduction.
- Measures of continuous improvement – suggestions/employee, number of problem-solving teams, savings accruing per worker, cumulative savings, etc.

There is also scope for measuring some of the influential conditions supporting or inhibiting the process – for example, the 'creative climate' of the organization or the extent to which strategy is clearly deployed and communicated.[16,17] And there is value in considering *inputs* to the process – for example, percentage of sales committed to R&D, investments in training and recruitment of skilled staff, etc.[18]

13.3 Auditing Innovation Management

A great deal of research effort has been devoted to the questions of what and how to measure in innovation. The risk is that we become so concerned with these questions that we lose sight of the practical objective which is to reflect upon and improve the *management* of the process. Having established in this book some of the factors which appear to influence success and failure in innovation in the experience of others, we can begin to develop a tool for assessing and developing innovative performance in organizations. We could construct a simple checklist of factors and assign a score to each of them so as to develop a profile of innovation performance. So, for example, an organization with no clear innovation strategy, with limited technological resources and no plans for acquiring more, with weak project management, with poor external links and with a rigid and unsupportive organization would be unlikely to succeed in innovation. By contrast, one which was focused on clear strategic goals, had developed long-term links to support technological development, had a clear project management process which was well supported by senior management and which operated in an innovative organizational climate would have a better chance of success.

An example of such a scale might be:

1 = innovation not even thought about, rarely happens;
2 = some awareness but random and occasional responses, informal systems;
3 = awareness and formal systems in place – but could still be improved;
4 = highly developed and effective systems including provision for improvement and development.

What to Look For

In carrying out auditing of this kind there is clearly no such thing as an absolute score. None the less it is possible to develop a number of indicators which give some under-pinning to what will otherwise be rather subjective judgments about the innovation management capability of a company. For example, a firm which spends 10% of its turnover on R&D is likely to be better on resourcing innovation than one which does no R&D at all.

Box 13.2 gives an example of an outline 'innovation audit' which could be used to focus attention on some of the issues flagged in the book and help begin the process of auditing innovation management capability. The responses to these questions describe 'the way we do things around here' – the pattern of behaviour which describes how the organization handles the question of innovation. These represent the tip of an iceberg but can help focus attention on areas where there is room for further development and where more detailed questions need to be asked.

Using the Framework

Box 13.2 provides a framework and a brief checklist of questions which might enable an assessment of innovation management to be undertaken. It is not exhaustive, but it does indicate the balance of facts and subjective judgments which would need to be considered to make a realistic response to the question, 'How well does this organization manage innovation?'

The format of the particular tool is not important; what is needed is the ability to use it to make a wide-ranging review of the factors affecting innovation success and failure, and how management of the process might be improved. Some of the uses to which it could be put are as:

BOX 13.2
HOW WELL DO WE MANAGE INNOVATION?

This simple self-assessment tool focuses attention on some of the important areas of innovation management. Below you will find statements which describe 'the way we do things around here' – the pattern of behaviour which describes how the organization handles the question of innovation. For each statement simply put a score between 1 (= not true at all) to 7 (= very true).

Statement	*Score 1 = Not true at all to 7 = Very true*

1. People have a clear idea of how innovation can help us compete
2. We have processes in place to help us manage new product development effectively from idea to launch
3. Our organization structure does not stifle innovation but helps it to happen
4. There is a strong commitment to training and development of people
5. We have good 'win–win' relationships with our suppliers
6. Our innovation strategy is clearly communicated so everyone knows the targets for for improvement
7. Our innovation projects are usually completed on time and within budget
8. People work well together across departmental boundaries
9. We take time to review our projects to improve our performance next time
10. We are good at understanding the needs of our customers/end-users
11. People know what our distinctive competence is – what gives us a competitive edge
12. We have effective mechanisms to make sure everyone (not just marketing) understands customer needs
13. People are involved in suggesting ideas for improvements to products or processes
14. We work well with universities and other research centres to help us develop our knowledge
15. We learn from our mistakes
16. We look ahead in a structured way (using forecasting tools and techniques) to try and imagine future threats and opportunities

Statement	Score 1 = Not true at all to 7 = Very true

17. We have effective mechanisms for managing process change from idea through to successful implementation
18. Our structure helps us to take decisions rapidly
19. We work closely with our customers in exploring and developing new concepts
20. We systematically compare our products and processes with other firms
21. Our top team have a shared vision of how the company will develop through innovation
22. We systematically search for new product ideas
23. Communication is effective and works top-down, bottom-up and across the organization
24. We collaborate with other firms to develop new products or processes
25. We meet and share experiences with other firms to help us learn
26. There is top management commitment and support for innovation
27. We have mechanisms in place to ensure early involvement of all departments in developing new products/processes
28. Our reward and recognition system supports innovation
29. We try to develop external networks of people who can help us – for example, with specialist knowledge
30. We are good at capturing what we have learned so that others in the organization can make use of it
31. We have processes in place to review new technological or market developments and what they mean for our firm's strategy
32. We have a clear system for choosing innovation projects
33. We have a supportive climate for new ideas – people don't have to leave the organization to make them happen
34. We work closely with the local and national education system to communicate our needs for skills
35. We are good at learning from other organizations
36. There is a clear link between the innovation projects we carry out and the overall strategy of the business
37. There is sufficient flexibility in our system for product development to allow small 'fast-track' projects to happen
38. We work well in teams
39. We work closely with 'lead users' to develop innovative new products and services
40. We use measurement to help identify where and when we can improve our innovation management

continues overleaf

BOX 13.2 (*continued*)

When you have finished, add the totals for the questions in the following way:

Question number				Scores					
1	. . .	2	. . .	3	. . .	5	. . .	4	. . .
6	. . .	7	. . .	8	. . .	10	. . .	9	. . .
11	. . .	12	. . .	13	. . .	14	. . .	15	. . .
16	. . .	17	. . .	18	. . .	19	. . .	20	. . .
21	. . .	22	. . .	23	. . .	24	. . .	25	. . .
26	. . .	27	. . .	28	. . .	29	. . .	30	. . .
31	. . .	32	. . .	33	. . .	34	. . .	35	. . .
36	. . .	37	. . .	38	. . .	39	. . .	40	. . .

Total
Divide by 8

— — — — — — — — — —

Your score for . . .   **Strategy**   **Processes**   **Organization** **Linkages**   **Learning**

Now plot a profile for the five dimensions on the next page.

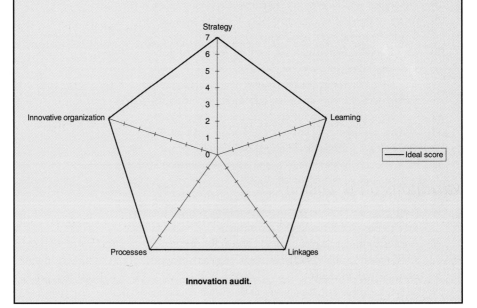

Innovation audit.

- an audit framework to see what you did right and wrong in the case of particular innovations or as a way of understanding why things happened the way they did;
- a checklist to see if you are doing the right things;
- a benchmark to see if you are doing them as well as others;
- a guide to continuous improvement of innovation management;
- a learning resource to help acquire knowledge and provide inspiration for new things to try;
- a way of focusing on sub-systems with particular problems and then working with the owners of those processes and their customers and suppliers to see if the discussion cannot improve on things.

(The website contains further detail to help you interpret your particular scores and think about what you might do next in terms of organizational development for innovation.)

13.4 What Kind of Innovator is Your Organization?

We can represent the different positions resulting from such a simple audit in diagrammatic form as in Figure 13.2 and identify a number of 'archetypes' of innovation capability.

Such an approach gives the possibility of a quick 'snapshot' to focus attention and create the commitment to improvement. In the end the question is not of how well you scored on the self-assessment audit but rather of using the information to help the learning process for improved innovation management.

Variations on a Theme

Throughout the book we have stressed that whilst the challenge in innovation management is generic there are specific issues around which specific responses need to be configured. Box 13.3, for example, looks at the case of service innovation and suggests some additional audit questions which might be relevant in thinking about managing such innovation.

Similarly we have been arguing that there are conditions – beyond the steady state – where we need to take a different approach to managing innovation and to introduce new or at least complementary routines to those helpful in dealing with 'steady state'

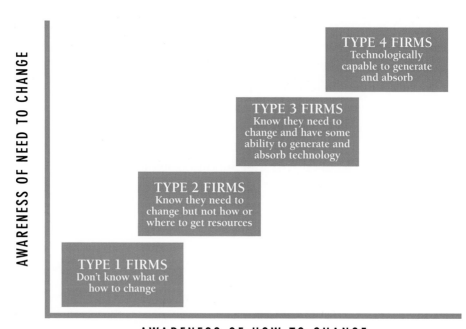

FIGURE 13.2 Distribution of innovation capability

BOX 13.3

MEASURING SERVICE INNOVATION

The organization and management of new service development and delivery can be assessed by five components: Strategy, Process, Organization, Tools/Technology and System (SPOTS). This framework has been developed and tested by analysing more than 100 firms in the USA and UK, and validated during the course of conducting a total of 27 cases studied from 18 companies.

Each of the five factors plays a different role in the performance of service innovation. *Strategy* provides focus; *process* provides control; *organization* provides co-ordination of people; *tools* and technologies provide transformation/transaction capabilities, and *system* provides integration. Performance is analysed as a total index and as three subscales: (1) innovation and quality, (2) time compression in development and cost reduction in development/delivery, and (3) service delivery.

The first two factors roughly correspond to generic strategic alternatives, differentiation vs. cost. The third factor is conceptually important because it distinguishes the service delivery process from product features. Delivery processes often comprise a significant proportion of value added by services, especially if interpersonal exchanges are involved.

The scores and comparisons with those of other companies in the database allows a company to identify its strengths and weaknesses. For example:

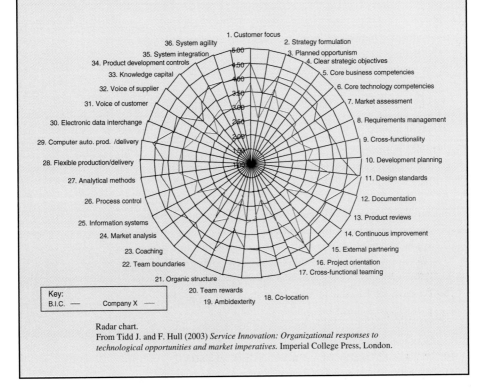

Radar chart.
From Tidd J. and F. Hull (2003) *Service Innovation: Organizational responses to technological opportunities and market imperatives.* Imperial College Press, London.

innovation. Box 13.4 offers some additional audit questions to help facilitate this kind of reflection.

We have repeatedly said that innovation is complex, uncertain and almost (but not quite) impossible to manage. That being so, we can be sure that there is no such thing as the perfect organization for innovation management; there will always be opportunities for experimentation and continuous improvement. As we have suggested throughout the book the challenge is to constantly review and reconfigure in the light of changing circumstances – whether discontinuous 'beyond the steady state' innovation

BOX 13.4
HOW WELL DO WE MANAGE DISCONTINUOUS INNOVATION?

As with the 'steady-state' audit, simply put a score between 1 (= not true at all) to 7 (= very true) against each statement. The same score sheet and profile can be used.

Around here . . .

Statement	*Score 1 = Not true at all to 7 = Very true*

1. We deploy 'probe and learn' approaches to explore new directions in technologies and markets
2. We actively explore the future, making use of tools and techniques like scenarios and foresight
3. Our organization allows some space and time for people to explore 'wild' ideas
4. We make connections across industry to provide us with different perspectives
5. We make regular use of formal tools and techniques to help us think 'out of the box'
6. We have alternative and parallel mechanisms for implementing and developing radical innovation projects which sit outside the 'normal' rules and procedures
7. We have capacity in our strategic thinking process to challenge our current position – we think about 'how to destroy the business'!
8. We have mechanisms to bring in fresh perspectives – for example, recruiting from outside the industry
9. We make use of formal techniques for looking and learning from outside our sector
10. We focus on 'next practices' as well as 'best practices'
11. We have mechanisms for managing ideas that don't fit our current business – for example, we license them out or spin them off
12. We use some form of technology scanning/intelligence-gathering – we have well developed technology antennae
13. We have mechanisms to identify and encourage 'intrapreneurship' – if people have a good idea they don't have to leave the company to make it happen
14. We have extensive links with a wide range of outside sources of knowledge – universities, research centres, specialized agencies – and we actually set them up even if not for specific projects

Statement	Score 1 = Not true at all to 7 = Very true

15. We make use of simulation, etc. to explore different options and delay commitment to one particular course
16. We work with 'fringe' users and very early adopters to develop our new products and services
17. We allocate a specific resource for exploring options at the edge of what we currently do – we don't load everyone up 100%
18. We have reward systems to encourage people to offer their ideas
19. We have well-developed peripheral vision in our business
20. We use technology to help us become more agile and quick to pick up on and respond to emerging threats and opportunities on the periphery
21. We have 'alert' systems to feed early warning about new trends into the strategic decision-making process
22. We have strategic decision-making and project selection mechanisms which can deal with more radical proposals outside of the mainstream
23. We value people who are prepared to break the rules
24. We practice 'open innovation' – rich and widespread networks of contacts from whom we get a constant flow of challenging ideas
25. We learn from our periphery – we look beyond our organizational and geographical boundaries
26. We are organized to deal with 'off-purpose' signals (not directly relevant to our current business) and don't simply ignore them
27. We deploy 'targeted hunting' around our periphery to open up new strategic opportunities
28. We have high involvement from everyone in the innovation process
29. We have an approach to supplier management which is open to strategic dalliances
30. We are good at capturing what we have learned so that others in the organization can make use of it
31. We have processes in place to review new technological or market developments and what they mean for our firm's strategy
32. Management create 'stretch goals' that provide the direction but not the route for innovation
33. Peer pressure creates a positive tension and creates an atmosphere to be creative
34. We have active links into a long-term research and technology community – we can list a wide range of contacts
35. We create an atmosphere where people can share ideas through cross-fertilization

continues overleaf

BOX 13.4 *(continued)*

Statement	Score 1 = Not true at all to 7 = Very true

36. There is sufficient flexibility in our system for product development to allow small 'fast-track' projects to happen
37. We are not afraid to 'cannibalize' things we already do to make space for new options
38. Experimentation is encouraged
39. We recognize users as a source of new ideas and try and 'co-evolve' new products and services with them
40. We regularly challenge ourselves to identify where and when we can improve our innovation management

or in the context of 'open innovation where the challenge is working' beyond the boundaries'. In the end innovation management is not an exact or predictable science but a craft, a reflective practice in which the key skill lies in reviewing and configuring to develop dynamic capability.

References

1 Shingo, S. (1983) *A Revolution in Manufacturing: The SMED system*. Productivity Press, Cambridge, Mass.
2 Kaplinsky, R., F. den Hertog and B. Coriat (1995) *Europe's Next Step*. Frank Cass, London.
3 Garvin, D. (1991) 'How the Baldrige award really works', *Harvard Business Review*, November/December, 80–93.
4 Paulk, M. *et al.* (1993) *Capability Maturity Model for Software*. Software Engineering Institute, Carnegie-Mellon University.
5 Voss, C. (1994) *Made in Britain*. London Business School.
6 Voss, C. (1999) *Made in Europe 3: The small company study*. London Business School/IBM Consulting, London.
7 Chiesa, V., P. Coughlan and C. Voss (1996) 'Development of a technical innovation audit', *Journal of Product Innovation Management*, **13** (2), 105–136.
8 Design Council (2002) *Living Innovation*. Design Council/Department of Trade and Industry, London. website: www.livinginnovation.org.uk.
9 Francis, D. (2001) 'Developing Innovative Capability', PhD thesis, University of Brighton, Brighton.
10 Oliver, N. and M. Blakeborough (1997) 'Innovation networks: the view from the inside', in Grieve Smith, J. and Michie, J. (eds), *Innovation, Co-operation and Growth*. Oxford University Press, Oxford.

11 Bessant, J. and S. Caffyn (1997) 'High involvement innovation', *International Journal of Technology Management*, **14** (1), 7–28.

12 Tidd, J. (ed.) (2000) *From Knowledge Management to Strategic Competence: Measuring technological, market and organizational innovation*. Imperial College Press, London.

13 Luchs, B. (1990) 'Quality as a strategic weapon', *European Business Journal*, **2** (4), 34–47.

14 Zairi, M. (1999) *Process Innovation Management*. Butterworth Heinemann, London.

15 Kay, J. (1993) *Foundations of Corporate Success: How business strategies add value*. Oxford University Press, Oxford, 416.

16 Ekvall, G. (1990) 'The organizational culture of idea management', in Henry, J. and Walker, D. (eds), *Managing Innovation*. Sage, London, 73–80.

17 Amabile, T. (1998) 'How to Kill Creativity', *Harvard Business Review*, September/October, 77–87.

18 DTI (2004) *Innovation Review*. Department of Trade and Industry, London.

Index